QUANTUM SYSTEMS IN CHEMISTRY AND PHYSICS

Basic Problems and Model Systems

Progress in Theoretical Chemistry and Physics

Volume 2

Quantum Systems in Chemistry and Physics
Volume 1

Basic Problems and Model Systems
Granada, Spain, 1998

Edited by

Alfonso Hernández-Laguna
Estación Experimental del Zaidín (C.S.I.C.), Granada, Spain

Jean Maruani
Laboratoire de Chimie Physique (C.N.R.S.), Paris, France

Roy McWeeny
Universita di Pisa, Pisa, Italy

and

Stephen Wilson
Rutherford Appleton Laboratory, Oxfordshire, United Kingdom

KLUWER ACADEMIC PUBLISHERS
DORDRECHT / BOSTON / LONDON

Library of Congress Cataloging-in-Publication Data is available

ISBN 0-7923-5969-0
ISBN (set) 0-7923-5971-2

Published by Kluwer Academic Publishers
PO Box 17, 3300 AA Dordrecht, The Netherlands

Sold and distributed in North, Central and South America
by Kluwer Academic Publishers,
101 Philip Drive, Norwell, MA 02061, USA

In all other countries, sold and distributed
by Kluwer Academic Publishers,
PO Box 322, 3300 AH Dordrecht, The Netherlands

Printed on acid-free paper

Printed and bound in Great Britain by Cromwell Press, Trowbridge, Wilts

Progress in Theoretical Chemistry and Physics

A series reporting advances in theoretical molecular and material sciences, including theoretical, mathematical and computational chemistry, physical chemistry and chemical physics

Aim and Scope

Science progresses by a symbiotic interaction between theory and experiment: theory is used to interpret experimental results and may suggest new experiments; experiment helps to test theoretical predictions and may lead to improved theories. Theoretical Chemistry (including Physical Chemistry and Chemical Physics) provides the conceptual and technical background and apparatus for the rationalisation of phenomena in the chemical sciences. It is, therefore, a wide ranging subject, reflecting the diversity of molecular and related species and processes arising in chemical systems. The book series *Progress in Theoretical Chemistry and Physics* aims to report advances in methods and applications in this extended domain. It will comprise monographs as well as collections of papers on particular themes, which may arise from proceedings of symposia or invited papers on specific topics as well as initiatives from authors or translations.

The basic theories of physics – classical mechanics and electromagnetism, relativity theory, quantum mechanics, statistical mechanics, quantum electrodynamics – support the theoretical apparatus which is used in molecular sciences. Quantum mechanics plays a particular role in theoretical chemistry, providing the basis for the valence theories which allow to interpret the structure of molecules and for the spectroscopic models employed in the determination of structural information from spectral patterns. Indeed, Quantum Chemistry often appears synonymous with Theoretical Chemistry: it will, therefore, constitute a major part of this book series. However, the scope of the series will also include other areas of theoretical chemistry, such as mathematical chemistry (which involves the use of algebra and topology in the analysis of molecular structures and reactions); molecular mechanics, molecular dynamics and chemical thermodynamics, which play an important role in rationalizing the geometric and electronic structures of molecular assemblies and polymers, clusters and crystals; surface, interface, solvent and solid-state effects; excited-state dynamics, reactive collisions, and chemical reactions.

Recent decades have seen the emergence of a novel approach to scientific research, based on the exploitation of fast electronic digital computers. Computation provides a method of investigation which transcends the traditional division between theory and experiment. Computer-assisted simulation and design may afford a solution to complex problems which would otherwise be intractable to theoretical analysis, and may also provide a viable alternative to difficult or costly laboratory experiments. Though stemming from Theoretical Chemistry, Computational Chemistry is a field of research in its own right, which can help to test theoretical predictions and may also suggest improved theories.

The field of theoretical molecular sciences ranges from fundamental physical questions relevant to the molecular concept, through the statics and dynamics of isolated molecules, aggregates and materials, molecular properties and interactions, and the role of molecules in the biological sciences. Therefore, it involves the physical basis for geometric and electronic structure, states of aggregation, physical and chemical transformations, thermodynamic and kinetic properties, as well as unusual properties such as extreme flexibility or strong relativistic or quantum-field effects, extreme conditions such as intense radiation fields or interaction with the continuum, and the specificity of biochemical reactions.

Theoretical chemistry has an applied branch – a part of molecular engineering, which involves the investigation of structure–property relationships aiming at the design, synthesis and application of molecules and materials endowed with specific functions, now in demand in such areas as molecular electronics, drug design or genetic engineering. Relevant properties include conductivity (normal, semi- and supra-), magnetism (ferro- or ferri-), optoelectronic effects (involving nonlinear response), photochromism and photoreactivity, radiation and thermal resistance, molecular recognition and information processing, and biological and pharmaceutical activities, as well as properties favouring self-assembling mechanisms and combination properties needed in multifunctional systems.

Progress in Theoretical Chemistry and Physics is made at different rates in these various research fields. The aim of this book series is to provide timely and in-depth coverage of selected topics and broad-ranging yet detailed analysis of contemporary theories and their applications. The series will be of primary interest to those whose research is directly concerned with the development and application of theoretical approaches in the chemical sciences. It will provide up-to-date reports on theoretical methods for the chemist, thermodynamician or spectroscopist, the atomic, molecular or cluster physicist, and the biochemist or molecular biologist who wish to employ techniques developed in theoretical, mathematical or computational chemistry in their research programmes. It is also intended to provide the graduate student with a readily accessible documentation on various branches of theoretical chemistry, physical chemistry and chemical physics.

Contents

Part III. Relativistic Formulations

Part IV. Valence Theory

Preface

These two volumes collect forty-four selected papers from the scientific contributions presented at the Third European Workshop on Quantum Systems in Chemistry and Physics, held in Granada (Spain), April 19–22, 1998. Ninety-nine scientists from Bulgaria, Columbia, Cuba, Denmark, Finland, France, Germany, Hungary, Israel, Italy, Mexico, Netherlands, Norway, Poland, Russia, Slovakia, Spain, Sweden, United Kingdom, Uruguay and Venezuela attended the workshop, discussing the state of the art, new trends, and future evolution of the methods and applications.

The workshop took place at the 'Los Alixares' Hotel, where 45 lectures were given by prominent members of the scientific community; in addition, 49 posters were presented in two very animated sessions. The success of this workshop is due, without doubt, to the excellent tradition initiated at the previous workshops, organised by Prof. R. McWeeny in San Miniato, Pisa (Italy), 1996, and by Prof. S. Wilson in Oxford (United Kingdom), 1997. These workshops create occasions for meetings and discussions on the current state of the art, emerging methods and applications and new trends in this area of science. The three meetings were sponsored and partially supported by the European Union (EU) in the frame of the Cooperation in Science and Technology (COST) chemistry actions.

Quantum Systems in Chemistry and Physics is a broad area of science in which scientists of different extractions and aims jointly place special emphasis on quantum theory. Several topics were presented in the sessions of the symposia, namely: 1: Density matrices and density functionals; 2: Electron correlation effects (many-body methods and configuration interactions); 3: Relativistic formulations; 4: Valence theory (chemical bonds and bond breaking); 5: Nuclear motion (vibronic effects and flexible molecules); 6: Response theory (properties and spectra; atoms and molecules in strong electric and magnetic fields); 7: Condensed matter (crystals, clusters, surfaces and interfaces); 8: Reactive collisions and chemical reactions, and 9: Computational chemistry and physics.

The first topic has an important role in the interpretation and calculation of atomic and molecular structures and properties. It is needless to stress the importance of electronic correlation effects, a central topic of research in quantum chemistry. The relativistic formulations are of great importance not only from a formal viewpoint, but also for the increasing number of studies on atoms with high Z values in molecules and materials. Valence theory deserves special attention since it improves the electronic description of molecular systems and reactions with the point of view used by most laboratory chemists. Nuclear motion constitutes a broad research field of great importance to account for the internal molecular dynamics and spectroscopic properties.

Also very broad, complex and of great importance in physics and chemistry is the sixth topic, where electric and magnetic fields interact with matter. Condensed matter is a field where theoretical studies are performed from few-atom clusters to crystals, materials and interfaces; the theory becomes more and more complex and new scientific ideas and models are sought. The theory with which to study chemical reactions and collisions is the cornerstone of chemistry; traditional concepts such as potential-energy surfaces or transition complexes appear to become insufficient.

The last topic is an innovation of this third workshop. To start with, the new EU COST-Chemistry D-9 action is devoted to 'Advanced Computational Chemistry of Increasingly Complex Systems', aiming at enlarging the scope, theory, techniques and algorithms of computational chemistry to perform more realistic modelling of chemical systems; in addition, computational chemistry has become an advanced field of research very relevant to science and technology in chemistry, biology, pharmacology, materials science and related fields. We recall with pleasure the 1998 chemistry Nobel prize awarded to two scientists for their important contributions to computational chemistry. Finally, we should mention the increasing power of modern computers to assist the development of more powerful theories and algorithms and more realistic modelling for complex systems. Nowadays, computational chemistry is finally close to that level envisioned by R. S. Mulliken who wrote, at the conclusion of his Nobel prize lecture: '. . . chemists will go to computer centres more frequently than to laboratories'.

We acknowledge the support from the following institutions: COST-Chemistry D-9, 'Dirección General de Enseñanza Superior e Investigación Científica', Spanish National Research Council (C.S.I.C.), Andalucian Government, University of Granada, City Hall of Granada and 'Caja Rural de Granada'. The kind attentions of the staff of Porcel's Hotels are also acknowledged. Finally, it is a pleasure to thank the work and dedication of the other members of the local organising team (Prof. D. Portal-Olea, Dr. C.I. Sainz-Diaz, Dr. J.A. Dobado, Dr. Z. Cruz-Rodríguez, Dr. H. Martínez-García, Lda. L. Alfonso-Méndez and Lda. M. Daza) and, last but not least, the interest and help of Dr. Gérard Rivière at the European Commission.

A. Hernández-Laguna and J. Molina-Molina
Granada, 1999

Part I

Density Matrices and Density Functionals

Three-Body Correlation Effects in Third-Order Reduced Density Matrices

C. Valdemoro[a], L.M. Tel[b] and E. Pérez-Romero[b]
[a]*Instituto de Matemáticas y Física Fundamental, Consejo Superior de Investigaciones Científicas, Serrano 123, Madrid 28006, Spain*
[b]*Dpto. de Química Física, Universidad de Salamanca, Salamanca 37008, Spain*

Abstract

A description of the different terms contributing to the correlation effects in the third order reduced density matrix taking as reference the Hartree Fock results is given here. An analysis of the approximations of these terms as functions of the lower order reduced density matrices is carried out for the linear BeH_2 molecule. This study shows the importance of the role played by the *homo*'s and *lumo*'s of the symmetry-shells in the correlation effect. As a result, a new way for improving the third order reduced density matrix, correcting the error of the basic approximation, is also proposed here.

1. Introduction

By integrating the Schrödinger Equation, Nakatsuji [1] and Cohen and Frishberg [2] obtained in 1976 a two electron integro-differential equation, which they named *density equation*. An equivalent matrix equation was reported in 1985 by Valdemoro [3]; and, since it had been generated by application of a contracting mapping acting on the matrix representation in the N-electron space of the Schrödinger equation, it was called Contracted Schrödinger Equation (2-*CSE*). This equation is undetermined [4] since its solution depends not only on the second order Reduced Density Matrix (2-*RDM*) but also on the 3-*RDM* and 4-*RDM*. The possibility of building high order *RDM*'s as function of the lower order ones –*construction procedure*– with a satisfactory degree of approximation [5–7] opened the way to an iterative solution of the 2-*CSE* [8–14]. This promising approach relies on the optimization of the construction procedure and therefore this question is dealt with here. Thus, the aim of this paper is to report our latest results in this line of research focusing our attention on the manner in which the electron correlation is taken into account in the construction procedure for the $\alpha\alpha\beta$ block of the 3-*RDM*. In order to have an overall view of the previous work, we dedicate the following section to the presentation of our notation and of a summary of the main approaches to the construction procedure. Then (section three) the analysis of the correlation matrices obtained for the ground state of the linear BeH_2 molecule is reported. In section four, after commenting an approximation proposed by Nakatsuji and Yasuda [10,11] for estimating the three body correlation effects, an improving modification is proposed. Finally, some concluding remarks are given in this same section.

A. Hernández-Laguna et al. (eds.), Quantum Systems in Chemistry and Physics, Vol. 1: Basic Problems and Model Systems, 3–16.

2. General Background

2.1. Definitions

In what follows the number N of electrons of the system under study is a fixed number, and the dimension of the finite Hilbert subspace spanned by the orthonormal basis of spin-orbitals is $2K$.

In our formalism, the 2-*RDM* is:

$$ {}^{2}\mathrm{D}_{i_1 i_2; j_1 j_2} = \frac{1}{2!} \langle \mathcal{L} | \mathrm{b}^{\dagger}_{i_1} \mathrm{b}^{\dagger}_{i_2} \mathrm{b}_{j_2} \mathrm{b}_{j_1} | \mathcal{L} \rangle \tag{1} $$

Similarly, the 2-order Holes Reduced Density Matrix (2-*HRDM*) is:

$$ {}^{2}\overline{\mathrm{D}}_{i_1 i_2; j_1 j_2} = \frac{1}{2!} \langle \mathcal{L} | \mathrm{b}_{j_2} \mathrm{b}_{j_1} \mathrm{b}^{\dagger}_{i_1} \mathrm{b}^{\dagger}_{i_2} | \mathcal{L} \rangle \tag{2} $$

The convention followed here is that the row/column indices of the *RDM*'s coincide respectively with the creator/annihilator labels. The form of all kinds of higher order *RDM*'s can be inferred by extension of the definitions given above.

If an ordering is imposed on the indices, i.e. $i_1 < i_2 < \ldots$ and $j_1 < j_2 < \ldots$, the dimension of the matrices is reduced and the 2!, 3!, ... respective factors are omitted. Herafter we will refer to these two-, three-electron basis as *ordered basis*. Note that – in an *ordered basis* – only one of the many inter-related elements is considered.

The first order transition *RDM*'s (1-*TRDM*) also play an important role in this theory. They are of the form

$$ {}^{1}\mathrm{D}^{\mathcal{L}\mathcal{L}'}_{r_1; s_1} = \langle \mathcal{L} | \mathrm{b}^{\dagger}_{r_1} \mathrm{b}_{s_1} | \mathcal{L}' \rangle $$

In general $\mathcal{L}$ and $\mathcal{L}'$ are N-electron orthonormal correlated states.

2.2. Construction Procedures

The different approaches proposed in the literature for constructing the high order *RDM*'s as function of the lower order ones are summarized here, although some of their features will be later considered in more detail.

The expectation values of the anticommutator/commutator of p-electron operators lead to expressions of the type [5]

$$ \begin{matrix} {}^{p}\overline{\mathrm{D}} \\ \pm \\ {}^{p}\mathrm{D} \end{matrix} = \begin{matrix} \mathcal{F}\{{}^{p-1}\overline{\mathrm{D}},\ {}^{p-2}\overline{\mathrm{D}}, \ldots {}^{1}\overline{\mathrm{D}}\} \\ \pm \\ \mathcal{F}\{{}^{p-1}\mathrm{D},\ {}^{p-2}\mathrm{D}, \ldots {}^{1}\mathrm{D}\} \end{matrix} \tag{3} $$

By identifying the holes part of the *l.h.s.* of this relation with the holes part of the *r.h.s.* and similarly for the electrons, Valdemoro, Colmenero and Pérez del Valle [5–7] built up a *p-RDM* in terms of the lower order *RDM*'s. This construction procedure will be referred to in what follows as *VCP* and denoted in the formulae as $\breve{\mathrm{D}}$. Since the value of the norm of the *VCP* matrix was very close – but not exactly equal – to $\binom{N}{p}$, different procedures of renormalization were proposed [8,9,12]. This step should not be underestimated, since the accuracy of the approximation is very sensitive to it. In fact, it is a way of partially correcting the error of the approximation which is defined as

$$^{i}\Delta = {}^{i}\mathrm{D} - {}^{i}\breve{\mathrm{D}} \tag{4}$$

For the 3-*RDM*, which will center our attention in this paper, the *VCP* generates 15 terms which, as Mazziotti pointed out [13,14], may be expressed in a compact form as

$$^{3}\mathrm{D}^{VCP} = {}^{1}\mathrm{D} \wedge {}^{1}\mathrm{D} \wedge {}^{1}\mathrm{D} + 3^{2}\Delta \wedge {}^{1}\mathrm{D} \tag{5}$$

Thus the exact 3-*RDM* may be written as:

$$^{3}\mathrm{D} = {}^{3}\mathrm{D}^{VCP} + {}^{3}\Delta \tag{6}$$

Nakatsuji and Yasuda [10,11], on the basis of the asymptotic properties of the Green functions and found an analogy between the *VCP* algorithm and the perturbative expansion of these functions, and gave a diagramatic expression totally equivalent to (6). The *VCP*, the Nakatsuji and Yasuda (hereafter refered to as *NY*) and Mazziotti's approaches differ in two kinds of ways. The first difference lies in the arguments leading to (5) used by these authors. Thus, *VCP* exploits the symmetry between holes and particles, *NY* conjectures that an *RDM* may be developed by a perturbative expansion by analogy to the Green's function one; and, more recently – Mazziotti [13,14] – while keeping the essence of the many-body diagramatic expansions uses a generating function instead of exploiting the asymptotic properties of the Green functions.

The other difference refers to the way in which the different groups approximate $^{3}\Delta$. Thus, while *VCP* focus the attention on a renormalization of the approximated 3-*RDM* which uses the complementary holes matrix, *NY* again inspire their procedure on the assumption of an analogy with the Dyson equation. Finally, Mazziotti uses a self-consistent algorithm which may be sumarized as: The 4-*RDM* is calculated by means of an algorithm which coincides with the one proposed by *NY*, expressed in a compact form as:

$$^{4}\mathrm{D} = {}^{4}\mathrm{D}^{VCP} + 3^{2}\Delta \wedge {}^{2}\Delta \tag{7}$$

then this 4-*RDM* is contracted to obtain a new 3-*RDM* which is used for building a new $^{4}\mathrm{D}^{VCP}$; and this procedure is repeated until consistency.

The problem lies therefore on the interpretation and on how to estimate the error $^{3}\Delta$.

Although the base of all these approaches is rigorous, there are steps in all of them which can only claim to be reasonable hypotheses. On the other hand, an analytical decomposition, that may be applied to any high order *RDM* and which obviously generates another construction procedure has been reported by Valdemoro, de Lara-Castells, Pérez-Romero and Tel [12]. This procedure will be refered to here as *VLPT*.

Since the *VLPT* is formally exact, instead of looking for the best construction procedure for high order *RDM*'s, the problem is reformulated as *how to estimate the pure two-, three-, four-body correlation terms in the VLPT* or, equivalently, the $^{2}\Delta$, $^{3}\Delta$, $^{4}\Delta$ whose expressions we now know and which in general are closely related but are not respectively identical to the two-body, three-body and four-body correlation terms. This redefinition of the problem is more precise – and therefore valuable – although we are fully aware that it does not provide the solution by itself.

2.2.1. *Decomposition of the 3-RDM: the VLPT*

We have recently reported [12] an approach for decomposing the high order *RDM*'s in terms of the 1-*RDM* and 1-*TRDM*'s which is formally exact and which, as has been just mentioned, will be referred to here as *VLPT*.

The idea in this procedure is to permute the operators in

$$ {}^{3}\mathrm{D}_{i_1 i_2 i_3; j_1 j_2 j_3} = \frac{1}{3!} \langle \mathcal{L} | \mathrm{b}^{\dagger}_{i_1} \mathrm{b}^{\dagger}_{i_2} \mathrm{b}^{\dagger}_{i_3} \mathrm{b}_{j_3} \mathrm{b}_{j_2} \mathrm{b}_{j_1} | \mathcal{L} \rangle \tag{8} $$

in such a way that an alternating pattern $\mathrm{b}^{\dagger}\mathrm{b}\ \mathrm{b}^{\dagger}\mathrm{b} \ldots$ appears in all the terms. Then, the unit operator $\sum_{\mathcal{L}'} |\mathcal{L}'\rangle\langle\mathcal{L}'|$ is inserted between each pair of operators. There are thirty six equivalent ways of performing this decomposition. Our choice of pattern generates the following expression:

$$ \begin{aligned} 3!\,{}^{3}\mathrm{D}_{i_1 i_2 i_3; j_1 j_2 j_3} = {} & {}^{1}\mathrm{D}_{i_1;j_1}\, {}^{1}\mathrm{D}_{i_2;j_2}\, {}^{1}\mathrm{D}_{i_3;j_3} \\ & + 2\, {}^{2}\mathrm{D}_{i_1 i_2; j_2 j_3} \delta_{i_3;j_1} - 2\, {}^{2}\mathrm{D}_{i_1 i_3; j_2 j_3} \delta_{i_2;j_1} \\ & - {}^{1}\mathrm{D}_{i_1;j_1}\, {}^{1}\mathrm{D}_{i_2;j_3}\ \delta_{i_3;j_2} - \delta_{i_3;j_2} \sum_{\mathcal{L}' \neq \mathcal{L}} {}^{1}\mathrm{D}^{\mathcal{L}\mathcal{L}'}_{i_1;j_1}\, {}^{1}\mathrm{D}^{\mathcal{L}'\mathcal{L}}_{i_2;j_3} \\ & + {}^{1}\mathrm{D}_{i_1;j_1} \left(\sum_{\mathcal{L}' \neq \mathcal{L}} {}^{1}\mathrm{D}^{\mathcal{L}\mathcal{L}'}_{i_2;j_2}\, {}^{1}\mathrm{D}^{\mathcal{L}'\mathcal{L}}_{i_3;j_3} \right) \\ & + \left(\sum_{\mathcal{L}' \neq \mathcal{L}} {}^{1}\mathrm{D}^{\mathcal{L}\mathcal{L}'}_{i_1;j_1}\, {}^{1}\mathrm{D}^{\mathcal{L}'\mathcal{L}}_{i_2;j_2} \right) {}^{1}\mathrm{D}_{i_3;j_3} \\ & + \sum_{\mathcal{L}', \mathcal{L}'' \neq \mathcal{L}} {}^{1}\mathrm{D}^{\mathcal{L}\mathcal{L}'}_{i_1;j_1}\, {}^{1}\mathrm{D}^{\mathcal{L}'\mathcal{L}''}_{i_2;j_2}\, {}^{1}\mathrm{D}^{\mathcal{L}''\mathcal{L}}_{i_3;j_3} \end{aligned} \tag{9} $$

In order to avoid mistakes and to analyse more easily the structure of the different relations, we usually replace the formulae by graph relations. These graphs, which may be easily adapted according to need, i.e. for an orbital or a spin-orbital basis, etc., have been described previously and are recalled here in Table 1. Note that in all the relations that follow the *ordered basis* has been used.

Using the graph language, the 2-*RDM* expansion is expressed as:

$^{2}\mathrm{D}_{ij;kl} \rightarrow$

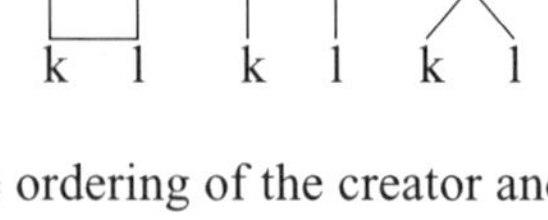

$+\, ^{2}\Delta_{ij;kl}$ (10)

As can be noticed, the same ordering of the creator and annihilator indices appears in all the graphs of a relation, unless otherwise explicitly stated.

Let us now compare the 2-*RDM* decomposition as given by (10) with the expression obtained for this same matrix with the *VLPT* decomposition:

Table 1 *Graph representation of elements of RDM's and related quantities*

Matrix element	Graphs
$\delta_{i;p}$	→ 
${}^1D_{i;p}$	→ 
${}^1\overline{D}_{i;p}$	→ 
${}^2D_{ij;pq}$	→ 
${}^3D_{ijr;pqs}$	→ 
$\sum_{\mathcal{L}'\neq\mathcal{L}} {}^1D^{\mathcal{L}\mathcal{L}'}_{i;p}\,{}^1D^{\mathcal{L}'\mathcal{L}}_{j;q}$	→ 
$\sum_{\mathcal{L}',\mathcal{L}''\neq\mathcal{L}} {}^1D^{\mathcal{L}'}_{i;p}\,{}^1D^{\mathcal{L}'\mathcal{L}''}_{j;q}\,{}^1D^{\mathcal{L}''\mathcal{L}}_{r;s}$	→ 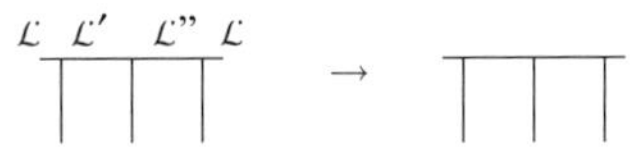

 (11)

since the Kronecker delta is equal to a sum of the 1-*RDM* and the 1-*HRDM* corresponding elements, one has:

 (12)

The result is an exact equation for the $^2\Delta$:

$^2\Delta = -$ 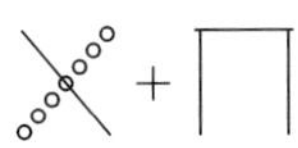 (13)

The first of these two terms cannot be considered a pure two-body term, therefore the $^2\Delta$ can only be considered as a connected diagram within the context of an antisymmetrized diagrammatical approach.

Similarly, an exact expression for the $^3\Delta$ may be easily obtained by comparing relation (9) with relation (6).

2.3. Hartree-Fock as a zero-correlation reference

An analysis of the structure of the electron correlation terms in which the reference was the antisymmetrized products of *FCI* 1-*RDM* elements was reported in [12]. The advantage of using correlated lower order matrices for building a high order reference matrix is that in an iterative process the reference is renewed in a natural way at each iteration. However, if the purpose is to analyse the structure of the electron correlation terms in an absolute manner that is, with respect to a fixed reference with no correlation, then the Hartree Fock *p*-*RDM*'s are the apropriate references. An important argument supporting this choice is that these *p*-*RDM*'s are well behaved N-representable matrices and, moreover, (as has been discussed in [15]) the set of 1-, 2-, and 3-Hartree Fock-*RDM* constitute a solution of the 1-*CSE*.

We will therefore examine in this section the different terms of the *VLPT* decomposition of a fully correlated 3-*RDM* when the set of Hartree-Fock *RDM*'s are taken as reference. This explicit reference to the Hartree Fock matrices, denoted in the formulae as D^*, can be introduced through a set of correlation matrices [15] defined as

$$ {}^{i}Y = {}^{i}D - {}^{i}D^* \tag{14} $$

In Table 2 we show the new graphs representing the Y matrices.

For $i = 1$ relation (14) takes the graph form:

$^{1}Y \rightarrow$ 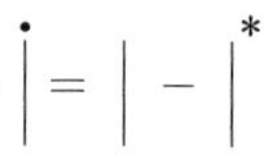 (15)

Table 2 *Graph representation of elements of the correlation matrices*

Matrix element		Graphs		
${}^{1}Y_{i;p}$	$\rightarrow$			
${}^{2}Y_{ij;pq}$	$\rightarrow$			
${}^{3}Y_{ijr;pqs}$	$\rightarrow$	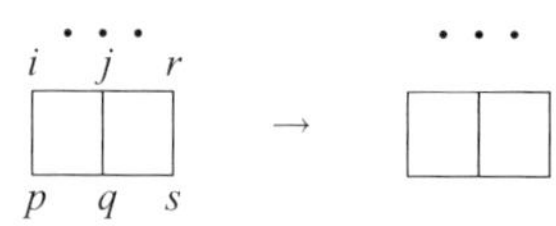		
$D^*_{i;p} - \overline{D}^*_{i;p}$	$\rightarrow$			

This relation renders explicit the averaged contribution of the N-electron correlation to the 1-*RDM*. That is, this matrix, ${}^1\mathrm{Y}$, is the one-body correlation matrix.

Similarly for $i = 2$, 3 one has:

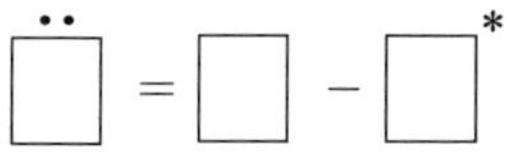 (16)

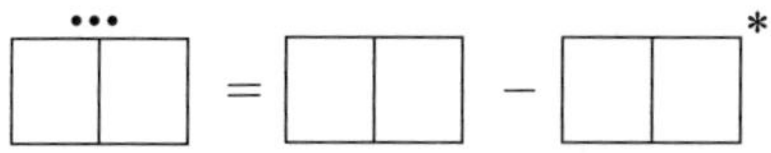 (17)

Replacing now in (11) the 1-*RDM* elements according to relation (15), and applying the definition (16), one has:

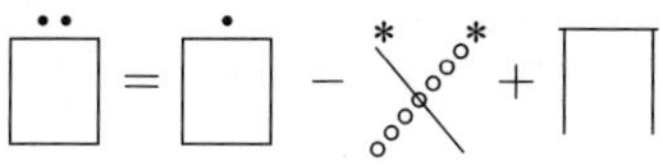 (18)

where 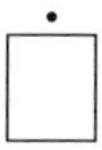 is the contribution of the one-body correlation matrix to the 2-*RDM*.

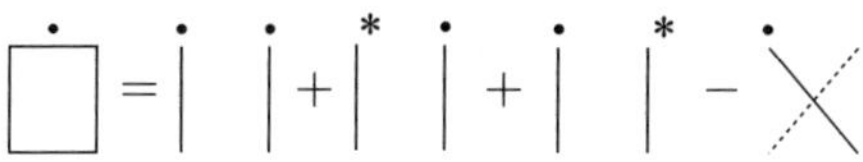 (19)

Taking now relation (9) as a starting point, and following a similar reasoning as for the second order case, one may write:

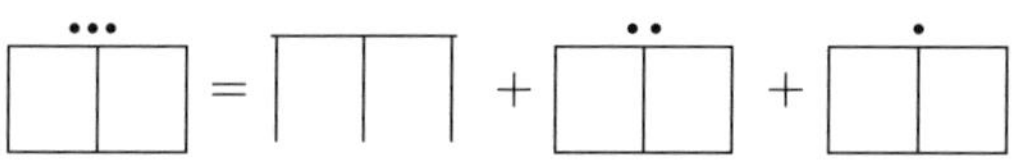 $+ {}^{1\to3}\overline{\mathrm{D}}^* + {}^{2\to3}\overline{\mathrm{D}}^*$ (20)

where

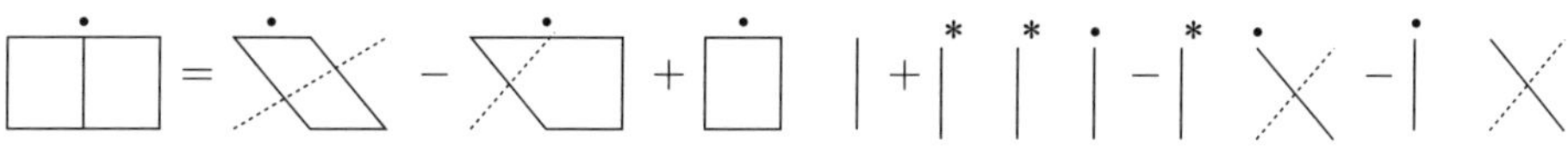

(21)

takes into account the effect on the 3-*RDM* of the one particle correlation and

 (22)

is the contribution of the pure 2-particle correlation on a 3-*RDM*.

In (20) the ${}^{1\to3}\overline{\mathrm{D}}^* + {}^{2\to3}\overline{\mathrm{D}}^*$ are terms involving the first and second order *HF HRDM*'s, which in principle contribute to the ${}^3\mathrm{Y}$. However, when using an ordered basis, the contribution of these terms to the elements explicitly considered is zero.

The decomposition just described shows clearly that the pure two-, three-, ... body correlation terms in the 2-, 3-, ... *RDM*'s are 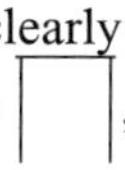, 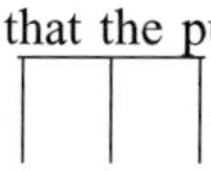, ... etc. These terms will be referred to hereafter as 2-*body*, 3-*body*, ...

Equation (9) is one of the 36 alternatives to decompose the 3-*RDM* as a sum of contributions; and the interpretation of these 2-*body*, 3-*body* terms is consistent with the option taken here.

Note that a given element of the 2-*body* and 3-*body* matrices may be compared to the corresponding element of ${}^2\Delta$ and ${}^3\Delta$ respectively; however, a global comparison of the whole matrices cannot be carried out. Thus, while the ${}^i\Delta$ possess the same symmetry properties than the corresponding *RDM*'s, the 2-*body* and 3-*body* correlation matrices do not have these symmetry properties.

3. An Analysis of the Correlation Matrices in the Linear BeH_2 Molecule

An analysis of the values taken by the different elements of the correlation matrices was recently reported [15] for the ground state of the Beryllium atom. This analysis suggested that the contributions of the 1-, 2- and 3-*body* correlation effects differed according to the kind of orbitals involved in a given element. In particular, the highest occupied (*homo*) and lowest empty (*lumo*) orbital of the *HF* configuration seemed to play an important role.

We have studied the FCI results for the ground, ${}^1\Sigma_g^+$, state of the linear, $D_{\infty h}$, BeH_2 molecule with a minimal basis set. The orbitals 1, 2, and 3 are of the σ_g type; 4 and 5 are σ_u; and 6 and 7 are degenerate π_u. The *HF* corresponds to a double occupancy of orbitals 1, 2 and 4. A bar over the orbital label indicates that its spin is β. It is important to note that 2 and 4 may be considered the *homo* of the σ_g and σ_u symmetry-shells respectively. Similarly, 3 and 5 are the respective *lumo* for the same symmetry-shells. In the following analysis we will denote generically the *homo*s by h_i and the *lumo*s by l_i.

This is a more complex system than the Beryllium atom (with a double zeta basis), retaining however a sufficient degree of simplicity for this kind of study. The number of elements which must be examined are nevertheless numerous, which is why we have limited our present study to the $\alpha\alpha\beta$ block of the 3-*RDM*. In the *ordered basis*, this matrix has 10 878 elements although the symmetry imposes a zero value to several blocks. The great majority of these elements is well estimated with the *VCP* since only 48 elements show an error (${}^3\Delta$) equal or larger than 0.001. The analysis that will now be reported is based on the study of the results for the whole matrix, although special attention has been given to these critical 48 elements. Both for the sake of clarity and conciseness, only the most significant examples (multiplied by $\times 10^2$) are explicitly reported in table 3, and all the other results are available upon request.

The elements can be classified into four groups. The first group includes those elements with non-negligible contributions from all the three kinds of correlation matrices which do not completly cancel each other. In the second group, the 1-*body* contribution is negligible. In the third group, both the 1- and the 2-*body* contributions are negligible. And finally, in the fourth group are included those elements which are rather well estimated by the *HF* method (the *HF* value is close to the *FCI* one) due to a global and nearly complete cancellation of the correlation errors. These results may be summarized as:

- All the first group elements are diagonal and their labels are of the type $h_1 h_2 l;\ h_1 h_2 l$ (the order in which these symbols appear in the element label is

Table 3 *Contribution* $(\times 10^2)$ *of the different correlation matrices to some relevant elements*

Matrix element	D − D*	$^3\Delta$	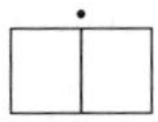	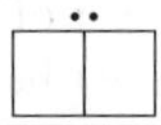	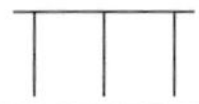
$23\,\bar{4};\,23\,\bar{4}$	0.177	0.217	1.395	−1.441	0.223
$23\,\bar{2};\,23\,\bar{2}$	0.680	0.300	1.392	−1.024	0.312
$34\,\bar{4};\,34\,\bar{4}$	0.331	0.670	1.398	−0.843	−0.224
$23\,\bar{3};\,23\,\bar{3}$	0.680	−0.300	0.020	0.965	−0.305
$23\,\bar{4};\,45\,\bar{4}$	−0.114	0.341	0.000	−0.455	0.341
$23\,\bar{2};\,34\,\bar{4}$	−0.461	−0.455	0.000	−0.558	0.097
$34\,\bar{2};\,45\,\bar{4}$	0.127	−0.384	0.000	0.518	−0.391
$24\,\bar{2};\,35\,\bar{3}$	−0.252	−0.218	0.000	−0.065	−0.187
$24\,\bar{4};\,35\,\bar{5}$	0.171	0.189	0.000	−0.044	0.215
$23\,\bar{3};\,45\,\bar{3}$	−0.347	−0.340	0.000	−0.007	−0.340
$25\,\bar{3};\,34\,\bar{3}$	0.261	0.255	0.000	0.006	0.255
$24\,\bar{2};\,56\,\bar{6}$	−0.405	−0.374	0.000	0.000	−0.405
$24\,\bar{4};\,36\,\bar{6}$	−0.709	−0.708	0.000	−0.000	**−0.708**
$25\,\bar{4};\,45\,\bar{2}$	0.196	0.194	0.000	0.000	0.196
$26\,\bar{2};\,26\,\bar{2}$	0.000	0.306	0.321	−0.631	0.310
$26\,\bar{6};\,27\,\bar{7}$	0.006	−0.320	0.000	0.326	−0.320
$26\,\bar{6};\,26\,\bar{6}$	0.000	−0.152	0.000	0.152	−0.153

variable). In all these elements there is a large but not complete cancellation of the 1-*body* and 2-*body* contributions among themselves and also, to some extent, of the 3-*body* part.

- As has been mentioned, in the second group the 1-*body* part does not contribute or its contribution is negligible. The element labels of this group are of the types $h_1 h_2 l;\ h_1 h_2 l$ or $l_1 l_2 h;\ l_1 l_2 h$.
- In the third group the non-negligible correlation effects are only due to the 3-body correlation effects. Most of these elements are the same type as the second group ones. The new types appearing in this group are $h_1 l_1 l_2;\ h_2 \pi\pi$ (where the order of the two *lumo*'s may differ in both elements) and $h_1 h_1 h_2;\ l_1 l_1 l_2$ where l_1 may also be one or the other of the π orbitals.
- As has been mentioned, the fourth group is formed by the elements where the cancellation of correlation effects is practically complete. In all these elements the orbital 2, which is the *homo* of the σ_g symmetry is either singly or doubly occupied in each trio of indices. The partners of 2 are all the empty orbitals but only the π orbitals may appear singly occupied as partners of the $2\bar{2}$.

The data presented in this section show how the symmetry, the degree of occupation of the orbitals as well as the partner orbitals determining the global label of a 3-*RDM* element, influence the relative contribution of the three kind of correlation matrices. Thus, it is now easy to understand why the *VCP* errors (the ${}^{i}\Delta$) are not of the same order for all the *RDM* elements [10]. A very positive consequence of this fact is that since the *VCP* errors affect only a rather small part of the matrix elements, and since it is possible to predict which kind of elements will be critical, we may limit ourselves to correcting these elements.

4. Correcting the Errors

The main purpose of this section is to analyse the *NY* correction and, as a come-out of this study, to suggest what we think is a more economical and effective modification of it (at least in the $\alpha\alpha\beta$-block case).

Before proceeding further, let us describe the main *NY* correction as we have adapted it for spin-orbitals for the $\alpha\alpha\beta$ 3-*RDM* block.

$$ {}^{3}\Delta_{ij\,\bar{l};mn\,\bar{q}} \approx \mathcal{A}\sum_{t} {}^{2}\Delta_{ij;mt}({}^{1}\mathrm{D}^{*}_{t;t} - {}^{1}\overline{\mathrm{D}}^{*}_{t;t})^{2}\Delta_{t\,\bar{l};n\,\bar{q}} \tag{23} $$

where $\mathcal{A}$ is the antisymmetrizer of the indices. In this case, this operator implies a sum of the eight following terms:

$$ A_1 = +\sum_{t}\{{}^{2}\Delta^{\alpha\alpha}_{ij;mt}({}^{1}\mathrm{D}^{*}_{t;t} - {}^{1}\overline{\mathrm{D}}^{*}_{t;t})^{2}\Delta^{\alpha\beta}_{t\,\bar{l};n\,\bar{q}}\} $$

$$ A_2 = -\sum_{t}\{{}^{2}\Delta^{\alpha\alpha}_{ij;nt}({}^{1}\mathrm{D}^{*}_{t;t} - {}^{1}\overline{\mathrm{D}}^{*}_{t;t})^{2}\Delta^{\alpha\beta}_{t\,\bar{l};m\,\bar{q}}\} $$

$$ A_3 = +\sum_{t}\{{}^{2}\Delta^{\alpha\beta}_{i\,\bar{l};m\,\bar{t}}({}^{1}\mathrm{D}^{*}_{\bar{t};\bar{t}} - {}^{1}\overline{\mathrm{D}}^{*}_{\bar{t};\bar{t}})^{2}\Delta^{\alpha\beta}_{j\,\bar{t};n\,\bar{q}}\} $$

$$ A_4 = -\sum_{t}\{{}^{2}\Delta^{\alpha\beta}_{i\,\bar{l};n\,\bar{t}}({}^{1}\mathrm{D}^{*}_{\bar{t};\bar{t}} - {}^{1}\overline{\mathrm{D}}^{*}_{\bar{t};\bar{t}})^{2}\Delta^{\alpha\beta}_{j\,\bar{t};m\,\bar{q}}\} $$

$$ A_5 = +\sum_{t}\{{}^{2}\Delta^{\alpha\beta}_{i\,\bar{l};t\,\bar{q}}({}^{1}\mathrm{D}^{*}_{\bar{t};\bar{t}} - {}^{1}\overline{\mathrm{D}}^{*}_{\bar{t};\bar{t}})^{2}\Delta^{\alpha\alpha}_{tj;mn}\} $$

$$ A_6 = -\sum_{t}\{{}^{2}\Delta^{\alpha\beta}_{j\,\bar{l};m\,\bar{t}}({}^{1}\mathrm{D}^{*}_{\bar{t};\bar{t}} - {}^{1}\overline{\mathrm{D}}^{*}_{\bar{t};\bar{t}})^{2}\Delta^{\alpha\beta}_{i\,\bar{t};n\,\bar{q}}\} $$

$$ A_7 = +\sum_{t}\{{}^{2}\Delta^{\alpha\beta}_{j\,\bar{l};n\,\bar{t}}({}^{1}\mathrm{D}^{*}_{\bar{t};\bar{t}} - {}^{1}\overline{\mathrm{D}}^{*}_{\bar{t};\bar{t}})^{2}\Delta^{\alpha\beta}_{i\,\bar{t};m\,\bar{q}}\} $$

$$ A_8 = -\sum_{t}\{{}^{2}\Delta^{\alpha\beta}_{j\,\bar{l};t\,\bar{q}}({}^{1}\mathrm{D}^{*}_{t;t} - {}^{1}\overline{\mathrm{D}}^{*}_{t;t})^{2}\Delta^{\alpha\alpha}_{ti;mn}\} \tag{24} $$

According to equation (13), the ${}^{2}\Delta^{\alpha\beta}$ is identical to the 2-*body* correlation term, while in the $\alpha\alpha$ one the contribution of the holes term also plays an important role.

In nearly all the cases the *NY* algorithm corrects the *VCP* error (before renormalization) within an order of magnitude. As an example, the most significant results obtained with the *NY* approximation are reported in the third column of table 4. The exceptions (in bold face) are no more than four elements in the whole matrix. They are coincident with rather large errors in the *VCP* approximation, which are not corrected by the *NY* algorithm that is being analysed here. Nakatsuji and Yasuda propose a different perturbative algorithm for estimating these elements whose label is of the type: *occupied, occupied, occupied; empty, empty, empty.* This point will, nevertheless, be discussed in a future publication.

An opposite situation appears in those elements (not given here) for which the *VCP* error is smaller than 0.001 where a pure 3-*body* contribution is well estimated with the simple *VCP*. To this cathegory belong the elements of the type: *occupied empty occupied; empty occupied empty*, where the *occupied* orbital need not be the *homo*.

When analysing the separate contributions to the total correction of each of the eight A_i terms of equation (24), we have found that only one term contributed non-negligibly to correcting each element. Moreover, the terms giving a non-negligible contribution were only those where the sum over the common index linked two ${}^2\Delta^{\alpha\beta}$.

The four terms, which have been found to be efficient for approximating ${}^3\Delta_{ij\bar{l};mn\bar{q}}$, may be given a pictorial image when expressed in our graph language (see last entry of table 2). They are:

$A_3 \rightarrow$ 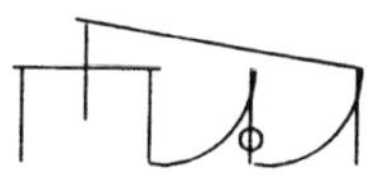

$A_4 \rightarrow -$ 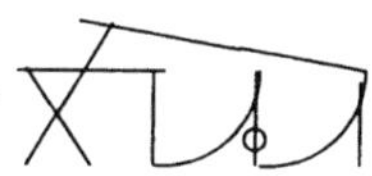

$A_6 \rightarrow -$ 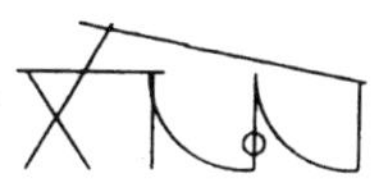

$A_7 \rightarrow$ 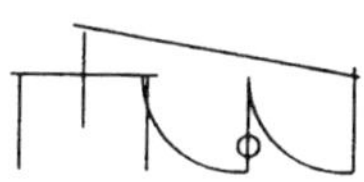

where the curved line represents a contraction; that is, the sum over all posible values of a common index $\bar{t}$. The values of these terms for the selected elements are reported in table 4 where all the quantities appear multiplied by 10^2. In this table we have omitted the negligible quantities in order to simplify their reading. As can be seen, in general, the value of one of these terms approximates better the ${}^3\Delta$ than the *NY* antisymmetrized sum. This is very convenient, since only these terms need be evaluated for the few critical elements that must be corrected. Another question which arises when examining the results given in table 4 is whether there is some kind of definite correspondence between each kind of graph and the type of each element. The answer is immediate since the following correspondence is easy to verify:

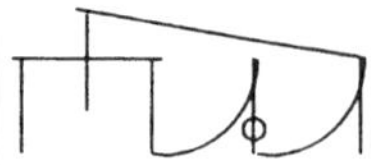	$\begin{matrix} h & l & \bar{h} \\ l & h & \bar{h} \end{matrix}$	or	$\begin{matrix} l & h & \bar{l} \\ h & l & \bar{l} \end{matrix}$
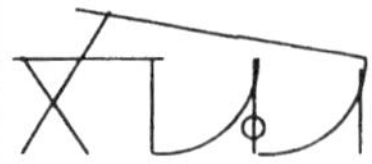	$\begin{matrix} h & l & \bar{h} \\ h & l & \bar{h} \end{matrix}$	or	$\begin{matrix} l & h & \bar{l} \\ l & h & \bar{l} \end{matrix}$
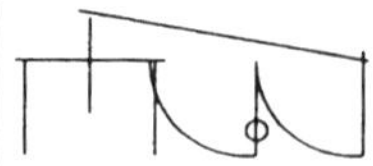	$\begin{matrix} l & h & \bar{h} \\ h & l & \bar{h} \end{matrix}$	or	$\begin{matrix} h & l & \bar{l} \\ l & h & \bar{l} \end{matrix}$
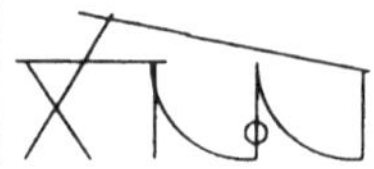	$\begin{matrix} l & h & \bar{h} \\ l & h & \bar{h} \end{matrix}$	or	$\begin{matrix} h & l & \bar{l} \\ h & l & \bar{l} \end{matrix}$

where h and l may be any of the *homo*'s and *lumo*'s respectively. The theoretical justification laying behind this correspondence remains now to be investigated. This table shows the great advantage of the graph representation which permits a quick analysis of the essential structural properties hidden behind the string of indices in the algebraic formulae.

We tried to find out if the performance of the procedure could be improved by replacing the $({}^1\text{D}^* - {}^1\overline{\text{D}}^*)$ in the *NY* algorithm by an equivalent one involving the *FCI* matrices. The results practically did not vary and some of the slight changes that this

Table 4 *Analysis of the NY approximation* $(\times 10^2)$ *in the* $\alpha\alpha\beta$*3-RDM: more significative elements*

Matrix element	${}^3\Delta$	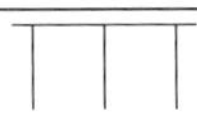	NY	A_3	A_4	A_7	A_6
23 $\bar{4}$; 23 $\bar{4}$	0.217	0.223	0.194	–	0.216	–	–
23 $\bar{2}$; 23 $\bar{2}$	0.300	0.312	0.283	–	0.309	–	–
34 $\bar{4}$; 34 $\bar{4}$	0.670	−0.224	0.630	–	–	–	0.675
23 $\bar{3}$; 23 $\bar{3}$	−0.300	−0.305	−0.284	–	–	–	−0.309
23 $\bar{4}$; 45 $\bar{4}$	0.341	0.341	0.328	–	0.344	–	–
23 $\bar{2}$; 34 $\bar{4}$	−0.455	0.097	−0.437	−0.457	–	–	–
34 $\bar{2}$; 45 $\bar{4}$	−0.384	−0.391	−0.364	–	–	−0.382	–
24 $\bar{2}$; 35 $\bar{3}$	−0.218	−0.187	**−0.020**	–	–	–	–
24 $\bar{4}$; 35 $\bar{5}$	0.189	0.215	**0.019**	–	–	–	–
23 $\bar{3}$; 45 $\bar{3}$	−0.340	−0.340	−0.327	–	–	–	−0.344
24 $\bar{2}$; 56 $\bar{6}$	−0.374	−0.405	**0.019**	–	–	–	–
25 $\bar{3}$; 34 $\bar{3}$	0.255	0.255	0.242	–	–	0.258	–
24 $\bar{4}$; 36 $\bar{6}$	−0.708	**−0.708**	**0.026**	–	–	–	–
25 $\bar{4}$; 45 $\bar{2}$	0.194	0.196	0.185	–	0.194	–	–
26 $\bar{6}$; 26 $\bar{6}$	−0.317	−0.153	−0.304	–	–	–	−0.310
26 $\bar{6}$; 27 $\bar{7}$	−0.320	−0.320	−0.306	–	–	–	−0.310
26 $\bar{2}$; 26 $\bar{2}$	0.306	0.310	0.295	–	0.310	–	–

modification introduced went in the wrong direction. However, this possibility exists and may prove useful in the case of excited states.

4.1. Concluding Remarks

The work reported here can only be considered as one part of a much larger task. Thus, while keeping as test samples the Beryllium atom and the BeH_2 molecule, a similar analysis must be carried out for the $\alpha\alpha\alpha$ 3-*RDM* and for all the spin blocks of the 4-*RDM*.

With a more extended basis set and for more complex systems, the number of element types will increase; but, hopefully, the performance of the *VCP* and that of the modified *NY* proposed here may probably be extrapolated. This working hypothesis will be tested in the future by extending the basis set as much as possible.

Another of our present aims is to continue investigating the theoretical properties of the 2- and 3-*body* terms. Thus, there are many exact relations linking the different 3-*body* terms arising from each of the 36 options of equation (9) as well as their holes counterparts [16]. These relations are interesting by themselves, because they widen our understanding of the problem and may be helpful for improving our approximations. Another important theoretical question is, as mentioned above, to investigate the reason why one A_i corrects the VCP error which in some cases is only due to the 3-*body* contribution but which may also involve other kind of correlation effects.

Let us finish by recalling that the final aim of this line of research is to build up an optimized program for solving iteratively the 2-*CSE*. In spite of the important questions which still remain to be solved before this aim is fulfilled, we expect this methodology to be of standard use in a not too distant future.

Acknowledgements

This work has been supported by the Dirección General de Investigación Científica y Técnica del Ministerio de Educación y Ciencia under project PB96-0891.

References

1. H. Nakatsuji, *Phys. Rev. A* **14**, 41 (1976).
2. L. Cohen and C. Frishberg, *Phys. Rev. A* **13**, 927 (1976).
3. C. Valdemoro, pg. 275 in *Density Matrices and Density Functionals*, Proceedings of the A.J. Coleman Symposium, Kingston, Ontario, 1985, edited by R. Erdahl and V. Smith (Reidel, Dordrecht, 1987).
4. J. E. Harriman, *Phys. Rev. A* **19**, 1893 (1979).
5. C. Valdemoro, *Phys. Rev. A* **45**, 4462 (1992).
6. F. Colmenero, C. Perez del Valle, C. Valdemoro, *Phys. Rev. A* **47**, 971 (1993).
7. F. Colmenero, C. Valdemoro, *Phys. Rev. A* **47**, 979 (1993).
8. F. Colmenero and C. Valdemoro, *Int. J. Quantum Chem.*, **51**(6), 369 (1994).
9. C. Valdemoro, L.M. Tel, E. Pérez-Romero, *Adv. Quant. Chem.*, **28**, 33, (1997).
10. H. Nakatsuji, K. Yasuda, *Phys. Rev. Let.* **76**, 1039 (1996).
11. K. Yasuda, H. Nakatsuji, *Phys. Rev. A* **56**, 2648 (1997).
12. C. Valdemoro, M.P. Lara-Castells, E. Pérez-Romero, L.M. Tel, *Adv. Quant. Chem.* **31**, 37 (1999).
13. D.A. Mazziotti, *Phys. Rev. A* **57**, 4219 (1998).

14. D.A. Mazziotti, *Chem. Phys. Lett.* **289**, 419 (1998).
15. C. Valdemoro, p. 187 in Correlation and Localization, volume in honour of Prof. E. Kapuy; ed. P. Surjan, publ. Springer Verlag in the series Topics in Current Chemistry. (1999)
16. L. Tel, E. Pérez-Romero, C. Valdemoro, unpublished results.

Part II

Electron Correlation Effects

Many-Particle Sturmians Applied to Molecules

John Avery and Stephan Sauer
H.C. Ørsted Institute, University of Copenhagen, Universitetsparken 5, DK-2100 Copenhagen, Denmark

Abstract

The method of many-electron Sturmian basis functions is applied to molecules. The basis potential is chosen to be the attractive Coulomb potential of the nuclei in the molecule. When such basis functions are used, the kinetic energy term vanishes from the many-electron secular equation, the matrix representation of the nuclear attraction potential is diagonal, the Slater exponents are automatically optimized, convergence is rapid, and a solution to the many-electron Schrödinger eqeuation, including correlation, is obtained directly, without the use of the self-consistent field approximation.

1. Introduction

Hydrogenlike orbitals can be written in the form:

$$\chi_{nlm}(\mathbf{x}) = R_{nl}(r) Y_{lm}(\theta, \phi) \tag{1}$$

where

$$R_{nl}(r) = \mathcal{N}_{nl}(2k_\mu r)^l e^{-k_\mu r} F(l+1-n|2l+2|2k_\mu r)$$

$$\mathcal{N}_{nl} = \frac{2k_\mu^{3/2}}{(2l+1)!}\sqrt{\frac{(l+n)!}{n(n-l-1)!}} \tag{2}$$

and where $F(a|b|z)$ is a confluent hypergeometric function. The functions $\chi_{nlm}(\mathbf{x})$ obey the following equations:

$$(\Delta - k_\mu^2)\chi_{nlm}(\mathbf{x}) = -\frac{2nk_\mu}{r}\chi_{nlm}(\mathbf{x}) \tag{3}$$

$$\int d^3x |\chi_{nlm}(\mathbf{x})|^2 \frac{1}{r} = \frac{k_\mu}{n} \tag{4}$$

and

$$\int d^3x |\chi_{nlm}(\mathbf{x})|^2 = 1 \tag{5}$$

where

$$\Delta \equiv \frac{\partial^2}{\partial x^2} + \frac{\partial^2}{\partial y^2} + \frac{\partial^2}{\partial z^2} \tag{6}$$

If $k_\mu = \zeta/n$, then the functions $R_{nl}(r)$ are just the familiar hydrogenlike radial functions, expressed in atomic units:

A. Hernandez-Laguna et al. (eds.), Quantum Systems in Chemistry and Physics, Vol I: Basic Problems and Model Systems, 19–39.

$$R_{10}(r) = 2\zeta^{3/2} e^{-\zeta r}$$

$$R_{20}(r) = \frac{\zeta^{3/2}}{\sqrt{2}} \left(1 - \frac{\zeta r}{2}\right) e^{-\zeta r/2}$$

$$R_{21}(r) = \frac{\zeta^{5/2}}{2\sqrt{6}} r e^{-\zeta r/2} \tag{7}$$

and so on. Early in the history of quantum chemistry, it was thought that it might be possible to use basis sets of the form shown in equation (7), with constant ζ, as building blocks for constructing the wave functions of many-electron systems. However, it soon became clear that such a set of basis functions is not complete unless the continuum is included. To remedy this situation, Shull and Löwdin [1] introduced basis sets of the form shown in equation (2), but with k_μ held constant, independent of the quantum number n. From equation (3), we can see that such a set of functions represents solutions to the Schrödinger equation for a weighted hydrogenlike potential, the weighting factor n being especially chosen in order to make all of the functions in the set correspond to the same energy, $\epsilon = -k_\mu^2/2$, regardless of the quantum numbers. If we let $t = k_\mu r$, then the first few basis functions in such a set are given by

$$R_{10}(t) = 2k_\mu^{3/2} e^{-t}$$

$$R_{20}(t) = 2k_\mu^{3/2}(1 - t)e^{-t}$$

$$R_{21}(t) = 2k_\mu^{3/2} \frac{1}{\sqrt{3}} t e^{-t} \tag{8}$$

It was shown by Shull and Löwdin that such basis sets with constant k_μ are complete without inclusion of the continuum. Later, Rotenberg [2] gave the name 'Sturmian' to this type of basis set, in order to emphasize its relationship with Sturm-Liouville theory. The weighted orthonormality relations for Sturmian basis sets were studied by Goscinski [3] and by Weniger [4]; and Weniger pointed out that such a set of functions forms the basis of a Sobolev space. In order to see that a Sturmian basis set obeys a weighted orthonormality relation, we can consider two different functions in the set, obeying

$$(\Delta - k_\mu^2)\chi_{nlm}(\mathbf{x}) = -\frac{2nk_\mu}{r}\chi_{nlm}(\mathbf{x})$$

$$(\Delta - k_\mu^2)\chi^*_{n'lm}(\mathbf{x}) = -\frac{2n'k_\mu}{r}\chi^*_{n'lm}(\mathbf{x}) \tag{9}$$

where $n \neq n'$ If we multiply these equations from the left respectively by $\chi^*_{n'lm}(\mathbf{x})$ and $\chi_{nlm}(\mathbf{x})$, integrate over the electron coordinates, and finally subtract the second equation from the first, we obtain:

$$0 = (n' - n)\int d^3x \chi^*_{n'lm}(\mathbf{x})\frac{1}{r}\chi_{nlm}(\mathbf{x}) \tag{10}$$

where we have made use of the Hermiticity of the operator $(\Delta - k_\mu^2)$. Thus when $n' \neq n$

the two functions are orthogonal with the weighting factor $1/r$. Combining equations (10) and (4), and making use of the orthonormality of the spherical harmonics, we obtain:

$$\int d^3x \chi^*_{n'l'm'}(\mathbf{x}) \frac{1}{r} \chi_{nlm}(\mathbf{x}) = \frac{k_\mu}{n} \delta_{n',n} \delta_{l',l} \delta_{m',m} \tag{11}$$

It is interesting to examine the momentum-space orthonormality relations of the Fourier transforms of hydrogenlike Sturmian basis sets. If we let

$$\chi^t_{nlm}(\mathbf{p}) \equiv \frac{1}{(2\pi)^{3/2}} \int d^3x e^{-i\mathbf{p}\cdot\mathbf{x}} \chi_{nlm}(\mathbf{x}) \tag{12}$$

then

$$\chi_{nlm}(\mathbf{x}) = \frac{1}{(2\pi)^{3/2}} \int d^3p e^{i\mathbf{p}\cdot\mathbf{x}} \chi^t_{nlm}(\mathbf{p}) \tag{13}$$

Inserting equation (13) into (3), we obtain:

$$\frac{1}{(2\pi)^{3/2}} \int d^3p e^{i\mathbf{p}\cdot\mathbf{x}} (p^2 + k_\mu^2) \chi^t_{nlm}(\mathbf{p}) = \frac{2nk_\mu}{r} \chi_{nlm}(\mathbf{x}) \tag{14}$$

and taking the Fourier transform of both sides gives:

$$(p^2 + k_\mu^2) \chi^t_{nlm}(\mathbf{p}) = \frac{2nk_\mu}{(2\pi)^{3/2}} \int d^3x e^{-i\mathbf{p}\cdot\mathbf{x}} \frac{1}{r} \chi_{nlm}(\mathbf{x}) \equiv 2nk_\mu \left[\frac{1}{r}\chi_{nlm}\right]^t \tag{15}$$

Since the scalar product of two functions in direct space is equal to the scalar product of their Fourier transforms in reciprocal space, we have

$$\int d^3x \chi^*_{n'l'm'}(\mathbf{x}) \frac{1}{r} \chi_{nlm}(\mathbf{x}) = \int d^3p [\chi^*_{n'l'm'}]^t \left[\frac{1}{r}\chi_{nlm}\right]^t \tag{16}$$

Finally, combining (15), (16) and (11), we obtain:

$$\int d^3p \left(\frac{p^2 + k_\mu^2}{2k_\mu^2}\right) \chi^{t*}_{n'l'm'}(\mathbf{p}) \chi^t_{nlm}(\mathbf{p}) = \delta_{n',n} \delta_{l',l} \delta_{m',m} \tag{17}$$

The momentum-space orthonormality relation for hydrogenlike Sturmian basis sets, equation(17), can be shown to be closely related to the orthonormality relation for hyperspherical harmonics in a 4-dimensional space. This relationship follows from the results of Fock [5], who was able to solve the Schrödinger equation for the hydrogen atom in reciprocal space by projecting 3-dimensional p-space onto the surface of a 4-dimensional hypersphere with the mapping:

$$
\begin{aligned}
u_1 &= \frac{2p_1 k_\mu}{p^2 + k_\mu^2} \\
u_2 &= \frac{2p_2 k_\mu}{p^2 + k_\mu^2} \\
u_3 &= \frac{2p_3 k_\mu}{p^2 + k_\mu^2} \\
u_4 &= \frac{p^2 - k_\mu^2}{p^2 + k_\mu^2}
\end{aligned} \tag{18}
$$

Fock was then able to show that the momentum-space Schrödinger equation for the hydrogen atom has properly normalized solutions of the form

$$
\chi^t_{nlm}(\mathbf{p}) = \frac{4k_\mu^{5/2}}{(p^2 + k_\mu^2)^2} Y_{n-1,l,m}(\mathbf{u}) \tag{19}
$$

where $Y_{n-1,l,m}(\mathbf{u})$ is a hyperspherical harmonic on the surface of the hypersphere onto which momentum space is mapped by Fock's transformation. The hyperangular solid angle element $d\Omega$ can be shown to be related to the volume element in momentum space by [6]

$$
d^3 p = \left(\frac{p^2 + k_\mu^2}{2k_\mu} \right)^3 d\Omega \tag{20}
$$

Substitution of (19) and (20) into equation (17) yields

$$
\int d\Omega\, Y^*_{n'-1,l',m'}(\mathbf{u}) Y_{n-1,l,m}(\mathbf{u}) = \delta_{n',n}\delta_{l',l}\delta_{m',m} \tag{21}
$$

which is just the orthonormality relation for the set of hyperspherical harmonics. Notice that for the relationship to be valid, k_μ must be held constant for all the members of the set.

2. Solutions to the One-Electron Many-Center Wave Equation

In 1965, Shibuya and Wulfman [7] extended Fock's momentum-space treatment of the hydrogen atom in such a way that they were able to solve the Schrödinger equation for an electron moving in the many-center attractive potential of a collection of nuclei. These authors made use of the properties of hyperspherical harmonics in their pioneering paper, but since most quantum chemists are unfamiliar with those properties, we shall try to review the theory of Shibuya and Wulfman in an alternative way, using the terminology of Fourier transforms: We begin by recalling that the Fourier transform of $1/r$ is given by

$$\left[\frac{1}{r}\right]^t \equiv \frac{1}{(2\pi)^{3/2}}\int d^3x e^{-i\mathbf{p}\cdot\mathbf{x}}\frac{1}{r} = \sqrt{\frac{2}{\pi}}\frac{1}{p^2} \tag{22}$$

Making use of equation (15) and (22), together with the Fourier convolution theorem, we can see that

$$(p^2 + k_\mu^2)\chi^t_{nlm}(\mathbf{p}) = \frac{nk_\mu}{2\pi^2}\int d^3p' \frac{1}{|\mathbf{p}-\mathbf{p}'|^2}\chi^t_{nlm}(\mathbf{p}') \tag{23}$$

The kernel of this integral equation can be represented by the expansion:

$$\frac{1}{2\pi^2}\frac{1}{|\mathbf{p}-\mathbf{p}'|^2} = \frac{(p^2+k_\mu^2)(p'^2+k_\mu^2)}{2k_\mu^2}\sum_{n'l'm'}\frac{1}{n'k_\mu}\chi^t_{n'l'm'}(\mathbf{p})\chi^{t*}_{n'l'm'}(\mathbf{p}') \tag{24}$$

Equation (24) can be derived from the theory of hyperspherical harmonics and Gegenbauer polynomials; but for readers unfamiliar with this theory, the expansion can be made plausible by substitution into the right-hand side of equation (23). With the help of the momentum-space orthonormality relations, (17), it can then be seen that right-hand side of (23) reduces to the left-hand side, which must be the case if the integral equation is to be satisfied. Let us now consider an electron moving in the attractive Coulomb potential of a collection of nuclei:

$$v(\mathbf{x}) = -\sum_a \frac{Z_a}{|\mathbf{x}-\mathbf{X}_a|} \tag{25}$$

Here Z_a and $\mathbf{X}_a$ represent the charge and position of the ath nucleus. The Schrödinger equation for such an electron can be written in a form closely analogous to equation (3):

$$[\Delta - k_\mu^2]\varphi_\mu(\mathbf{x}) = 2b_\mu k_\mu v(\mathbf{x})\varphi_\mu(\mathbf{x}) \tag{26}$$

If we let $b_\mu k_\mu = 1$, then (26) is just the Schrödinger equation for an electron moving in the potential $v(\mathbf{x})$, the energy of the electron being given by $\epsilon = -k_\mu^2/2$. However, in a later section of this paper, we shall use the functions $\varphi_\mu(\mathbf{x})$ as building blocks for constructing the many-electron wave function of a molecule; and for this purpose we shall relax the condition $b_\mu k_\mu = 1$, and instead we shall impose on the parameters b_μ and k_μ subsidiary conditions derived from the many-electron wave equation. The momentum-space Schrödinger equation corresponding to (26) is:

$$(p^2 + k_\mu^2)\varphi^t_\mu(\mathbf{p}) = -\frac{2b_\mu k_\mu}{(2\pi)^{3/2}}\int d^3p' v^t(\mathbf{p}'-\mathbf{p})\varphi^t_\mu(\mathbf{p}') \tag{27}$$

where

$$\varphi^t_\mu(\mathbf{p}) \equiv \frac{1}{(2\pi)^{3/2}}\int d^3x e^{-i\mathbf{p}\cdot\mathbf{x}}\varphi_\mu(\mathbf{x}) \tag{28}$$

and

$$v^t(\mathbf{p}-\mathbf{p}') = -\sqrt{\frac{2}{\pi}}\sum_a \frac{Z_a}{|\mathbf{p}-\mathbf{p}'|^2}e^{i(\mathbf{p}-\mathbf{p}')\cdot\mathbf{X}_a} \tag{29}$$

Substituting (24) and (29) into (27) and cancelling the common factor $(p^2 + k_\mu^2)$ from both sides of the resulting equation, we obtain:

$$\varphi_\mu^t(\mathbf{p}) = b_\mu \sum_{a'n'l'm'} \sqrt{\frac{Z_{a'}}{n'}} e^{i\mathbf{p}\cdot\mathbf{X}_{a'}} \chi_{n'l'm'}^t(\mathbf{p}) \times \int d^3p' \left(\frac{p'^2 + k_\mu^2}{2k_\mu^2}\right) \sqrt{\frac{Z_{a'}}{n'}} e^{-i\mathbf{p}'\cdot\mathbf{X}_{a'}} \chi_{n'l'm'}^{t*}(\mathbf{p}')\varphi_\mu^t(\mathbf{p}') \tag{30}$$

which can be rewritten in the form:

$$\varphi_\mu^t(\mathbf{p}) = b_\mu \sum_\tau \xi_\tau(\mathbf{p}) \int d^3p' \left(\frac{p'^2 + k_\mu^2}{2k_\mu^2}\right) \xi_\tau^*(\mathbf{p}')\varphi_\mu^t(\mathbf{p}') \tag{31}$$

where

$$\xi_\tau(\mathbf{p}) \equiv \sqrt{\frac{Z_a}{n}} e^{i\mathbf{p}\cdot\mathbf{X}_a} \chi_{nlm}^t(\mathbf{p}) \tag{32}$$

and where we have let the index τ stand for the set of indices, $\{a, n, l, m\}$. If we now expand the Fourier-transformed one-electron wave functions in terms of the basis functions shown in (32),

$$\varphi_\mu^t(\mathbf{p}) = \sum_\tau \xi_\tau(\mathbf{p}) C_{\tau,\mu} \tag{33}$$

we can see that the integral equation (31) will be fulfilled provided that

$$\sum_\tau [K_{\tau',\tau} - b_\mu^{-1}\delta_{\tau',\tau}]C_{\tau,\mu} = 0 \tag{34}$$

where

$$K_{\tau',\tau} \equiv \int d^3p \left(\frac{p^2 + k_\mu^2}{2k_\mu^2}\right) \xi_{\tau'}^*(\mathbf{p})\xi_\tau(\mathbf{p}) \tag{35}$$

Making use of equations (19), (20) and (32), we can rewrite (35) in the form:

$$K_{\tau',\tau} = \sqrt{\frac{Z_{a'}Z_a}{n'n}} S_{n'l'm'}^{nlm}(\mathbf{s}) \tag{36}$$

where

$$S_{n'l'm'}^{nlm}(\mathbf{s}) \equiv \int d\Omega e^{i\mathbf{p}\cdot\mathbf{R}} Y_{n'-1,l',m'}^*(\mathbf{u}) Y_{n-1,l,m}(\mathbf{u}) \tag{37}$$

and $\mathbf{s} \equiv k_\mu \mathbf{R} \equiv k_\mu(\mathbf{X}_a - \mathbf{X}_{a'})$. Several alternative methods are available for evaluation of the Shibuya-Wulfman integrals, $S_{n'l'm'}^{nlm}$. They can be evaluated by means of the coupling coefficients of the hyperspherical harmonics [7–11]. Alternatively [12–14], the Shibuya-Wulfman integrals can be evaluated by means of a theorem which states that

$$\frac{1}{2\pi^2}\int d\Omega e^{i\mathbf{p}\cdot\mathbf{R}} u_4^k h_l(u_j) = \mathcal{F}_{k,l}(s) h_l(s_j) \tag{38}$$

where

$$\mathcal{F}_{k,l}(s) \equiv \frac{i^l k! e^{-s}}{2^k(2l+1)!}\left[\sum_{q=0}^{[k/2]} \frac{2(k+2l+1-2q)!F(2q-k|2l+2|2s)}{q!(k+l+1-q)!(k-2q)!}\right.$$
$$\left. -(k+1)1 \sum_{q=0}^{[(k+1)/2]} \frac{(k+2l+2-2q)!F(2q-k-1|2l+2|2s)}{q!(k+l+2-q)!(k+1-2q)!}\right] \tag{39}$$

In (38), $h_l(u_j)$ is an harmonic polynomial of order l in u_1, u_2 and u_3, while $h_l(s_j)$ is the same harmonic polynomial with u_j replaced by s_j. The Shibuya-Wulfman integrals can then be calculated by resolving the product of hyperspherical harmonics in (37) into terms of the form $u_4^k h_l(u_j)$. To illustrate this second method for the evaluation of $S^{nlm}_{n'l'm'}(\mathbf{s})$, we can consider integrals involving the first few hyperspherical harmonics,

$$\sqrt{2\pi}Y_{0,0,0} = 1$$
$$\sqrt{2\pi}Y_{1,1,1} = i\sqrt{2}(u_1 + iu_2)$$
$$\sqrt{2\pi}Y_{1,1,0} = -i2u_3$$
$$\sqrt{2\pi}Y_{1,0,0} = 2u_4 \tag{40}$$

where the 4-dimensional unit vector, $\mathbf{u}$ is defined by equation (18). Then, for example,

$$S^{1,0,0}_{1,0,0}(\mathbf{s}) = \frac{1}{2\pi^2}\int d\Omega e^{i\mathbf{p}\cdot\mathbf{R}} = \mathcal{F}_{0,0}(s) = (1+s)e^{-s} \tag{41}$$

while

$$S^{1,0,0}_{2,0,0}(\mathbf{s}) = \frac{1}{2\pi^2}\int d\Omega e^{i\mathbf{p}\cdot\mathbf{R}} u_4 = \mathcal{F}_{1,0}(s) = -\frac{2}{3}s^2 e^{-s} \tag{42}$$

and

$$S^{1,1,1}_{1,0,0}(\mathbf{s}) = \frac{i\sqrt{2}}{2\pi^2}\int d\Omega e^{i\mathbf{p}\cdot\mathbf{R}}(u_1 + iu_2)$$
$$= i\sqrt{2}\mathcal{F}_{0,1}(s)(s_1 + is_2) = -\frac{\sqrt{2}}{3}(s_1 + is_2)(1+s)e^{-s} \tag{43}$$

To illustrate the Shibuya-Wulfman method, we can consider an electron moving in the attractive potential of two protons with internuclear separation R. In the lowest approximation, we can represent the ground state by a linear combination of the two basis functions:

$$\xi_1(\mathbf{p}) = e^{i\mathbf{p}\cdot\mathbf{X}_1}\chi^t_{1,0,0}(\mathbf{p})$$

$$\xi_2(\mathbf{p}) = e^{i\mathbf{p}\cdot\mathbf{X}_2}\chi^t_{1,0,0}(\mathbf{p}) \tag{44}$$

Making use of (36) and (41), we have:

$$K_{\tau',\tau} = \begin{pmatrix} 1 & (1+s)e^{-s} \\ (1+s)e^{-s} & 1 \end{pmatrix} \tag{45}$$

If we impose the condition $b_\mu k_\mu = 1$, the secular equation, (34), requires that

$$k_\mu = 1 \pm (1+s)e^{-s} \tag{46}$$

In the united-atom limit, $R = 0$, the positive root is $k_\mu = 2$, which corresponds to the exact ground-state energy of the He^+ ion (in Hartrees):

$$\epsilon = -k_\mu^2/2 = -2 \tag{47}$$

In the separated-atom limit, $R \to \infty$, both roots yield $k_\mu = 1$, which again gives the exact energy of the system, $\epsilon = -1/2$ Hartrees. For intermediate values of R, the energies given by our crude approximation differ appreciably from the exact energies, but extremely high precision (10-figure accuracy) can be obtained by using more basis functions, as was demonstrated by Koga and Matsuhashi [10].

3. Many-Electron Sturmians

The method of Shibuya and Wulfman has been extended and developed by a number of authors [6,8–16], and it gives us an elegant method for solving the Schrödinger equation for a single electron moving in a many-center Coulomb potential. However, in theoretical chemistry it is the many-electron many-center problem which really interests us. It is therefore tempting to ask whether there might be some way of extending the method of Shibuya and Wulfman to many-electron systems. If we wish to discuss this problem, we must shift notation slightly and introduce an index j to label the individual electrons of an N-electron system. Thus, for example, we must add the index j to the Laplacian operator for each of the electrons of the system and write

$$\Delta_j \equiv \frac{\partial^2}{\partial x_j^2} + \frac{\partial^2}{\partial y_j^2} + \frac{\partial^2}{\partial z_j^2} \tag{48}$$

The total kinetic energy operator of the N-electron system, in atomic units, will then be given by $-\Delta/2$, where Δ is the generalized Laplacian operator for the $d = 3N$ dimensional space:

$$\Delta \equiv \sum_{j=1}^{N} \Delta_j \tag{49}$$

Similarly we can let $\mathbf{x}_j$ and $\mathbf{p}_j$ represent the coordinate and momentum of the jth electron while $\mathbf{x}$ and $\mathbf{p}$ are d-dimensional vectors:

$$\mathbf{x} \equiv \{\mathbf{x}_1, \mathbf{x}_2, \ldots \mathbf{x}_N\}$$

$$\mathbf{p} \equiv \{\mathbf{p}_1, \mathbf{p}_2, \ldots \mathbf{p}_N\} \tag{50}$$

The N-electron Schrödinger eqeuation can then be written in the form:

$$[\Delta - p_0^2]\psi(\mathbf{x}) = 2V(\mathbf{x})\psi(\mathbf{x}) \tag{51}$$

where $V(\mathbf{x})$ is the potential experiennced by the electrons, including both nuclear attraction and interelectron repulsion terms, and where the total electronic energy is given by

$$E = -p_0^2/2 \tag{52}$$

We can try to build up solutions to (51) from a set of many-electron basis functions which satisfy the equations:

$$[\Delta - p_0^2]\phi_\nu(\mathbf{x}) = 2\beta_\nu V_0(\mathbf{x})\phi_\nu(\mathbf{x}) \tag{53}$$

where $V_0(\mathbf{x})$ is some potential for which (53) can be solved exactly, and where $\{\beta_\nu\}$ is a set of weighting factors especially chosen so that all of the basis functions $\phi_\nu(\mathbf{x})$ correspond to the same value of p_0 and hence the same energy regardless of their quantum numbers. Comparison of (53) with (3) shows that the two equations are analogous, and that a set of basis functions $\phi_\nu(\mathbf{x})$ with constant p_0 is analogous to a set of basis functions $\chi_{nlm}(\mathbf{x})$ with constant k_μ. Thus the functions $\phi_\nu(\mathbf{x})$ might be called a 'many-electron Sturmian basis set'. Functions of this type were introduced a number of years ago by Herschbach and Avery [6,17], the 'basis potential' $V_0(\mathbf{x})$ being the d-dimensional analogue of the hydrogen atom potential. More recently, Aquilanti and Avery [17–19] have shown that it is possible to construct basis sets of this type using the actual nuclear attraction potential experienced by the electrons in an atom or molecule as the 'basis potential', $V_0(\mathbf{x})$, i.e. letting

$$V_0(\mathbf{x}) = \sum_{j=1}^{N} \upsilon(\mathbf{x}_j) \tag{54}$$

where $\upsilon(\mathbf{x}_j)$ is defined by equation (25). As we shall see, this more realistic choice of the basis potential brings the basis functions much closer to the solutions of the actual Schrödinger equation, (51), and thus it leads to a much more rapid convergence of the expansion

$$\psi(\mathbf{x}) = \sum_\nu \phi_\nu(\mathbf{x})B_\nu \tag{55}$$

Goscinski's treatment of the orthonormality relations of Sturmian basis sets [3] is easy to generalize; and we can see by an argument analogous to equations (9) and (10) that when $\beta_{\nu'} \neq \beta_\nu$,

$$\int dx\phi_{\nu'}^*(\mathbf{x})V_0(\mathbf{x})\phi_\nu(\mathbf{x}) = 0 \tag{56}$$

where dx represents the d-dimensional volume element. Thus the many-particle

Sturmian basis set obeys a potential-weighted orthogonality relationship analogous to equation (10). This still does not tell us how to normalize the functions, and in fact the choice is arbitrary. However, it will be convenient to choose the normalization in such a way that in momentum space the orthonormality relations become:

$$\int dp \left(\frac{p^2 + p_0^2}{2p_0^2} \right) \phi_{\nu'}^{t*}(\mathbf{p}) \phi_{\nu}^{t}(\mathbf{p}) = \delta_{\nu',\nu} \tag{57}$$

(where dp represents the d-dimensional volume element in momentum space), which is analogous to equation (17). From an argument analogous to equations (11)–(17), it then follows that the direct-space orthonormality relations will have the form:

$$\int dx \phi_{\nu'}^{*}(\mathbf{x}) V_0(\mathbf{x}) \phi_{\nu}(\mathbf{x}) = -\frac{p_0^2}{\beta_\nu} \delta_{\nu',\nu} \tag{58}$$

It should be noted that since we are dealing with a many-dimensional space, the index ν represents a set of quantum numbers. We can divide these into 'grand principal quantum numbers', on which the value of β_ν depends, and the remaining 'minor' quantum numbers. In equation (17), orthogonality with respect to the minor quantum numbers l and m does not follow from (10) but from the properties of the spherical harmonics. Similarly, in (57) and (58), orthogonality with respect to the minor quantum numbers does not follow from (56) but must be constructed or proved in some other way, for example by means of symmetry arguments. Having discussed the orthonormality relations of the many-particle Sturmian basis functions, we are now in a position to derive a secular equation for the N-electron system: If we substitute (55) into (51), multiply from the left by $\phi_{\nu'}^{*}(\mathbf{x})$, and integrate over dx, we obtain:

$$\sum_{\nu} \int dx \phi_{\nu'}^{*}(\mathbf{x}) [\beta_\nu V_0(\mathbf{x}) - V(\mathbf{x})] \phi_\nu(\mathbf{x}) B_\nu = 0 \tag{59}$$

where we have also made use of equation (53). If we also use the potential-weighted orthonormality relation (58), and if we introduce the notation:

$$T_{\nu',\nu} \equiv -\frac{1}{p_0} \int dx \phi_{\nu'}^{*}(\mathbf{x}) V(\mathbf{x}) \phi_\nu(\mathbf{x}) \tag{60}$$

we can rewrite the secular equation (59) in the form:

$$\sum_{\nu} [T_{\nu',\nu} - p_0 \delta_{\nu',\nu}] B_\nu = 0 \tag{61}$$

The definition of the matrix $T_{\nu',\nu}$ in equation (60) requires some explanation: The minus sign is motivated by the fact that $V(\mathbf{x})$ is assumed to be an attractive potential. Division by p_0 is motivated by the fact that for Coulomb systems, when $T_{\nu',\nu}$ is so defined, it turns out to be independent of p_0, as we shall see below. The Sturmian secular equation (61) has several remarkable features: In the first place, the kinetic energy has vanished! Secondly, the roots are not energy values but values of the parameter p_0, which is related to the electronic energy of the system by equation (52). Finally, as we shall see below, the basis functions depend on p_0, and therefore they are not known until solution of the secular equation is complete. This corresponds, as we shall see, to an automatic optimization of the Slater exponents of the basis functions. The remarkable features just mentioned are shared by the one-particle secular equation shown in (34).

4. Construction of Many-Electron Sturmians

Now suppose that we have solved the one-electron equations (26); and suppose that the subsidiary conditions:

$$k_\mu^2 + k_{\mu'}^2 + k_{\mu''}^2 + \ldots = p_0^2 \tag{62}$$

and

$$k_\mu b_\mu = k_{\mu'} b_{\mu'} = k_{\mu''} b_{\mu''} = \ldots = \beta_\nu \tag{63}$$

are also fulfilled. Then

$$\phi_\nu(\mathbf{x}) = \varphi_\mu(\mathbf{x}_1)\varphi_{\mu'}(\mathbf{x}_2)\ldots\varphi_{\mu''}(\mathbf{x}_N) \tag{64}$$

will satisfy equation (53), since

$$\begin{aligned}
[\Delta - p_0^2]\phi_\nu(\mathbf{x}) &= [\Delta_1 - k_\mu^2 + \Delta_2 - k_{\mu'}^2 + \ldots]\varphi_\mu(\mathbf{x}_1)\varphi_{\mu'}(\mathbf{x}_2)\ldots \\
&= [2b_\mu k_\mu v(\mathbf{x}_1) + 2b_{\mu'} k_{\mu'} v(\mathbf{x}_2) + \ldots]\varphi_\mu(\mathbf{x}_1)\varphi_{\mu'}(\mathbf{x}_2)\ldots \\
&= 2\beta_\nu[v(\mathbf{x}_1) + v(\mathbf{x}_2) + \ldots]\varphi_\mu(\mathbf{x}_1)\varphi_{\mu'}(\mathbf{x}_2)\ldots \\
&= 2\beta_\nu V_0(\mathbf{x})\phi_\nu(\mathbf{x})
\end{aligned} \tag{65}$$

It can easily be seen that if the antisymmetrizer is applied to the function shown in equation (64), the resulting function will still satisfy (53). The total potential of the system is given by

$$V(\mathbf{x}) = V_0(\mathbf{x}) + V'(\mathbf{x}) \tag{66}$$

where

$$V_0(\mathbf{x}) = -\sum_{j=1}^{N}\sum_a \frac{Z_a}{|\mathbf{x}_j - \mathbf{X}_a|} \tag{67}$$

represents the nuclear attraction potential, while

$$V'(\mathbf{x}) = \sum_{i>j}^{N}\sum_{j=1}^{N} \frac{1}{|\mathbf{x}_i - \mathbf{x}_j|} \tag{68}$$

is the interelectron repulsion term. Correspondingly, we can divide the matrix $T_{\nu',\nu}$ of equation (60) into two parts, the nuclear attraction part being:

$$T^0_{\nu',\nu} \equiv -\frac{1}{p_0}\int dx\,\phi^*_{\nu'}(\mathbf{x})V_0(\mathbf{x})\phi_\nu(\mathbf{x}) \tag{69}$$

Making use of the potential-weighted orthonormality relation (58) and the subsidiary conditions (62) and (63), we obtain:

$$T^0_{\nu',\nu} = \frac{p_0}{\beta_\nu}\delta_{\nu',\nu} = \left(\frac{1}{b^2_\mu} + \frac{1}{b^2_{\mu'}} + \frac{1}{b^2_{\mu''}} + \dots\right)^{1/2} \delta_{\nu',\nu} \tag{70}$$

Thus the secular equation, (61), requires that

$$\det\left|\left(\frac{1}{b^2_\mu} + \frac{1}{b^2_{\mu'}} + \frac{1}{b^2_{\mu''}} + \dots\right)^{1/2} \delta_{\nu',\nu} + T'_{\nu',\nu} - p_0\delta_{\nu',\nu}\right| = 0 \tag{71}$$

where

$$T'_{\nu',\nu} \equiv -\frac{1}{p_0}\int dx \phi^*_{\nu'}(\mathbf{x}) V'(\mathbf{x})\phi_\nu(\mathbf{x}) \tag{72}$$

where $V'(\mathbf{x})$ is the interelectron repulsion potential shown in equation (68). From a calculational point of view, the evaluation of the Coulomb and exchange integrals shown in equation (72) is the most difficult step in the method which we have outlined. However, there has recently been very important progress in the evaluation of multi-center two-electron integrals involving exponential-type orbitals [20–22]. We are extremely grateful to Professor J. Fernández Rico and Dr. R. López of the Departimento de Física Aplicada, Universidad Autónoma de Madrid, 28049 Madrid, Spain, for allowing us to use the integral packages which they have developed for this purpose.

5. An Illustrative Example

In order to illustrate the discussion given above, let us consider the simplest possible example - the H_2 molecule with a basis consisting of a single two-electron Sturmian:

$$\begin{aligned}\phi_1(\mathbf{x}) &= |\varphi_g \varphi_{\overline{g}}| \\ &= \mathcal{N}\varphi_g(\mathbf{x}_1)\varphi_g(\mathbf{x}_2)\frac{1}{\sqrt{2}}[\alpha(1)\beta(2) - \alpha(2)\beta(1)]\end{aligned} \tag{73}$$

To make the example still more simple, let us use only two atomic functions (one on each center) in constructing $\varphi_g(\mathbf{x}_j)$:

$$\varphi^t_g(\mathbf{p}_j) = e^{i\mathbf{p}\cdot\mathbf{X}_1}\chi^t_{1,0,0}(\mathbf{p}_j)C_{1,g} + e^{i\mathbf{p}\cdot\mathbf{X}_2}\chi^t_{1,0,0}(\mathbf{p}_j)C_{2,g} \tag{74}$$

The matrix $K_{\tau',\tau}$ will then be given in terms of the parameter $s = k_g R$ by equation (45), and the one-electron problem closely resembles the example discussed in connection with this equation, the only difference being that we now relax the condition $k_\mu b_\mu = 1$, and we instead impose the subsidiary conditions (62) and (63). These respectively require that

$$2k_g^2 = p_0^2 \tag{75}$$

and

$$\beta_1 = b_g k_g \tag{76}$$

so that $k_g = p_0/\sqrt{2}$, where the subscript g stands for *gerade*. The one-electron secular equation, (34), then requires that

$$b_g^{-1} = 1 + (1+s)e^{-s} \tag{77}$$

since the root with the minus sign correspond to the *ungerade* solution. Because we are using only a single two-electron Sturmian in our basis, the two-electron secular equation becomes trivial, and equation (71) simply requires that

$$p_0 = \frac{\sqrt{2}}{b_g} + T'_{1,1} = \sqrt{2}[1 + (1+s)e^{-s}] + T'_{1,1} \tag{78}$$

From the one-electron secular equation, it follows that the *gerade* solution (in momentum space) has the form:

$$\varphi_g^t(\mathbf{p}_j) = \sqrt{\frac{\mathcal{N}}{2}}[\xi_1(\mathbf{p}_j) + \xi_2(\mathbf{p}_j)] \tag{79}$$

where the basis functions $\xi_a(\mathbf{p}_j)$ are defined by equation (44) and where $\mathcal{N}$ is a normalization constant which must be determined from the orthonormality condition (57) or from (58). It turns out to be most convenient to impose the normalization condition in momentum space, and thus we make use of (57), which requires that

$$\int dp \left(\frac{p^2 + p_0^2}{2p_0^2}\right)|\phi_1^t(\mathbf{p})|^2 = 1 \tag{80}$$

Using the subsidiary condition (75) and the fact that $p^2 = p_1^2 + p_2^2$, we can rewrite this requirement in the form:

$$\mathcal{N}^2 \int d^3p_1 \left(\frac{p_1^2 + k_g^2}{2k_g^2}\right)|\varphi_g^t(\mathbf{p}_1)|^2 \int d^3p_2 |\varphi_g^t(\mathbf{p}_2)|^2 = 1 \tag{81}$$

The first of the integrals in this product is easy to evaluate, since

$$\begin{aligned}\int d^3p_1 \left(\frac{p_1^2 + k_g^2}{2k_g^2}\right)|\varphi_g^t(\mathbf{p}_1)|^2 &= \frac{1}{2}\int d^3p_1 \left(\frac{p_1^2 + k_g^2}{2k_g^2}\right)|\xi_1(\mathbf{p}_1) + \xi_2(\mathbf{p}_1)|^2 \\ &= \frac{1}{2}[K_{11} + K_{12} + K_{21} + K_{22}] \\ &= [1 + (1+s)e^{-s}] \end{aligned} \tag{82}$$

The second integral in the product shown in equation (81) is given by

$$\begin{aligned}\int d^3p_2 |\varphi_g^t(\mathbf{p}_2)|^2 &= \frac{1}{2}\int d^3p_2 |\xi_1(\mathbf{p}_1) + \xi_2(\mathbf{p}_1)|^2 \\ &= \frac{1}{2}[M_{11} + M_{12} + M_{21} + M_{22}] \end{aligned} \tag{83}$$

where

$$M_{a',a} \equiv \int d^3 p_2 \xi^*_{a'}(\mathbf{p}_2)\xi_a(\mathbf{p}_2) \tag{84}$$

The integrals $M_{a',a}$ can be evaluated in direct space using ellipsoidal coordinates, or alternatively in momentum space using methods discussed in reference [17]. The result is:

$$\begin{aligned} M_{11} &= M_{22} = 1 \\ M_{12} &= M_{21} = \left(1 + s + \frac{s^2}{3}\right) e^{-s} \end{aligned} \tag{85}$$

Thus we obtain the normalization constant:

$$\mathcal{N} = \left[\{1 + (1+s)e^{-s}\}\left\{1 + \left(1 + s + \frac{s^2}{3}\right)e^{-s}\right\}\right]^{-1/2} \tag{86}$$

In the united-atom limit, $s = 0$, this gives $\mathcal{N} = 1/2$ while in the separated-atom limit, $s \to \infty$, we have $\mathcal{N} \to 1$. Beginning with a series of s-values, we can now immediately generate the corresponding values of b_g^{-1} and $\mathcal{N}$, as shown in Table 1. Interestingly, the interelectron repulsion matrix element, T'_{11}, which seems at first sight to depend independently on the two parameters k_g and R, can be shown to depend only on their product, $s = k_g R$, (Appendix 1). The approximate functional dependence of this matrix element on s is given [19] by

$$T'_{11} \approx -\frac{1}{2\sqrt{2}}\left[\frac{5}{8} + f(s)\right] \tag{87}$$

where

$$f(s) = \frac{1}{s} - \left(\frac{24 + 33s + 18s^2 + 4s^3}{24s}\right)e^{-2s} \tag{88}$$

Approximate values of T'_{11}, derived from equations (87) and (88), are shown in Table 1, compared with more exact values, calculated using the integral package of Professor Fernández Rico and Dr. López and co-workers [20–22].

Table 1 H_2 *molecule with 1 a.o. per atom*

s	b_g^{-1}	$\mathcal{N}$	Eq.(87)	T'_{11}	p_0	R
0	2.00000	0.50000	−0.44194	−0.44194	2.38649	0
1	1.73575	0.55678	−0.41702	−0.44730	2.00744	0.70449
2	1.40601	0.66956	−0.37157	−0.42026	1.56813	1.80370
3	1.19915	0.78639	−0.33404	−0.37273	1.32312	3.20655
4	1.09158	0.87768	−0.30849	−0.33001	1.21371	4.66079
5	1.04043	0.93621	−0.29153	−0.30115	1.17023	6.04244
6	1.01735	0.96888	−0.27987	−0.28359	1.15516	7.34554
7	1.00730	0.98550	−0.27147	−0.27275	1.15178	8.59497

The next step in improving the crude approximation discussed above might be to use more atomic orbitals in solving the one-electron equation, (34). If we use more Sturmian atomic orbitals of the type defined in equation (32), then the dimension of the matrix $K_{\tau',\tau}$ will become larger, and the roots of the secular equation, (34), will correspond to increasingly accurate solutions to the one-electron part of the problem. In Tables 2 and 3, the ground-state root, b_g^{-1} is shown, with respectively 3 atomic orbitals per center and 15 atomic orbitals per center, as functions of the parameter $s = k_g R$. When 15 atomic orbitals are used, the curve of the ground-state root closely approximates the highly-precise results of Koga and Matsuhashi, as illustrated in Figure 1.

When more atomic orbitals are used in solving the one-electron part of the problem, the normalization constant, $\mathcal{N}$, must of course be recalculated. Instead of equation (83), we then have

$$\begin{aligned}\int d^3 p_2 |\varphi_g^t(\mathbf{p}_2)|^2 &= \sum_{\tau',\tau} C^*_{\tau',g} C_{\tau,g} \int d^3 p_2 \xi^*_{\tau'}(\mathbf{p}_2)\xi_\tau(\mathbf{p}_2) \\ &= \sum_{\tau',\tau} C^*_{\tau',g} M_{\tau',\tau} C_{\tau,g} \equiv \sigma_g \qquad (89)\end{aligned}$$

Table 2 H_2 *molecule with 3 a.o.'s per atom*

s	b_g^{-1}	$\mathcal{N}$	(T'_{11})	p_0	R
0	2.00000	0.50000	−0.44194	2.38649	0
1	1.83419	0.50885	−0.44730	2.14664	0.65880
2	1.61510	0.55086	−0.42026	1.86383	1.51753
3	1.39588	0.62187	−0.37273	1.60134	2.64943
4	1.20614	0.71788	−0.33001	1.37573	4.11189
5	1.08271	0.82670	−0.30115	1.23003	5.74868
6	1.02829	0.91573	−0.28359	1.17063	7.24846
7	1.00972	0.96356	−0.27275	1.15520	8.56944

Table 3 H_2 *molecule with 15 a.o's per atom*

s	b_g^{-1}	$\mathcal{N}$	(T'_{11})	p_0	R
0	2.00000	0.50000	−0.44194	2.38649	0
1	1.84702	0.51266	−0.44730	2.16478	0.65328
2	1.64335	0.54763	−0.42026	1.90379	1.48568
3	1.47329	0.59251	−0.37273	1.71082	2.47989
4	1.34131	0.64199	−0.33001	1.56689	3.61025
5	1.24190	0.69333	−0.30115	1.45516	4.85930
6	1.16885	0.74555	−0.28359	1.36941	6.19629
7	1.11153	0.80024	−0.27275	1.29919	7.61974

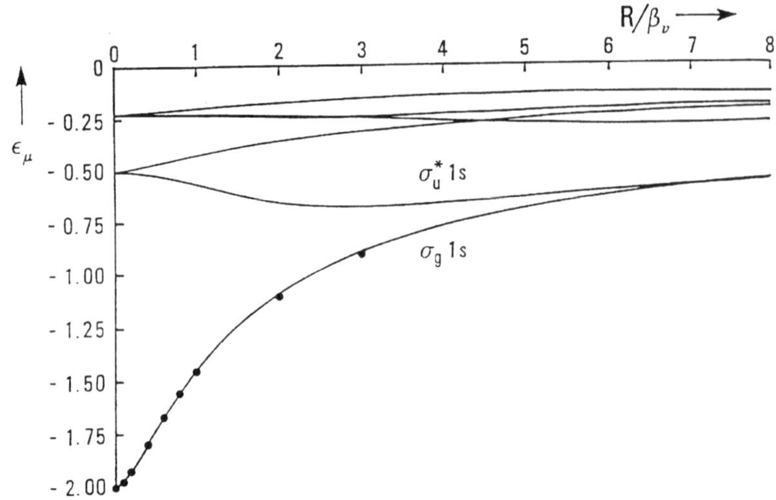

Fig. 1. This figure shows the electron energies of the H_2^+ ion calculated using the Shibuya-Wulfman technique, with 15 orbitals on each center. The nearly-exact values of Koga and Matsuhashi are indicated by dots for the ground state. Excited states with $m = 0$ are also shown.

where, as before, the index τ stands for the set of indices $\{a, n, l, m\}$, and where the coefficients $C_{\tau,g}$ represent the ground-state eigenvector of $K_{\tau',\tau}$. Similarly, the integral shown in equation (82) becomes:

$$\int d^3p_1 \left(\frac{p_1^2 + k_g^2}{2k_g^2}\right) |\varphi_g^t(\mathbf{p}_1)|^2 = \sum_{\tau',\tau} C_{\tau',g}^* K_{\tau',\tau} C_{\tau,g} = b_g^{-1} \tag{90}$$

Combining equations (81), (89) and (90), we obtain:

$$\mathcal{N} = \frac{1}{\sqrt{b_g^{-1}\sigma_g}} \tag{91}$$

The normalization constants corresponding respectively to 3 and 15 atomic orbitals per center are shown in Tables 2 and 3. The improved values of b_g^{-1} result in improved values for the electronic energy and the total energy of the H_2 molecule even when the crude interelectron repulsion values ($T'_{1,1}$) from Table 1 are used, as illustrated in Figures 2 and 3. In these figures it can be seen that for small values of R, the curves obtained in this way closely approximate the results of the benchmark calculation of Kolos and Wolniewicz. For large values of R, the agreement is less good because, as is well known, a second configuration, $|\varphi_u\varphi_{\overline{u}}|$, is needed to give correct dissociation of the molecule, φ_u being the *ungerade* solution to the 1-electron problem. We hope to present multiconfigurational calculations on molecules in a future publication.

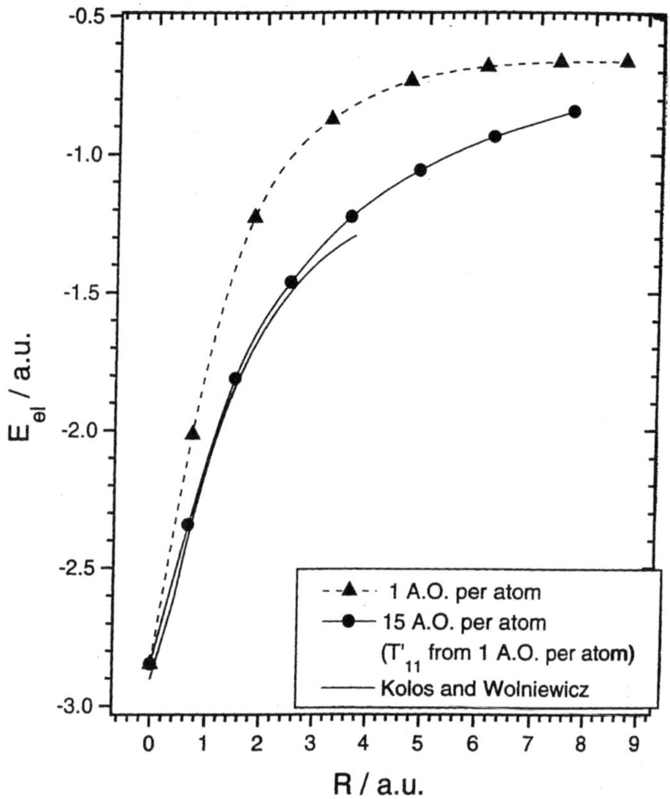

Fig. 2. This figure shows the electronic energy of the ground state of H_2 molecule, calculated in a crude approximation using only one configuration. The benchmark calculation of Kolos and Wolniewicz is exhibited for comparison. Accuracy can be seen to be improved by using more atomic orbitals even when a rough approximation is used for the interelectron repulsion matrix element.

6. Discussion

The results presented in this paper seem to indicate that it will be possible to apply successfully the method of many-electron Sturmians to molecules. Momentum-space methods, pioneered by Shibuya and Wulfman [7], seem very well suited to solving the one-electron part of the problem. When the 'basis potential' used in constructing the many-electron Sturmian basis set is taken to be the nuclear attraction potential experienced by the electrons in the molecule, the method of many-electron Sturmians has the following advantages:

1. The matrix representation of the nuclear attraction potential is diagonal.
2. The kinetic energy term vanishes from the secular equation.
3. The Slater exponents of the atomic orbitals are automatically optimized.
4. Convergence is rapid.

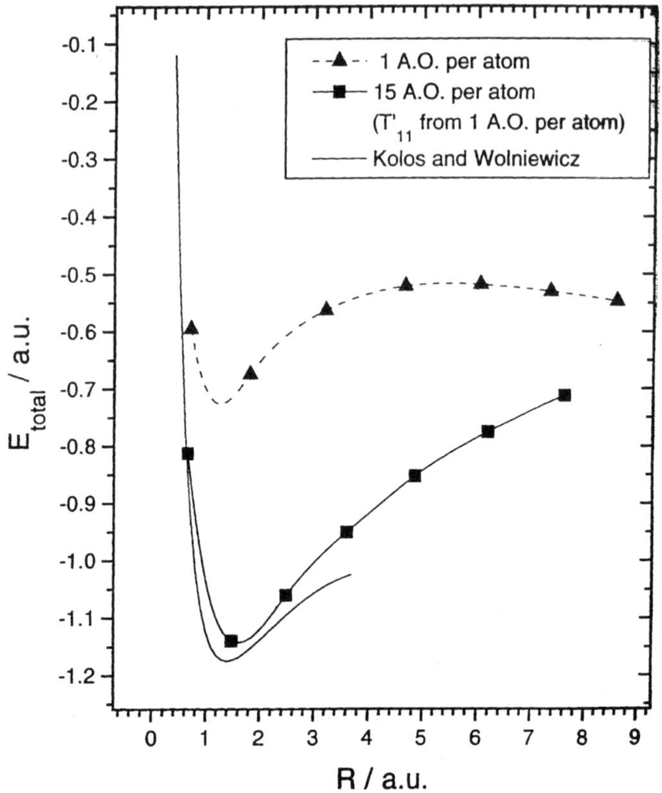

Fig. 3. This figure shows the total energy of the H_2 molecule as a function of internuclear distance, calculated from the electronic energies shown in Figure 2. For small values of R, the calculation using 15 atomic orbitals per atom agrees well with the values of Kolos and Wolniewicz; but another configuration would be needed for agreement at large R.

5. A solution to the many-electron problem, including correlation, can be obtained directly, without the use of the SCF approximation.

Basis sets of the type discussed in this paper can only be applied to bound-state problems. It is interesting to ask whether it might be possible to construct many-electron Sturmian basis sets appropriate for problems in reactive scattering in an analogous way, using hydrogenlike continuum functions as building-blocks. We hope to explore this question in future publications.

7. Appendix A: Properties of $T'_{\nu',\nu}$

In this appendix we shall try to show that interelectron repulsion matrix element T'_{11} of equation (78) does not depend independently on $k_\mu = k_g$ and R, but depends only on their product, $s = k_g R$: If we take the Fourier transform of equation (79), we obtain:

$$\varphi_g(\mathbf{x}_j) = \sqrt{\frac{\mathcal{N}}{2}}[\chi_{1,0,0}(\mathbf{x}_j - \mathbf{X}_1) + \chi_{1,0,0}(\mathbf{x}_j - \mathbf{X}_2)] \tag{A1}$$

Letting $\mathbf{R} \equiv \mathbf{X}_2 - \mathbf{X}_1$ and taking the origin at of our coordinate system the point $\mathbf{X}_1$, we can rewrite this as

$$\varphi_g(\mathbf{x}_j) = \sqrt{\frac{\mathcal{N}}{2}}[\chi_{1,0,0}(\mathbf{x}_j) + \chi_{1,0,0}(\mathbf{x}_j - \mathbf{R})] \tag{A2}$$

Since the hydrogenlike Sturmian basis functions form a complete set, the term $\chi_{1,0,0}(\mathbf{x}_j - \mathbf{R})$ can be represented as a single-center expansion in terms of functions localized at the origin:

$$\chi_{1,0,0}(\mathbf{x}_j - \mathbf{R}) = \sum_{nlm} \chi_{nlm}(\mathbf{x}_j) C_{nlm}^{1,0,0} \tag{A3}$$

More generally we can write:

$$\chi_{n'l'm'}(\mathbf{x}_j - \mathbf{R}) = \sum_{nlm} \chi_{nlm}(\mathbf{x}_j) C_{nlm}^{n'l'm'} \tag{A4}$$

As has been pointed out by Shibuya, Wulfman [7] and Aquilanti [18], the expansion coefficients in equation (A4) are just the integrals shown in equation (37), i.e., the Shibuya-Wulfman integrals. To see this, we can take the Fourier transform of (A4), which gives us the relationship

$$e^{i\mathbf{p}_j\cdot\mathbf{R}} \chi_{n'l'm'}^t(\mathbf{p}_j) = \sum_{nlm} \chi_{nlm}^t(\mathbf{p}_j) C_{nlm}^{n'l'm'} \tag{A5}$$

We now multiply both sides of (A5) by $(p^2 + k_\mu^2/2k_\mu^2)\chi_{n''l''m''}^{t*}(\mathbf{p}_j)$ and integrate over $d^3 p_j$. Then, making use of the orthonormality relationship shown in equation (17), we obtain:

$$\int d^3 p_j \left(\frac{p^2 + k_\mu^2}{2k_\mu^2}\right) e^{i\mathbf{p}_j\cdot\mathbf{R}} \chi_{n''l''m''}^{t*}(\mathbf{p}_j) \chi_{n'l'm'}^t(\mathbf{p}_j) = C_{n''l''m''}^{n'l'm'} \tag{A6}$$

With the help of equations (19) and (20), the left-hand side of (A6) can be seen to reduce to an integral of the type shown in equation (37); and thus we can write:

$$\chi_{n'l'm'}(\mathbf{x}_j - \mathbf{R}) = \sum_{nlm} \chi_{nlm}(\mathbf{x}_j) S_{nlm}^{n'l'm'}(\mathbf{s}) \tag{A7}$$

Thus

$$T_{11}'' \equiv -\frac{1}{p_0}\int dx \frac{1}{r_{12}} |\varphi_g(\mathbf{x}_1)|\varphi_g(\mathbf{x}_2)|^2 \tag{A8}$$

will consist of a sum of one-center integrals multiplied by the Shibuya-Wulfman integrals, $S_{nlm}^{n'l'm'}(\mathbf{s})$, which are functions only of the product, $\mathbf{s} = k_\mu \mathbf{R}$, and not of k_μ and R independently. The one-center integrals are sums of terms of the form

$$(J_l)_{j',j} \equiv -\frac{k_\mu}{p_0}\int_0^\infty dt_1 t_1^{j'+2} e^{-t_1}\int_0^\infty dt_2 t_2^{j+2} e^{-t_2}\frac{t_<^l}{t_>^{l+1}}$$

$$= -\frac{k_\mu}{p_0}\frac{\Gamma(j+j'+5)}{4^{j+j'+3}}\left[\frac{{}_2F_1(1, j+j'+5; j+l+4; 1/2)}{j+l+3}\right.$$

$$\left. + \frac{{}_2F_1(1, j+j'+5; j'+l+4; 1/2)}{j'+l+3}\right] \qquad (A9)$$

where $t_1 = k_\mu r_1$, $t_2 = k_\mu r_2$ and where j, j' and l are integers, while ${}_2F_1(a, b; c; d)$ is a hypergeometric function. In other words, after the integrations have been performed, the one-center integrals are pure numbers multiplied by the ratio k_μ/p_0. But this ratio too is a pure number whose value is determined by the subsidiary conditions, equations (62) and (63). In our illustrative example, $k_g/p_0 = 1/\sqrt{2}$. Similar arguments can be used to show in the general case that the matrix elements $T'_{\nu',\nu}$ depend on $\mathbf{s} = k_\mu \mathbf{R}$; but they do not depend on k_μ and $\mathbf{R}$ independently, nor are they dependent on p_0. This theoretical prediction is confirmed by numerical results which we have obtained using the integral package of Professor Fernández Rico and Dr. López.

8. Appendix B: Second-Iterated Solutions

Koga and others [10] have shown that the loss of accuracy resulting from truncation of the basis set can be reduced by replacing the one-electron secular equation, (34), with an equation based on a second iteration of the integral equation (31). The second-iterated integral equation has the form:

$$\varphi_\mu^t(\mathbf{p}) = b_\mu^2 \sum_{\tau,\tau'} \xi_\tau(\mathbf{p})\int d^3p' \left(\frac{p'^2 + k_\mu^2}{2k_\mu^2}\right)\xi_\tau^*(\mathbf{p}')\xi_{\tau'}(\mathbf{p}')\int d^3p'' \left(\frac{p''^2 + k_\mu^2}{2k_\mu^2}\right)\xi_{\tau'}^*(\mathbf{p}'')\varphi_\mu^t(\mathbf{p}'') \qquad (B1)$$

If we expand $\varphi_\mu^t(\mathbf{p})$ in terms of the basis functions $\xi_\tau(\mathbf{p})$ defined in equation (32), equation (B1) becomes:

$$\sum_{\tau''} \xi_{\tau''}(\mathbf{p})C_{\tau'',\mu} = b_\mu^2 \sum_{\tau,\tau''} \xi_\tau(\mathbf{p})K^2_{\tau,\tau''}C_{\tau'',\mu} \qquad (B2)$$

where

$$K^2_{\tau,\tau''} \equiv \sum_{\tau'} K_{\tau,\tau'}K_{\tau',\tau''} \qquad (B3)$$

with $K_{\tau,\tau'}$ defined by equation (35). If the constants $C_{\tau'',\mu}$ and b_μ are solutions of the secular equations

$$\sum_{\tau''}[K^2_{\tau,\tau''} - b_\mu^{-2}\delta_{\tau,\tau''}]C_{\tau'',\mu} = 0 \qquad (B4)$$

then equation (B2) becomes an identity. The increased accuracy obtained by replacing the one-electron secular equation (34) by (B4) comes from the fact that in calculating the matrix $K^2_{\tau,\tau''}$ defined by (B3), sum rules can sometimes be found which allow the sum $\sum_{\tau'}$ to run over functions which are not included in the truncated basis set.

References

1. Shull, H. and Löwdin, P.O., J. Chem. Phys. **30**, 617 (1959).
2. Rotenberg, M., Adv. At. Mol. Phys. **6**, 233 (1970).
3. Goscinski, O., *Preliminary Research Report No. 217*, Quantum Chemistry Group, Uppsala University, (1968).
4. Weniger, E.J., J. Math. Phys. **26**, 276 (1985).
5. Fock, V.A., Z. Phys. **98**, 145 (1935); Bull. Acad. Sci. USSR, (1935).
6. Avery, J., *Hyperspherical Harmonics; Applications in Quantum Theory*, Kluwer Academic Publishers, Dordrecht, Netherlands, (1989).
7. Shibuya, T. and Wulfman, C.E., Proc. Roy. Soc. **A286**, 376 (1965).
8. Judd, B.R., *Angular Momentum Theory for Diatomic Molecules*, Academic Press, New York, (1975).
9. Monkhorst, H. and Jeziorski, B., J. Chem. Phys. **71**, 5268 (1979).
10. Koga, T., and Matsuhashi, T., J. Chem. Phys. **89**, 983 (1988).
11. Wen, Z.-Y. and Avery, J., J. Math. Phys. **26**, 396, (1985).
12. Avery, J., in *Conceptual Trends in Quantum Chemistry*, E.S. Kryachko and J.L. Calais (eds.), Kluwer Academic Publishers, Dordrecht, Netherlands, (1994), pp. 135-169.
13. Avery, J., in *New Methods in Quantum Theory*, C.A. Tsipis, V. Popov, D.R. Herschbach and J.S. Avery (eds.), Kluwer Academic Publishers, Dordrecht, Netherlands, (1996), pp. 281-296.
14. Avery, J., and Hansen, T.B., Int. J. Quantum Chem. **60**, 201 (1996).
15. Aquilanti, V., Cavalli, S., Coletti, C., De Fazio, D. and Grossi, G., in *New Methods in Quantum Theory*, C.A. Tsipis, V.S. Popov, D.R. Herschbach and J.S. Avery editors, Kluwer, (1996).
16. Avery, J. and Herschbach, D.R., Int. J. Quantum Chem. **41**, 673 (1992).
17. Avery, J., Adv. Quantum Chem. (1998) (in press).
18. Aquilanti, V. and Avery, J., Chem. Phys. Letters, **267**, 1 (1997).
19. Avery, J., J. Math. Chem. **21** 285, (1997).
20. López, R., Ramírez, G., García de la Vega, J.M., and Fernández Rico, J., J. Chim. Phys. **84**, 695 (1987).
21. Fernández Rico, J., Ramírez, G., López, R., and Fernández Alonso, J.I., Collect. Czech. Chem. Comm. **53**, 2250, (1987).
22. Fernández Rico, J., López, R., Ema, I., and Ramírez, G., preprints, (1997).
23. Kolos, W. and Wolniewicz, L., J. Chem Phys., **41**, 3663 (1964); 49, 404 (1968).

Treatment of Electron Correlation in Localized Representation

Cornelia Kozmutza[a], Ede Kapuy[b] and László Udvardi[c]
[a]*Department of Theoretical Physics, TU Budapest, Budafoki út 8, Budapest, H–1521, Hungary*
[b]*Department of Theoretical Physics, University József Attila, Aradi vrt. 1, Szeged, H–6720, Hungary*
[c]*MTA, TKI, Condensed Matter Research Group at TU Budapest, Budafoki út 8, Budapest, H–1521, Hungary*

1. Introduction

The scope of the quantum chemistry is manifold. It provides numerical solutions obtained by the use of equations of the quantum mechanical theory. It enables to calculate bond and dissociation energies, characteristics of spectral transitions, force constants, electron and spin densities, polarizabilities. The properties referring to atoms, molecules and other systems can be theoretically determined without knowledge of empirical data of these systems (ab initio level).

The various methods used in quantum chemistry make it possible to compute equilibrium intermolecular distances, to describe intermolecular forces and chemical reactions too. The usual way to calculate these properties is based on the independent particle model: this is the Hartree–Fock method. The expansion of one–electron wave–functions (molecular orbitals) in practice requires technical work on computers. It was believed for years and years that ab initio computations will become a 'routine' task even for large molecules. In spite of the enormous increase and development in computer technique, however, this expectation has not been fulfilled. The treatment of large, extended molecular systems still needs special theoretical background. In other words, some approximations should be used in the methods which describe the properties of molecules of large size and/or interacting systems. The further approximations are to be chosen carefully: this caution is especially important when going beyond the HF level. The inclusion of the electron correlation in the calculations in a convenient way is still one of the most significant tasks of quantum chemistry.

Quantum chemistry aims to understand a large variety of chemical facts. In some systems an interesting feature was obtained whose study and whose application can help to reduce the computational effort considerably: this is the transferability. Transferability can be interpreted in several ways. The orbitals, on the one hand, may be considered transferable in the case when certain properties of these orbitals are close to each other to a certain extent (Rothenberg, 1971). The transferability of orbitals can be discussed 'directly' on the other hand too. Orbitals of small molecules can be used for constructing the wave-function of related, larger molecules. This can be done with or without further optimizations. In this interpretation the orbitals are transferable if the molecular properties calculated with and without optimizations are close to each other (O'Leary *et al.*, 1975). The transferability of orbitals for cyclic hydrocarbons was discussed exhaustively (Edmiston *et al.*, 1963).

A. Hernández-Laguna et al. (eds.), Quantum Systems in Chemistry and Physics, Vol. 1: Basic Problems and Model Systems, 41–71.

The transferability of several molecular properties may lead to advantages: for the aliphatic hydrocarbons, e.g., the bond energies can be approximated in advance as they are known to be transferable to a good accuracy (Allen, 1959).

In the framework of the independent particle model the one–electron properties can be written as the sum of contributions from the orbitals. In other words, the transferability of the one–electron properties is implied by the transferability of the individual orbitals. This holds also for the total energy at the equilibrium geometry. The solutions of the HF equations, the canonical molecular orbitals (CMO) are non transferable since they are delocalized over the whole system. Therefore theoreticians have been looking for a special kind of orbitals for a long time in order to establish correspondence to chemical intuition. An important step in this direction was the introduction of the concept of localized molecular orbitals (LMO). The first idea (England *et al.*, 1975) was followed by further ones. All of them aim to give a description more or less analogous to the intuitive Lewis' hypothesis (Lewis, 1916). The use of LMO have reached considerable success at the HF level first for diatomic molecules and later for larger, closed shell systems as well. It turned out furthermore that the use of LMOs may lead to certain advantages in the framework of methods calculating electron correlation effects.

The scope of this work is to deal with the possible treatments of electron correlation in a localized representation. Several methods will be discussed in detail elaborated by present authors. Special attention will be payed to the analysis of the transferability of certain correlation energy contributions. The use of their transferability will be discussed for extended systems: series of hydrocarbons and polyenes will be investigated. The transferable properties of the contributions to the correlation energy, furthermore, turned out to be useful in the study of weakly interacting intermolecular systems. A detailed description of this procedure will be given in the present work.

2. Electron Correlation by Localization

2.1. Electron correlation effects

A usual approach to the theoretical treatment of the electronic structure is a decomposition of the many–electron problem into one–electron problem. The Hartree-Fock method is the one most commonly applied for the above purpose. Although using the HF method the total energy of the system studied can be computed to a quite good accuracy the remaining error is comparable to some energies, like bonding energy, also of interest. The error is due to the correlation of the electrons. Since the correlation effects are not observable they can not be directly compared to experiments. The treatment of the electron correlation effects is thus inevitable for the calculation of certain energetic quantities as bonding energies etc. This also holds for the interaction energy of weakly bonded systems. In systems where at least several tens of electrons are to be accounted for the description of the electron correlation effects is not a simple task. Several theories and methods have been elaborated for treating the electron correlation effects (Wilson, 1984). Most of the commonly used computational methods suffer from serious difficulty when dealing with large systems (Saebo *et al.*, 1985) since the increasing number of electrons to be treated requires a significant computational effort. The computational work is inevitably to be reduced if larger and larger systems are to be dealt with. A promising way to overcome this barrier is the use of localized representation in the correlation methods. Beside the reduction in computational time the application of LMOs in correlation procedures may yield information on the molecular electronic structure, too. Some localized orbital methods are quite successful in the description of interaction energies in van der Waals systems as well (Duijneveldt *et al.*, 1994).

2.2. Localization methods

A single–determinant wave-function of closed shell molecular systems is invariant against any unitary transformation of the molecular orbitals apart from a phase factor. The transformation can be chosen in order to obtain LMOs. Starting from CMOs a number of localization procedures have been proposed to get LMOs: the most commonly used methods are as given by the authors of (Edmiston *et al.*, 1963) and (Boys, 1966), while the procedures provided by (Pipek *et al.*, 1989) and (Saebo *et al.*, 1993) are also of interest. It could be stated that all the methods yield comparable results. Each LMO densities are found to be relatively concentrated in some spatial region. They are, furthermore, expected to be determined mainly by that part of the molecule which occupies that given region and its nearby environment rather than by the whole system.

It could thus be stated that a pair description analogous to the Lewis' picture is quite conveniently yielded by localized molecular orbitals. In other words, to the extent that the one–electron (orbital) method is in principle capable of faithful description of molecular electronic structure, the orbital analysis may provide a useful rationale of some molecular characteristics. Having generated LMOs, using any localizability criteria, the degree of spatial localization obtained may also be inquired.

It may be of interest the extent to which the orbital density distribution (a certain percentage of the density) occupy non overlapping region of the given molecule. The orbital 'tails' population, whose overlap destroy the direct transferability of orbitals, calculated in some ten–electron hydrides (Kapuy *et al.*, 1976) was found not negligible, almost 10 % of the total. The principal lobes, on the other hand, containing almost 90 % of the population suggest a transferability of the LMO densities to a good approximation.

A special kind of molecular systems are the molecular complexes. The interaction energies in these systems can be calculated in some cases more precisely than the experimental data available. Several approximations have been elaborated during the period of the last decades for the theoretical study both in weakly and strongly interacting systems (Duijneveldt *et al.*, 1994). One of the most commonly used method for investigating weakly interacting systems (in other terminology they are called van der Waals complexes) is the super-molecule approach. The super-molecule method both at HF and higher levels of the correlation can be applied in canonical and localized representation. (See later in Chapter 5) In order to localize the orbitals of the super-system, a special requirement is to be fulfilled: it is advantageous if the orbitals in the monomers both in the occupied and virtual subspaces are separated as much as possible. A new localization method for this purposes was suggested in (Kapuy *et al.*, 1994). Its use will be discussed in Chapter 6 in detail.

3. The Localized Many-Body Perturbation Theory

The localized many-body perturbation theory (LMBPT) applies localized HF orbitals which are unitary transforms of the canonical ones in the diagrammatic many-body perturbation theory. The method was elaborated on models of cyclic polyenes in the Pariser-Parr-Pople (PPP) approximation. These systems are considered as not well localized so they are suitable to study the importance of non local effects. The description of LMBPT follows the main points as it was first published in 1984 (Kapuy *et al.*, 1983).

In the many-body perturbation theory the non-relativistic Hamiltonian $\hat{H}$ is partitioned in the following way:

$$\hat{H} = \hat{H}_0 + \hat{W} \tag{1}$$

The occupied single-particle functions ϕ_i and the virtual single-particle functions ϕ_a are solutions of the corresponding canonical HF equations

$$\hat{F}\psi_m = \varepsilon_m \psi_m \qquad m = 1, 2, \ldots M \tag{2}$$

The $\hat{F}$ operator is

$$\hat{F} = \hat{H}(1) + \sum_{j=1}^{N} \langle j | r_{12}^{-1}(1 - \hat{P}_{12}) | j \rangle \tag{3}$$

while $\hat{H}_0$ is chosen as

$$\hat{H}^{(0)} = \sum_{i=1}^{N} \hat{F}(i) \tag{4}$$

The perturbation thus is as follows:

$$\hat{W} = \frac{1}{2} \sum_{\substack{i,j=1 \\ i \neq j}} r_{ij}^{-1} - \sum_{i=1}^{N} \sum_{j=1}^{N} \langle j | r_{12}^{-1}(1 - \hat{P}_{i2}) | j \rangle_i \tag{5}$$

Transforming the occupied and the virtual single-particle functions separately by unitary transformations U and V, respectively, the localized single-particle functions ϕ_m will satisfy the following (non-diagonal) HF equations

$$\hat{F}\phi_i = \sum_{j=1}^{N} \varepsilon_{ij} \phi_j \qquad i, j = 1, 2 \ldots N \tag{6}$$

$$\hat{F}\phi_a = \sum_{b=N+1}^{M} \varepsilon_{ab} \phi_b \qquad a, b = N + 1, N + 2 \ldots M \tag{7}$$

Choosing a new HF operator (according to (Amos *et al.*, 1971))

$$\hat{F}^{loc} = \hat{F} - \sum_{\substack{i,j \\ i \neq j}} |i\rangle \varepsilon_{ij} \langle j| - \sum_{\substack{a,b \\ a \neq b}} |a\rangle \varepsilon_{ab} \langle b| \tag{8}$$

The eigenfunctions of this operator are the transformed single-particle functions ϕ_m,

$$\hat{F}^{loc} \phi_i = \varepsilon_i \phi_i \qquad i = 1, 2, \ldots N \tag{9}$$

$$\hat{F}^{loc} \phi_a = \varepsilon_a \phi_a \qquad i = N+1, N+2, \ldots M \tag{10}$$

The zeroth order Hamiltonian and the perturbation are, therefore, the following:

$$\hat{H}^{loc}(0) = \sum_{i=1}^{N} \hat{F}^{loc}(i) \tag{11}$$

$$\hat{W}^{loc} = \hat{W} + \sum_{i=1}^{N} \left(\sum_{\substack{k,l=1 \\ k \neq l}}^{N} |\phi_k(i)\rangle \varepsilon_{ij} \langle \phi_j(i)| - \sum_{\substack{a,b=N+1 \\ a \neq b}}^{M} |\phi_a(i)\rangle \varepsilon_{ab} \langle \phi_b(i)| \right) \tag{12}$$

It is an important consequence that certain non-zero off-diagonal elements (Lagrangian multipliers) ε_{ij}, ε_{ab} as extra terms enter the perturbation expansion at third and higher orders.

The terms of the perturbation series are represented by graphs in the diagrammatic formulation. The 'mixed' Hugenholtz-Feynman representation (proposed in (Brandow, 1967) and (Bartlett *et al.*, 1975)) is used in the LMBPT. An algorithm that first constructs the Hugenholtz diagrams and then the Feynman diagrams and automatically selects the conjugate pairs, equivalent diagrams, and so on was developed for this purpose. Through fourth order, all diagrams using canonical orbitals can be found in the literature (Bartlett *et al.*, 1975): each Hugenholtz diagram can be represented by a Feynman graph with antisymmetrized vertices

$$\langle kl | r_{12}^{-1} | mn \rangle = \langle kl \| mn \rangle = \langle kl | mn \rangle - \langle kl | nm \rangle \tag{13}$$

The rules for translating graphs into formulas can also been found in many places in the literature (Brandow, 1967; Bartlett *et al.*, 1975; Hubac *et al.*, 1978).

Representing terms due to the non-zero off-diagonal Fock matrix element can be called localization diagrams (in contrast to the others which are canonical diagrams).

Using canonical HF orbitals the diagrams representing the second and third order energy corrections are shown in Figure 1. In the fourth order we have 39 Hugenholtz diagrams.

Using localized HF orbitals, new terms enter the perturbation corrections in third and higher orders. These terms are called localization corrections. In the diagrams representing the localization terms the off-diagonal Lagrangian multipliers are denoted by crosses in circle. As the occupied and the virtual orbitals are localized separately, no element connects particle and hole states. The localization diagrams can be derived thus from the canonical ones by inserting crosses in the hole and/or in the particle lines.

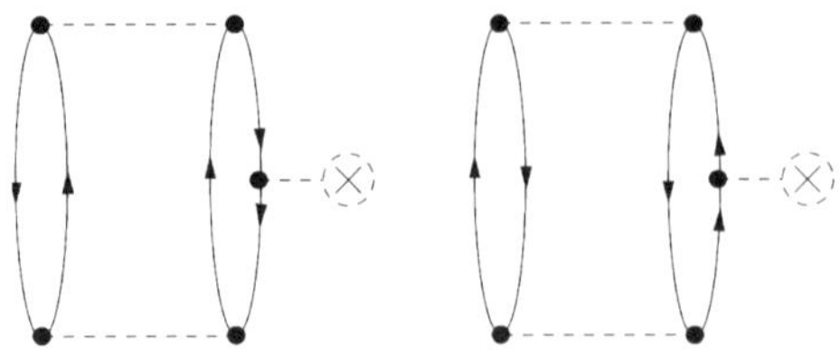

Fig. 1. Second and third order antisymmetrized Feynman diagrams.

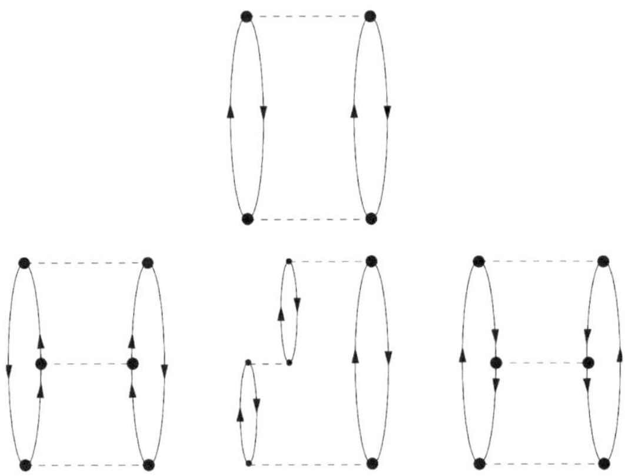

Fig. 2. Third order localization diagrams.

Putting one cross in the second order diagram the two third order localization diagrams, can be obtained (Figure 2.)

Inserting two crosses in the second order diagram or one cross in the third order diagrams the fourth order localization diagrams can be constructed (Figure 3). In the fourth order one does have 22 antisymmetrized Feynman diagrams among them 18 form 9 conjugate pairs i.e. 13 are different.

Due to the localization terms entering the localized representation, an extra computational work is necessary. It represents only a small fraction of the total computing time, in a given order, because the number of indices to be summed up are always less in the localization terms than in the canonical ones.

The method used for the localization of the orbitals is to be carefully chosen. It is natural to expect that if the orbitals are localized into different spatial regions, for the matrix elements $\langle ij|kl\rangle$ the zero differential approximation can be applied: all terms containing at least one factor $\langle ij|kl\rangle$ in which $\psi_i\psi_j$ and/or $\psi_k\psi_l$ are localized to different spatial regions can be neglected. Thus the summation in a closed loop in evaluating a perturbation correction should only be extended over indices of orbitals which are localized into the same region of space.

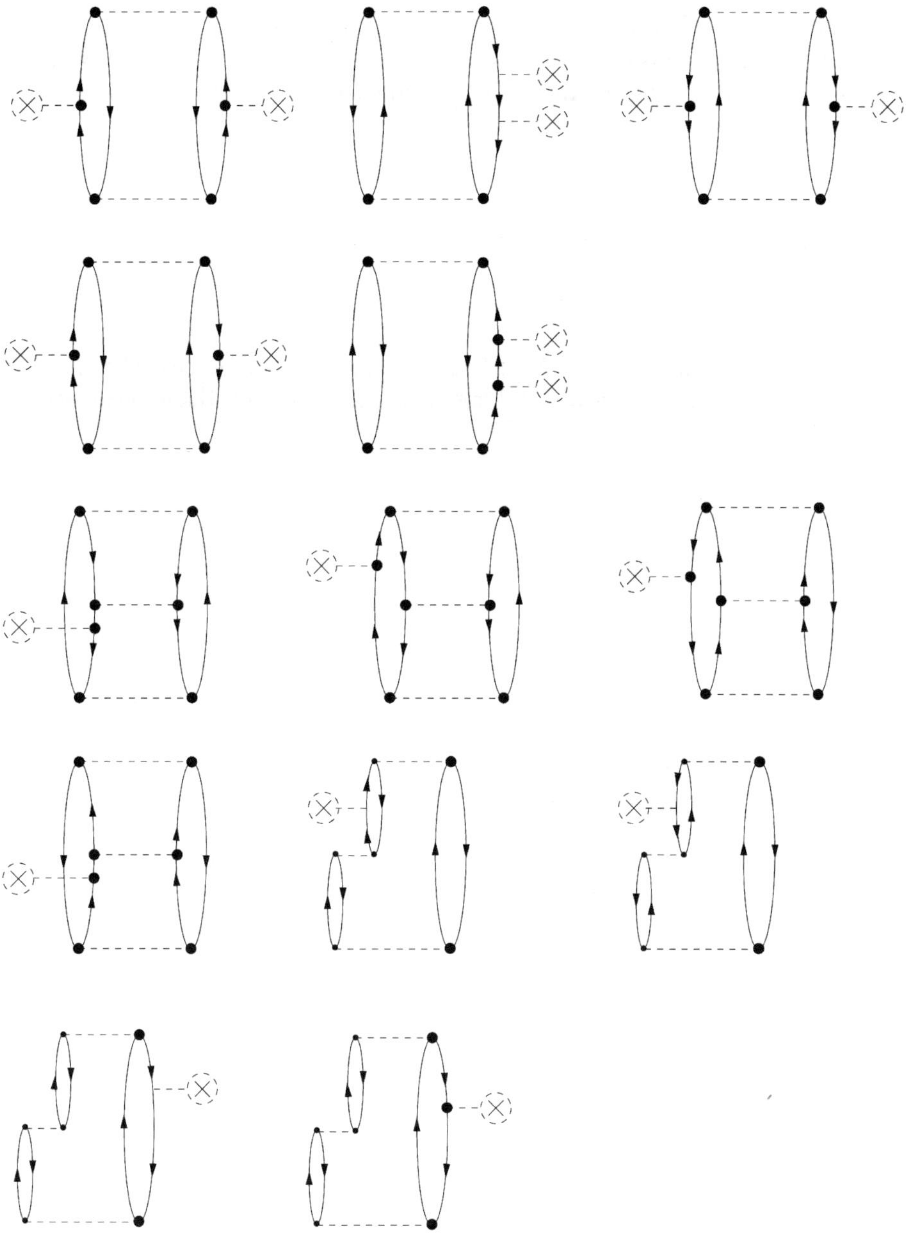

Fig. 3. Fourth order localization diagrams.

As to the localization of occupied orbitals the conventional procedures (Edmiston-Ruedenberg (Edmiston *et al.*, 1963), Boys (Boys, 1966) might not be the most suitable because they do not restrict the magnitude of the off-diagonal Fock matrix elements.

Regarding the localization of virtual orbitals, they cannot be localized uniquely into the spatial regions of the (previously) localized occupied orbitals. There are certain exceptions: in the case of some small bases (single zeta, double zeta, etc.), or for model systems where the localized orbitals are determined by symmetry. Special cases represent the weakly interacting systems, see later in Chapter 6. A suitable localization procedure has been elaborated for their convenient study.

4. Application of LMBPT for Studying Transferability

4.1. Calculation for some polyenes

The first actual calculations using LMBPT have been carried out for the cyclic polyenes C_6H_6, $C_{10}H_{10}$ and $C_{14}H_{14}$ in the PPP approximation. The PPP Hamiltonian can be described as follows:

$$\hat{H}^{PPP} = \beta \sum_{\substack{\mu,\nu,\sigma \\ \mu \neq \nu}} a^+_{\mu\sigma} a_{\nu\sigma} + \sum_{\substack{\mu,\nu \\ \sigma,\tau}} \gamma_{\mu\nu} a^+_{\mu\sigma} a^+_{\nu\sigma} a_{\nu\tau} a_{\mu\tau} \tag{14}$$

where μ, ν and σ, τ are (atomic) orbital and spin indices, respectively.

The Mataga-Nishimoto parameterization (Mataga *et al.*, 1957) is used for the matrix elements. The CC bond length was taken to be 1.4 Å. The so-called β^{-1} was regarded as the coupling constant of the interaction of electrons. A set of localized orbitals corresponding to a Kekule structure was singled out.

In the canonical representation the total energy correction of a given order n is denoted by $\zeta^{(n)}_{can}$. In the localized representation $\zeta^{(n)}_{loc}$ consists of two terms $\zeta^{(n)}_{cl}$, $\zeta^{(n)}_{ll}$. The former is the total contribution of the canonical diagrams (with localized orbitals) the latter is the total contributions from the localization terms. The total energy corrections are denoted as follows (through a given order n):

$$E^{(n)}_{can} = \sum_{i=2}^{n} \zeta^{(i)}_{can} \tag{15}$$

$$E^{(n)}_{loc} = \sum_{i=2}^{n} \zeta^{(i)}_{loc} \tag{16}$$

The actual calculation were performed through fourth order, the perturbation corrections in both the canonical and the localized representation for the interval $-2 > \beta \geqslant -10$. The results for C_6H_6 and $C_{10}H_{10}$ were compared to those obtained by full CI. It was shown (Kapuy *et al.*, 1984; Kapuy *et al.*, 1988) that the perturbation theory recovers a fraction of the total correlation correction. The localization correction are relatively small (compared to the canonical ones) in the localized representation.

The zero differential overlap approximation can be applied in the localized representation. This was demonstrated by calculating for C_6H_6, $C_{10}H_{10}$ and $C_{14}H_{14}$, respectively the total energy corrections and the pair correlation energies through second and third order in different approximations. When the strongly local contributions were only retained the results could be compared to those obtained by the separated pair approximation: the results are quite similar (Kapuy *et al.*, 1983).

Taking into account that the electrons of the systems investigated are only weakly localizable the results are not discouraging: the intra-pair correlation energies are not much affected. It is shown that even for these weakly localizable systems more than 50% of the correlation energy consist of (strictly) local contributions (Kapuy *et al.*, 1984).

The following important conclusion can be drown, that the diagrammatic many-body perturbation theory can be used in a localized representation.

The main advantage suggested by the use of the localized many-body perturbation theory (LMBPT) is that the local effects can be separated from the non-local ones. The summations in the corrections at a given order can be truncated. As to the practical applicability of the localized representation, a localization (separation) method, satisfying a double requirement is highly desired. Well-localized (separated) orbitals with small off-diagonal Lagrangian multipliers are required (Kapuy *et al.*, 1983).

The cyclic polyenes (in PPP approximation) have been examined in (Kapuy *et al.*, 1984) too. The correlation energy contributions obtained for the all−*trans* polyenes further were analyzed both in canonical and localized representation, respectively (Kapuy *et al.*, 1994). The results are in agreement with those found for the smaller cyclic polyenes in the series.

The results obtained in localized representation for the cyclic polyenes unambiguously suggest, that local and non-local contributions at the correlated level can be separated. Separating these effects it is possible to deal with larger systems (that cannot or is difficult to be treated by using the canonical version (Pipek *et al.*, 1986). The transferability of the correlation energy quantities for some all-trans conjugated polyenes was demonstrated by using STO-3G and 6–3IG* basis sets at the ab initio level (Kapuy *et al.*, 1991). The pair-correlation energies at the second order level of LMBPT were remarkably satisfactory (see Table 1). Both second and third order

Table 1 *Transferability of intra-pair correlation energy contriutions in some all-trans conjugated polyenes*

	Intra-pair for a CH-bond	Inter-pair for a CC-bond
Basis set: STO-3G		
C_2H_4	−0.01049220	−0.00873641
C_4H_6	−0.01048325	−0.00843519
C_6H_8	−0.01048305	−0.00838256
Basis set: 6-31G*		
C_2H_4	−0.01552858	−0.01385117
C_4H_6	−0.01537345	−0.01372821
C_6H_8	−0.01522244	−0.01368269

Values are given in Hartree.

perturbation corrections were calculated and the degree of the single-particle perturbation was augmented up to 8. The convergence was found to be rapid, the local and non-local effects can be remarkably separated (Kapuy *et al.*, 1994).

It should be noted that the transferability of certain molecular properties has also been demonstrated for the same systems and for their larger more extended relatives, too (Kapuy *et al.*, 1991).

4.2. Study of related systems at the ab initio level

The application of the LMBPT on the ab initio level to some normal saturated hydrocarbons was first performed using STO-3G basis set (Kapuy *et al.*, 1987). Similar calculations have been carried out by the use of MINI basis set. The transferability of some correlation energy contributions in systems CH_4, C_3H_8, C_5H_{12} and C_7H_{16}, respectively was demonstrated (at the MP2 level, Boys' localization method was used throughout). It was also shown (similarly to the series of the polyenes), that the partitioning of the pair correlations can be done in a straightforward manner according to the order of neighborhood (see Table 2 and Table 3 in ref (Kapuy *et al.*, 1987). There, it is shown that more than 94% the intra-pair correlation is due to the zero order 'neighborhood' i.e. the effects of third and higher order can be neglected. For the inter-pair correlation the deviation is larger: the zero order terms overestimate the correlation energy by more than 10%. The zero order terms give more than 93% of the total correlation energy. Including the first and second neighbor effects more than 99% can be obtained. The second neighbor contributions are important. The transferability properties of the pair-correlation energies are demonstrated in Table 2.

Table 2 *Transferability of intra- and inter-pair correlation energy contributions in some normal saturated hydrocarbons (for a CH-bond at the end of the molecules and between two neighboring CH-bonds, respectively)*

	Intra-pair	Inter-pair
Basis set: STO-3G		
CH_4	−0.01090022	−0.00203261
C_3H_8	−0.01083636	−0.00214123
C_5H_{12}	−0.01083376	−0.00213967
C_7H_{16}	−0.01083356	−0.00213928
Basis set: MINI		
CH_4	−0.00853457	−0.00247570
C_3H_8	−0.00812241	−0.00253746
C_5H_{12}	−0.00809347	−0.00252513
C_7H_{16}	−0.00809118	−0.00252732

Values are given in Hartree.

Table 3 *Transferability of intra-pair correlation energy contributions in some related systems containing CH- and CC-bonds, respectively*

	Intra-pair for a CH-bond	Intra-pair for a CC-bond
Basis set: MINI		
CH_3OH	−0.008647	–
CH_3CH_2OH	−0.008432	−0.007736
CH_3NH_2	−0.008671	–
$CH_3CH_2NH_2$	−0.008503	−0.007631
Basis set: 6-31G*		
CH_3OH	−0.01609	–
CH_3CH_2OH	−0.01578	−0.01236
CH_3NH_2	−0.01593	–
$CH_3CH_2NH_2$	−0.01574	−0.01185

Values are given in Hartree.

The method to partition the correlation energy corrections according to the topological and geometrical structures of the molecules is thus apparently a useful one. It should be noted, however, that due to restricted size of the basis, the correlation energies calculated are less than a third of the 'experimental'. The leading terms on the other hand, give the largest contributions when using systematically extended basis sets. By using larger basis sets the importance of localization corrections seem to increase. This was illustrated by the results obtained for CH_4 with different basis sets, e.g. The partitioning of the correlation energy corrections could be carried out for all standard basis sets applied until now.

The partitioning of the pair-correlations, could thus be done according to a so-called law or order of 'neighborhood'. This low is further applied in (Kapuy *et al.*, 1990). respectively. The application of this law of neighborhood is useful for studying the transferable properties of the correlation energy contributions. The results clearly show, that the local and non-local effects can be separated. The contributions of distant (virtual) orbitals can be omitted to a rather good approximation.

Two important classes of related systems are also to be mentioned. One of these series is obtained by replacing a terminal CH_3 group in the normal saturated hydrocarbons by a functional group: this was an OH or an NH_2 group in our study. The correlation energy partitioning – according to the several pair contributions – for some systems are given in Table 3. The transferability of the CH and CC intra-pair correlation energy quantities could be noticed in each system, let it containing OH- or NH-bond.

Another field, were the transferability property of certain correlation energy contributions can be demonstrated is in the study of weakly interacting systems in localized representation. Although the interaction energy between the systems will be discussed later (Chapter 5), it should be emphasized here that the transferable property of the total correlation energy contributions from each of the monomers (which take part in the interaction) is of importance. The existence of this transferability property suggests that the interaction studied is weak. Another kind of transferability between certain pair–correlation energy contribution may also hold from the non-interacting monomers to the so-called super-molecule. This transferability could also be pointed out for several super-molecules containing two or more interacting monomers. (See chapter 6.) The transferability may significantly reduce the computational cost and work.

Some other systems – beside the series of relative ones – have also been studied in localized representation, using the LMBPT scheme. Some of these systems have importance in biology. Two kinds of molecular systems have been investigated, each of them is related to the CH_2O molecule which is now believed to be one of the most important small molecular species in human life.

In a paper of Pipek (Pipek *et al.*, 1986) and Ladik an ab initio electron correlation study was described for some molecules using localized representation. Long-range and local correlation effects could be separated in a mathematically well-defined manner. The effect of various atomic basis sets furthermore, was also investigated. The results show, that the role of the localization corrections in the LMBPT is important. When the long-range correlation is neglected, only a fraction of the integrals is to be taken into account. The energy loss, on the other hand, (about 7%) is not significant.

Summarizing this chapter, one can state that the transferability of the correlation energy contributions is a quite general property of several molecular systems. In the framework of LMBPT method, that property can be used to **a**/ come closer to chemical intuition, **b**/ demonstrate the importance of some local and/or non-local effects and **c**/ reduce some of the computational work.

5. The Super Molecule Approach by using LMBPT

The study of the interaction energies between atoms and/or molecules has been of interest for a long time. One of the most commonly used approach for calculating the interaction energies between the molecules is the super molecule method (Duijneveldt *et al.*, 1994). In the framework of this method the same theoretical model is employed for the determination of the total energy of the super molecule as well as that of the subsystems. The calculations of weak interactions are often performed using the super molecule approach. The computations in these cases are suffering from the basis set superposition error (BSSE). The BSSE, namely, manifests itself as an artificial energy lowering. A correction procedure was proposed by Boys and Bernardi (Boys *et al.*, 1971): using their counter-poise (CP) method an essential part of the superposition error can be taken into account. The application of localized orbitals for describing intermolecular interactions was elaborated as early as in the 70s (Schaefer *et al.*, 1971). The use of the local correlation method (Pulay *et al.*, 1986), furthermore, leads to certain computational savings. A recent review summarizes the main features of the methods using localized orbitals within the super molecule approach (Saebo *et al.*, 1993). The calculations at the HF level do not give rise to significant work when applying the CP-corrections. Acceptable values can be obtained for the interaction energy, provided that large enough basis sets are used. The treatment at the correlated level, on the other hand, is more difficult. A detailed analysis on this subject can be found in (Duijneveldt *et al.*, 1994) (a discussion is also given in ref (Kozmutza *et al.* 1996; Kozmutza *et al.*, 1998). Here the correlated level will only be investigated.

5.1. The use of spatial symmetry

The localized representation can be applied in the study of simple interacting systems like diatomic molecules. A series of diatomics has been studied by using the LMBPT method at the MP2 level (Kozmutza *et al.*, 1993). The localized orbitals for the calculations where obtained using a slightly modified Boys' method: the spatial symmetry, namely, was taken into consideration. Taking into account the spatial symmetry the virtual orbitals could also be separated in a straightforward manner. The canonical space has been partitioned into certain subgroups: these groups of CMOs were localized separately without any difficulties. Some significant results are summarized in Table 4. The results clearly show that **a**/ the canonical and localized representation produce comparable values for the correlation energies and **b**/ the interaction energies at the correlated level (all are CP-corrected) do not differ seriously

Table 4 *Total and interaction energies obtained at the correlated MP2 level for some studied dimers in canonical and localized representation*

System, basis set and inter-atomic distance (Bohr)	Total correlation Canonical	Total correlation Localized	Interaction Canonical	Interaction Localized
Ne_2 (6-31G*) 5.0	−0.30408	−0.30373	−0.48-3	−0.28-3
Ne_2 (6-31G*) 7.0	−0.30357	−0.30345	−0.89-5	−0.27-4
Ar_2 (6-31G) 5.0	−0.07227	−0.07226	+0.78-3	−0.21-3
Ar_2 (6-31G) 7.0	−0.07253	−0.07218	+0.50-4	−0.19-4
Ar_2 (6-31G) 9.0	−0.07249	−0.07211	−0.72-4	−0.88-6
Ar_2 (6-31G*) 7.0	−0.08313	−0.08306	−0.01-4	−0.72-4
Be_2 (6-31G) 5.0	−0.05119	−0.05048	−0.43-2	−0.53-2
Be_2 (6-31G) 6.0	−0.04868	−0.04803	−0.19-2	−0.26-2
Be_2 (6-31G) 7.0	−0.04780	−0.04777	−0.11-2	−0.12-2
Be_2 (6-31G*) 5.0	−0.06309	−0.06286	−0.50-2	−0.55-2
Be_2 (6-31G*) 6.0	−0.05885	−0.05879	−0.30-2	−0.27-2
Be_2 (6-31G*) 9.0	−0.05609	−0.05606	−0.43-3	−0.36-3
Be_2 (6-31G*) 10.0	−0.05592	−0.05590	−0.32-3	−0.27-3
Mg_2 (6-31G) 5.0	−0.04808	−0.03171	−0.48-2	−0.36-2
Mg_2 (6-31G) 7.0	−0.04418	−0.02943	−0.16-2	−0.13-2
Mg_2 (6-31G) 9.0	−0.04303	−0.02885	−0.79-3	−0.46-3
Mg_2 (6-31G*) 7.0	−0.06211	−0.05142	−0.32-2	−0.33-2

Values are given in Hartree.

from each other in the most cases, however, some discrepancies can be found but only for Ar_2 system.

5.2. *Studies by using Boys' localized orbitals*

The first application of the LMBPT method for studying weakly interacting systems was published in 1990 (Kozmutza *et al.*, 1990). A detailed description on the possibility of treating dimer (weakly interacting) molecules using Boys' localized orbitals (LMOs) is given in (Kapuy *et al.* 1991). The dimer of water was subjected to a careful study in the above-mentioned work. The investigations have clearly shown that the use of LMOs is convenient for the calculation of the various correlation energy components of the interaction energy at the MP2 level around the intermolecular energy minimum. The correlation energy could be separated into intra- and inter-part – and this result is very similar to that obtained for related systems (see results given in Chapter 4). The results thus affirm that the use of LMOs in the LMBPT scheme have certain advantages in analyzing weakly interacting systems. This conclusion does not hold exclusively for the dimer of water: other dimers have also been investigated in ref. (Kapuy *et al.*, 1991). The interaction energy in several dimers composed from different monomers are also of interest: a $BH_3 + H_2O$ complex is one super molecule of this type. Performing a Boys' localization for the monomers' canonical orbitals as well as for the whole super molecule's CMOs, the obtained LMOs were used in the LMBPT for calculating the correlation energy components in that complex.The results for the intra- and inter-parts of the contributions to the correlation energy at the MP2 level in the $BH_3 + H_2O$ complex is given in Table 5. The values clearly show that **a/** transferability holds between the contributions obtained for the subsystems' and the super-system's LMOs and **b/** certain contributions of the inter-parts are significantly responsible for the interaction energy between the monomers. The use of Boys' localized orbitals, on the other hand, has some disadvantages (Bartha *et al.*, 1990). In order to overcome certain difficulties when working with Boys' LMOs, a new scheme was elaborated especially to produce orbitals in order to treat weakly interacting atoms and/or molecules.

6. The Use of SMOs in the Framework of LMBPT

The convenient study of weakly interacting systems in the LMBPT procedure requires the following criteria:

- the convergence of the perturbation series (i.e. that of the localized corrections) should be fast,
- both the occupied and virtual orbitals are to be separated as much as possible,
- the separation procedure should be simple.

The method which produces the separated molecular orbitals (SMOs) was first published in ref. (Kozmutza *et al.*, 1994) and then later applied successfully for dimer molecules (Kozmuta *et al.*, 1994; Kozmutza *et al.*, 1995) and also for several interacting systems (Kozmutza *et al.*, 1996; Kozmutza *et al.*, 1994).

Let us now summarize:

Table 5 *Intra- and inter-pair correlation energy contributions for the localized molecular orbitals in the super-molecule $BH_3 + H_2O$*

	B–O distance (in bohr)		
	3.29	3.48	3.67
Intra-pair contributions			
BH bond-pair (1)	−0.172992-1	−0.173972-1	−0.174961-1
BH bond-pair (2)	−0.175154-1	−0.176019-1	−0.176817-1
OH bond-pair	−0.199002-1	−0.199028-1	−0.199125-1
O lone-pair (1)	−0.165466-1	−0.165349-1	−0.165319-1
O lone-pair (2)	−0.176695-1	−0.176178-1	−0.175161-1
Total inter-pair contributions			
BH(1)/OH bond 1	−0.619673-3	−0.511201-3	−0.419456-3
BH(1)/OH bond 2	−0.556679-3	−0.468625-3	−0.390640-3
BH(1)/O lone (1)	−0.552541-3	−0.461226-3	−0.382323-3
BH(1)/O lone (2)	−0.278069-2	−0.236630-2	−0.196980-2
BH(2)/OH bonds	−0.514322-3	−0.432267-3	−0.359953-3
BH(2)/O lone (1)	−0.575586-3	−0.468549-3	−0.380511-3
BH(2)/O lone (2)	−0.267981-2	−0.227560-2	−0.189010-2
Net inter-pair contributions			
BH bond(1)/OH bond (1)	−0.401010-3	−0.334417-3	−0.271178-3
BH bond(1)/OH bond (2)	−0.316304-3	−0.265544-3	−0.217674-3
BH bond(1)/O lone (1)	−0.291279-3	−0.263864-3	−0.216215-3
BH bond(1)/O lone (2)	−0.945971-3	−0.815977-3	−0.635466-3
BH bond (2)/OH bonds	−0.301603-3	−0.270688-3	−0.229020-3
BH bond(2)/O lone (1)	−0.337492-3	−0.300217-3	−0.245644-3
BH bond(2)/O lone (2)	−0.984463-3	−0.766881-3	−0.572497-3

Basis = 6–31G*, values are given in Hartree. The definition for the total and net terms are given in Eq. (29) and (30), respectively.

- the essence of the main feature of SMOs,
- the treatment of the second, third and further order of the correlated level (MP2, MP3 and MP4 respectively),
- the scheme for separating the correlation energy contributions at each level of correlation.

The procedure for obtaining SMOs starts from the occupied and virtual canonical molecular orbitals (CMOs) of the super-system: it uses a bridge in order to transform the CMOs towards a prescribed set of orbitals. The criterion prescribes, that the overlap between the constructed orbitals ψ_i of a super molecule (SM) should be maximal with the initial, canonical orbitals ϕ_i of the non–interacting molecules, the bridge thus implies an overlap criterion:

$$\sum_{i=1}^{N} |\langle\phi_i|\psi_i\rangle|^2 = max. \tag{17}$$

At the correlated level the many–body perturbation theory is applied, the localized version of which (LMBPT) has already proven to be useful in the study of molecular electronic structure. The LMBPT is a double perturbation theory, and the perturbational correction are calculated as:

$$E_{loc}^{corr} = \sum_N \sum_M E_{loc}^{(N,M)} \tag{18}$$

$$E_{can}^{(N)} = \sum_{M=1}^{\infty} E_{loc}^{(N,M)} \tag{19}$$

where N is the order of the two–particle perturbation and M is the order of the single–particle perturbation. Summing to infinite order over the index M, the canonical correlation energy corrections are obtained.

For two–particle perturbations we can easily separate the energy components of the interacting system into different terms. The expressions for the second and third order MBPT correction has the form of:

$$E^{(2)} = \sum_{M_1 M_2 M_3 M_4} e(M_1 M_2 M_3 M_4) \tag{20}$$

where

$$e(M_1 M_2 M_3 M_4) = \sum_{h_1 \in M_1^h} \sum_{h_2 \in M_2^h} \sum_{p_1 \in M_3^p} \sum_{p_2 \in M_4^p} \langle p_1 p_2 | h_1 h_2 \rangle c(h_1 h_2 p_1 p_2) \tag{21}$$

$$E^{(3)} = \sum_{M_1 M_2 M_3 M_4 M_5 M_6} (e^1(M_1 M_2 M_3 M_4 M_5 M_6) + e^2(M_1 M_2 M_3 M_4 M_5 M_6) + e^3(M_1 M_2 M_3 M_4 M_5 M_6)) \tag{22}$$

where

$$e^1(M_1 M_2 M_3 M_4 M_5 M_6) = \sum_{h_1 \in M_1^h} \sum_{h_2 \in M_2^h} \sum_{p_1 \in M_3^p} \sum_{p_2 \in M_4^p} \sum_{p_1' \in M_5^p} \sum_{p_2' \in M_6^p} c(h_1 h_2 p_1 p_2) \times \frac{\langle p_1 p_2 | p_1' p_2' \rangle \langle p_1' p_2' | h_1 h_2 \rangle}{\epsilon_{h_1} + \epsilon_{h_2} - \epsilon_{p_1'} - \epsilon_{p_2'}} \tag{23}$$

$$e^2(M_1 M_2 M_3 M_4 M_5 M_6) = \sum_{h_1 \in M_1^h} \sum_{h_2 \in M_2^h} \sum_{h_1' \in M_3^h} \sum_{h_2' \in M_4^h} \sum_{p_1 \in M_5^p} \sum_{p_2 \in M_6^p} c(h_1 h_2 p_1 p_2) \times \frac{\langle p_1 p_2 | h_1' h_2' \rangle \langle h_1' h_2' | h_1 h_2 \rangle}{\epsilon_{h_1'} + \epsilon_{h_2'} - \epsilon_{p_1} - \epsilon_{p_2}} \tag{24}$$

$$e^3(M_1M_2M_3M_4M_5M_6) = \sum_{h_1\in M_1^h}\sum_{h_2\in M_2^h}\sum_{h_1'\in M_3^h}\sum_{p_1\in M_4^p}\sum_{p_2\in M_5^p}\sum_{p_2\in M_6^p} c(h_1h_2p_1p_2)$$

$$\times\left(-2\frac{\langle h_1'p_2|h_1p_1'\rangle\langle p_1p_1'|h_1'h_2\rangle}{\epsilon_{h_1'}+\epsilon_{h_2}-\epsilon_{p_1'}-\epsilon_{p_2}} - 2\frac{\langle h_1'p_1|p_1'h_1\rangle\langle p_2p_1'|h_1'h_2\rangle}{\epsilon_{h_1'}+\epsilon_{h_2}-\epsilon_{p_1'}-\epsilon_{p_2}}\right.$$

$$\left.+\,2\frac{(2\langle h_1'p_1|p_1'h_1\rangle - \langle h_1'p_1|h_1p_1'\rangle)\langle p_1'p_2|h_1'h_2\rangle}{\epsilon_{h_1'}+\epsilon_{h_2}-\epsilon_{p_1'}-\epsilon_{p_2}}\right) \tag{25}$$

Here h_1, h_2, h_1', h_2' (p_1, p_2, p_1', p_2') denote the occupied/virtual orbitals (holes and particles), M_i^h, (M_i^p) are the set of the occupied/virtual orbitals belonging to the i-th monomer, and

$$c(h_1h_2p_1p_2) = \frac{2\langle h_1h_2|p_1p_2\rangle - \langle h_1h_2|p_2p_1\rangle}{\epsilon_{h_1}+\epsilon_{h_2}-\epsilon_{p_1}-\epsilon_{p_2}} \tag{26}$$

The definition of the one–body, two–body, three–body and four–body correlation energy terms at several levels of correlation, respectively, as follows:

$$E^2(A,\ total) = \sum_{\substack{M_1M_2M_3M_4\\ S(1,2)=\{A\}}} e(M_1M_2M_3M_4) \tag{27}$$

$$E^2(A,\ net) = \sum_{\substack{M_1M_2M_3M_4\\ S(1,2)=S(3,4)=\{A\}}} e(M_1M_2M_3M_4) \tag{28}$$

$$E^2(A,\ B,\ total) = \sum_{\substack{M_1M_2M_3M_4\\ S(1,2)=\{A,B\}}} e(M_1M_2M_3M_4) \tag{29}$$

$$E^2(A,\ B,\ net) = \sum_{\substack{M_1M_2M_3M_4\\ S(1,2)=S(3,4)=\{A,B\}}} e(M_1M_2M_3M_4) \tag{30}$$

$$E^3(A,\ total) = \sum_{\substack{M_1M_2M_3M_4M_5M_6\\ S(1,2)=\{A\}}} e^1(M_1M_2M_3M_4M_5M_6)$$

$$+\sum_{\substack{M_1M_2M_3M_4M_5M_6\\ S(1,2,3,4)=\{A\}}} e^2(M_1M_2M_3M_4M_5M_6)$$

$$+\sum_{\substack{M_1M_2M_3M_4M_5M_6\\ S(1,2,3)=\{A\}}} e^3(M_1M_2M_3M_4M_5M_6) \tag{31}$$

$$E^3(A,\ net) = \sum_{\substack{M_1M_2M_3M_4M_5M_6\\ S(1,2,3,4,5,6)=\{A\}}} e^1(M_1M_2M_3M_4M_5M_6) + \sum_{\substack{M_1M_2M_3M_4M_5M_6\\ S(1,2,3,4,5,6)=\{A\}}} e^2(M_1M_2M_3M_4M_5M_6) + \sum_{\substack{M_1M_2M_3M_4M_5M_6\\ S(1,2,3,4,5,6)=\{A\}}} e^3(M_1M_2M_3M_4M_5M_6) \tag{32}$$

$$E^3(A,\ B,\ total) = \sum_{\substack{M_1M_2M_3M_4M_5M_6\\ S(1,2)=\{A,B\}}} e^1(M_1M_2M_3M_4M_5M_6) + \sum_{\substack{M_1M_2M_3M_4M_5M_6\\ S(1,2,3,4)=\{A,B\}}} e^2(M_1M_2M_3M_4M_5M_6) + \sum_{\substack{M_1M_2M_3M_4M_5M_6\\ S(1,2,3)=\{A,B\}}} e^3(M_1M_2M_3M_4M_5M_6) \tag{33}$$

$$E^3(A,\ B,\ net) = \sum_{\substack{M_1M_2M_3M_4M_5M_6\\ S(1,2)=S(3,4,5,6)=\{A,B\}}} e^1(M_1M_2M_3M_4M_5M_6) + \sum_{\substack{M_1M_2M_3M_4M_5M_6\\ S(1,2,3,4)=S(5,6)=\{A,B\}}} e^2(M_1M_2M_3M_4M_5M_6) + \sum_{\substack{M_1M_2M_3M_4M_5M_6\\ S(1,2,3)=S(4,5,6)=\{A,B\}}} e^3(M_1M_2M_3M_4M_5M_6) \tag{34}$$

$$E^3(A,\ B,\ C,\ total) = \sum_{\substack{M_1M_2M_3M_4M_5M_6\\ S(1,2,3,4)=\{A,B,C\}}} e^2(M_1M_2M_3M_4M_5M_6) + \sum_{\substack{M_1M_2M_3M_4M_5M_6\\ S(1,2,3)=\{A,B,C\}}} e^3(M_1M_2M_3M_4M_5M_6) \tag{35}$$

$$E^3(A,\ B,\ C,\ net) = \sum_{\substack{M_1M_2M_3M_4M_5M_6\\ S(1,2,3)=S(4,5,6)=\{A,B,C\}}} e^3(M_1M_2M_3M_4M_5M_6) \tag{36}$$

$$E^3(A,\ B,\ C,\ D,\ total) = \sum_{\substack{M_1M_2M_3M_4M_5M_6\\ S(1,2,3,4)=\{A,B,C,D\}}} e^2(M_1M_2M_3M_4M_5M_6) \tag{37}$$

$$E^{(2)} = \sum_{A} E^2(A,\ total) + \sum_{\substack{A,B \\ A>B}} E^2(A,\ B,\ total) \tag{38}$$

$$E^{(3)} = \sum_{A} E^3(A,\ total) + \sum_{\substack{A,B \\ A>B}} E^3(A,\ B,\ total) + \sum_{\substack{A,B,C \\ A>B>C}} E^3(A,\ B,\ C,\ total)$$

$$+ \sum_{\substack{A,B,C,D \\ A>B>C>D}} E^3(A,\ B,\ C,\ D,\ total) \tag{39}$$

7. Some Results Obtained by the SMO–LMBPT Formalism

In the previous chapters it was demonstrated that the separated molecular orbitals (as their name suggests) are originally devoted to describe the interaction energy in weakly interacting systems in a straightforward manner. In cases when only the interaction energy is studied it is evident that the separation of the CMOs of the super-molecule is performed accordingly to the CMOs of the non–interacting monomers. This procedure corresponds to a CMO $\rightarrow$ SMO transformation (see Eq. 17). It may also happen that some local correlation energy contributions are of interest: in this case in the transformation the CMOs of the super-molecule are separated accordingly to the LMOs (provided previously by the Boys' procedure e.g.) of the non–interacting monomers. This procedure corresponds to a LMO $\rightarrow$ SMO transformation. The above kinds of schemes (CMO $\rightarrow$ SMO and LMO $\rightarrow$ SMO, respectively) can thus be used depending on the subject of the work. There are some cases where the use of one or other scheme could be more advantageous. It is yet an important task of the present chapter to point out the main features of the schemes described. These are as follows:

1. The deviation of the diagonal Lagrangian–multipliers (see Eq. 6, 7) obtained for the orbitals after the given transformation from the canonical diagonal Fock–matrix elements.
2. The rapidity of the convergence of the correlation corrections in Eq. 19 using the orbitals after the given transformation.
3. The transferability of the correlation energy contributions: the comparison are to be made in this case in a different way when using the different transformation. In the case of the CMO $\rightarrow$ SMO procedure the obtained energy terms in whole super-system according to the subsystems are expected to be close to the corresponding contributions of the initial monomers. When a LMO $\rightarrow$ SMO scheme is applied, on the other hand, the transferability of the local contributions are worthwhile discussing.

The calculations were performed for a large variety of systems. The results are selected according to the above mentioned points of view.

The diagonal Lagrangian–multipliers in the case of water–dimer are depicted on Fig. 4. One can see that the difference between the diagonal Fock–matrix element and the corresponding diagonal Lagrangian–multipliers is considerably smaller for the

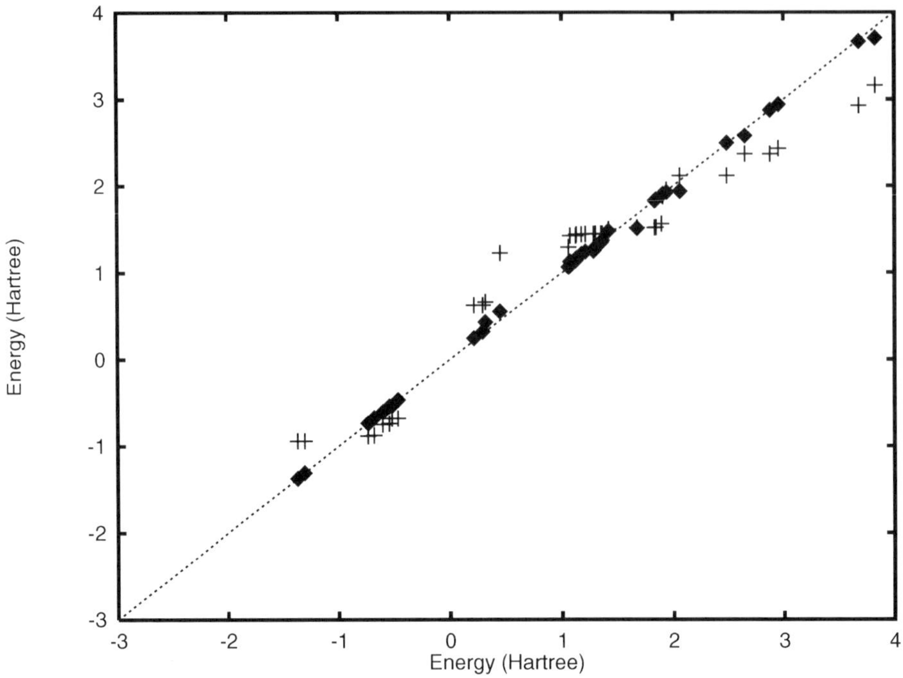

Fig. 4. Diagonal Lagrandgian-multipliers versus canonical one-particle energies for water dimer. Values corresponding to CMO→SMO and LMO→SMO denoted by diamonds and crosses, respectively.

CMO → SMO transformation than in the case of the LMO → SMO procedure. This fact is reflected also in the better convergence of the localization contributions for CMO → SMOs.

The results obtained for a model system, the $BH_3 + H_2O$ complex by both of the CMO → SMO and LMO → SMO schemes, respectively are compared. The results given in Table 5 and Table 6 reflect some important characteristics or differences of the two procedures. The geometry for the $BH_3 + H_2O$ system is shown by Figure 5. The values obtained are in a good agreement with the expectations (see point 3. above) concerning the LMO → SMO and CMO → SMO procedures, respectively.

Several studies demonstrated that the canonical contributions at the correlated level (MP2, MP3, MP4) can be approximated to a very good extent by their corresponding values obtained in localized representation. The transferability of the intra-parts can be further used: this makes it also possible to discuss conveniently the interaction energy of weakly bounded-systems in terms of SMOs.

It is expected, furthermore, that the difference between the E(intra) quantities and the corresponding values obtained from the CP corrected calculations offers a possibility for studying the effect of the CP technique. The first series of calculation were performed for the He dimer: it was studied at the experimental equilibrium

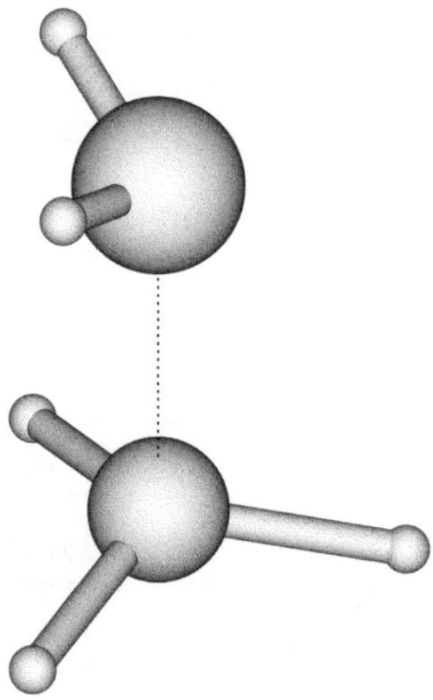

Fig. 5. $BH_a + H_2O$ complex

(R = 5.67 bohr). The Ne-dimer was also investigated at several inter-nuclear distances. The trimer and tetramer systems were calculated at certain inter-nuclear distances. Well-known basis sets were used.

The calculated intra-correlated energy terms for the He and Ne dimers at several inter-nuclear distances- are given in Table 7. The results demonstrate, that the values do not change significantly with the variation of the inter-nuclear distances. The E(intra/total) terms obtained for some linear trimer- and tetramer systems are also demonstrated in Table 8: a very significant transferability holds for these terms in each of the systems investigated.

Defining the several parts of the correlation energy, some significant terms are collected for the He tetramer studied in Table 9. The results unambiguously show that not only the values follow the differences in their geometrical position, but a transferability property also holds for both the one- and two-body terms, respectively, as well at each level of correlation.

One of the advantage of the SMO-LMBPT formalism is that we can separate the two-, three-, four-body interaction terms from those of the corresponding components of the intra-terms. In this way it is possible to exclude the effects which are caused by the deviation from the additivity of the intra-terms. Some significant results are collected in Table 10. It should be noted, that for the higher order terms there are not many references in the literature. This might be the case because these terms can not be calculated in an easy way. Our results affirm, that the higher-order interaction energies are small but not negligible.

In order to investigate the validity of the CP-recipe, a systematic study was performed for the *He* dimer. The values calculated at different inter-nuclear distances are collected in Table 11. The values for the E(intra/total) quantities were calculated

Table 6 *Diagonal Lagrangian multipliers in the $BH_3 + H_2O$ system*

Orbital No.	CMOs in super-molecule	CMOs in monomers	CMOs in super-molecule
1.	−20.630400	−7.621740	−7.573758
2.	−7.574843	−0.707125	−0.660250
3.	−1.436929	−0.493712	−0.449613
4.	−0.794581	−0.491035	−0.453883
5.	−0.718814	−20.549446	−20.630116
6.	−0.622607	−1.359485	−1.432649
7.	−0.562078	−0.721472	−0.792422
8.	−0.451724	−0.559283	−0.650717
9.	−0.444055	−0.503155	−0.592623
10.	0.105597	0.057849	0.492300
11.	0.183464	0.198212	0.229113
12.	0.205086	0.241012	0.274776
13.	0.232628	0.241769	0.273197
14.	0.266475	0.331774	0.379129
15.	0.269223	0.331955	0.383541
16.	0.367746	0.426786	0.539160
17.	0.373778	0.474439	0.570469
18.	0.492427	0.675953	0.730036
19.	0.548924	0.699655	0.735561
20.	0.552551	0.747823	0.791530
21.	0.626635	2.040553	2.151963
22.	0.730326	2.152540	2.199657
23.	0.735561	2.154511	2.197560
24.	0.789481	2.543609	2.610917
25.	1.002980	2.667696	2.716142
etc.			

Basis 6-31G*, values are given in Hartree, the B-O distance equals 3.48 bohr.

including the CP recipe, too. The results obtained at several levels of correlation, show that **a/** the intra-terms in the super-systems are in each case between the values calculated with and without the CP technique and **b/** comparing the values obtained for the non corrected and for the CP−corrected systems, one can notify, that the CP corrected E(intra/total) components are closer to those obtained by the use of the SMO−LMBPT scheme. The results thus unambiguously suggest, that the E(intra) values obtained using the CP recipe can be replaced by those resulted in the calculations performed by our scheme.

The 'beneficial' effect of the counter-poise method is thus taken into account in our method without the need of additional (i.e. CP corrected) calculations.

The dimer of H_2O and *HF*, respectively, have been the subject of theoretical studies by several authors. We also performed a work on these dimers (Kapuy *et al.*, 1998).

Table 7 *Correlated energy components, the E(intra/total) values calculated for some studied dimers*

	MP2 [hartree]	MP3 [hartree* 10^{-2}]	MP4 [hartree* 10^{-2}]
He + He			
Basis: 8s3p			
Monomer	−0.0294556	−0.532739	−0.09913
Distance (bohr)			
5.67	−0.0294578	−0.527007	−0.09639
7.67	−0.0294559	−0.529093	−0.09819
9.67	−0.0294556	−0.532741	−0.09913
Basis: 6-311G*			
Monomer	−0.0304800	−0.536942	−0.09871
Distance (bohr)			
5.67	−0.0304813	−0.536976	−0.09884
7.67	−0.0304806	−0.536955	−0.09885
9.67	−0.0304802	−0.536947	−0.09873
Ne + Ne			
Basis: 6-31G*			
Monomer	−0.151729	+0.015098	−0.45097
Distance (bohr)			
5.0	−0.151781	+0.015162	−0.45103
7.0	−0.151769	+0.015109	−0.45109
9.0	−0.151730	+0.015100	−0.45097
Basis: 6-311G*			
Monomer	−0.189088	+0.242123	−0.67011
Distance (bohr)			
5.0	−0.189204	+0.240067	−0.66872
7.0	−0.189089	+0.242134	−0.67031
9.0	−0.189088	+0.242124	−0.67011

Basis set 8s3p is from S. Huzinaga: J. Chem. Phys. 50, 1374 (1963).

That study at the MP2 correlated level for the dimers investigated are summarized in Table 7. In Table 7, we use the following notations: SM denotes the values obtained in the super-molecule, while CP denotes those obtained in the counter-poise corrected calculations. The results clearly show, that the E(intra/net) values are close to each other (using any basis sets) for the monomers in both of SM and CP systems. This suggests that the SMO-LMBPT scheme takes into account the 'benefit' effect of the basis set superposition. The deviations found in the intra terms between the SM and CP systems are explained recently (Kapuy *et al.*, 1998) in detail.

The conclusion of that paper states, that concerning the scepticism over the CP recipe might be unfounded. The differences in the SM and CP values of the E(intra) components for the two monomers are presumably due to the different basis set

Table 8 *Correlated energy components, the E(intra/total) values for some linear trimers and tetramers. Two different values are obtained in each case*

	MP2 [hartree]	MP3 [hartree* 10]	MP4 [hartree* 10]a
	He + He + He	Basis: 8s3p	
Distance:			
5.67 bohr	−0.0294578	−0.526901	−0.09684
	−0.0294565	−0.527711	−0.09617
7.67 bohr	−0.0294559	−0.535671	−0.09717
	−0.0294561	−0.534058	−0.09711
	He + He + He + He	Basis: 8s3p	
Distance:			
5.67 bohr	−0.0294571	−0.526871	−0.09675
	−0.0294563	−0.527813	−0.09637
7.67 bohr	−0.0294547	−0.535547	−0.09718
	−0.0294549	−0.534951	−0.09716
	Ne + Ne + Ne	Basis: 6-311G*	
Distance:			
5.0 bohr	−0.189320	+0.240750	−0.67903
	−0.189300	+0.240322	−0.67582
7.0 bohr	−0.189095	+0.242035	−0.67132
	−0.189090	+0.242017	−0.67121
	Ne + Ne + Ne + Ne	Basis: 6-311G*	
Distance:			
5.0 bohr	−0.189326	+0.237717	−0.66899
	−0.189209	+0.240010	−0.66901
7.0 bohr	−0.189090	+0.241931	−0.67082
	−0.189089	+0.241817	−0.67083

Basis 8s3p is from S. Huzinaga: J. Chem. Phys. 50, 1374 (1963).

dependence of the so-called charge-transfer components for the different monomers. It is expected that no increase of the basis set can cause a systematic trend in the calculated value for the interaction energy at the correlated level when using CP corrections. Without taking into account the effect of BSSE the results are incorrect. The effect studied, on the other hand, can be located in the SMO-LMBT formalism easily.

The results given in Table 12 also suggest, that the SMO-LMBPT formalism not only eliminates the basis set super-position error but also takes into account the beneficial part of the BSSE.

A third example was the advantageous application of the SMO-LMBPT model to a zeolite type structure: $AlH_2OH + H_2O + NH_3$. Ab initio modeling of zeolite structures

Table 9 *Correlation energy decomposition for the one- and two-body terms in a linear tetramer He-cluster. Values are given in mHartree*

	E (total)	E (net)
MP2 correlated level		
He(1) and He(4)	−29.458	−29.321
He(2) and He(3)	−29.456	−29.184
He(12) and He(34)	−0.0351	−0.0346
He(31) and He(42)	−0.0005	−0.0045
He(32)	−0.0352	−0.0347
He(41)	−0.0001	−0.0001
MP3 correlated level		
He(1) and He(4)	−5.2692	−5.8499
He(2) and He(3)	−5.1763	−5.1376
He(12) and He(34)	−0.0057	−0.0051
He(31) and He(42)	−0.0001	−0.0001
He(32)	−0.0055	−0.0049
He(41)	−0.0000	−0.0000
MP4 correlated level		
He(1) and He(4)	−0.9635	−0.9586
He(2) and He(3)	−0.9354	−0.9259
He(12) and He(34)	−0.0004	−0.0003
He(31) and He(42)	−0.0000	−0.0000
He(32)	−0.0003	−0.0002
He(41)	−0.0000	−0.0000

Basis 8s3p is from S. Huzinaga: J. Chem. Phys. 50, 1374 (1963).

was discussed in several works. It is known, that a study of weak dative complexes can be performed for structures of two components using extended basis sets. To come closer to the reality, i.e. the extension of the zeolite structures implies the use of small or medium sized basis sets. This occurs, for example, when modeling of ammonia or water complexation: usually minimal basis sets (STO-3G, MINI etc) are used.

The inclusion of correlation effects, as known, is inevitable. The importance of correlation has been pointed out for these structures as well. The geometry parameters used in the calculations were taken from an optimization procedure at the SCF level ((Kozmutza *et al.*, 1995)). The complex is given in Figure 6.

According to the formulae used in the SMO-LMBPT procedure (see those concerning the correlated level as given in Chapter 6), the following quantities are calculated: the intra-correlated part as well as the inter-correlated part of the energy terms at the second (MP2), third (MP3) and the fourth (MP4) level of correlation. The MINI basis

Table 10 *Some of the correlated parts of the many-body interaction energy terms obtained for the studied systems using different basis sets. The values are given in Hartree. (Distances: He-He = 5.67 bohr, Ne-Ne = 5.0 bohr)*

System Basis set	Two-body MP2 level	Three-body MP3 level	Four-body MP4 level
He + He			
Basis I.	−34.99	–	–
Basis III.	−34.39	–	–
He + He + He			
Basis I.	−34.77	−0.0258	–
Basis III.	−34.32	−0.0227	–
He + He + He + He			
Basis I.	−34.67	−0.0247	−0.00027
Basis III.	−34.30	−0.0223	−0.00030
Ne + Ne			
Basis II.	−321.30	–	–
Basis III.	−387.58	–	–
Ne + Ne + Ne			
Basis II.	−327.70	−64.92	–
Basis III.	−392.60	−0.4277	–
Ne + Ne + Ne + Ne			
Basis II.	−329.31	−69.75	−0.5632
Basis III.	−397.38	−0.4167	−0.0374

Basis I.: 8s3p (S. Huzinaga: J. Chem. Phys. 50, 1374 (1963).
Basis II.: 6-31G*, Basis III.: 6-311G*.

set was used in the calculations. The values were calculated both in the whole complex system as well as in some counter-poise (CP) corrected systems. The CP correction calculations were performed for the ALH_2OH structure: in one case the charges of the nuclei on the NH_3, in the second case those of the nuclei on the H_2O were taken equal zero. The total and the net contributions were calculated in each case for the intra contributions, at each level of correlation. The results obtained are presented in Table 13 and Table 14, respectively.

The main conclusions are as follows:

- according to the systems there is a convenient separation of the net contributions at each level in the complex system;
- the values reflect suitably the different magnitude of the energy components at each level of correlation.

Table 11 *The E(intra/total) values in some He dimers. CP corrected values are also calculated*

	MP2 [hartree]	MP3 [hartree* 10^{-2}]	MP4 [hartree* 10^{-2}]
Basis: 8s3p			
Monomer	−0.0294556	−0.532739	−0.09913
Distance (bohr)			
5.67	−0.0294578	−0.527007	−0.09639
CP corr.	−0.0294591	−0.527613	−0.09640
7.67 bohr	−0.0294559	−0.529093	−0.09819
CP corr.	−0.0294559	−0.529103	−0.09822
9.67 bohr	−0.0294556	−0.532741	−0.09913
CP corr.	−0.0294556	−0.532741	−0.09914
Basis: 6-311G*			
Monomer	−0.0304800	−0.536942	−0.09871
Distance (bohr)			
5.67 bohr	−0.0304813	−0.536976	−0.09884
CP corr.	−0.0304820	−0.536992	−0.09893
7.67 bohr	−0.0304806	−0.536955	−0.09885
CP corr.	−0.0304807	−0.536955	−0.09885
9.67 bohr	−0.0304802	−0.536947	−0.09873
CP corr.	−0.0304802	−0.536947	−0.09873

Table 12 *E(intra) values calculated at the MP2 correlated level for the dimer of water and dimer of hydrogen-fluoride. Values are given in Hartree*

		E(intra/total)	E(intra/net)
(H_2O)$_2$			
Basis set: MINI			
Monomer 1	SM	−0.0387	−0.0317
	CP	−0.0371	−0.0321
Monomer 2	SM	−0.0338	−0.0308
	CP	−0.0338	−0.0318
Basis set: 6-31G*			
Monomer 1	SM	−0.1891	−0.1838
	CP	−0.1890	−0.1839
Monomer 2	SM	−0.1873	−0.1858
	CP	−0.1883	−0.1856
Basis set: 6-311G*			
Monomer 1	SM	−0.2290	−0.2229
	CP	−0.2292	−0.2232
Monomer 2	SM	−0.2270	−0.2248
	CP	−0.2280	−0.2250
(HF)$_2$			
Basis set: 13s7p/4s contracted to 4s2p/2s			
Monomer 1	SM	−0.1277	−0.1218
	CP	−0.1272	−0.1219

Table 12 (*continued*)

		E(intra/total)	E(intra/net)
Monomer 2	SM	−0.1243	−0.1205
	CP	−0.1239	−0.1191
Basis set: 6-31G*			
Monomer 1	SM	−0.1831	−0.1775
	CP	−0.1834	−0.1775
Monomer 2	SM	−0.1728	−0.1789
	CP	−0.1808	−0.1763
Basis set: 6-311G*			
Monomer 1	SM	−0.2147	−0.2102
	CP	−0.2149	−0.2106
Monomer 2	SM	−0.2154	−0.2120
	CP	−0.2142	−0.2107

SM denotes the SMOs determined in the super-molecule.
CP denotes the SMOs obtained in the counter-poise calculations.
Geometries are as given in E. Kapuy et al.: J. Mol. Structure (Theochem) 422, 143 (1998).

Table 13 *The intra-correlated part of the energy at several levels of MBPT for the $AlH_2OH + H_2O$ structure (including CP corrections). Values are given in mhartree*

	Total contributions	Net contributions
MP2 level		
In the complex system for		
AlH_2OH	−55.50	−52.79
H_2O	−42.94	−27.10
In the CP-corrected system for		
AlH_2OH	−56.74	−54.51
H_2O	−43.68	−29.32
MP3 level		
In the complex system for		
AlH_2OH	−7.81	−8.26
H_2O	−5.37	−6.05
In the CP-corrected system for		
AlH_2OH	−7.88	−8.13
H_2O	−6.09	−6.74
MP4 level		
In the complex system for		
AlH_2OH	−7.02	−6.41
H_2O	−2.91	−1.65
In the CP-corrected system for		
AlH_2OH	−7.19	−6.98
H_2O	−2.89	−1.85

Table 14 *The intra-correlated part of the energy at several levels of MBPT for the $AlH_2OH + NH_3$ structure (including CP corrections). Values are given in mhartree*

	Total contributions	Net contributions
MP2 level		
In the complex system for		
AlH_2OH	−55.50	−52.79
NH_3	−52.63	−44.58
In the CP-corrected system for		
AlH_2OH	−56.31	−52.28
NH_3	−50.61	−45.06
MP3 level		
In the complex system for		
AlH_2OH	−7.81	−8.26
NH_3	−8.81	−9.53
In the CP-corrected system for		
AlH_2OH	−7.86	−8.27
NH_3	−9.31	−9.77
MP4 level		
In the complex system for		
AlH_2OH	−7.02	−6.41
NH_3	−3.32	−2.68
In the CP-corrected system for		
AlH_2OH	−8.53	−8.01
NH_3	−2.82	−2.75

A similar conclusion holds for the inter-correlated terms (see the values collected in Table 15). The separation of the energy terms is obvious, although the values obtained at each level of correlation are suffering from the restricted size of the basis set (Kozmutza *et al.*, 1998). As to the counter-poise calculations, furthermore, it is to be emphasized, that for these values there is a systematic agreement compared to the quantities obtained in the whole complex at each level of correlation. These results suggests (similarly to that found in one of our earlier studies on homo-nuclear dimer, trimer and tetramer systems (Kozmutza *et al.*, 1996)) and the dimers of $(H_2O)_2$ and $(HF)_2$, respectively that the SMO-LMBPT scheme is able to taken into account the effect of the CP recipe in weakly interacting system: this conclusion thus seems to hold for interacting systems containing three interacting molecules of different types.

Acknowledgements

We would like to thank E. M. Evleth and E. Tfirst for the useful discussions and for the careful reading. The work was supported by the OTKA grant no.: T019741, T025622, T021228, T024137, Budapest, Hungary.

Table 15 *The inter-correlated part of the energy at several levels of MBPT in the studied complices. Values are given in mhartree*

	Total contributions	Net contributions
MP2 level		
In the complex system for		
$AlH_2OH - H_2O$	−1.494	−0.610
$AlH_2OH - NH_3$	−0.981	−0.394
In the CP-corrected system for		
$AlH_2OH - H_2O$	−1.550	−0.720
$AlH_2OH - NH_3$	−0.959	−0.473
MP3 level		
In the complex system for		
$AlH_2OH - H_2O$	−0.111	−0.037
$AlH_2OH - NH_3$	−0.020	−0.024
In the CP-corrected system for		
$AlH_2OH - H_2O$	−0.102	−0.047
$AlH_2OH - NH_3$	−0.039	−0.048
MP4 level		
In the complex system for		
$AlH_2OH - H_2O$	−0.128	−0.050
$AlH_2OH - NH_3$	−0.151	−0.079
In the CP-corrected system for		
$AlH_2OH - H_2O$	−0.126	−0.069
$AlH_2OH - NH_3$	−0.125	−0.105

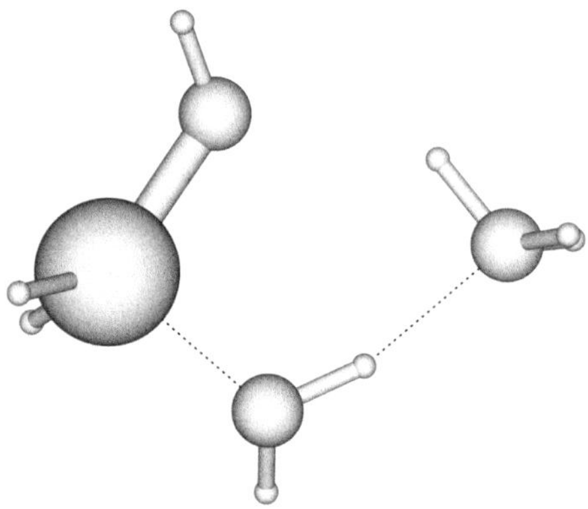

Fig. 6. Al H_2 OH + H_2O + NH_3 complex.

References

S. Rothenberg, J. Am. Chem. Soc. **93**, 68 (1971).
B. O'Leary, B.J. Duke, J.E. Eilers: Advan. Quant. Chem. **9**, 1 (1975).
C. Edmiston and K. Ruedenberg, Rev. Mod. Phys. **35**, 457 (1963).

T.L. Allen, J. Chem. Phys. **31**, 1039 (1959).
A.T. Amos, J.I. Musher, J. Chem. Phys. **54**, 2380 (1971).
W. England, M.J. Gordon, K. Ruedenberg, Theoret. Chim. Acta (Berl.) **37**, 177 (1975).
G.N. Lewis, J. Am. Chem. Soc. **33**, 762 (1916).
S. Wilson: Elektron correlation in molecules (The international series of monograhs on chemistry No. 11) Clarendon Press, Oxford, 1984. (1984).
S. Saebo, P. Pulay, Chem. Phys. Letters **113**, 13 (1985).
F.B. van Duijneveldt, J.G.M. van Duijneveldt–van de Rijdt, J.H. van Lenthe, Chem. Rev. **94**, 1873 (1994).
J. Pipek, P.G. Mezey, J. Chem. Phys. **90**, 4916 (1989).
S. Saebo, P. Pulay, Ann. Rev. Phys. Chem. **44**, 213 (1993).
E. Kapuy, C. Kozmutza, M.E. Stephens, Theoret Chim. Acta (Berl.) **43**, 175 (1976).
E. Kapuy, C. Kozmutza, Chem. Phys. Letters, **226**, 484 (1994).
B.H. Brandow, Rev. Mod. Phys. **39**, 771 (1967).
R.J. Bartlett and D.M. Silver, Int. J. Quant. Chem. Quantum, Chem. Symp. **9**, 183 (1975).
S.F. Boys, in Quantum Theory of Atoms, Molecules and Solid State, P.O. Lowdin Ed. (Academic, New York, 1966) pp. 253–262.
I. Hubac and P. Carsky, Topics Current Chem. **75**, 99 (1978).
S.F. Boys, F. Bernardi, Mol. Phys. **19**, 553 (1971).
N. Mataga, K. Nishimoto, Z. Phys. Chem. **13**, 140 (1957).
E. Kapuy, Z. Csépes and C. Kozmutza, Croatica Chem. Acta **57**, 855 (1984).
E. Kapuy, Z. Csepes, F. Bartha, F. Bogar and C. Kozmutza, Collection Czechoslovak Chem. Commun. **53**, 2073 (1988).
E. Kapuy, Z. Csepes, and C. Kozmutza, Int. J. Quant. Chem. **23** 981 (1983).
E. Kapuy, Z. Csepes, J. Pipek, Acta Phys. Hung. **55**, 365 (1984).
E. Kapuy, F. Bogar, E. Tfirst, Int. J. Quant. Chem. **52**, 127 (1994).
E. Kapuy, P.G. Szalay, R. Daudel, F. Bogar. F. Bartha, C. Kozmutza, J. Mol. Structure (Theochem) **226**, 351 (1991).
E. Kapuy, F. Bartha, F. Bogar, C. Kozmutza, Theor. Chim. Acta **72**, 337 (1987).
E. Kapuy, F. Bartha, C. Kozmutza, F. Bogar, J. Mol. Structure (Theochem) **170**, 59 (1988).
E. Kapuy, F. Bartha, F. Bogar, Z. Csepes, C. Kozmutza, Int. J. Quant. Chem. **38**, 139 (1990).
J. Pipek, J. Ladik, Chem. Phys. **102**, 445 (1986).
H.F. Schaefer III, D.R. McLaughlin, F.E. Harris, B.J. Alder, Phys. Rew. Letters **25**, 988 (1971).
P. Pulay, S. Saebo, Theoret. Chim. Acta (Berl.) 69, 357 (1986).
C. Kozmutza, E. Tfirst, E. Kapuy, Mol. Phys. **87**, 569 (1996).
C. Kozmutza, E. Kapuy, Int. J. Quant. Chem. **38**, 665 (1990).
E. Kapuy, C. Kozmutza, J. Chem. Phys. **94**, 5565 (1991).
F. Bartha, F. Bogar, E. Kapuy, Int. J. Quant. Chem. **38**, 215 (1990).
C. Kozmutza, E. Tfirst, E. Kapuy, Mol. Phys. **82**, 343 (1994).
C. Kozmutza, E. Kapuy, E.M. Evleth, E. Kassab, J. Mol. Structure (Theochem) **332**, 141 (1995).
C. Kozmutza, E. Tfirst, E. Kapuy, Mol. Phys. **80** 1059 (1993).
C. Kozmutza, E. Tfirst, L. Udvardi, J. Mol. Structure (Theochem) (submitted).
E. Kapuy, C. Kozmutza, L. Ovari, J. Mol. Structure (Theochem) **422**, 143 (1998).

Comparing $(SC)^2$CAS-SDCI and Externally Corrected CCSD Methods

G. Peris[a,b], J.-P. Malrieu[b] and J. Planelles[a]

[a]*Departament de Ciències Experimentals, Universitat Jaume I, Apartat 224, E-12080, Castelló, Spain*

[b]*I.R.S.A.M.C., Laboratoire de Physique Quantique (URA 505 du CNRS), Université Paul-Sabatier, 118 route de Narbonne, 31062 Toulouse Cedex, France*

Abstract

The performance of *size-consistent self-consistent complete active space singles and doubles configuration interaction* ($(SC)^2$CAS-SDCI) and *externally corrected CCSD* methods based on the same determinantal space is compared. Both methods may be considered as effective potential methods provided that they include, in a reduced model space, the effect of a complementary space. In the case of $(SC)^2$CAS-SDCI this is carried out by dressing, iteratively, the CI matrix with its own eigenvectors. In corrected CCSD, the CAS-SDCI wavefunction is used as a source for approximate T_3 and T_4 cluster amplitudes that are included as constant terms when decoupling the CCSD equations from the complete CCA chain. The reliability of both methods is displayed in situations of stretched bonds, that are tipically outside of the domain of good performance of the standard CCSD method. In particular, HF (single bond), H_2O (two single bonds) and N_2 (triple bond) molecules are considered.

1. Introduction and Theoretical Background

It is well-known that there is no way to find out exact eigenvalues and eigenfunctions of a differential hamiltonian corresponding to many interacting electrons in an external field. In order to reach an approximate information of a given part of the spectrum of this operator, $\mathcal{H}$, its eigenvalue equation is solved within the framework of a given model space $\mathcal{S}_M$, *i.e.*, its projection onto such model space $\mathcal{S}_M$, $H = P_M \mathcal{H} P_M$, is diagonalized. This procedure is equivalent to solve a set of N_M linear equations, N_M being the dimension of the model space $\mathcal{S}_M$.

The most commonly used model space in quantum chemistry is the so-called full configuration interaction (FCI) space. It is the antisymmetric and spin-adapted N-fold tensorial power, $V_A^{\otimes N}$, of the $2k$-dimensional spin-orbital one-electron space V. The dimension of the FCI space is given by the Weyl-Paldus formula [1]

$$D(S, N, K) = \frac{2S+1}{k+1} \binom{k+1}{\frac{N}{2} - S} \binom{k+1}{\frac{N}{2} + S + 1}.$$

For the sake of practical simplicity (the use of Slater rules and a two-slope space geometry [2]) one may employ the larger, just antisymmetric but non-spin-adapted space spanned by $\binom{2k}{N}$ Slater determinants.

Usually, despite the reduction of the original problem up to this model space, still a formidable problem remains. If so, smaller and carefully defined subspaces are used.

A. Hernández-Laguna et al. (eds.), Quantum Systems in Chemistry and Physics, Vol. 1: Basic Problems and Model Systems, 73–85.

Perhaps the most widely considered in the field of quantum chemistry is the complete active space (CAS) [3]. In a CAS space some holes (occupied orbitals) and particles (empty orbitals) are frozen. The remaining orbitals and electrons are used to build up a FCI space of the active electrons and orbitals. The dimension of a CAS space is, generally, much lower than the dimension of a FCI space. Unfortunately, this space disregards most of the so-called dynamical correlation energy [4]. One may account for the most important part of this correlation energy by enlarging the CAS space. The new space includes, additionally, all single and double excitations from holes to particles for every reference N-electron function of the CAS space. The resulting space is referred to as CAS-SDCI. Its dimension d is still much lower than FCI, in general.

Solving a hamiltonian $\mathcal{H}$ in a CAS-SDCI space (avoiding this way the often untractable FCI dimension) is equivalent to solve d linear equations, disregarding the rest of the FCI space. All the same, one may keep the low dimensionality but try to account for the influence of the outer space in the eigenvalues yielded by the CAS-SDCI (or whatever) subspace. To achieve this goal one may use the technique of effective hamiltonians [5]. Basically, one deals with the hamiltonian

$$\overline{H} = P_M(H + \Delta)P_M \tag{1}$$

where $\overline{H}$ is a projection onto the subspace $\mathcal{S}_M$ of a dressed hamiltonian, Δ being a dressing operator and H the hamiltonian. The eigenvalue equation of this effective hamiltonian reads,

$$P_M(H + \Delta)P_M|\Psi_0\rangle = \varepsilon_0 P_M|\Psi_0\rangle. \tag{2}$$

If we consider $|\Psi_0\rangle$ in an intermediate normalization form, and refering the complementary space of $\mathcal{S}_M$ as $\mathcal{S}_E$, we have:

$$|\Psi_0\rangle = |\phi_0\rangle + \sum_{\substack{i\in\mathcal{S}_M\\ i\neq 0}} c_i|\phi_i\rangle + \sum_{\alpha\in\mathcal{S}_E} c_\alpha|\phi_\alpha\rangle. \tag{3}$$

Multiplying equation (2) by $\langle\phi_i| \in \mathcal{S}_M$ and carrying out integrations, we have:

$$H_{i0} + \sum_{\substack{j\neq 0,i\\ j\in\mathcal{S}_M}} c_j H_{ij} + c_i(H_{ii} - \varepsilon_0) + \sum_{j\in\mathcal{S}_M} c_j\Delta_{ij} = 0. \tag{4}$$

The equivalent equation of the FCI chain reads:

$$H_{i0} + \sum_{\substack{j\neq 0,i\\ j\in\mathcal{S}_M}} c_j H_{ij} + c_i(H_{ii} - \varepsilon_0) + \sum_{\alpha\in\mathcal{S}_E} c_\alpha H_{i\alpha} = 0. \tag{5}$$

Therefore, we may define the matrix elements of the dressing operator according to

$$\sum_{j\in\mathcal{S}_M} c_j\Delta_{ij} = \sum_{\alpha\in\mathcal{S}_E} c_\alpha H_{i\alpha}. \tag{6}$$

Provided there are only d variables, we have some freedom when defining Δ (see [6–8] for different definitions of Δ). For example, we may ensure hermiticity of Δ by choosing a diagonal dressing, *i.e.* $\Delta_{ij} = \Delta_{ii}\delta_{ij}$. Then,

$$\Delta_{ii} = \frac{1}{c_i}\sum_{\alpha\in\mathcal{S}_E} c_\alpha H_{i\alpha}. \tag{7}$$

Of course, the c_α coefficients are unknown unless the FCI hamiltonian is diagonalized, but one may try different ways to estimate c_α from the set of model space coefficients c_i (for a wide description on how this can be done, see e.g. [9]). In particular, if we write:

$$c_\alpha = c_i c_j^{(2)}, \tag{8}$$

we ensure the requirement of separability of two non-interacting systems [10]. In equation (8), $c_j^{(2)}$ is the coefficient of the double excited determinant $|\phi_j\rangle$ (associated to the excitation operator $G_j^{(2)}$), fulfilling:

$$|\phi_\alpha\rangle = G_j^{(2)}|\phi_i\rangle. \tag{9}$$

Within this approach (called (SC)² because it is size-consistent if localized orbitals are used and the coefficients are, in practice, yielded by a recursive self-consistent procedure) the diagonal elements Δ_{ii} of the dressing operator become:

$$\Delta_{ii} = \sum_{\substack{j \\ G_j^{(2)}|\phi_i\rangle\neq 0\notin\mathcal{S}_M}} c_j^{(2)} H_{0j} \tag{10}$$

where,

$$\langle\phi_i|H|\phi_\alpha\rangle = \langle\phi_i|H|G_j^{(2)}\phi_i\rangle = H_{0j}. \tag{11}$$

The diagonal element of the dressing operator may be rewritten as:

$$\Delta_{ii} = \sum_{\substack{j \\ G_j^{(2)}|\phi_0\rangle\in\mathcal{S}_M}} c_j^{(2)} H_{0j} - \sum_{\substack{j \\ G_j^{(2)}|\phi_i\rangle = 0}} c_j^{(2)} H_{0j} - \sum_{\substack{j \\ G_j^{(2)}|\phi_i\rangle\neq 0\in\mathcal{S}_M}} c_j^{(2)} H_{0j}. \tag{12}$$

Here, the first term represents the correlation energy while the second one stands for the exclusion principle violating (EPV) terms which involve a forbidden double excitation over $|\phi_i\rangle$ (*i.e.*, those fulfilling $G_j^{(2)}|\phi_i\rangle = 0$). The third term excludes the effect of all higher than double excitations which are already included in the determinantal space $\mathcal{S}_M$ (they should not be considered when describing in Δ the effect of the outer space). Then, we may write equation (12) as,

$$\Delta_{ii} = E_{corr} - E_i^{EPV} - \sum_{\substack{j \\ G_j^{(2)}|\phi_i\rangle\neq 0\in\mathcal{S}_M}} c_j^{(2)} H_{0j}. \tag{13}$$

This method, (SC)² CAS-SDCI, in its actual *minimal dressing* implementation, may be considered as an approximation to the state-selective multireference* Coupled Cluster (SS-MRCC) [11,12]. A full dressing, including all remaining linked terms in Δ, would yield a method equivalent to SS-MRCC.

A different way to approximate SS-MRCC is the so-called externally corrected CCSD (*ec*-CCSD) [13–17]. The *ec*-CCSD method is based on the Coupled Cluster Approach (CCA). In CCA the exact wavefunction is written in an exponential form,

$$|\Psi_0\rangle = e^T|\phi_0\rangle, \tag{14}$$

where T is the so-called *cluster operator*. In the *ec*-CCSD method, this operator is approximated as:

$$T = T_1 + T_2 + T_3^0 + T_4^0 \tag{15}$$

where T_1, T_2 are unknowns to be determined and T_3^0, T_4^0 are three- and four-body fixed cluster operators previously determined from an external CI or VB calculation (in this paper a CAS-SDCI wavefunction). T_3^0 and T_4^0 are actually obtained by means of a cluster analysis of the external CI wavefunction, using the well-known formulas

$$T_1 = C_1 \tag{16}$$

$$T_2 = C_2 - \frac{1}{2}C_1^2 \tag{17}$$

$$T_3 = C_3 - C_1C_2 + \frac{1}{3}C_1^3 \tag{18}$$

$$T_4 = C_4 - C_1C_3 + C_1^2C_2 - \frac{1}{2}C_2^2 - \frac{1}{4}C_1^4. \tag{19}$$

Once T_3^0 and T_4^0 are determined, they are introduced as constant terms when decoupling CCSD equations from the full CC chain of equations (standard CCSD assumes T_3^0 and T_4^0 to be zero).

We may now stress some formal similarities and differences between both approaches when the model space used in $(SC)^2$CAS-SDCI and the source of corrections in *ec*-CCSD is the same (a more mathematically oriented comparison will be showed in the next section).

- $(SC)^2$CAS-SDCI is a size-extensive method, while *ec*-CCSD is almost size-extensive (CC procedure is itself size-extensive, but the corrections are not, provided they come from a truncated variational CI wavefunction).
- $(SC)^2$CAS-SDCI may be viewed as a set of d equations to be solved simultaneously (actually, the implementation of the method is a recursive diagonalization process). In contrast, *ec*-CCSD implies a unique diagonalization and a further resolution of a rather small set of non-linear equations.

---

*SS-MRCC is not a genuine multireference approach. In a genuine multireference approach several electronic states are simultaneously studied in a single calculation. In SS-MRCC a single electronic state is studied in a single calculation. It is actually a single reference CC method. Nevertheless, it includes most meaningful T_n ($n > 2$), the ones corresponding to a CAS-SDCI expansion.

- The *minimal dressing* of (SC)^{2}CAS-SDCI yields a size-extensive method deleting all unlinked terms in CAS-SDCI, but does not include any other linked term. One might carry out a *full dressing* including all pertinent linked terms. In such a case *dressed*-CAS-SDCI becomes equivalent to SS-MRCC. There is no natural way to turn *ec*-CCSD into SS-MRCC (except a formal correction using T_3^0 and T_4^0 obtained from the SS-MRCC we attempt to mimic).
- (SC)^{2}CAS-SDCI is, in general, computationally heavier than *ec*-CCSD. A *full dressing* method (equivalent to SS-MRCC) would be much heavier.
- Both methods may use a reduced optimally selected CI instead of a CAS-SDCI (for example, a selected CI using a CIPSI procedure [10,18]). In such a case, the computational cost of *ec*-CCSD falls down dramatically, while the final energies hardly change [35].

2. Mathematical Aspects

For a detailed comparison of (SC)^{2}CAS-SDCI and *ec*-CCSD methods, equations corresponding to single and double excitations should be written explicitly for both approaches. For the former one, we can substitute eq. (10) in (4) and, using operator notation, we have

$$\langle \phi_{I_1}^{(1)} | H_N (1 + C_1 + C_2 + C_3^{int} + [C_1 C_2]_{I_1}^{ext}) | \phi_0 \rangle = c_{I_1}^{(1)} E_{corr} \tag{20}$$

$$\langle \phi_{I_2}^{(2)} | H_N \left(1 + C_1 + C_2 + C_3^{int} + C_4^{int} + \left[\frac{1}{2} C_2^2 \right]_{I_2}^{ext} \right) | \phi_0 \rangle = c_{I_2}^{(2)} E_{corr} \tag{21}$$

where

$$[C_1 C_2]_{I_1}^{ext} = \left[\sum_{I_2} c_{I_1}^{(1)} c_{I_2}^{(2)} G_{I_1}^{(1)} G_{I_2}^{(2)} \right]^{ext} \tag{22}$$

$$\left[\frac{1}{2} C_2^2 \right]_{I_2}^{ext} = \left[\sum_{I_2'} c_{I_2}^{(2)} c_{I_2'}^{(2)} G_{I_2}^{(2)} G_{I_2'}^{(2)} \right]^{ext} \tag{23}$$

and superscripts *int* and *ext* refer to excitations belonging to $\mathcal{S}_M$ or $\mathcal{S}_E$, respectively. H_N is the normal form of the hamiltonian operator [19].

On the other hand, by substituting Eq. (14) in the Schrödinger equation and projecting it onto single and double excitations, we get,

$$\langle \phi_{I_1}^{(1)} | H_N \left(1 + T_1 + T_2 + T_3 + \frac{1}{2} T_1^2 + T_1 T_2 + \frac{1}{6} T_1^3 \right) | \phi_0 \rangle = c_{I_1}^{(1)} E_{corr} \tag{24}$$

$$\langle \phi_{I_2}^{(2)} | H_N \left(1 + T_1 + T_2 + T_3 + T_4 + \frac{1}{2} T_1^2 + T_1 T_2 + \frac{1}{2} T_2^2 + T_1 T_3 \right.$$
$$\left. + \frac{1}{6} T_1^3 + \frac{1}{3} T_1^2 T_2 + \frac{1}{24} T_1^4 \right) | \phi_0 \rangle = c_{I_2}^{(2)} E_{corr} \tag{25}$$

where no approximations are involved. By replacing T_3 and T_4 by T_3^0 and T_4^0 we get *ec*-CCSD. In order to compare these set of equations with the corresponding (SC)^{2}CAS-SDCI ones, one may use the equivalence between CI and CC operators (16)–(19) and write (24)–(25) as

$$\langle \phi_{I_1}^{(1)} | H_N \left(1 + C_1 + C_2 + (C_3^L)^0 + C_1 C_2 - \frac{1}{3} C_1^3 \right) | \phi_0 \rangle = c_{I_1}^{(1)} E_{corr} \tag{26}$$

$$\langle \phi_{I_2}^{(2)} | H_N \left(1 + C_1 + C_2 + (C_3^L)^0 + (C_4^L)^0 + \frac{1}{2} C_2^2 + C_1 C_2 + C_1 (C_3^L)^0 \right.$$
$$\left. - \frac{1}{12} C_1^4 - \frac{1}{3} C_1^3 \right) | \phi_0 \rangle = c_{I_2}^{(2)} E_{corr} \tag{27}$$

where $(C_i^L)^0 = T_i^0$ is the connected part of the i excitation operator. Comparing eqs. (20)–(21) and (26)–(27), we can see that *ec*-CCSD method includes all linked T_1 and T_2 contributions and linked approximate T_3 and T_4. (SC)^2 CAS-SDCI, though eliminates all unlinked effects that ensure size-extensivity, does not include additional linked terms. This is the main difference between both methods.

Furthermore, *ec*-CCSD only considers triple and quadruple excitations appearing in the CI wavefunction (*i.e.*, those 3- and 4-body excitations included in $\mathcal{S}_M$) and they are treated non-iteratively. (SC)^{2}CAS-SDCI uses the same subset of triples and quadruples in an iterative process, and, additionally, an approximation to the disconnected components of the external triple and quadruple excitations (*i.e.*, those belonging to $\mathcal{S}_E$). These terms are labelled as *ext* in eqs. (20) and (21). This is, indeed, an approximation, because the exact disconnected $C_1 C_2$ and $\frac{1}{2} C_2^2$ terms should be written as:

$$C_1 C_2 = \sum_{J_1} \sum_{I_2} c_{J_1}^{(1)} c_{I_2}^{(2)} G_{J_1}^{(1)} G_{I_2}^{(2)} \tag{28}$$

and

$$\frac{1}{2} C_2^2 = \sum_{J_2 < I_2'} c_{J_2}^{(2)} c_{I_2'}^{(2)} G_{J_2}^{(2)} G_{I_2'}^{(2)} \tag{29}$$

while in eqs. (20) and (21) only the cases $J_1 = I_1$ and $J_2 = I_2$ are included. Furthermore, the disconnected triples are only considered in monoexcited equations, while disconnected quadruples appear only in double excited equations.

Finally, (SC)^{2}CAS-SDCI involves, iteratively, excitations C_i^{int}, $i > 4$ and also approximations to high order disconnected external terms.

The present paper is devoted to display the performance of both procedures in situations of streched bonds. These areas of the molecular energy hypersurface are typically placed outside of the domain of good performance of the standard CCSD method [19–23]. In particular, we consider the dissociation of HF, H_2O and N_2 molecules (*i.e*, a single bond, two single bonds and a triple bond).

3. Applications

3.1. *Computational details*

HF molecule: we use the Dunning's contraction $[4s2p/2s]$ [24] of Huzinaga's $(9s5p/4s)$ [25] primitive set, augmented with a d polarization function with exponent 1.6 for F, and a set of p functions for H with exponent 0.75. We run different calculations for the equilibrium distance ($R_e = 1.733$ bohr), 1.5 R_e, 2.0 R_e and 3.0 R_e. In this case, Cartesian Gaussians with 6d-functions are used. Comparison is made with near FCI (nFCI, up to sextuple excitation, *i.e.* CISDTQQS) [26].

H_2O: we also use the Dunning's contraction of Huzinaga's basis. Polarization exponent for oxygen is 1.2, while for the set of p orbitals of H $\alpha = 0.8$. Water molecule is stretched symetrically, with the equilibrium geometry R_e defined in [27]. Calculations are made at R_e, 1.5 R_e and 2.0 R_e. Spherical Gaussians (5d-orbitals) were used.

N_2: TZ basis optimized by Dunning [28]. FCI energies and distances are taken from Evangelisti *et al* [29].

In all calculations core orbitals were frozen for sake of comparison with FCI results of the literature.

3.2. *Choice of active spaces*

The choice of active orbitals is crucial in our calculations. Orbital energies are not the most important factor when choosing the active space, but rather their shape and character.

For the HF molecule, the active space two electron in two orbitals [2e/2o] consisting of σ bonding and σ^* antibonding orbitals, is the minimum required for a proper description of the bond breaking. Both orbitals are quasidegenerate in the dissociative region, being the highest occupied MO (HOMO) and the lowest unoccupied MO (LUMO), respectively, although for small nuclear separation, the σ bonding is, energetically, much lower than the two π non-bonding orbitals (lone pairs). The inclusion of F_{2p_x}, F_{2p_y} and F_{2s} lone pairs and one correlating antibonding equivalent for each of them constitutes a better space ([8e/8o]) to describe the dissociation process. Provided F^*_{2s} is, actually, energetically very high and does not contribute significantly to the description of the dissociation, we disregard it and choose, instead, the smaller space [8e/7o] in our calculations.

For the H_2O molecule, the same philosophy lead us to select the two σ_{OH} bonding orbitals and their partner antibonding σ^*_{OH} orbitals, yielding the [4e/4o] space. As in the HF molecule, we can enlarge this space with the valence occupied orbitals and their antibonding partners (except for the O_{2s} orbitals) and use a [8e/7o] space.

Finally, for the N_2 molecule, the smallest reasonable space is [6e/6o], involving three bonding orbitals ($\sigma\pi_x\pi_y$) and their antibonding partners. Provided the two remaining occupied valence orbitals lie very close, a better space would be [10e/8o], involving all valence orbitals.

3.3. *Results and discussion*

For sake of comparison, in all studied cases, we run calculations for those geometries and basis sets with a FCI (or near FCI) available. The methods we deal with are CCSD, CAS-SDCI, $(SC)^2$CAS-SDCI and *ec*-CCSD corrected from both CAS-SDCI and $(SC)^2$CAS-SDCI. The performance of the methods is examined from two aspects: the total energy and the quality of the potential energy surface (PES), being this quality measured by the so-called "non-parallelity error" (NPE). For a given set of calculations in a dissociative curve, the NPE is defined as the difference between the maximal and minimal deviation from the exact FCI PES.

We compare the performance of a variational CAS-SDCI with its size-extensive version $(SC)^2$CAS-SDCI and the standard size-extensive CCSD. Additionally, we correct CCSD using the non size-extensive T_3^0 and T_4^0 amplitudes yielded by a cluster analysis of the CAS-SDCI wavefunction and also using the size-extensive T_3^0 and T_4^0 amplitudes coming from $(SC)^2$CAS-SDCI.

The molecular integral transformations were obtained employing a modified version of MOTRA program from the Molcas 4.0 package [30], adapted for the IDDCI series of programs [31], while for CAS-SDCI and $(SC)^2$CAS-SDCI we used a new set of programs [32]. Cluster analysis and corrections to CCSD equations were run with our own programs, and CC calculations are based on a orthogonally spin-adapted (OSA) implementation by Piecuch and Paldus [33].

3.3.1. *HF molecule*

We start dealing with the dissociation of a single bond system. HF has been broadly used to test the performance of new methods describing the breaking of a single bond. The resulting total energies as well as the differences relative to nFCI benchmark, representing an almost exact solution for the basis set employed, are listed in Table 1. The last column in these tables gives the NPE. Although we generate only a few points on the PES, so that the NPE given in our tables is based on this limited information, it nevertheless characterizes the parallelism of the approximate and exact FCI PESs over the examined interval of internuclear separations, since these curves are smooth and seldom intersect. The first column describes the active space used. Its dimension is given in the third column (Roman numbers for Slater determinants and italics for configuration spin functions (CSF), depending on the program used or the reference in the literature).

The results in Table 1 for [2e/2o] space clearly indicate that at the equilibrium geometry both, CAS-SDCI and $(SC)^2$CAS-SDCI, do not reach the good performance of standard CCSD (errors: 14.6 mHa, 10.9 mHa and 3.1 mHa, respectively). Additionally, we see that by dressing CAS-SDCI we quite improve the energy. On the other hand, *ec*-CCSD takes the best from both, CCSD and the external source, yielding the most accurate result (only 2.4 mHa away from nFCI, irrespective of the external source used).

The behaviour reverses when the bond is stretched. At $R = 3R_e$ the good performance of CCSD deteriorates due to the increasing importance of T_3 and T_4 amplitudes

Table 1 *Calculations on HF with DZ+P basis. The total energies are reported as* $-(E+99)$ *hartree. Dimension should be understood as number of determinants (number of spin-adapted configurations in italics). NPE (non-parallelity error) is the difference between the maximal and minimal deviation from FCI*

Active Space	Method	Dim	$-(E+99)$ (hartree)			E-nFCI (mhartree)			NPE mhartree
			R_e	$2R_e$	$3R_e$	R_e	$2R_e$	$3R_e$	
	CCSD	*1,023*	1.250421	1.073203	1.023577	3.088	10.269	21.640	18.552
[2e/2o]	CAS-SDCI	3,784	1.238920	1.068293	1.029656	14.589	15.179	15.561	0.972
	ec-CCSD		1.251110	1.080369	1.041546	2.399	3.103	3.671	1.272
	$(SC)^2$CAS-SDCI		1.242643	1.071989	1.032799	10.866	11.483	12.418	1.552
	ec-CCSD		1.251134	1.080289	1.041269	2.375	3.183	3.948	1.573
[8e/7o]	CAS-SDCI	64,768	1.244648	1.075327	1.037278	8.861	8.145	7.939	0.922
	ec-CCSD		1.252827	1.083022	1.044726	0.682	0.450	0.491	0.232
	$(SC)^2$CAS-SDCI		1.246244	1.076824	1.038586	7.265	6.648	6.631	0.634
	ec-CCSD		1.252830	1.083002	1.044674	0.679	0.470	0.543	0.209
	nFCI	*512,003*	1.253509	1.083472	1.045217	0.000	0.000	0.000	0.000

that are assumed to be zero in standard CCSD. We see at $3R_e$ that CCSD differs from nFCI by 21.6 mHa, while CAS-SDCI and $(SC)^2$CAS-SDCI differ only by 15.6 mHa and 12.4 mHa, respectively (again, the size-consistent approach performes better than CAS-SDCI). It is due to the fact that both, CAS-SDCI and $(SC)^2$CAS-SDCI, include the most important tri- and tetraexcitation. As a consequence, *ec*-CCSD, that employs T_3^0 and T_4^0 coming from these external sources, yields, again, extremely accurate results. Indeed, we see from Table 1 that *ec*-CCSD differs from nFCI by 3.7 mHa (CAS-SDCI as external source) or 3.9 mHa ($(SC)^2$CAS-SDCI as external source).

We can also see from Table 1 that very accurated values of NPE are yielded by CAS-SDCI (1.0 mHa), *ec*-CCSD(CAS-SDCI) (1.3 mHa), $(SC)^2$CAS-SDCI (1.6 mHa) and *ec*-CCSD($(SC)^2$CAS-SDCI) (1.6 mHa as well). It is not the case of standard CCSD which failure at stretched geometries yields 18.6 mHa NPE error.

The same trends are observed with the larger space [8e/7o]. The differences between CAS-SDCI and nFCI range from 7.9 mHa at $3R_e$ up to 8.9 mHa at R_e. In the case of $(SC)^2$CAS-SDCI it goes from 6.6 mHa to 7.3 mHa, while CCSD, as stated above, goes from 21.6 mHa to 3.1 mHa. Finally, *ec*-CCSD error lies between 0.5 mHa and 0.7 mHa (roughly the same with the two external sources employed).

As a whole, we can see that a *minimal dressing* of CAS-SDCI, that just ensures its size-extensivity, improves the absolute value of the energies and does not substantially change the NPE (using the space [8e/7o] it is improved by 0.29 mHa, while using the smaller space [2e/2o] it is worsen by 0.58 mHa). We may get a significant improvement using *ec*-CCSD. The computational cost of this last procedure is equivalent to a CI plus a standard CCSD. If we would need to ensure size-extensivity, we could use a $(SC)^2$type wavefunction as the external source. All the same, it can be seen in our examples that using CAS-SDCI as the external source, *ec*-CCSD yields an almost size-extensive result.

3.3.2. H_2O molecule

We consider a situation where two single bonds are simultaneously broken, modelled by symmetrically stretching the two OH bonds in H_2O molecule. As we did before, two spaces have been considered, [4e/4o] and [8e/7o]. The bonds are stretched up to twice the equilibrium distance R_e, and the results are compared to FCI as benchmark. Results are collected in Table 2.

In the case of the small space [4e/4o] and equilibrium geometry R_e, CAS-SDCI differs 9.7 mHa from the benchmark while (SC)^{2}CAS-SDCI improves this result up to 4.9 mHa, roughly the same as CCSD (4.1 mHa). All the same, *ec*-CCSD shows a smaller difference of 2.4 mHa, using either external correcting source. When stretching symmetrically the bonds up to $2R_e$, standard CCSD clearly deteriorates (21.4 mHa), while CAS-SDCI and (SC)^{2}CAS-SDCI slightly improve (8.0 mHa and 4.1 mHa, respectively), being *ec*-CCSD the method more insensitive to changes (2.4 mHa or 2.6 mHa, depending on the external source used). As a consequence, the NPEs are extremely good: CAS-SDCI 1.7 mHa, (SC)^{2}CAS-SDCI 0.9 mHa and *ec*-CCSD less than 0.52 mHa. Meanwhile CCSD has, as it could be expected, quite a poor NPE (17.3 mHa). The same trends are observed with the larger space, being the absolute values of energy closer to FCI and NPEs of the same order.

3.3.3. N_2 molecule

The most challenging problem for any ab initio correlated method is the reliable description of the dissociation of multiple bonds. A typical case is the nitrogen

Table 2 *Same as Table 1 for H_2O with DZ+P basis. The total energies are reported as* $-(E+75)$ *hartree*

Active Space	Method	Dim	$-(E+75)$ (hartree)			E-FCI (mhartree)			NPE mhartree
			R_e	$1.5R_e$	$2R_e$	R_e	$1.5R_e$	$2R_e$	
	CCSD	2,349	1.252502	1.061248	0.930865	4.122	10.158	21.404	17.282
[4e/4o]	CAS-SDCI	37,600	1.246972	1.062644	0.944312	9.652	8.762	7.957	1.695
	ec-CCSD		1.254214	1.069390	0.949845	2.410	2.016	2.424	0.408
	(SC)^{2}CAS-SDCI		1.251684	1.067078	0.948218	4.940	4.328	4.051	0.889
	ec-CCSD		1.254240	1.069326	0.949678	2.384	2.080	2.591	0.511
[8e/7o]	CAS-SDCI	149,914	1.250885	1.067303	0.948933	5.739	4.103	3.336	2.403
	ec-CCSD		1.255274	1.070648	0.951644	1.350	0.758	0.625	0.725
	(SC)^{2}CAS-SDCI		1.254058	1.069836	0.951042	2.566	1.570	1.227	1.339
	ec-CCSD		1.255285	1.070624	0.951502	1.339	0.782	0.767	0.570
	CCSDT		1.256091	1.069581	0.954440	0.533	1.825	−2.171	3.996
	FCI	28,233,466	1.256624	1.071406	0.952269	0.000	0.000	0.000	0.000

molecule, which requires up to six-fold excitations in order to describe the dissociation of its molecular ground state into the ground state of separated atoms. One may predict the failure of any single-reference based methods as, for example, CCSD. *ec*-CCSD is also, formally, a single-reference based method. However, as it was stated in the introduction, it may also be described as an approximate state-selective non genuine multireference coupled cluster approach (SS-MRCC). Indeed, our previous [14–17] and present calculations neatly show that this description is appropiate. CAS-SDCI and (SC)^{2}CAS-SDCI are free from this problem, unless an unreasonable CAS space is defined.

The results for N_2 are summarized in Table 3. The bond is streched from 2.15 a.u. up to 3.00 a.u. Opposite to the previous molecules, even at the equilibrium geometry CCSD does not perform as well as the other approaches. For the [6e/6o] active space, it differs from FCI by 12.9 mHa, while CAS-SDCI does it by 8.3 mHa, (SC)^{2}CAS-SDCI by 3.9 mHa and *ec*-CCSD by 2.0 mHa or 1.8 mHa, depending on the external correcting source. Again, one may observe that the minimal dressing of CAS-SDCI not only makes it size-extensive, but also improves the absolute value of the yielded energy.

As the bond is streched up to 3.00 a.u., CAS-SDCI and (SC)^{2}CAS-SDCI hardly change (differences with FCI being 8.4 mHa and 4.0 mHa, respectively). *ec*-CCSD behaves in a similar way (errors 2.7 mHa and 2.6 mHa). As it was expected, standard CCSD dramatically fall down (up to 36.8 mHa at 3.0 a.u.). The PES quality factor NPE is quite good in any case: 0.4 mHa (CAS-SDCI), 0.1 ((SC)^{2}CAS-SDCI) and 0.8 (*ec*-CCSD). When using the larger [10e/8o] space, the results are quite similar for near equilibrium distances. As the bond is streched, we get better results for the energy with all the employed methods. This effect works slightly improving NPE for *ec*-CCSD method and slightly deteriorating it for CAS-SDCI and (SC)^{2}CAS-SDCI.

Table 3 *Same as Table 1 for* N_2 *with TZ basis. The total energies are reported as* $-(E + 108)$ *hartree*

Active Space	Method	Dim	$-(E+108)$ (hartree)			E-FCI (mhartree)			NPE mhartree
			2.15$a.u.$	2.50$a.u.$	3.00$a.u.$	2.15$a.u.$	2.50$a.u.$	3.00$a.u.$	
	CCSD	*726*	1.141157	1.085796	0.962355	12.858	21.247	36.809	23.951
[6e/6o]	CAS-SDCI	192,852	1.145749	1.098403	0.990769	8.266	8.640	8.395	0.374
	ec-CCSD		1.152057	1.104645	0.996450	1.958	2.238	2.714	0.756
	(SC)^{2}CAS-SDCI		1.150077	1.102981	0.995144	3.938	4.062	4.020	0.124
	ec-CCSD		1.152176	1.104805	0.996551	1.839	2.238	2.613	0.774
[10e/8o]	CAS-SDCI	268,908	1.145955	1.098908	0.992348	8.060	8.135	6.816	1.319
	ec-CCSD		1.152060	1.104700	0.996824	1.955	2.343	2.340	0.388
	(SC)^{2}CAS-SDCI		1.150114	1.103030	0.995777	3.901	4.013	2.287	0.626
	ec-CCSD		1.152164	1.104838	0.996908	1.851	2.205	2.256	0.405
	FCI	225,854,028	1.154015	1.107043	0.999164	0.000	0.000	0.000	0.000

4. Conclusions

We have studied the performance of the so-called $(SC)^2$CAS-SDCI and *ec*-CCSD methods versus standard CAS-SDCI method on one hand, and CCSD on the other hand, in quasidegenerate situations that arise when exploring dissociation channels of simple molecules. In particular, we dissociate a single bond, two single bonds simultaneously and a triple bond.

The first conclusion we may draw is that the minimal dressing of a CI matrix (CAS-SDCI in our study) not only ensures the importanty property of size-extensivity, but also improves the absolute values of the yielded energies. This can be stated in all studied cases: single bond breaking, two single bond breaking and triple bond breaking.

As a second conclusion we may say that at least some of the linked T_3 and T_4 terms, not included in the minimal dressing, and approximately accounted in *ec*-CCSD, reveal themselves very important for an accurate description of the wavefunction in streched geometries.

CAS-SDCI and $(SC)^2$CAS-SDCI methods seem to be rather insensitive to changes in the geometry (as the NPE reveals).

ec-CCSD confirms itself as a very workable method, with an extremely good ratio performance/cost. Our present calculations reveal, in particular, an almost identical result when correcting from CAS-SDCI or its size-extensive partner $(SC)^2$CAS-SDCI. In other words, *ec*-CCSD looks almost size-extensive despite being corrected with a variational non size-extensive external source. We believe that if an arbitrary selected CI is used as a external source, having a higher lack of size-extensivity, the *ec*-CCSD would reveal non size-extensivity in a major extension than it does in our present calculations. An obvious way to rise the ratio performance/cost of *ec*-CCSD is the use of some kind of reduced, but optimally selected CI, instead of CAS-SDCI (e.g. a selected CI using a CIPSI procedure). This wavefunction would have a great lack of size-extensivity. In such a case a $(SC)^2$-selected CI [34] appears to be the natural external correcting source [35].

Acknowledgements

One of us (G.P.) wishes to express his thanks to Professor J.P. Malrieu for inviting him to visit the I.R.S.A.M.C., where this work was carried out with a grant given by the Generalitat Valenciana. Support from the Fundació Caixa de Castelló, Project No. P1B97-23 is gratefully acknowledged. G.P. would also like to thank O. Castell for all his help and useful discussions.

References

1. J. Paldus: J. Chem. Phys. 61, 5321 (1974).
2. W. Duch: RMS or graphical representations of model spaces, Lecture Notes in Chemistry, Vol. 42 (Springer, Berlin 1986).
3. B.O. Roos in: Ab initio methods in quantum chemistry, 2, K.P. Lawley (Ed.), Advan. Chem. Phys. 69 (1987) 63, and references therein.

4. See, e.g., R.J. Bartlett and J.F. Stanton in: Reviews in Computational Chemistry, Vol. V, K.B. Lipkowitz znd D.B. Boyd (Eds.), VCH Publishers, Inc. New York, 65 (1994).
5. J.P. Malrieu, Ph. Durand and J.P. Daudey: J. Phys. A 18, 809 (1984).
6. J.P. Malrieu, I. Nebot-Gil and J. Sánchez-Marín: J. Chem. Phys. 100, 1440 (1994).
7. I. Nebot-Gil, J. Sánchez-Marín, J.P. Malrieu, J.L. Heully and D. Maynau: J. Chem. Phys. 103, 2576 (1995).
8. J.L. Heully and J.P. Malrieu: Chem. Phys. Lett. 199, 545 (1992).
9. J. Sánchez-Marín, I. Nebot-Gil, J.P. Malrieu, J.L. Heully and D. Maynau: Theor. Chim. Acta 95, 215 (1997).
10. J.P. Daudey, J.L. Heully and J.P. Malrieu: J. Chem. Phys. 99, 1240 (1993).
11. N. Oliphant and L. Adamowicz: J. Chem. Phys. 94, 1229 (1991).
12. N. Oliphant and L. Adamowicz: J. Chem. Phys. 96, 3739 (1992).
13. J. Paldus and J. Planelles: Theor. Chim. Acta 89, 13 (1994).
14. J. Planelles, J. Paldus, and X. Li: Theor. Chim. Acta 89, 33, 59 (1994).
15. G. Peris, J. Planelles and J. Paldus: Int. J. Quantum Chem. 62, 137 (1997).
16. X. Li, G. Peris, J. Planelles, F. Rajadell and J. Paldus: J. Chem. Phys. 107, 90 (1997).
17. G. Peris, F. Rajadell, X. Li, J. Planelles and J. Paldus: Mol. Phys. 94 (1) 235 (1998).
18. A. Povill, R. Caballol, J. Rubio and J.P. Malrieu: Chem. Phys. Lett. 209, 126 (1993).
19. J. Paldus in: S. Wilson, G.H.F. Diercksen (Eds.), Methods in computational molecular physics, NATO ASI Series, Series B, Vol. 293. Plenum, New York, 99. (1992).
20. R.J. Bartlett: Annu. Rev. Phys. Chem. 32, 359 (1981); J. Phys. Chem. 93, 1697 (1989).
21. R.J. Bartlett, C.E. Dykstra, and J. Paldus, in: Advanced Theories and Computational Approaches to the Electronic Structure of Molecules, C.E. Dykstra, Ed. (Reidel, Dordrecht, 1984) 127.
22. M. Urban, I. Černušák, V. Kellö, and J. Noga, in: Methods in Computational Chemistry Vol. 1, Electron Correlation in Atoms and Molecules, S. Wilson, Ed. (Plenum, New York, 1987), 117.
23. R.J. Bartlett, in: Modern Electronic Structure Theory, Part I, D.R. Yarkony, Ed. (World Scientific, Singapore, 1995), 1047.
24. T.H. Dunning: J. Chem. Phys. 53, 2823 (1970).
25. S. Huzinaga: J. Chem. Phys. 42, 1293 (1965).
26. X. Li and J. Paldus: J. Chem. Phys. 108, 637 (1998).
27. C.W. Bauschlicher and P.R. Taylor: J. Chem. Phys. 85, 2779 (1986).
28. T.H. Dunning: J. Chem. Phys. 90, 1264 (1991).
29. S. Evangelisti and G.L. Bendazzoli: Il Nuovo cimento della societa italiana di fisica, 17, 289 (1995).
30. Molcas Version 4. K. Andersson, M.R.A. Blomberg, M.P. Flscher, G. Karlstrm, R. Lindh, P.-E. Malmqvist, P. Neogrdy, J. Olsen, B.O. Roos, A.J. Sadlej, M. Schtz, L. Seijo, L. Serrano-Andrs, P. E. M. Siegbahn, and P.-O. Widmark, Lund University, Sweden (1997); Modified version of MOTRA written by: O. Castell.
31. Computer Programs are MR-DDCI: O. Castell; SCIEL: O. Castell, R. Caballol, J.P. Daudey, and J.P. Malrieu, NATURALS; O. Castell.
32. N. Ben Amor and D. Maynau: Chem. Phys. Lett. 286, 211-220 (1998).
33. P. Piecuch and J. Paldus, Int. J. Quantum Chem. 36, 429 (1989).
34. R. Caballol and J.P. Malrieu, Chem. Phys. Letters 188, 543 (1992)
35. G. Peris, J. Planelles, J.P. Malrieu and J. Paldus: J. Chem. Phys. 110 (24) 11708 (1999).

The Size-Consistent Self-Consistent SDCI Method for Excited States and Ionization Potentials

J. Pitarch-Ruiz[a], J. Sánchez-Marín[a]*, I. Nebot-Gil[a], N. Ben Amor[b] and D. Maynau[b]

[a]*Departament de Química Física, Facultat de Ciencies Químiques, Dr. Moliner 50, E-46100, Burjassot (Valencia), Spain*

[b]*Laboratoire de Physique Quantique, IRSAMC, Université Paul-Sabatier, 118 route de Narbonne, 31062 Toulouse Cedex, France*

Abstract

The size-consistent self-consistent operator $(SC)^2$ provides a general solution to the well-known problem of performing calculations that remain in an SDCI logic whilst obtaining size-extensive results. The size-extensivity correction affects both the energies and the wavefunctions and can be applied to any single-reference (SR) and multi-reference (MR) singles and doubles CI. Of course, the method is no longer a CI, but proceeds through matrix diagonalization and deals with determinant coefficients instead of excitation amplitudes. The method is based on an iterative application of the matrix dressing formalism of the Intermediate Hamiltonians Theory. It can be applied to any state provided that its CI expansion is dominated by a single determinant. This leading determinant can be either closed or open shell and examples of both applications are shown in the present work. The examples deal with some valence and Rydberg excited states of CO, as well as the outer valence ionization potentials. Mean absolute errors (MAE) of less than 0.1 eV relative to experiment have been obtained for both the vertical excitation energies (VEE) and the IPs using small MR spaces for the SDCI to be dressed.

1. Introduction

The use of CI expansions to descrive excited states is very familiar to the molecular scientists and provides chemists with a very useful and intuitive schema [1]. Most representations of photochemical reactivity and electronic spectra descriptions find valuable help in the dominant excitations that characterize each state [2]. These excitations are generally referred to by means of the one-electron functions (usually MOs) that contain the holes and particles that define the excitations starting from a reference function (e.g., the closed shell Hartree Fock determinant). It is generally recognized that, from a theoretical viewpoint, the major drawback of the CI method is its lack of size-consistency and size-extensivity [3–7]. From a technical point of view, an additional complications arises in the application of the variational procedure to very large CI spaces [3]. In spite of this, the conceptual simplicity of the method has led to huge efforts in the search for efficient algorithms that yield the lower roots and eigenvalues at a reasonable computational cost [3,5–7]. Additional features of the method that make it very popular are its great flexibility and generality [3]. It can be

*To whom correspondence should be addressed.

A. Hernández-Laguna et al. (eds.), Quantum Systems in Chemistry and Physics, Vol. 1: Basic Problems and Model Systems, 87–100.

applied to single-reference (SR) or multi-reference(MR) dominated states, whatever their multiplicity.

The MR-SDCI method is suitable for dealing with states or conformations that have important non-dynamic correlation effects, i.e., those that require multiconfigurational wavefunctions. The dynamic correlation effects, i.e., those mainly related to the coupled movement of the electrons, are usually taken into account through inclusion in the CI of all possible single and double substitutions on the reference space (i.e., generating the SDCI space on the reference). In spite of the large CI matrices that this procedure generates, the treatment of the dynamic correlation is still limited, and very important effects that rely on the coupling with the outer space (i.e., the space beyond the SD on the reference) are still being neglected [3,8]. Especially important among these terms are those which would be required to satisfy the size-extensivity property that ensures the correct scaling of the energy with the number of correlated electrons. These are effects that can be analyzed in terms of many body perturbation theory (MBPT) and correspond to terms appearing at third and higher orders of the perturbation relative to the reference determinants [9,10]. Such terms are known as non-linked MBPT diagrams and are canceled order by order in the Rayleigh-Schrödinger perturbative treatments [10].

Several attempts to obtain size-consistency from the CI eigenvalues are represented by the various Davidson-like correction proposals [11,12], which include a generalized Davidson correction for the MR-CI case [3,13]. More elaborate proposals that improve both energies and eigenvectors lead to the Coupled Pair Functional (CPF) methods [14,15]. Other approaches that are usually named as Electron Pair Approximations, and that include the different versions of CEPA [16–18] or the Quadratic CI [19], can be considered to be somewhere in between the non size-extensive but linear eigenvalue equations of CI and the non-linear but size-extensive equations of the Coupled Cluster ansatz. The CC methods may be regarded as the most efficient ones in terms of their recovery of correlation effects (e.g., those expressed in terms of MBPT diagrams) that are taken into account for a given dimension of the CI space [20–24]. However, as with the perturbative methods, they face important formal difficulties in the case of multi-configurational problems [25].

Very efficient methods for the calculation of excited state energies can be obtained from a clever combination of variational and perturbative treatments [26–28]. The most successful methods such as CASPT2 combine a MCSCF (e.g., a CASSCF) treatment [29] of the non-dynamic correlation effects followed by a second order perturbation [30,31] and are nowadays widely used [32]. Another approach for combining perturbation and eigenvalue-like calculations is provided by the dressing matrix methods. Here, the CI matrices are corrected (dressed) with the addition of terms that describe the effects that would be lacking in a bare CI diagonalization [33–35]. These effects can provide the size-extensivity corrections or even the most complete (and computationally demanding) ones that are required to obtain the CC solutions [35].

In this work we are concerned with the dressing method known as Size-Consistent Self-Consistent Singles and Doubles Configuration Interaction, usually abbrebiated to $(SC)^2$ SDCI [8]. This method can be considered as a generalization of the CEPA-3 formalism, as has been shown elsewhere [8,35]. It provides size-extensive roots and vectors for the SDCI built on a Single-Reference (SR-SDCI) or a Multi-Reference (MR-SDCI) space of determinants. The principles and details of the method have been thoroughly discussed elsewhere, and applied both to ground and excited states [8,36]. However, in the case of excited states it can be applied in one of two ways, depending on the state from which the information required to built the dressing operators is to be taken. In the methods section we present different practical approaches that can be used to deal with excitation energies, including the case where the dominant determinant of a state function is an open shell. To illustrate these procedures, some results are shown in the next section that concern the excited states of the CO molecule. These results include the application of the method to doublet states for the calculation of the outer valence Ionization Potentials of the same molecule.

2. Method

A brief discussion of the principles and terminology related to the Matrix Dressing technique and the $(SC)^2$ SDCI method seems to be in order here to make clearer the different approaches that can be employed. It is worth noting that the $(SC)^2$ method is a very flexible one. It has been conceived in the framework of the Intermediate Hamiltonian Theory (IHT) [37,38]. On the one hand, IHT can be chosen on a single-state or a multi-state basis. On the other, both the IHT and the SDCI space can be taken as single-reference (SR) or multi-reference (MR). The theory and a few test applications of (single or multi-state) multireference-$(SC)^2$ have been developed by the Toulouse group [39]. They are usually denoted as MR-$(SC)^2$ and will not be further referred to here. To clarify things from the outset, the reader should bear in mind that the present paper shall only deal with *single state Intermediate Hamiltonians* and *single reference* formulations of the *intermediate* (effective) *Hamiltonian*. Throughout this work, we deal with the $(SC)^2$ SDCI dressing operator, denoted Δ^{m}. Whenever we refer to SR or MR calculations, we are merely indicating the nature of the reference space of Slater determinants on which the SDCI space has been built. Hence, the dimensions of such SR-SDCI or MR-SDCI spaces determine the dimensions of the matrix whose lower or lowest roots must be found. Of course, MR-SDCI matrices are much better suited for dealing with multiconfigurational states than SR-SDCI ones.

The Intermediate Hamiltonian Theory is a generalization of the Effective Hamiltonian Theory. The full CI space of Slater determinants can be divided into three parts,

a) The principal or S_P space, that includes the reference function (or functions) of the SDCI, ϕ_I^P.

b) The intermediate space, S_Q, made of all the singles and doubles substitutions on the S_P space. The union of S_P and S_Q is called the model space S_{PQ}.

c) The outer space, S_E, including all other excitations.

For the closed shell ground state, S_P includes ϕ_0. Furthermore, for the same state, but in the MR-SDCI cases, other ϕ_I^P are included in S_P, but their weight in the wave function will always remain lower than unity. Hence, we can write, in the intermediate normalization

$$\Psi_0 = \phi_0 + \sum_{\substack{i \in S_{PQ} \\ i \neq 0}} c_i^0 \phi_i + \sum_{\alpha \in S_E} c_\alpha^0 \phi_\alpha \tag{1}$$

The imposition of a closed shell state condition is not mandatory provided that the wavefunction of the state being sought can be written as in eq. (1). In this case, ϕ_0 should be replaced by the dominant determinant ϕ_I^m and hence the eq. (1) should read as

$$\Psi_m = \phi_I^m + \sum_{\substack{i \in S_{PQ} \\ i \neq I}} c_i^m \phi_i + \sum_{\alpha \in S_E} c_\alpha^m \phi_\alpha \tag{2}$$

Note that it is not required that all the c_i^m coefficients be very different from unity, and our experience is that even in the cases of having a second coefficient equal to unity (e.g., for a state dominated by two single excited spin partners), the method does not diverge, though the accuracy of the method is reduced. The problem of the applicability of single reference dressing methods to states which are not actually dominated by a single determinant has been discussed elsewhere [41].

Suppose now that we are interested in a given state M that satisfies the condition stated in eq. (2). As we deal just with one eigenvector, i.e., the set of c_i^M coefficients, the dressing matrix need not be composed entirely of non-zero elements. In fact, the single state $(SC)^2$ operator matrix is diagonal [8]. Its elements are calculated for each $\phi_i \in S_Q$ as

$$\Delta_{ii}^M = \sum_{\substack{j \\ D_j^+\phi_i \neq 0 \\ D_j^+\phi_I^M \neq 0}} c_j^M \langle \phi_i | H | D_j^+ \phi_i \rangle = \sum_{\substack{j \\ D_j^+\phi_i \neq 0 \\ D_j^+\phi_I^M \neq 0}} c_j^M \langle \phi_0 | H | D_j^+ \phi_0 \rangle \tag{3}$$

where D_j^+ is the diexcitation operator that creates the diexcited function ϕ_j from ϕ_0 and c_j^M is the coefficient of ϕ_j in eq. (1) or eq. (2). Note the very important restrictions that apply to the j index. The first ensures that the disubstitution D_j^+ is possible on the ϕ_i determinant whose H_{ii} matrix element is being dressed, and the second that D_j^+ is possible on the reference determinant ϕ_I^M. The second condition is equivalent to saying that the D_j^+ operation is disconnected with the excitation operation that would generate ϕ_I^M from the closed shell ground state reference ϕ_0. As H is a two-electron operator, the equivalence between the summations in eq. (3) follows from the diexcitation nature of the D_j^+ operator.

The E_M root of $H + \Delta^M$ depends on the eigenvector coefficients c_j^M through eq. (3), and can be obtained by means of a conventional eigenvalue solving algorithm (e.g., by diagonalization). An iterative procedure is required to achieve self-consistency between the set of coefficients used to build the dressing terms defined in eq. (3) and the set of coefficients of the Ψ_M eigenvector. The converged eigenvector will be denoted as $\tilde{\Psi}_M$ and the corresponding eigenvalue $\tilde{E}_M$. Hence, we can write, once the convergence has been achieved,

$$\tilde{\Psi}_M = \phi_I^M + \sum_{\substack{i \in S_M \\ i \neq I}} \tilde{c}_i^M \phi_i. \tag{4}$$

Two important remarks are in order here. The first is that eq. (3) does not impose any limitation on the degree of excitation of ϕ_i, so that Δ^M can be calculated whatever the composition and symmetry of S_Q. In this way, even if eqs. (1), (2) and (4) assume a single reference determinant, the SDCI space that constitutes S_Q can be built as a MR-SDCI and the procedure described above can be applied. The second is that care must be taken to remove the possible redundancies that could be present if $D_j^+\phi_i$ belongs to the model space S_{PQ}. In such a case, the physical effects included in each term $c_j^M\langle\phi_i|H|D_j^+\phi_i\rangle$ of eq. (3) would be included twice in the diagonalization of $H + \Delta^M$.

The selected state procedure described above can be applied to the ground state and to each excited state of interest. The excitation energies are then obtained by substraction as in conventional CI methods, so that

$$VEE_M = \tilde{E}_M - \tilde{E}_0 \tag{5}$$

We can call this one state at a time procedure as *one-state one-root* $(SC)^2$ dressing.

However, this is in general a quite impractical method, as it requires the iterative series of dress-then-diagonalize steps to get convergence for each individual state. Moreover, even if no particular numerical problems arise in the case of states that are not dominated by one particular excitation, the method does not seem to be well-adapted from a formal point of view to such cases because they do not satisfy the condition stated in eq. (2). Notwithstanding, this procedure can be practical for the calculation of outer-valence ionization potentials of closed-shell molecules. In such cases, one must to deal with the doublet states of the cation that are well dominated by a unique Koopmans' determinant.

Fortunately, there is another way of obtaining excitation energies in the framework of $(SC)^2$ SDCI theory. As indicated above, the main role played by the terms in the summation of eq. (3) is to include in the CI matrix the so-called disconnected terms that are required to cancel the non-linked MBPT diagrams at the time of performing the diagonalization. It can be demonstrated from perturbation theory arguments [40], that the part of these non-linked contributions that goes through the external space determinants is rather state independent. Hence, no great error is introduced if one assumes that the $(SC)^2$ converged ground state coefficients $\tilde{c}_0^j$ can be used in eq (3) instead of the state specific $\tilde{c}_j^M$. In this way, only the ground state root and wavefunction must be calculated by means of the iterative $(SC)^2$ SDCI or $(SC)^2$ MR-SDCI procedure. Once a set of $\{\tilde{c}_j^0\}$ coefficients is available the excitation energies are obtained as follows:

i) The SR-SDCI or MR-SDCI model space S_{PQ} of the required spatial and spin symmetry is built. We call H_{PQ} the corresponding CI matrix.

ii) The appropriate diagonal dressing matrix element is calculated for each $\phi_i \in S_{PQ}$ as

$$\Delta_{ii} = \sum_{\substack{j \\ \forall\phi_j = D_j^+\phi_l^0}} \tilde{c}_j^0 H_{0j} - \sum_{\substack{k \\ D_k^+\phi_i = 0}} \tilde{c}_k^0 H_{0k} - \sum_{\substack{l \\ D_l^+\phi_i \in S_{PQ}}} \tilde{c}_l^0 H_{0l} \tag{6}$$

This is a generalization of eq. (3). Note, however, that whilst the coefficients of eq (4) are used and the superscript M has been written as 0, this does not necesarily mean that the ground state must be a closed shell. In eq. (6), we have written H_{0j} as a shorthand for $\langle\phi_0|H|D_j^+\phi_0\rangle$. The three summations of eq. (6) are introduced for practical reasons and the technical details relative to them have been extensively discussed elsewhere [41].

iii) A few roots of $H_{PQ} + \Delta_{PQ}$ are then obtained and the excitation energies are calculated as energy differences in the same way as in conventional CI

$$VEE_m = E_m - E_0 \tag{7}$$

Now, the choice of E_0 depends on the nature of the S_{PQ} space. The following cases occur:

a) $E_0 = \tilde{E}_0$ from the single reference (SC)^2 SDCI. This choice has been used for single reference CI in all symmetries and for MR-SDCI in symmetries other than that of the ground state when the MR space is very small and contains determinants of a single symmetry.

b) $E_0 = \tilde{E}_0$ from each (SC)^2-MR-SDCI. This has been used for MR-SDCI in the symmetry of the ground state and for other symmetries in those cases where the MR space includes determinants of different symmetries (e.g., a CAS of a few electrons in a few active orbitals).

The procedure just described can be termed *one-state several-roots* (SC)^2 dressing. It has been shown to be very convenient for the study of vertical excitation energies of closed-shell systems [36,42,43]. Note that the *one-state several-roots* (SC)^2 dressing method is somehow related to the EOM-CCSD method [58,59] in the sense that, in EOM calculations, the ground state CCSD amplitudes are assumed in the cluster operator T before diagonalization of the disconnected-diagrams-free effective Hamiltonian $\overline{H} = \exp(-T)H\exp(T)$. A significant difference, but by no means the only one, comes from the single-reference nature of the CCSD calculations.

The computational details concerning basis sets and geometries are given in the following section. The MOs have been calculated with the MOLCAS series of programs [44]. The (SC)^2-MR-SDCI have been calculated with the CASDI series of programs [41,45] as adapted to run on IBM RS/6000 workstations and on SGI multiprocessor computers.

3. Results and Discussion

Second period diatomics are excellent representative systems for assesing the suitability of methods that are under development. A large number of vertical excitation energies are available. The size of these molecules may be small but the physical contents of the states can be very diverse. The condition required for extrapolating the mean errors to greater systems is the size-extensivity of the methods. All of the calculations in the present work have been performed on the CO molecule. Both the VEE and the IP of this system have been of great interest to theorists because this molecule, as well as its isoelectronic N_2, is a very correlated system described in its ground state as having a triple bond [46–48]. The CO molecule is the most efficient chemical energy retainer in terms of energy per atom, and the triple bond in CO shows the greatest molar bond dissociation enthalpy [49]. The study of the excited states in CO extends from the highly approximate mean value description by Nesbet [50] to the recent time-dependent density-functional response theory calculation of Casida *et al.* [51], and goes through a large number of representative papers [52–57].

Concerning the VEE, we have chosen, as standard references for comparison, the EOM-CCSD results from Stanton *et al.* [56] and from Comeau and Bartlett [57]. Results from other very efficient state-of-the-art methods such as Exponentially Generated CI or the now widely used CASPT2 could also have been used. In general, these are methods from which mean errors in VEE less than 0.2 eV might be expected provided that the basis set used is at least of split-valence plus polarization quality and it is augmented in a well-conditioned way to account for Rydberg states.

The basis sets have been taken according to the reference calculations. We have used Dunning's aug-cc-pVDZ [60, 61] and a 5s4p2d basis set augmented with two sets of diffuse s and p centered at the mass centroid of the molecule. Further details about this basis set can be obtained from reference [57].

All of the calculations have been performed at the experimental equilibrium distance $R_e = 1.128$ Å, in order to enable a proper comparison with the EOM-CCSD reference. In so far as there are neither largely interacting excited states nor special reasons for expecting a breakdown of the Born Oppenheimer approximation, great changes in the MAE are not expected if one takes the $(SC)^2$ SDCI ground state equilibrium value for R_e, which is $R_e' = 1.140$ Å (very close to the CCSD value, as expected: Cfr. table I). We have performed a separate calculation of the whole set of VEE with the aug-cc-pVDZ basis set at the R_e' distance, in any case. The results have not been included in table II for the sake of clarity, but the total MAE values where 2.34 eV for the MR-SDCI and 0.17 eV for $(SC)^2$MR-SDCI.

The vertical IPs of CO deserve special attention because carbon monoxide is a reference compound for the application of photoelectron spectroscopy (PES) to the study of adsorption of gases on metallic surfaces. Hence, the IP of free CO is well-known and has been very accurately measured [62]. A number of very efficient theoretical methods specially devoted to the calculation of ionization energies can be found in the literature. Most of these are related to the so-called random phase approximation (RPA) [63]. The most common formulations result in the equation-of-motion coupled-cluster (EOM-CC) equations [59] and the one-particle Green's function equations [64,65] or similar formalisms [65,66]. These are powerful ways of dealing with IP calculations because the ionization energies are directly obtained as roots of the equations, and the repolarization or relaxation of the MOs upon ionization is implicitly taken into account [59]. In the present work we remain close to the CI procedures so that a separate calculation is required for each state of the cation and of the ground state of the neutral to obtain the IP values.

one-state several-roots (SC)^2 SDCI results

Some spectroscopic properties of the GS (X $^1\Sigma^+$) and the two first excited states a$^3\Pi$ and A$^1\Pi$ of CO are shown in Table 1. The data have been obtained from the curves calculated near the bottom of the potential wells and covering a range of about 0.8 Å around the equilibrium point. This range has been considered sufficient to take into account the anharmonicity effects in the lower region. The SDCI was taken as SR for the ground state. For the excited states, a very small MR was used for the SDCI, made of the first two single excitations of the π symmetry. Both the vertical and adiabatic excitation energies are included in Table 1. The important corrections that the (SC)^2 procedure provides to the SDCI or MR-SDCI results can be clearly seen. These corrections always result in a better fit to available experimental values. However, the basis set is very limited because it was chosen to compare with the published CCSD data from reference [56]. The basis set is not big enough to give accurate data for CO and it would be difficult to separate errors due to the basis set from errors related to the theoretical model. Data on spectroscopic properties of excited states and AEE calculated with (SC)^2 SDCI are still scarce. For the moment we cannot say more than that they appear promising.

The results reported in Table 2 are much more significant because they correspond to a larger manifold of states. VEE are reported for seven low-lying valence states (up to about 10 eV) and five Rydberg states lying in the 10–12 eV region. The mean of the

Table 1 *Spectroscopic data[a] and Excitation Energies[b] of the ground,* a$^3\Pi$ *and* A$^1\Pi$ *states of CO calculated with the aug-cc-pVDZ basis set*

		R_e	B_e	ω_e	$\omega_e x_e$	α_e	$\bar{D}_e*10^6$	EEV	EEA
X$^1\Sigma^+$									
	SDCI	1.134	1.9101	2219.3	11.11	0.01568	5.7		
	(SC)^2-SDCI	1.140	1.8907	2160.7	12.01	0.01665	5.8		
	CCSD	1.138		2183[c]					
	Exp[d]	1.128	1.9313	2169.8	13.29	0.01750	6.1		
a$^3\Pi$									
	MR-SDCI	1.214	1.6678	1916.7	20.79	0.02120	5.1	6.53	6.16
	(SC)^2-MR-SDCI	1.214	1.6683	1829.4	13.99	0.01518	5.5	6.27	6.01
	Exp[d]	1.206	1.6912	1743.4	14.36	0.01904	6.4	6.32[e]	
A$^1\Pi$									
	MR-SDCI	1.245	1.5846	1633.0	25.92	0.01954	6.0	8.97	8.44
	(SC)^2-MR-SDCI	1.241	1.5947	1578.1	20.75	0.01980	6.5	8.57	8.16
	EOM-CCSD	1.242		1516[c]				8.64	7.94
	Exp[d]	1.235	1.6115	1518.2	19.40	0.02325	7.33	8.51[e]	8.07[f]

[a] R_e in Å. Other parameters in cm^{-1}.
[b] In eV.
[c] Harmonic frequency.
[d] From ref. [68].
[e] From ref. [57].
[f] From ref. [56].

Table 2 *Vertical Excitation Energies (VEE) in eV for the excited states of CO*

States	Excitation	Basis Set aug-cc-pVDZ[c]					Basis Set (5s4p2d//2s2p)[c]						Exp[a]
		SDCI	$(SC)^2$-SDCI	MR-SDCI[d]	(SC)2 MR-SDCI[d]		SDCI	$(SC)^2$-SDCI	MR-SDCI[d]	$(SC)^2$ MR-SDCI[d]		EOM-CCSD	
Valence States													
$a^3\Pi$	5σ-2π	11.56	5.70	8.77	2S	6.38	12.14	5.88	10.99	2S	6.25	6.46	6.32
$A^1\Pi$	5σ-2π	13.64	7.81	10.10	2S	8.64	14.21	8.01	11.50	2S	8.57	8.79	8.51
$a'^3\Sigma^+$	1π-2π	12.91	7.51	9.91	G2S	8.51	13.44	7.53	9.15	G2S	8.60	8.38	8.51
$e^3\Sigma^-$	1π-2π	14.10	9.00	11.27	2S	9.92	14.62	9.16	10.46	2S	9.91	9.94	9.88
$d^3\Delta$	1π-2π	13.67	8.55	10.71	2S	9.34	14.20	8.71	9.92	2S	9.40	9.36	9.36
$I^1\Sigma^-$	1π-2π	14.25	9.15	11.45	2S	10.02	14.77	9.30	10.57	2S	10.06	10.16	9.88
$D^1\Delta$	1π-2π	14.39	9.30	11.53	2S	10.21	14.91	9.45	10.82	2S	10.26	10.29	10.23
M.A.E[b] (A)		4.55	0.81	1.58	0.06		5.09	0.66	1.53	0.07		0.14	
Rydberg States													
$b^3\Sigma^+$	5σ-6σ	16.64	10.48	11.16	G2S	10.55	17.17	10.41	11.86	G1S	10.41	10.51	10.40
$B^1\Sigma^+$	5σ-6σ	17.04	11.00	11.48	G2S	11.03	17.53	10.83	11.89	G1S	10.83	10.98	10.78
$C^1\Sigma^+$	5σ-7σ	17.73	11.57	17.65	G2S	11.58	18.25	11.49	17.57	G1S	11.44	11.58	11.40
$j^3\Sigma^+$	5σ-7σ	17.48	11.30	17.33	G2S	11.31	18.12	11.35	17.19	G1S	11.31	11.45	11.30
$E^1\Pi$	5σ-3π	17.82	12.05	16.03	2S	11.79	18.27	11.59	15.52	2S	11.55	11.67	11.53
M.A.E[b] (B)		6.26	0.20	3.65	0.17		6.79	0.05	3.7	0.03		0.16	
M.A.E[b] (T)		**5.26**	**0.55**	**2.44**	**0.11**		**5.79**	**0.41**	**2.44**	**0.05**		**0.14**	

[a]Experimental VEE taken from ref [69].
[b]M.A.E: Mean Absolute Errors:(A) Valence states. (B) Rydberg states. (T) All states.
[c]See text for references concerning the basis sets.
[d]The size of the MR space is given: G stands for the closed shell determinant. nS stands for n single excitations (two spin partners each).

absolute values of the errors relative to experimental data (MAE) are also indicated. SR and MR-SDCI results, as well as the respective $(SC)^2$ ones, are shown for the two basis sets, and a number of observations can be made. On average, the correction of the size-extensivity for the SR-SDCI amounts to several eV. The correction is also significant but much smaller in the MR cases, which is in line with the usual view that MR-CI is less sensitive to the size-extensivity error [3]. Of course, the larger the MR-CI, the closer to the FCI and to the size-extensive exact solution for the given basis set. The MAE for the Rydberg states is smaller than for the valence states in the $(SC)^2$ SR-SDCI. This is an unexpected result, especially for the aug-cc-pVDZ basis set. This basis set, in similar calculations for the N_2 molecule, gave errors of about 0.7 eV [43]. The MAE results for the $(SC)^2$ MR-SDCI are also better than expected, ranging about 0.1–0.05 eV, which can be considered to be excellent results for this level of the theory. It is important, however, to bear in mind that the basis set is clearly small, so that the reported accuracy may be simply due to a cancellation of errors.

Note that the selection criteria for the CI reference for each state was directed by the $(SC)^2$ MR-SDCI wavefunction. As a consequence, the MR spaces are not necessarily the same as they would be if the MR-SDCI directed the selection. It follows that the MAE values for MR-SDCI reported in Table 2 should be considered as upper bounds.

one-state one-root $(SC)^2$ SDCI results

The state-specific iterative $(SC)^2$ dressing on each individual state has been applied to the vertical outer valence IPs of CO. These calculations represent a first application of this dressing technique to doublet open shells. All IP calculations have been performed with the (5s4p2d//2s2p) basis set at the experimental geometry of neutral CO. The results are reported in Table 3, along with a number of reference calculations.

In a first approximation, we have taken the canonical HF MO's from the neutral CO as a common set of one-electron functions for both the cationic and the neutral systems. This procedure is thus simply taking the ionization as a limiting excitation case and treating it in the same formal way as the excited states of CO. The only major difference comes from the fact that each dressed state has been iterated separately. Both the SR-SDCI and the MR-SDCI have been considered. For programming convenience [41], instead of a small MR space, a CAS-CI has been used as reference for the SDCI. Such a change should result in a slight improvement of the results in calculations at equilibrium geometry according to our experience with excitation energies in similar systems [43].

The simplest $(SC)^2$ SDCI calculations give a very important improvement over the Koopman's IPs for the same basis set. It should not be expected that a CI takes into account properly the repolarization effects of the MOs of the cation relative to the neutral molecule. However, the MAE is reduced from 1.3 eV (KT) to 0.23 (SDCI) or 0.18 eV ($(SC)^2$ SDCI). A further improvement of the results can be obtained with CAS-SDCI. The calculations have been performed in the C_{2v} point symmetry group, so that we indicate the active spaces as $(n_1 n_2 n_3 n_4)$ corresponding respectively to the irreducible representations $(a_1\ b_1\ b_2\ a_2)$. The CAS for the ground state of CO was 8 electrons in (2220). For the 5σ and 4σ cations, the CAS was 7 electrons in (2220) also, but for the second excited state of the same symmetry (4σ cation), the second vector was dressed. The π cation gave good results with a smaller CAS of 3 electrons in (0220). The MAE

Table 3. *Vertical ionization potentials for CO in eV.* (5s4p2d//2s2p) Basis Set. Ground State Energy. SCF = −112.7798 a.u.

	KT	SR-CI (NMO)		CAS-CI (NMO)		CAS-CI (CASMO)		Exp[a]	EKT[b] ⟨5s3p1d⟩	MCSTEP[b] pVTZ	MBPT(2)-GF[c] ⟨7s4p2d⟩	CCSD-GF[c] ⟨7s4p2d⟩
		CI	$(SC)^2$	CI	$(SC)^2$	CI	$(SC)^2$					
5σ	15.13	14.08	13.81	13.88	13.82	13.86	13.87	14.01	14.10	13.99	14.15	14.09
	(+1.12)	(+.07)	(−.20)	(−.13)	(−.19)	(−.15)	(−.14)		(+.09)	(−.02)	(+.14)	(+.08)
1π	17.43	16.99	16.93	16.97	16.91	17.07	16.96	16.90	17.36	17.15	16.99	16.96
	(+.53)	(+.09)	(+.03)	(+.07)	(+.01)	(+.17)	(+.06)		(+.46)	(+.25)	(+.09)	(+.06)
4σ	21.93	20.22	20.00	19.59	19.56	19.54	19.59	19.68	22.14	19.70	19.64	19.69
	(+2.25)	(+.54)	(+.32)	(−.09)	(−.12)	(−.14)	(−.09)		(+2.46)	(+.02)	(−.04)	(+.01)
M.A.E	1.30	.23	.18	.10	.11	.15	.10		1.00	.10	.09	.05

[a]Ref. [62] and references therein.
[b]Ref. [66] and references therein.
[c]Ref. [70].

of the IPs improved to near 0.1 eV, but two remarks are in order. Firstly, the effect of the $(SC)^2$ dressing was not always favorable in this case. Secondly, and more importantly, the selection of the active space is a very sensitive issue. The results shown in Table 3 in the second set of columns correspond to a CAS where the virtual π orbitals were the 3π set instead of the HOMO 2π. This was the choice indicated by the simpler $(SC)^2$ SR-SDCI wavefunctions. However, if one takes the 2π virtuals in the CAS, the first calculated IP is 14.04 eV, very close to the experimental value of 14.01 eV.

The last set of results in Table 3 correspond to using a set of CASSCF natural orbitals (NOs) for the neutral molecule. Of course, these orbitals are not expected to provide a good set for each state of the cation. However, they have the advantage that one can use the occupation numbers of the NOs of each symmetry as a guide for selecting the CAS at the CI step. The CAS for the CASSCF step was 8 electrons in (4440). The results are slightly better, and we consider them lying close to the limit possibilities of the method using a common set of MOs, as was the case in the reference calculations. Note that in this case, again, the $(SC)^2$ dressing has a clearly favorable effect. The MAE lies about 0.1 eV, and it seems that the two Σ states are responsible for this error not being smaller. J.V. Ortiz [67] has reported the difficulty offered by these states due to the shake up coupling with π to π^* excitations. We must also indicate that the dressing method used in this work does not take account of the unlinked effects of the singles (cf. eq. (3)). Such a dressing could result in an improvement of orbital relaxation effects.

4. Conclusions

The *one-state several-roots* $(SC)^2$ dressing method for the calculation of excitation energies that consists of adding a closed shell single-state single-reference $(SC)^2$ dressing operator to a SDCI matrix has been succesfully applied to the CO molecule. Single reference (ϕ_0) as well as small MR-CI spaces have been used as a reference for the SDCI. In spite of the limited extension of the basis set used, the mean absolute error for the VEE was in the range 0.1–0.05 eV for singlet and triplet valence and Rydberg states.

The method can be also applied to open-shell CI references. It has been applied for the first time to the calculation of the outer valence IPs of CO. This is a classic but by no means simple problem of theoretical studies of PES. The formalism used was the *one-state one-root* $(SC)^2$ dressing approach. Small MR-SDCI have been used, along with common sets of MOs for both the neutral and cationic systems. The results are also good, and we can reasonably expect to obtain improved results for similar problems in the future using MOs previously adapted to each ionized state. The selection of small sets of active MOs for the CAS is important to avoid very large SDCI matrices, but the results can be very sensitive to the choice of the active space.

Acknowledgements

This work has been supported by the European Commission through the TMR network ERB FMRX-CT 96-0079 (QUCEX). The authors acknowledge also financial support of the Spanish DGCYT (Project No. PB94-993) and SEUI (Grant No. OP90-0042). JVP acknowledges the Spanish MEC for a FPU grant. The authors are indebted to Dr. J.L. Heully for some FORTRAN code concerning the coupling between the MOLCAS and CASDI codes and MR-SDCI dressing.

References

1. A. Szabo and N.S. Ostlund: *Modern Quantum Chemistry.*, Macmillan, New York, (1990).
2. N.J. Turro: *Modern Molecular Photochemistry.*, University Science Books, Sausalito, CA, (1991).
3. M. Urban, I. Cernusák, V. Kellö and J. Noga: Electron Correlation in Molecules., in S. Wilson (ed.) *Methods in Computational Chemistry, Vol 1.*, Plenum Press, New York, p. 117 (1987).
4. J. Karwowski: Configuration Interaction., in S. Fraga (ed.) *Computational Chemistry. Structure, Interactions, and Reactivity.*, Elsevier, Amsterdam, p. 197 (1992).
5. I. Shavitt: The Method of Configuration Interaction., in H. F. S. III (ed.) *Methods of Electronic Structure Theory.*, Plenum Press, New York, pp 189 (1977).
6. P.E.M. Siegbahn: The Direct-CI Method., in G.H.F. Diercksen and S. Wilson (ed.) *Methods in Computational Molecular Physics.*, D. Reidel Publising Company, Dordrecht, p. 189 (1983).
7. B.O. Roos and P.E.M. Siegbahn: The Direct Configuration Interaction Method from Molecular Integrals., in H. F. S. III (ed.) *Methods of Electronic Structure Theory.*, Plenum, New York, p. 277 (1977).
8. J.P. Daudey, J.L. Heully and J.P. Malrieu: *J. Chem. Phys.* **99**, 1240 (1993).
9. J. Goldstone: *Proc. R. Soc. London* **A 239**, 269 (1957).
10. I. Lindgren and J. Morrison: *Atomic Many-Body Theory.*, Springer-Verlag, Berlin, (1986).
11. S.R. Langhoff and E.R. Davidson: *Int. J. Quantum Chem.* **8**, 61 (1974).
12. R.J. Bartlett and I. Shavitt: *Int. J. Quantum Chem., Quantum Chem. Symp.* **11**, 165 (1977).
13. P.G. Burton, R.J. Buenker, P.J. Bruna and S.D. Peyerimhoff: *Chem. Phys. Letters.* **95**, 379 (1983).
14. R. Ahlrichs, P. Scharf and C. Ehrhardt: *J. Chem. Phys.* **82**, 890 (1985).
15. R. Ahlrichs and P. Scharf: The Coupled Pair Aproximation., in K. P. Lawley (ed.) *Ab Initio Methods in Quantum Chemistry. Part I.*, J. Wiley & Sons, Chichester, (1987).
16. H.P. Kelly and M.A. Sessler: *Phys. Rev. A* **132**, 2091 (1963).
17. W. Meyer: *Int. J. Quantum. Chem.* **5**, 341 (1971).
18. W. Meyer: *J. Chem. Phys.* **58**, 1017 (1973).
19. J.A. Pople, M. Head-Gordon and K. Raghavachari: *J. Chem. Phys.* **87**, 5968 (1987).
20. F. Coester: *Nucl. Phys.* **1**, 421 (1958).
21. J. Cizek: *J. Chem. Phys.* **45**, 4256 (1966).
22. J. Paldus, J. Cizek and I. Shavitt: *Phys. Rev. A* **5**, 50 (1972).
23. G.D. Purvis and R.J. Bartlett: *J. Chem. Phys.* **76**, 1910 (1982).
24. R.J. Bartlett: *J. Phys. Chem.* **93**, 1697 (1989).
25. D. Mukherjee and (Ed.): *Applied Many-Body Methods in Spectroscopy and Electronic Structure.*, Plenum Press, New York, (1992).
26. B. Huron, P. Rancurel and J.P. Malrieu: *J. Chem. Phys.* **58**, 5745 (1973).
27. R.J. Buenker and S.D. Peyerimhoff: *Theor. Chim. Acta* **35**, 33 (1974).
28. R.J. Buenker and S.D. Peyerimhoff: *Theor. Chim. Acta* **39**, 217 (1975).
29. A.C. Wahl and G. Das: The Multiconfiguration Self-Consistent Field Method., in H. F. S. III (ed.) *Methods of Electronic Structure Theory.*, Plenum Press, New York, pp 51–78 (1977).
30. K. Andersson, P.-Å. Malmqvist, B.O. Roos, A.J. Sadlej and K. Wolinski: *J. Phys. Chem.* **94**, 5483 (1990).
31. K. Andersson, P.-Å. Malmqvist and B.O. Roos: *J. Chem. Phys.* **96**, 1218 (1992).
32. B.O. Roos, M. Fülscher, P.A. Malmqvist, M. Merchán and L. Serrano-Andrés: Theoretical Studies of the Electronic Spectra of Organic Molecules., in S.R. Langhoff (ed.) *Quantum Mechanical Electronic Structure Calculations with Chemical Accuracy.*, Kluwer Academic Publishers, Dordrecht, (1995).
33. P. Durand and J.P. Malrieu: Effective Hamiltonians and Pseudo-Operators as Tools for Rigorous Modelling., in K.P. Lawley (ed.) *Ab Initio Methods in Quantum Chemistry.*, J. Wiley & sons, New York, pp 321–412 (1987).

34. J.L. Heully and J.P. Malrieu: *Chem. Phys. Letters* **199**, 545 (1992).
35. J. Sánchez-Marín, I. Nebot-Gil, J.P. Malrieu, J.L. Heully and D. Maynau: *Theor. Chim. Acta* **95**, 215 (1997).
36. J.L. Heully, J.P. Malrieu, I. Nebot-Gil and J. Sánchez-Marín: *Chem. Phys. Letters* **256**, 589 (1996).
37. J.P. Malrieu, P. Durand and J.P. Daudey: *J. Phys. A, Math. Gen.* **18**, 809 (1985).
38. D. Mukhopadhyay, B. Datta and D. Mukherjee: *Chem. Phys. Letters* **197**, 236 (1992).
39. J. Meller, J.P. Malrieu and J.L. Heully: *Chem. Phys. Letters* **244**, 440 (1995).
40. N. Ben-Amor, D. Maynau, J. Sánchez-Marín, I. Nebot-Gil and S. Evangelisti: *J. Chem. Phys.* **109**, 8275 (1998).
41. N. Ben-Amor and D. Maynau: *Chem. Phys. Letters* **286**, 211 (1998).
42. M. Reguero, R. Caballol, J.L. Heully and J.P. Malrieu: *Chem. Phys. Letters* **265**, 621 (1997).
43. J. Pitarch-Ruiz, J. Sánchez-Marín, I. Nebot-Gil, and N. Ben Amor: *Chem. Phys. Letters* **291**, 407 (1998).
44. K. Andersson, M.R.A. Blomberg, M.P. Fülsher, V. Kellö, R. Lindh, P.-Å. Malmqvist, J. Noga, J. Olsen, B.O. Roos, A. J. Sadlej, P.E.M. Siegbahn, M. Urban and P.-O. Widmark. MOLCAS-3. University of Lund, Sweden. (1994)
45. J.L. Heully. Personal communication.
46. F. Hund: *Z. Physik* **42**, 93 (1927).
47. R.S. Mulliken: *Phys. Rev.* **32**, 186 (1928).
48. R.S. Mulliken: *Rev. Mod. Phys.* **4**, 1 (1932).
49. R.C. Weast (Ed.): *Handbook of Chemistry and Physics.*, CRC Press, Boca Raton, (1984).
50. R.K. Nesbet: *J. Chem. Phys.* **43**, 4403 (1965).
51. M.E. Casida, C. Jamorski, K.C. Casida and D.R. Salahub: *J. Chem. Phys.* **108**, 4439 (1998).
52. H. Lefebvre-Brion and C.M. Moser: *J. Chem. Phys.* **34**, 1950 (1961).
53. J. Rose, T. Shibuya and V. Mckoy: *J. Chem. Phys.* **58**, 74 (1973).
54. S.V. O'Neil and H. F. Schaefer III: *J. Chem. Phys.* **53**, 3994 (1970).
55. W. Coughran, J. Rose, T. Shibuya and V. Mckoy: *J. Chem. Phys.* **58**, 2699 (1973).
56. J.F. Stanton, J. Gauss, N. Ishikawa and M. Head-Gordon: *J. Chem. Phys.* **103**, 4160 (1995).
57. D.C. Comeau and R.J. Bartlett: *Chem. Phys. Letters* **207**, 414 (1993).
58. J.F. Stanton and R.J. Bartlett: *J. Chem. Phys.* **98**, 7029 (1993).
59. R.J. Bartlett and J.F. Stanton: Applications of Post-Hartree-Fock Methods: A Tutorial., in K.B. Lipkowitz and D.B. Boyd (eds.) *Reviews in Computational Chemistry, Vol. V.*, VCH Publishers, Inc., New York, pp 65–169 (1994).
60. T.H. Dunning: *J. Chem. Phys.* **90**, 1007 (1989).
61. R.A. Kendall, T.H. Dunning and R.J. Harrison: *J. Chem. Phys.* **96**, 6796 (1992).
62. S. Hüfner: *Photoelectron Spectroscopy.*, Springer-Verlag, Berlin, (1995).
63. J. Oddershede: Introductory Polarization Propagator Theory., in G. H. F. Diercksen and S. Wilson (eds.) *Methods in Computational Molecular Physics.* D. Reidel Publishing Company, Dordrecht, pp 249–268 (1983).
64. W. von Niessen, J. Schirmer and L.S. Cederbaum: On a Green's Function Method for the calculation of Ionization Spectra in the Outer and Inner Valence Region., in G.H.F. Diercksen and S. Wilson (ed.) *Methods in Computational Molecular Physics.*, D. Reidel Publishing Company, Dordrecht, pp 227–248 (1983).
65. M. Nooijen and J.G. Snijders: *Int. J. Quantum Chem.* **48**, 15 (1993).
66. D. Heryadi, D.L. Yeager, J.T. Golab and J.A. Nichols: *Theor. Chim. Acta* **90**, 273 (1995).
67. J.V. Ortiz: *J. Chem. Phys.* **104**, 7599 (1996).
68. K.P. Huber and G. Herzberg: *Constants of Diatomic Molecules.*, Van Nostrand Reinhold, New York, (1979).
69. E.S. Nielsen, P. Jorgensen and J. Oddershede: *J. Chem. Phys.* **73**, 6238 (1980).
70. M. Nooijen and J.G. Snijders: *J. Chem. Phys* **102**, 1681 (1995).

Ab Initio Summation Over States/SCI for Static and Dynamic Hyperpolarizabilities of Small Molecules

Milena Spassova[a], Valentin Monev[a], Ivo Kanev[a†], Benoît Champagne[b*], David H. Mosley[b] and Jean-Marie André[b]

[a] *Institute of Organic Chemistry, Bulgarian Academy of Sciences, 1113 Sofia, Bulgaria*
[b] *Laboratoire de Chimie Théorique Appliquée, Facultés Universitaires Notre-Dame de la Paix, 61, rue de Bruxelles, B-5000 Namur, Belgium*

Abstract

A Summation Over States (SOS) scheme for calculating the linear and nonlinear optical properties is implemented *ab initio* within the Configuration Interaction procedure including Single excitations (CIS). The potential of this SOS/CIS approach is assessed by considering the basis set dependence of the polarizability and first hyperpolarizability of a set of eight reference molecules, their comparison with Hartree-Fock and post Hartree-Fock techniques as well as with available experimental data. We emphasize the strength of this approach for analyzing the responses in terms of excited state contributions well-suited to the deduction of structure-property relationships. Secondly, the SOS/CIS technique is applied to an investigation of the three isomers of nitroaniline.

1. Introduction

The design of new materials for nonlinear optical (NLO) applications requires molecules possessing large hyperpolarizabilities leading, at the macroscopic scale, to large nonlinear susceptibilities. In addition, these materials have to be transparent to the ingoing and outgoing waves and be stable with respect to processing conditions and laser irradiation. Quantum chemistry is closely involved in the understanding of the physical phenomena of NLO responses of molecules [1,2] due to its ability to predict the relationships between the structures and their properties.

(Hyper)polarizabilities are defined as the coefficients in the Taylor series expansion of the dipole moment – or the energy – in the presence of static and/or oscillating electric fields:

$$\mu_\xi(\omega_\sigma) = \mu_\xi^0 + \sum_\eta \alpha_{\xi\eta}(-\omega_\sigma;\,\omega_1)\mathrm{E}_\eta(\omega_1) + \frac{1}{2}\mathrm{K}^{(2)} \sum_{\eta\zeta} \beta_{\xi\eta\zeta}(-\omega_\sigma;\,\omega_1,\,\omega_2)\mathrm{E}_\eta(\omega_1)\mathrm{E}_\zeta(\omega_2) + \ldots \quad (1)$$

where the $\alpha_{\xi\eta}$ and $\beta_{\xi\eta\zeta}$ components define the linear and first nonlinear responses to the external field and $\omega_\sigma = \sum_i \omega_i$. The subscripts ξ, η, and ζ correspond to the Cartesian molecular axes while the factor $\mathrm{K}^{(2)}$ is required in order that all frequency-dependent β converge towards the same static limit [$\mathrm{K}^{(2)} = 1$ for $\beta(0;\,0,\,0)$]. $\beta(-2\omega;\,\omega,\,\omega)$ desig-

---

*To whom correspondence should be addressed.
†Deceased August 2nd 1999

A. Hernández-Laguna et al. (eds.), Quantum Systems in Chemistry and Physics, Vol. 1: Basic Problems and Model Systems, 101–125.

nates the process of second harmonic generation (SHG) whereas $\beta(-\omega; \omega, 0)$ is associated with the static electric field-induced phase shift or birefringence, also called the linear electrooptic or dc-Pockels effect.

Sum Over States [3] (SOS) approaches constitute one of the most commonly used class of methodologies for theoretical estimation of hyperpolarizabilities. The strength of this approach is related to the fact that for many compounds of interest, only a few excited states make a major contribution. The simplest scheme, proposed by Oudar and Chemla [4–5] to analyze variations of β among push-pull conjugated materials, restricts the summation to a unique excited state. In this resulting two-state approximation (TSA), the static longitudinal electronic first hyperpolarizability, β_L, is given by:

$$\beta_L = 6\frac{\Delta\mu_{ge}\mu_{ge}^2}{\Delta E_{ge}^2} \tag{2}$$

where μ_{ge} is the longitudinal dipole moment transition between the ground and excited states, ΔE_{ge} is the corresponding excitation energy and $\Delta\mu_{ge}$ is the variation of the longitudinal dipole moment between the ground and excited states. The TSA has the merit of establishing a direct link between β and spectroscopic quantities associated with the main low-lying charge transfer transition, providing a simple scheme for interpretation. The charge transfer excited state of these donor/acceptor (D/A) systems is predominantly described by the excitation of an electron from an occupied to an unoccupied spinorbital (often the HOMO $\rightarrow$ LUMO). For a large variety of D/A conjugated systems, the TSA overestimates β_L by roughly a factor of 2 whereas the inclusion of a second excited state corrects the estimate to about 80% of the full-SOS value [6].

As reviewed by Kanis *et al.* [6], most of these SOS-based investigations of second-order NLO materials are carried out at various semiempirical levels. Generally, the ground and excited wavefunctions and energies are determined within a Configuration Interaction (CI) scheme limited to the single excitations (CIS) which has only effects on the excited states (Brillouin's theorem). However, extra electron correlation effects are indirectly accounted for in the parametrization of these semiempirical procedures. As a consequence, the semiempirical SOS/CIS approaches – although widely used for D/A compounds – can hardly be improved by including higher-order excitations in the CI expansion because this would require a re-parametrization. Effects of including higher-order excitations have nevertheless been shown to be quantitatively significant and dependent on the chemical nature of the compounds [7–8]. On the other hand, to our knowledge, an *ab initio* SOS/CIS treatment has only been adopted by Tsunekawa and Yamaguchi [9] to investigate the effects of incorporating nitrogen atoms in the conjugated bridges of push-pull compounds. In addition to these CI-based techniques, Tomonari and coworkers [10–13] have employed *ab initio* the SOS approach where the wavefunctions and energies are Slater determinants and their associated Hartree-Fock energies. This approach, which is equivalent to an uncoupled Hartree-Fock treatment [14–15], significantly overestimates the excitation energies and therefore underestimates β.

This paper concerns the use of *ab initio* CI methods to calculate the (hyper)polarizabilities within the SOS scheme. This work starts by adopting the simplest CI scheme, *i.e.* involving only the single excitations, and assesses different theoretical aspects of computing α and β. In the following Section we briefly summarize the theoretical background of the SOS/CIS approach. Eight small molecules – HF, H_2O, NH_3, HCl, H_2S, PH_3, H_2CO, and H_2CS – have been chosen to characterize the SOS/CIS method as most of them have already given rise to a large amount of theoretical and experimental investigations. Section 3 describes and discusses the results on i) the basis set dependence of the α and β values computed at the SOS/CIS level; ii) the comparison between the SOS/CIS approach, the coupled-perturbed (CPHF) or time-dependent (TDHF) Hartree-Fock schemes as well as other techniques including electron correlation effects; iii) the analysis of α and β in terms of the excited state contributions as well as by using the missing state analysis; and iv) the study of the first hyperpolarizability of the three isomers of nitroaniline.

2. Methodology and Computational Procedure

When a molecular system is placed in static and/or dynamic external electric fields, a perturbation term has to be added to the unperturbed time independent Hamiltonian, H_0:

$$H = H_0 + H_1 = H_0 - \vec{\mu}.\vec{E} \tag{3}$$

where $\vec{\mu}$ is the dipole moment operator and $\vec{E}$ is the vector sum of the different electric fields. Although the contributions originating from the field-induced nuclear relaxation can be non negligible [16], the present study only considers the effects of the fields on the electronic distribution. Therefore using the SOS perturbation theory expressions of Orr and Ward [3] we can write the expressions for the electronic α and β tensor components:

$$\alpha_{\xi\eta}(-\omega_\sigma;\omega_1) = \sum P_{-\sigma,1} \sum_{m}{}' \frac{\langle 0|\mu_\xi\rangle|m\rangle\langle m|\mu_\eta|0\rangle}{(\omega_m - \omega_\sigma)} \tag{4}$$

$$\beta_{\xi\eta\zeta}(-\omega_\sigma;\omega_1,\omega_2) = \sum P_{-\sigma,1,2} \sum_{m,n}{}' \frac{\langle 0|\mu_\xi|m\rangle\overline{\langle m|\mu_\eta|n\rangle}\langle n|\mu_\zeta|0\rangle}{(\omega_m - \omega_\sigma)(\omega_n - \omega_1)} \tag{5}$$

where the summations run over all the electronic excited states $|m\rangle$ and $|n\rangle$ of energy E_m and E_n. $|0\rangle$ is the ground state wavefunction of energy E_0, $\overline{\langle m|\mu_\eta|n\rangle} = \langle m|\mu_\eta|n\rangle - \langle 0|\mu_\eta|0\rangle\delta_{mn}$ and $\hbar\omega_n = \omega_n = E_n - E_0$. $\sum P_{-\sigma,1}(\sum P_{-\sigma,1,2})$ is the sum over the 2 (6) permutations of the pairs $(-\omega_\sigma, \xi)$, (ω_1, η), and (ω_2, ζ). These are the exact (hyper)polarizability expressions which generate a hierarchy of approximations classified according to the method employed to evaluate the wavefunctions and energies. Once the excitation energies and dipole transition elements are computed, these SOS expressions can be evaluated for as many frequency values as wanted at a very low additional computational cost.

If Slater determinants obtained from the Hartree-Fock procedure are used in equations (4) and (5), we obtain the uncoupled Hartree-Fock (UCHF) scheme because the field effects upon the electron-electron interactions are not taken into account [14–15]. To go beyond this crude approximation, the wavefunctions are built as linear combinations of the ground state Hartree-Fock wavefunction with singly-, doubly-, ... excited wavefunctions; the expansion coefficients are obtained via CI energy minimization procedures. The simplest way consists of adopting the CI-singles scheme [17], where the configuration space is limited to the single excitations. Since Brillouin's theorem states that there is no direct interaction between the ground state and the singly-excited states, the excited state wavefunction *ansatz* is given by the expression:

$$\Psi_{\mathrm{CIS}} = \sum_{\mathrm{i}} \sum_{\mathrm{a}} \mathrm{c}_{\mathrm{ia}} \Psi_{\mathrm{ia}} \tag{6}$$

Ψ_{ia} is the Slater determinant in which the ith occupied spinorbital is replaced by the ath unoccupied one. Setting this *ansatz* in the Schrödinger equation, one obtains a set of linear equations

$$\sum_{\mathrm{i}} \sum_{\mathrm{a}} \mathrm{c}_{\mathrm{ia}} \langle \Psi_{\mathrm{j}}^{\mathrm{b}} | \mathrm{H} | \Psi_{\mathrm{i}}^{\mathrm{a}} \rangle = \mathrm{E}\mathrm{c}_{\mathrm{jb}} \tag{7}$$

which reduces the problem to the evaluation of the eigenvectors and eigenvalues of the Hamiltonian, the matrix elements of which are,

$$\mathrm{H}_{\mathrm{jb,ia}} = (\varepsilon_{\mathrm{a}} - \varepsilon_{\mathrm{i}})\delta_{\mathrm{ij}}\delta_{\mathrm{ab}} - \langle \mathrm{bi}|\mathrm{aj}\rangle - \langle \mathrm{bi}\|\mathrm{ja}\rangle \tag{8}$$

ε_{a} and ε_{i} are the spinorbital energies and $\langle \mathrm{bi}\|\mathrm{ja}\rangle$ is a two-electron integral between spinorbitals in the physicist's notation. Due to the singlet nature of the dipole moment operator and because we are dealing with closed-shell molecules, only spin-adapted singlet excitations are considered. These CIS excited wavefunctions, energies, transition dipole moments between ground and excited states have been computed by the Gaussian94 series of program [18]. From these quantities (Eqs. 4 and 5), the static and frequency-dependent α and β tensor elements are computed by using the $\alpha\beta$-SOS code [19].

In addition to the nonzero α and β tensor components, we also consider the isotropic average polarizability $\langle\alpha\rangle$ and the polarizability anisotropy $\Delta\alpha$ defined as:

$$\langle\alpha\rangle = \frac{1}{3}\sum_{\xi} \alpha_{\xi\xi} \tag{9}$$

$$\Delta\alpha = \frac{1}{\sqrt{2}}[(\alpha_{\mathrm{xx}} - \alpha_{\mathrm{yy}})^2 + (\alpha_{\mathrm{xx}} - \alpha_{\mathrm{zz}})^2 + (\alpha_{\mathrm{yy}} - \alpha_{\mathrm{zz}})^2]^{1/2} \tag{10}$$

since these quantities are most commonly determined experimentally. For β we evaluate its projection on the dipole moment,

$$\beta_{\mathrm{vec}} = \sum_{\xi} \frac{\mu_{\xi}\beta_{\xi}}{|\mu|} \tag{11}$$

where $|\mu|$ is the norm of the dipole moment and

$$\beta_{\xi} = \frac{1}{3}\sum_{\eta} \beta_{\xi\eta\eta} + \beta_{\eta\xi\eta} + \beta_{\eta\eta\xi} \tag{12}$$

$\beta_{\|} = 3/5\beta_{vec}$, the vector component of the β tensor in the direction of the dipole moment is also a commonly determined experimental quantity.

At the Time-Dependent Hartree-Fock (TDHF) level, the polarizability and first hyperpolarizabilities are evaluated as the first and second derivatives of the dipole moment with respect to the external fields [20]. This procedure includes the field-induced electron reorganizational effects self-consistently in terms of the average Coulomb and Pauli potentials. It is equivalent to the Random Phase Approximation (RPA) [21]. The CIS or Tamm Dancoff Approximation (TDA) [22] is a simplification of the RPA in which the interactions between the ground state and doubly excited states have been neglected. Although one can dispute upon the correlated nature of the RPA (or TDHF) scheme, the THDF/RPA method is a fully-relaxed Hartree-Fock procedure. These TDHF calculations have been performed by using the GAMESS program [23] or the GAUSSIAN94 package if considering the static limit. In this event, one uses the Coupled-Perturbed Hartree-Fock (CPHF) or TDHF schemes which determine the first and second derivatives of the dipole moment with respect to the electric fields.

In order to assess the basis set dependence on α and β tensor components, we have used a series of standard basis sets from the minimal, STO-3G [24], split-valence, 6-31G [25], to basis sets including polarization functions, 6-31G* [26], 6-31G** [27], and 6-311G** [28]. For nitroaniline, our study is limited to the 6-31G basis set. We have also employed basis sets with diffuse functions, the 6-31G + SPD (HF, H_2O, NH_3, and H_2CO), 6-31G* + SPD (HCl, H_2S, and PH_3) and 6-311G* + SPD (H_2CO and H_2CS) and 6-311G** + SPD (HF, H_2O, NH_3, HCl, H_2S, and PH_3) basis sets where diffuse *s* functions are added to the hydrogen atoms whereas a set of diffuse *p* and *d* functions of equal exponent is added on the other atoms. The *s*, *p* and *d* diffuse functions have been shown to be of importance to provide semi-quantitative hyperpolarizabilities [29]. They have been chosen to be equal to 0.040, 0.050, 0.0715, 0.100, 0.135, 0.040, 0.050, and 0.065 for the H C, N, O, F, P, S, and Cl atoms, respectively. For the hydrogen and carbon atoms these exponents, α_C and α_H, are identical to those suggested by Hurst *et al.* [29] whereas the other exponents have been derived by mapping the $\alpha_X/\alpha_C(X = N, 0, \ldots)$ ratios of the diffuse function exponents within the polarized basis set of Sadlej [30] and the aug-cc-pVTZ of Dunning [31–32]. The 6-311G** + SPD basis is considered to be reasonably extended in order to provide semi-quantitative α and β values. This is confirmed by comparing our CPHF/TDHF results with those obtained with much larger basis sets (see Section 3.2). This basis set is therefore considered as our reference.

The geometries of HF, H_2O, NH_3 and H_2S were taken from Ref. 33, of H_2CO from Ref. 34, and of HCl and PH_3 from Ref. 35. The molecular geometry of H_2CS (the nitroanilines) have been optimized at the RHF/6-31G** (RHF/6-31G) level by using the GAUSSIAN94 program (see Scheme 1 for their orientation in the Cartesian frame).

3. Results and Discussion

3.1. Basis set dependence

Figures 1–3 display the basis set dependence of the average polarizability, $\langle\alpha\rangle$, the polarizability anisotropy, $\Delta\alpha$, and the vector component of the first hyperpolarizability,

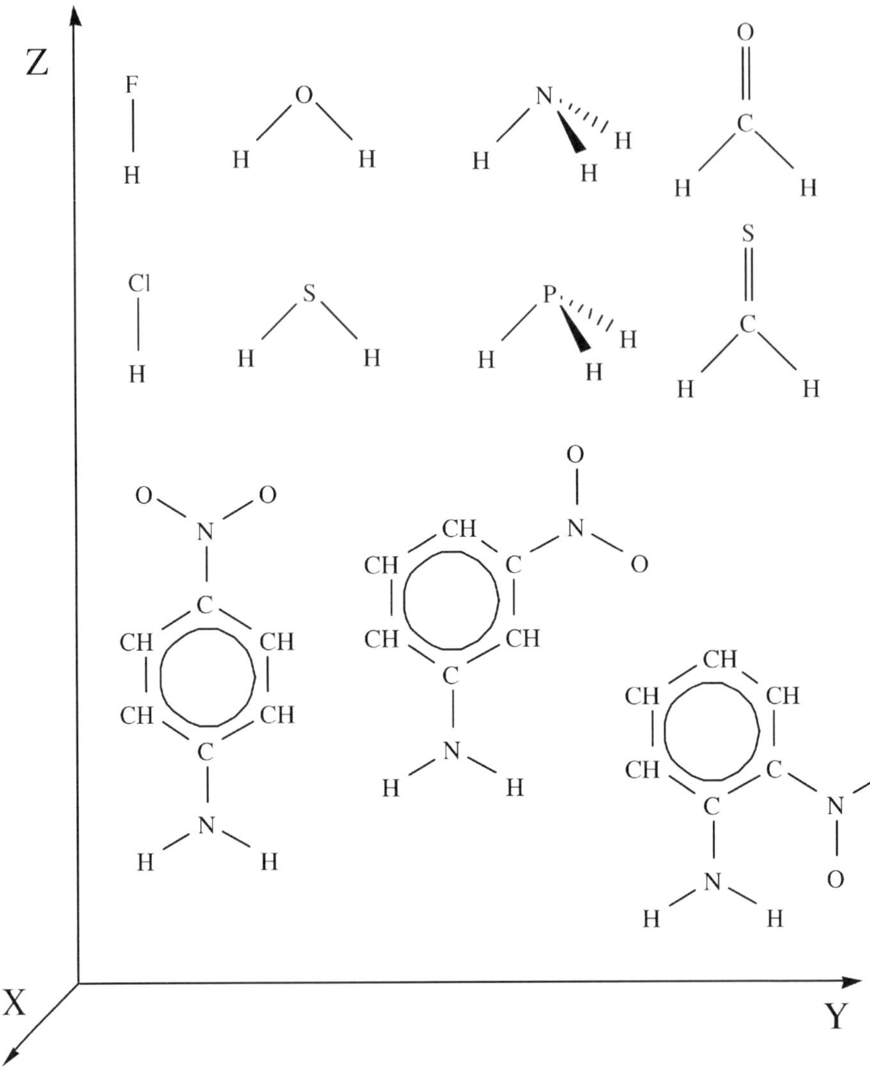

Scheme 1. 3D Cartesian representation of the molecules under investigations.

$\beta_{//}$, respectively for the eight reference molecules. These values have been obtained at the SOS/CIS level of approximation by including a maximum of 120 excited states (100 for H_2CO and H_2CS); larger number of excited states is beyond our current computational facilities. As we will comment in paragraph 3.3., we have assumed that higher-energy excited states present negligible contributions. Small basis sets underestimate $\langle\alpha\rangle$ but overestimate $\Delta\alpha$ because the lack of diffuse and/or polarization functions underestimates mainly the non-axial tensor elements (α_{xx} and α_{yy}). Adding *d*-polarization functions on third-row elements has a much significant impact on their $\Delta\alpha$ than for

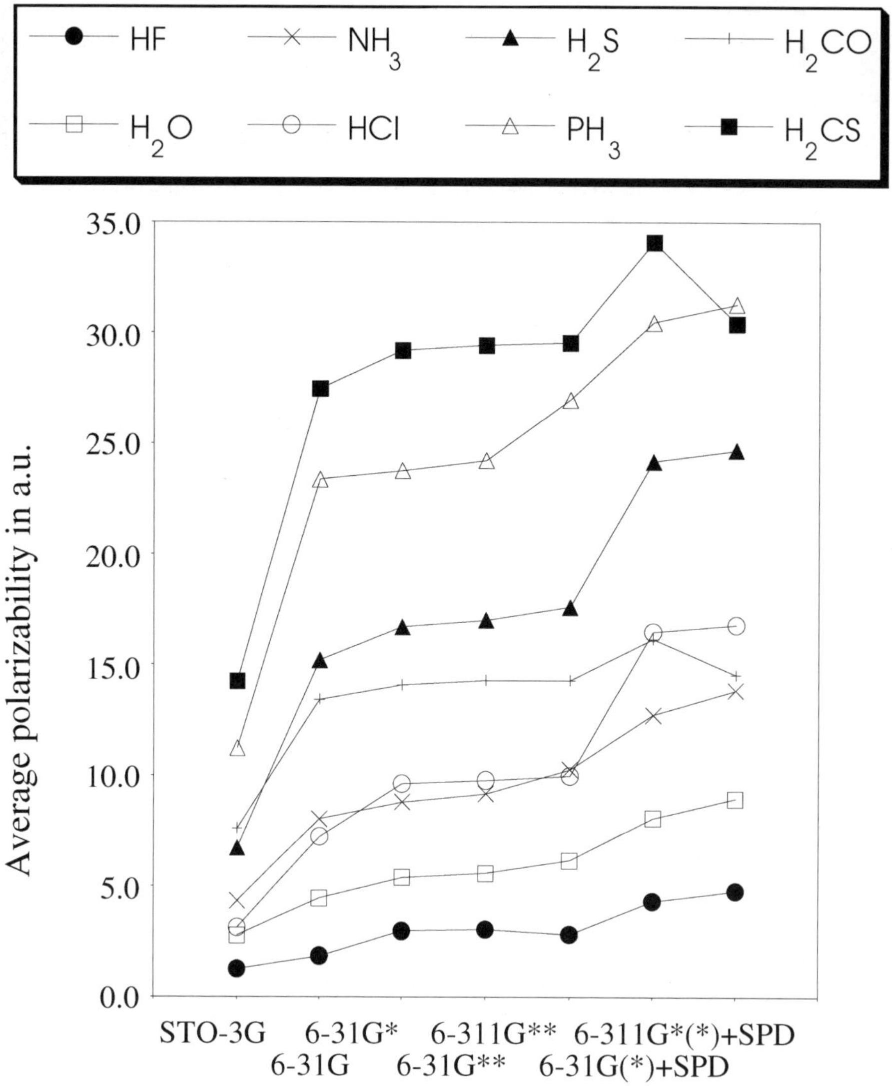

Fig. 1. Basis set dependence of the average static polarizability (in a.u., 1 a.u. of polarizability = 1.6488 10^{-41} C^2 m^2 J^{-1} = 0.14818 Å3) for the eight reference molecules evaluated at the SOS/CIS level of approximation.

molecules containing only hydrogen and second-row atoms. For $\beta_{//}$, variations with basis set are proportionally larger and also non-monotonic in the sense that adding diffuse and/or polarization functions can either increase or decrease $\beta_{//}$. Similar basis set dependencies are obtained when adopting the TDHF scheme [36].

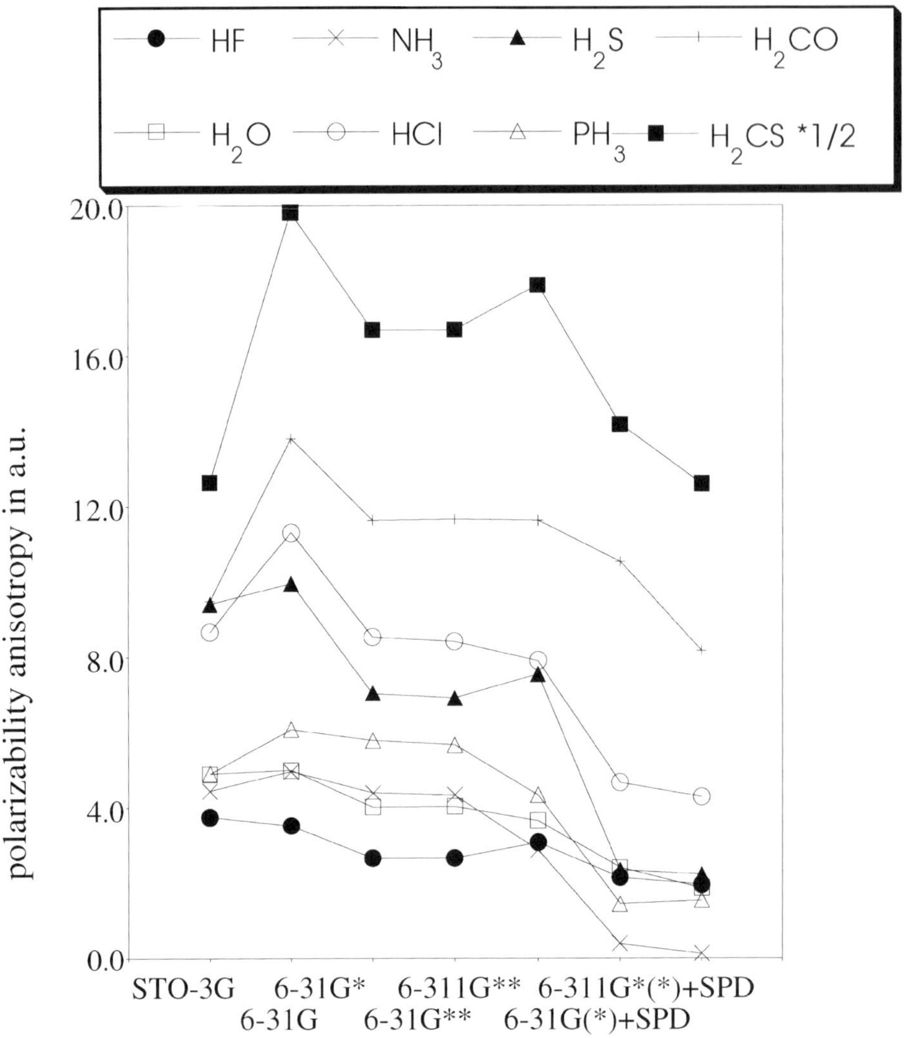

Fig. 2. Basis set dependence of the static polarizability anisotropy (in a.u.) for the eight reference molecules evaluated at the SOS/CIS level of approximation.

3.2. Comparison with experiment and other theoretical methods

In Tables 1–6, the SOS/CIS/6-311G** + SPD (6-311G* + SPD for H_2CO and H_2CS) α and β tensor elements, and related quantities are compared on one hand with available experimental data and on the other hand with reference theoretical values. A maximum of 120 excited states (100 for H_2CO and H_2CS) has been considered. Using the same basis set, we have evaluated the corresponding TDHF/CPHF values. However, it should be stressed that by default, GAUSSIAN94 (SOS/CIS) uses pure d-functions whereas

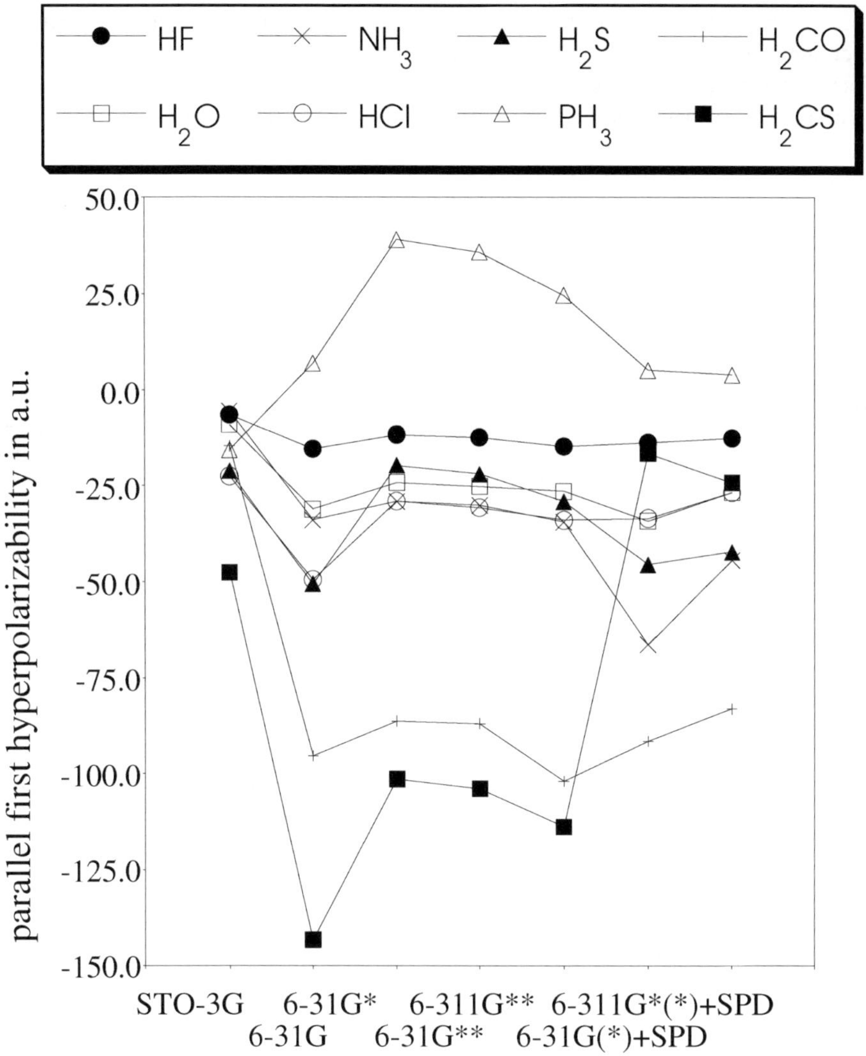

Fig. 3. Basis set dependence of the vector component of the first hyperpolarizability (in a.u., 1.0 a.u. of first hyperpolarizability = 3.2063×10^{-53} $C^3\,m^3\,J^{-2}$ = 8.641×10^{-33} esu) for the eight reference molecules evaluated at the SOS/CIS level of approximation.

GAMESS (CPHF/TDHF) adopts the Cartesian functions. From comparing the static α and β quantities with 5 and 6 *d*-functions, it turns out that this effect is negligible. In order to illustrate the consequences of the lack of electron correlation on the α and β values, we have made a selection of authoritative investigations [37–42] where, using very large basis sets, various levels of approximations – including TDHF/CPHF – have been adopted to evaluate both static and dynamic quantities. In addition to these works, many other investigations have considered the electron correlation effects of the (hyper)polarizabilities of one or more of these eight reference molecules, either with conventional approaches [43–52] or with density functional techniques [53–54].

For α the agreement between these CPHF/TDHF results performed with near-Hartree-Fock limit basis sets and our corresponding results (3rd columns of Tables 1–6) obtained with our largest basis set is rather good. It proves the adequacy of adding just one set of diffuse functions to reach the basis set saturation for these properties.

Table 1 *Comparison of static and dynamic SOS/CIS results for α and β* of HF molecule *(in a.u., 1 a.u. of polarizability = 1.6488 10^{-41} C^2 m^2 J^{-1} = 0.14818 Å^3; 1.0 a.u. of first hyperpolarizability = 3.2063 × 10^{-53} C^3 m^3 J^{-2} = 8.641 × 10^{-33} esu) with literature values*

1) HF

	Present work		Sekino and Bartlett [37]			Experiment
	SOS/CIS	TDHF	TDHF	MP2	CCSD(T)	
$\alpha_{xx}(0;0) = \alpha_{yy}(0;0)$	4.14	3.88	4.51	5.40	5.34	5.10[b], 4.96[c]
$\alpha_{zz}(0;0)$	6.10	5.50	5.76	6.43	6.44	6.59[b], 6.29[c]
$\langle\alpha(0;0)\rangle$	4.79	4.42	4.93	5.74	5.71	5.60[b], 5.40[c]
$\Delta\alpha(0;0)$	1.96	1.43	1.25	1.04	1.10	1.49[b], 1.33[c]
$\beta_{xxz}(0;0,0) = \beta_{yyz}(0;0,0)$	2.32	0.28	0.28	1.25	1.27	
$\beta_{zzz}(0;0,0)$	16.39	9.31	8.40	9.10	9.62	
$\beta_{//}(0;0,0)$	−12.62	−5.93	−5.38	−6.96	−7.30	
$\beta_{//}(-2\omega;\omega,\omega)$[a]	−13.43	−6.44	−5.85	−7.60	−8.00	−11.0 ± 1.0[d]
$\beta_{//}(0;\omega,-\omega)$[a]	−12.87	−6.10	−5.52	−7.20	−7.50	

[a] ω = 0.0656 a.u. (λ = 694.3 nm).
[b] Extrapolated static values, see Table 2 of Ref. 43 and references therein.
[c] Extrapolated static values, see Table 3 of Ref. 47 and references therein.
[d] Results of J.W. Dudley II and J.F. Ward, J. Chem. Phys. **82**, 4673 (1985) rescaled to the new γ(He) values due to D.M. Bishop and J. Pipin, J. Chem. Phys. **91**, 3549 (1989) by D.P. Shelton and J. Rice, Chem. Rev. **94**, 3 (1994).

Table 2 *Comparison of static and dynamic SOS/CIS results for α and β of H_2D molecule (in a.u.)*

2) H_2O

	Present work		Sekino and Bartlett [37]			Maroulis [38]	Experiment
	SOS/CIS	TDHF	TDHF	MP2	CCSD(T)	SDQ-MP4	
$\alpha_{xx}(0;0)$	7.87	7.26	7.96	9.75	9.64	9.84	9.26 ± 0.09[b]
$\alpha_{yy}(0;0)$	10.03	8.84	9.16	10.04	10.02	9.18	10.01 ± 0.09[b]
$\alpha_{zz}(0;0)$	8.95	8.01	8.53	9.82	9.73	9.49	9.62 ± 0.02[b]
$\langle\alpha(0;0)\rangle$	8.95	8.04	8.55	9.87	9.79	9.50	9.64[b], 9.81[c]
$\Delta\alpha(0;0)$	1.87	1.36	1.05	0.25	0.34	0.57	0.65[b], 0.66[c]

Table 2 (*Continued*)

2) H_2O

	Present work		Sekino and Bartlett [37]			Maroulis [38]	Experiment
	SOS/CIS	TDHF	TDHF	MP2	CCSD(T)	SDQ-MP4	
$\beta_{xxz}(0; 0, 0)$	6.89	1.52	1.4	5.9	6.2	9.8	
$\beta_{yyz}(0; 0, 0)$	16.17	10.65	9.4	9.4	10.2	4.9	
$\beta_{zzz}(0; 0, 0)$	21.37	7.73	7.1	13.7	13.7	13.3	
$\beta_{//}(0; 0, 0)$	−26.66	−11.94	−10.8	−17.5	−18.0	−16.8	
$\beta_{//}(-2\omega; \omega, \omega)^a$	−29.90	−14.01	−12.6	−20.4	−21.1		-22.0 ± 0.9^d
$\beta_{//}(0; \omega, -\omega)^a$	−27.65	−12.57	−11.3	−18.4	−19.0		

$^a\omega = 0.0656$ a.u.

^{b}I.G. John, G.B. Bacskay, and N.S. Hush, Chem. Phys. **51**, 49 (1980) from the $\langle\alpha\rangle$ (static value) of G.D. Zeiss and W.J. Meath, Mol. Phys. **33**, 1155 (1977) and $\Delta\alpha$ (514.5 nm) of W.F. Murphy, J. Chem. Phys. **67**, 5877 (1977).

^{c}see Table 3 of Ref. 47 and references therein.

^{d}J.F. Ward and C.K. Miller, Phys. Rev. A **19**, 826 (1984) rescaled to the new γ(He) values due to D.M. Bishop and J. Pipin, J. Chem. Phys. **91**, 3549 (1989) by D.P. Shelton and J. Rice, Chem. Rev. **94**, 3 (1994).

Table 3 *Comparison of static and dynamic SOS/CIS results for α and β of NH_3 molecule (in a.u.)*

3) NH_3

	Present work		Sekino and Bartlett [37]			Rice and Handy [39]	Experiment
	SOS/CIS	TDHF	TDHF	MP2	CCSD(T)	MP2	
$\alpha_{xx}(0; 0) = \alpha_{yy}(0; 0)$	13.80	12.18	12.77	13.78	13.71		
$\alpha_{zz}(0; 0)$	13.93	12.67	13.35	15.83	15.71		
$\langle\alpha(0; 0)\rangle$	13.84	12.34	12.96	14.46	14.38		14.56^d
$\Delta\alpha(0; 0)$	0.13	0.49	0.58	2.06	2.01		
$\langle\alpha(-\omega; \omega)\rangle^a$	14.13	12.60					
$\Delta\alpha(-\omega; \omega)^a$	0.32	0.70					1.94^e
$\beta_{xxz}(0; 0, 0) = \beta_{yyz}(0; 0, 0)$	13.25	8.90	7.0	8.1	8.8		
$\beta_{zzz}(0; 0, 0)$	47.57	18.72	11.1	37.8	39.6		
$\beta_{//}(0; 0, 0)$	−44.44	−21.91	−15.1	−32.5	−34.3	−29.7	
$\beta_{//}(-2\omega; \omega, \omega)^b$	−57.70	−31.17	−22.0	−46.6	−49.1		-48.9 ± 1.2^f
$\beta_{//}(0; \omega, -\omega)^b$	−48.15	−24.43	−16.7	−36.3	−38.3		
$\beta_{//}(-2\omega; \omega, \omega)^c$	−46.70	−23.44				−32.0	
$\beta_{//}(0; \omega, -\omega)^c$	−45.17	−22.40				−30.4	

$^a\omega = 0.0719$ a.u. ($\lambda = 632.8$ nm), $^b\omega = 0.0656$ a.u., $^c\omega = 0.0300$ a.u. ($\lambda = 1517$ nm).

^{d}static value from G.D. Zeiss and W.J. Meath, Mol. Phys. **33**, 1155 (1977).

^{e}dynamic value at 632.8 nm, N.J. Bridge and A.D. Buckingham, Proc. R. Soc. London Ser. A **295**, 334 (1966).

^{f}J.F. Ward and C.K. Miller, Phys. Rev. A **19**, 826 (1984) rescaled to the new γ(He) values due to D.M. Bishop and J. Pipin, J. Chem. Phys. **91**, 3549 (1989) by D.P. Shelton and J. Rice, Chem. Rev. **94**, 3 (1994).

The SOS/CIS scheme leads to larger α tensor components than the TDHF/CPHF approach and, in most cases, to larger polarizability anisotropy. This can easily be understood by refereeing to the differences between the RPA and TDA schemes. In the latter which is equivalent to SOS/CIS, the excitation energies are smaller because the ground state is not stabilized by interactions with doubly-excited states [21–22]. This

Table 4 *Comparison of static and dynamic SOS/CIS results for α and β of HCl molecule (in a.u.)*

4) HCl

	Present work		Hammond and Rice [40]			Experiment
	SOS/CIS	TDHF	TDHF	MP2	CCSD(T)	
$\alpha_{xx}(0;0) = \alpha_{yy}(0;0)$	15.38	13.87	15.98	16.74	16.65	
$\alpha_{zz}(0;0)$	19.67	16.49	17.91	18.46	18.42	
$\langle\alpha(0;0)\rangle$	16.81	14.74	16.62	17.32	17.24	
$\Delta\alpha(0;0)$	4.29	2.31	1.93	1.72	1.77	1.56[d]
$\langle\alpha(-\omega;\omega)\rangle^{a}$	17.14	15.05				17.40 ± 0.03[e]
$\beta_{xxz}(0;0,0) = \beta_{yyz}(0;0,0)$	2.12	−4.08	−2.30	1.11	0.47	
$\beta_{zzz}(0;0,0)$	40.34	14.97	10.79	11.27	11.11	
$\beta_{//}(0;0,0)$	−26.74	−4.09	−3.72	−8.09	−7.23	
$\beta_{//}(-2\omega;\omega,\omega)^{b}$	−27.44	−4.61	−4.14	−9.33		−9.9 ± 1.2[f]
$\beta_{//}(0;\omega,-\omega)^{b}$	−26.97	−4.23	−3.69	−8.47		
$\beta_{//}(-2\omega;\omega,\omega)^{c}$	−29.52	−4.20	−3.58	−8.34		
$\beta_{//}(0;\omega,-\omega)^{c}$	−27.58	−4.13	−3.71	−8.18		

[a] $\omega = 0.0719$ a.u., [b] $\omega = 0.0656$ a.u., [c] $\omega = 0.0345$ a.u. ($\lambda = 1319$ nm).
[d] static value of D.W. Hohnson and N.F. Ramsey, J. Chem. Phys. **67**, 941 (1977).
[e] dynamic value at 632.8 nm, N.J. Bridge and A.D. Buckingham, Proc. R. Soc. London Ser. A **295**, 334 (1966).
[f] Results of J.W. Dudley II and J.F. Ward, J. Chem. Phys. **82**, 4673 (1985) rescaled to the new γ(He) values due to D.M. Bishop and J. Pipin, J. Chem. Phys. **91**, 3549 (1989) by D.P. Shelton and J. Rice, Chem. Rev. **94**, 3 (1994).

Table 5 *Comparison of static and dynamic SOS/CIS results for α and β of H_2S molecule (in a.u.)*

5) H_2S

	Present work		Sekino and Bartlett [37]			Experiment
	SOS/CIS	TDHF	TDHF	MP2	CCSD(T)	
$\alpha_{xx}(0;0)$	23.44	21.24	23.76	24.41	24.35	25.22[c]
$\alpha_{yy}(0;0)$	26.01	21.73	23.84	25.35	25.36	25.92[c]
$\alpha_{zz}(0;0)$	24.60	21.03	23.43	24.45	24.40	25.31[c]
$\langle\alpha(0;0)\rangle$	24.68	21.33	23.68	24.74	24.70	25.48[c], 24.67[d]
$\Delta\alpha(0;0)$	2.23	0.62	0.37	0.92	0.99	0.67[c]
$\langle\alpha(-\omega;\omega)\rangle^{a}$	25.37	21.97				

Table 5 (*Continued*)

5) H_2S

	Present work		Sekino and Bartlett [37]			Experiment
	SOS/CIS	TDHF	TDHF	MP2	CCSD(T)	
$\beta_{xxz}(0; 0, 0)$	22.62	3.72	−5.6	5.5	4.4	
$\beta_{yyz}(0; 0, 0)$	29.64	17.70	9.9	9.2	9.5	
$\beta_{zzz}(0; 0, 0)$	18.21	−8.21	−7.9	−0.2	−1.1	
$\beta_{//}(0; 0, 0)$	−42.28	−7.92	2.2	−8.8	−7.7	
$\beta_{//}(-2\omega; \omega, \omega)$[b]	−51.07	−10.18	2.4	−10.1	−8.8	-10.1 ± 2.1[e]
$\beta_{//}(0; \omega, -\omega)$[b]	−44.74	−8.48	2.2	−9.4	−8.2	

[a] $\omega = 0.0719$ a.u.,

[b] $\omega = 0.0656$ a.u.

[c] dynamic values at 632.8 nm, M.P. Bogaard, A.D. Buckingham, and G.L.D. Ritchie, Chem. Phys. Lett. **90**, 183 (1982).

[d] static value, M.P. Bogaard, A.D. Buckingham, R.K. Pierens, and A.H. White, J. Chem. Soc. Faraday Trans. 1 **74**, 3008 (1978).

[e] J.F. Ward and C.K. Miller, Phys. Rev. A **19**, 826 (1984) rescaled to the new γ(He) values due to D.M. Bishop and J. Pipin, J. Chem. Phys. **91**, 3549 (1989) by D.P. Shelton and J. Rice, Chem. Rev. **94**, 3 (1994).

Table 6 *Comparison of static and dynamic SOS/CIS results for α and β of PH_3, H_2CO and H_2CS molecules (in a.u.)*

6)	**PH_3**			**H_2CO**				**H_2CS**	
	present work			present work		Post HF		present work	
	SOS/CIS	TDHF	Experiment	SOS/CIS	TDHF	methods	Experiment	SOS/CIS	TDHF
$\alpha_{xx}(0; 0)$	31.79	26.97		10.20	10.88			21.62	23.31
$\alpha_{yy}(0; 0)$	31.79	26.97		13.89	14.74			22.38	27.40
$\alpha_{zz}(0; 0)$	30.25	26.95		19.58	20.44	17.13[d]		47.21	45.50
$\langle\alpha(0; 0)\rangle$	31.28	26.96	32.6[c]	14.56	15.35			30.40	32.07
$\Delta\alpha(0; 0)$	1.54	0.02		8.18	8.33			25.22	20.45
$\langle\alpha(-\omega; \omega)\rangle$[a]				14.91			16.56[g]		
$\Delta\alpha(-\omega; \omega)$[a]				8.54			6.25[h]		
$\beta_{xxz}(0; 0, 0)$	−6.66	−4.89		3.41	2.95	3.2[d]		−5.48	−25.22
$\beta_{yyz}(0; 0, 0)$	−6.66	−4.89		34.89	21.08	40.2[d]		−45.59	−71.33
$\beta_{zzz}(0; 0, 0)$	6.77	−35.79		98.16	32.03	24.0[d], 16 ± 1[e]		91.32	2.64
$\beta_{//}(0; 0, 0)$	3.93	27.34		−83.08	−33.63	−40.4[d], −40.4[f]		−24.15	56.37
$\beta_{//}(-2\omega; \omega, \omega)$[b]				−93.23	−39.82	−49.7[f]			
$\beta_{//}(0; \omega, -\omega)$[b]				−85.36	−35.51	−43.0[f]			

[a] $\omega = 0.0772$ a.u. ($\lambda = 589.3$ nm), [b] $\omega = 0.0600$ a.u. ($\lambda = 759$ nm).

[c] static value, CRC Handbook of Chemistry and Physics, (CRC Press, Boca Raton, 1986–87), pp E70-E75 and references therein; see M.A. Spackman, J. Phys. Chem. **93**, 7594 (1989).

[d] static MP2 results of Velders and Feil (Ref. [41]),

[e] static CCSD(T) result from Fig. 3 of the Ref. [42].

[f] dynamic MP2 results at 759 nm from Ref. [39].

[g] J.A. Applequist, J.R. Carl, and K.K. Fung, J. Am. Chem. Soc. **94**, 2952 (1974).

[h] S. Parthasarathy, Indian J. Phys. **7**, 139 (1932).

phenomenon is magnified for β for which sign discrepancies between the SOS/CIS and TDHF/CPHF results can appear in the case of small β tensor elements.

The comparison with the correlated results provides a good idea of what improvements can be brought about by including double, triple, ... excitations in the wavefunction descriptions. For most molecules, we present MP2 [Møller-Plesset scheme limited to second order in electron-electron interactions] and CCSD(T) [coupled-cluster treatment including the single and double excitations as well as a perturbational estimate of the connected triple excitations] results. The former shows the importance of the leading electron correlation correction whereas the latter is a reference for higher-order electron correlation treatments which could serve to calibrate standards of measurements. With respect to these correlated treatments, the SOS/CIS procedure underestimates $\langle\alpha\rangle$ by less than 20% whereas $\Delta\alpha$ is in most cases largely overestimated – often by more than 50%. Although this is not the main source of discrepancy between the SOS/CIS and the experimental values, the vibrational contributions to the static α should also be considered to obtain a better agreement. $\beta_{//}$ is also overestimated, by less than 50% for H_2O and NH_3, but by a factor greater than 2 for HCl, H_2S and H_2CO. As a consequence a satisfactory agreement with experiment is only reached for $\langle\alpha\rangle$. For $\beta_{//}$, our SOS/CIS approach can predict the trends among HF, H_2O, and NH_3 but not with the molecules containing third-row atoms for which electron correlation corrections are required for a correct ordering. Clearly, one has to go further than the CIS scheme for these small molecules.

3.3. *Analysis of α and β in terms of excited state contributions*

The knowledge of the excited state contributions to α and β is of importance not only for understanding the origin of the responses but also for computational purposes. Indeed, for medium-size molecules like *p*-nitroaniline or small molecules studied with extended basis sets, the full configuration space becomes rapidly very huge and out of reach for a complete *ab initio* treatment. Therefore, if one needs to truncate the configuration space, one has to ensure that the most contributing excited states are present and correctly reproduced.

One way of considering the state contributions to α and β consists in limiting the summations over m and n in Eqs. (4–5) to the Nth lowest-energy excited states and in analyzing the evolution of α and β with N. Figures 4–5 display the variations of static $\langle\alpha\rangle$ and β_{vec} with the number of excited states for NH_3 and H_2CO, respectively. These 2 molecules are representative of the various behaviours that appear. The corresponding graphs for the 6 other molecules are available upon request to the authors. Since each state contribution in Eq. 4 is positive, $\langle\alpha\rangle$ increases monotonically with N. For NH_3, $\langle\alpha\rangle$ saturates when 80–90 states are included whereas for H_2CO there are 2–3 higher-energy excited states which contribute significantly. The evolution of $\langle\alpha\rangle$ with N for NH_3 presents many plateau's whereas for H_2CO the stairs are much smaller. $\langle\alpha\rangle$ of the HF, H_2O, HCl, H_2S, and PH_3 molecules present also a plateau-shape evolution which saturates for $N = 60$, 40, 80, 90, and 90, respectively. H_2CS presents a similar behavior as H_2CO and its $\langle\alpha\rangle$ saturates for $N = 70$. The situation is totally different for β_{vec} because the state contributions can be either positive or negative. Nevertheless, the

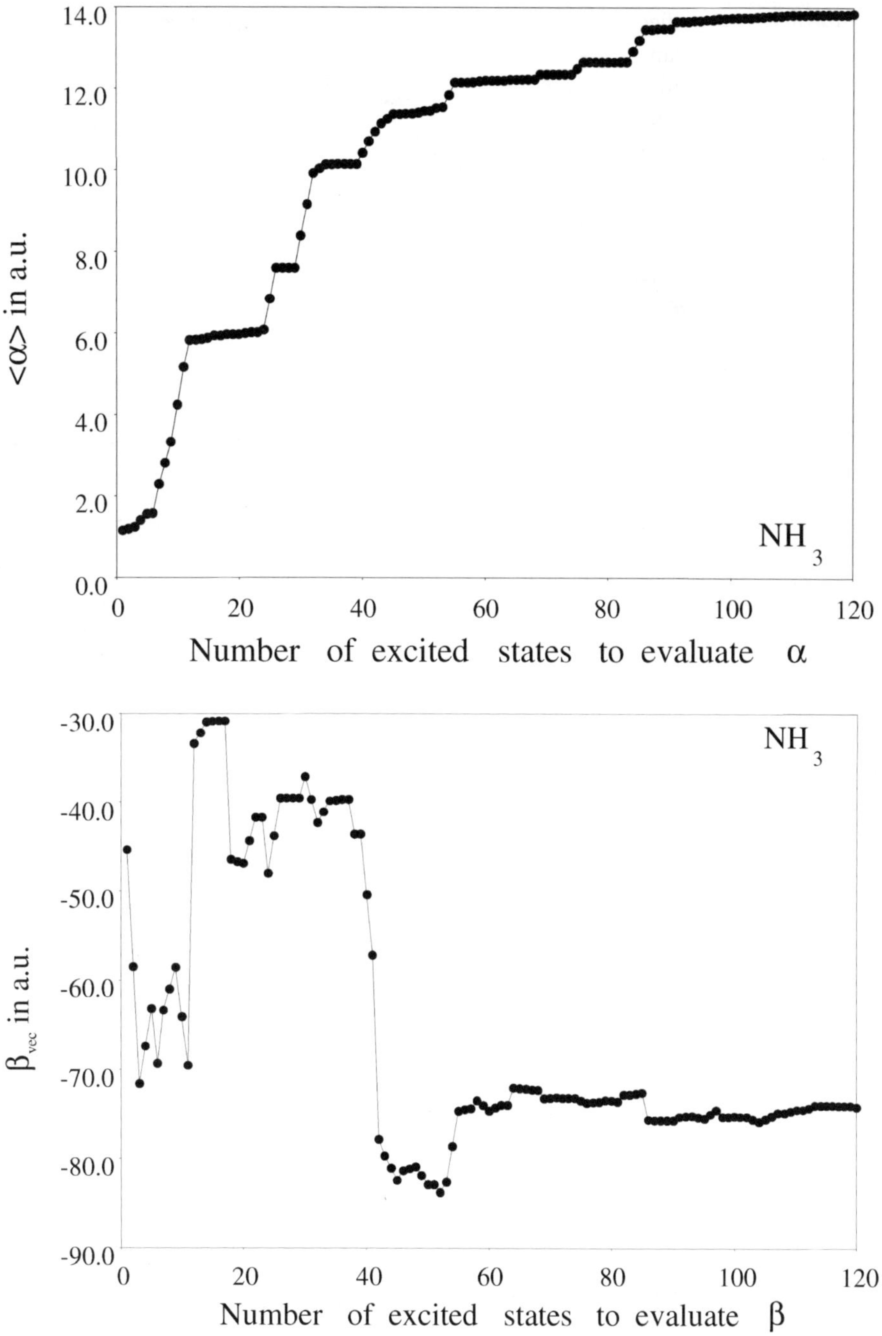

*Fig. 4. Evolution with the number of excited states considered in the SOS expressions of the $\langle\alpha(0;0)\rangle$ (a) and $\beta_{vec}(0;0,0)$ (b) of NH_3 evaluated at the SOS/CIS/6-311G** + SPD level of approximation. All the values are given in a.u..*

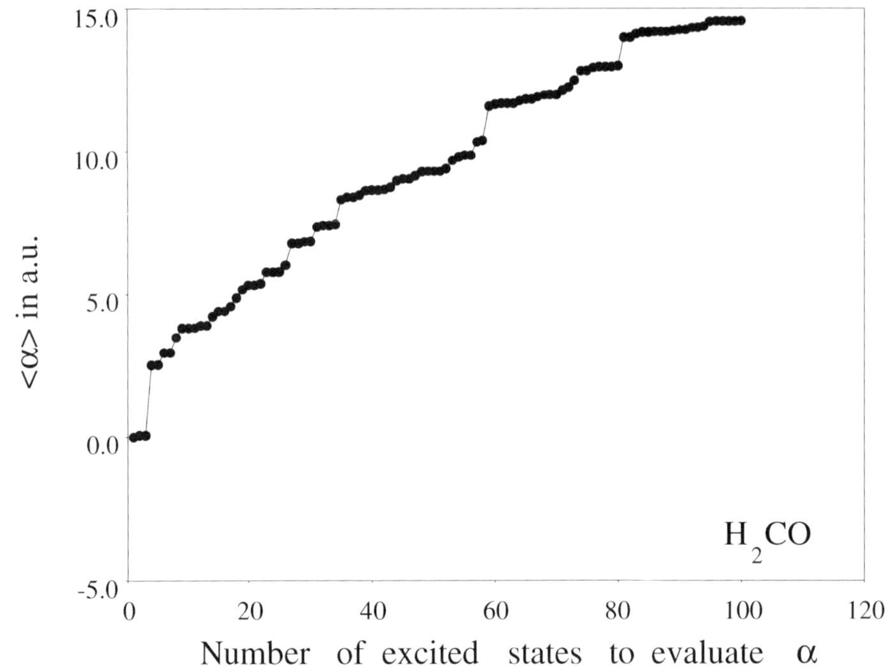

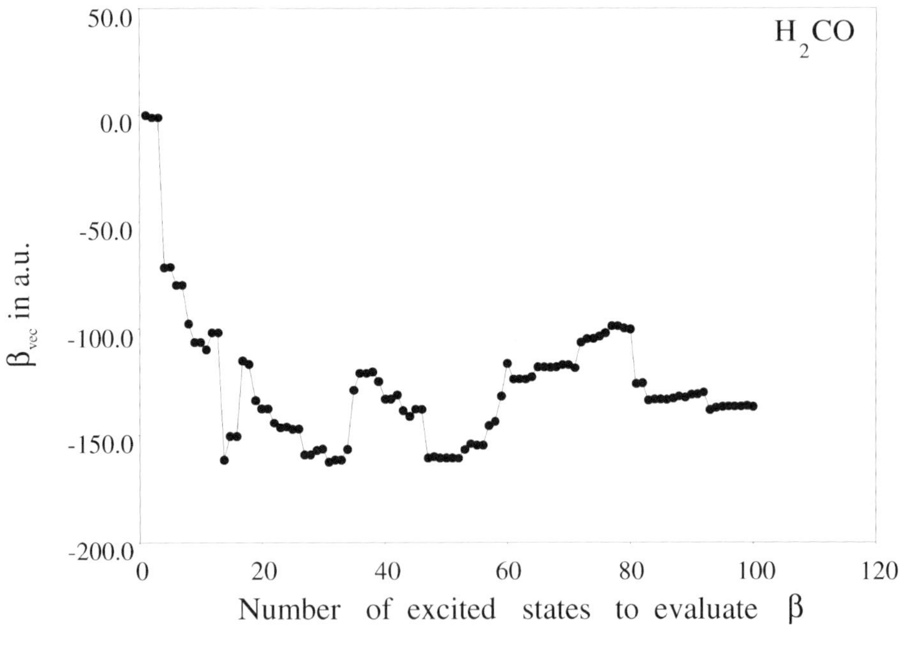

Fig. 5. Evolution with the number of excited states considered in the SOS expressions of the $\langle\alpha(0;0)\rangle$ *(a) and* $\beta_{vec}(0;0,0)$ *(b) of* H_2CO *evaluated at the SOS/CIS/6-311G* + SPD level of approximation. All the values are given in a.u..*

convergence behaviors are similar. Indeed, for HF, H_2O, NH_3, HCl, H_2S, and PH_3, β_{vec} attains a plateau for N = 60, 40, 60, 40, 40, and 60, respectively. This contrasts with H_2CO and H_2CS for which there remain scattered higher-energy contributing states.

Since for β there exist mixed contributions (m $\neq$ n Eq. (5)), a model which follows the evolution of β with the number of lowest-energy excited states cannot highlight the true state contributions. Dirk and Kuzik [55] have proposed the *missing-state analysis* which determines the relative importance of states by considering the effect of their absence. This relative importance is defined as

$$\sigma_n = \frac{\beta_{tot} - \beta_{-n}}{\beta_{tot}} \tag{13}$$

where β_{tot} has been evaluated with the *full* SOS expression which is limited to a maximum of 120 (or 100 for H_2CO and H_2CS) excited states. For evaluating β_{-n} the nth excited state is missing. If $0 < \sigma_n < 1$, σ_n determines the relative importance of state n, while $\sigma_n < 0$ indicates a negative contribution to β. Moreover, note that when $\sigma_n > 1$, β would have the opposite sign if state n is not present. Dirk and Kuzik have further refined their model in order to describe the couplings between states which lead to large mixed contributions. We illustrate this analysis scheme for three representative molecules (NH_3, HCl, and H_2CO) in Figures 6–8. Obviously, there are less than 10–20 excited states which need to be included. However, one can distinguish between three categories of molecules. In the first category (HF, H_2O, and NH_3), there are a few

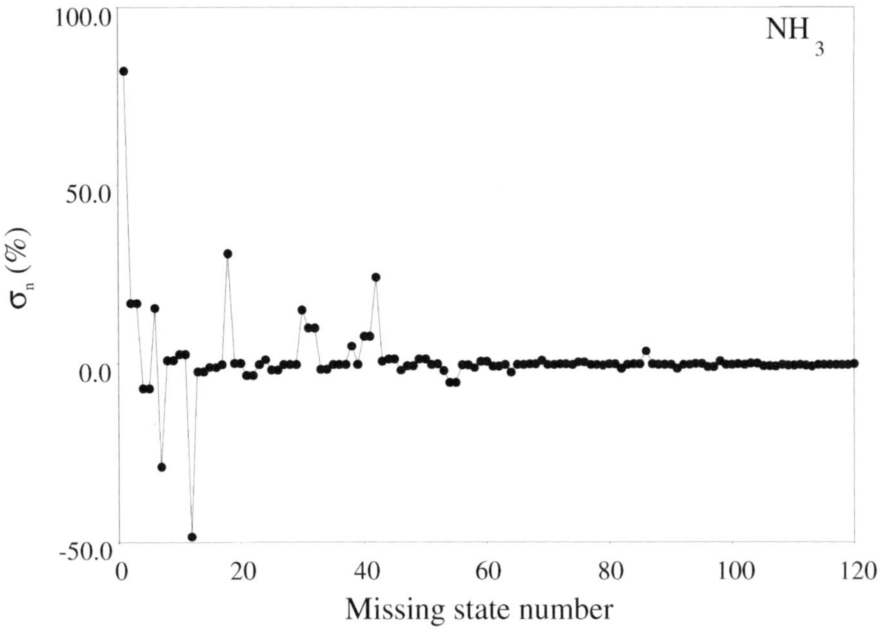

*Fig. 6. Missing state analysis for $\beta_{vec}(0; 0, 0)$ of NH_3 evaluated at the SOS/CIS/6-311G** + SPD level of approximation.*

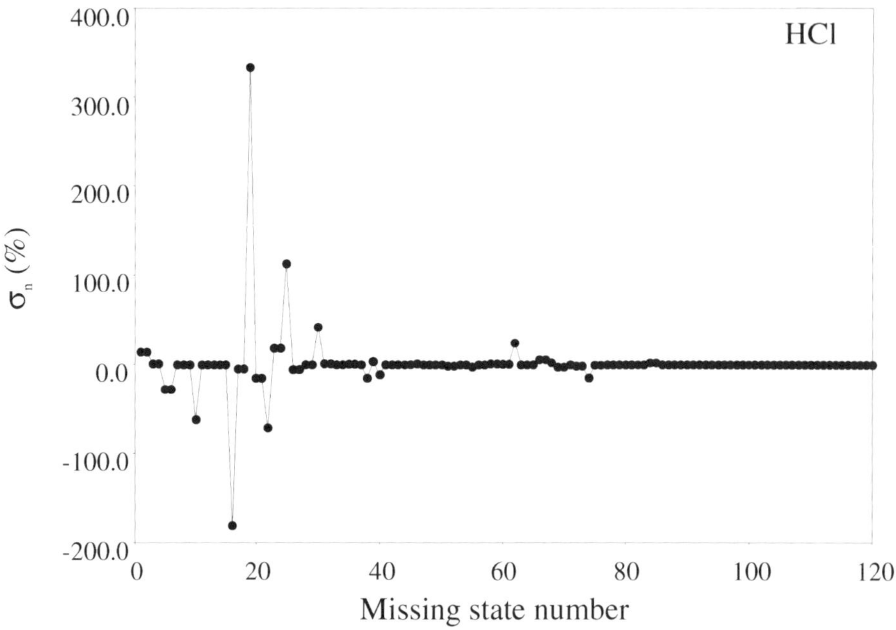

*Fig. 7. Missing state analysis for $\beta_{vec}(0; 0, 0)$ of HCl evaluated at the SOS/CIS/6-311G** + SPD level of approximation.*

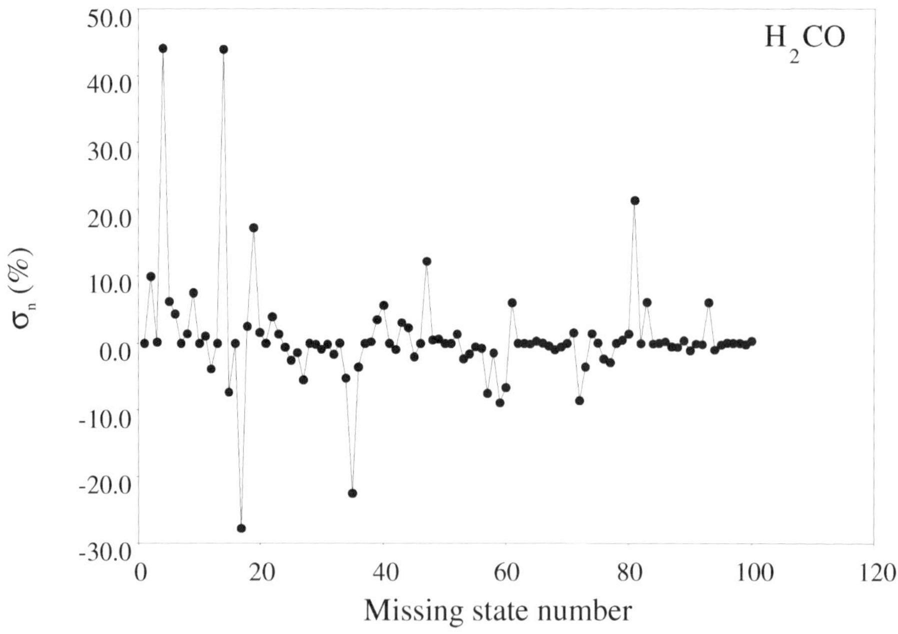

Fig. 8. Missing state analysis for $\beta_{vec}(0; 0, 0)$ of H_2CO evaluated at the SOS/CIS/6-311G + SPD level of approximation.*

dominant low-energy excited states associated with σ_n ranging between -1.0 and $+1.0$. For the second category associated with molecules containing third-row atoms (HCl, H_2S, and PH_3), there are 3, 2, and 10 low-energy states with $|\sigma_n|$ larger than unity as well as a dozen of important low-energy states with $1.0 \geqslant |\sigma_n| \geqslant 0.2$. On the basis of the presence/absence of third-row atoms, H_2CO and H_2CS could enter in these two classes but these two molecules are different due to non-negligible contributions of a few scattered higher-energy excited states.

In any case, our analysis has shown that for these eight small molecules more than 2 or 3 dominant excited states dictate the magnitude of β. The present behavior is clearly different from these of push-pull conjugated molecules. It is however cautious to note that truncation procedures which are validated by the fact that α or β value seem to have converged over a large range of N cannot guarantee no important high-energy excited state is missing. Similarly, one can never be sure that multiply-excited states are not of crucial importance for the ground state characterization.

3.4. *Linear and nonlinear responses of the three isomers of nitroaniline*

In order to put the *ab initio* SOS/CIS approach to test, the archetypical [56–65] donor-acceptor para-nitroaniline (*p*-NA) and its ortho- and meta isomers (*o*-NA and *m*-NA) are investigated. The CI space has been truncated to the 60 lowest-energy excited states and the split-valence 6-31G basis set has been used. For *p*-NA, at this SOS/CIS level, $\langle\alpha(0;0)\rangle = 40.0$ a.u. and $\beta_{vec}(0;0,0) = 1585$ a.u.. Sim and coworkers [62] have obtained at the CPHF(FF/MP2) level using a double-ζ quality basis set augmented with diffuse polarization functions [3s2p2d|2s] $\langle\alpha(0;0)\rangle = 97.8$ a.u. (106.9 a.u.) and $\beta_{vec}(0;0,0) = 1069$ a.u. (1934 a.u.). The RPA (equivalent to the CPHF/TDHF approach) and MCSCF $\beta_{vec}(0;0,0)$ values due to Mikkelsen *et al.* [64] are equal to 947 a.u. and 1373 a.u. whereas for $\langle\alpha(0;0)\rangle$, they obtained 93.8 a.u. and 93.0 a.u., respectively. The large underestimation of our $\langle\alpha(0;0)\rangle$ estimate with respect to other *ab initio* works has to be related to the basis set which can only reproduce with satisfaction the component of α along the molecular axis. On the other hand, our present SOS/CIS $\beta_{vec}(0;0,0)$ estimate is 50–60% larger than the SCF, 15% larger than the MCSCF but 20% smaller than the FF/MP2 values. This good agreement with correlated treatments does not prevent our result to remain one order of magnitude smaller than most of the experimental values. At $\lambda = 1064$ nm, Teng and Garito [56] have measured for *p*-NA in 1-, 4-dioxane a $\beta_{vec}(-2\omega;\omega,\omega)$ value of 11735 a.u. whereas in gas phase, the SOS/CIS $\beta_{vec}(-2\omega;\omega,\omega) = 2183$ a.u., the scaled FF/MP2 $\beta_{vec}(-2\omega;\omega,\omega)$ value due to Sim *et al.* [62] $= 2778$ a.u., and the two-state MCSCF $\beta_{vec}(-2\omega;\omega,\omega)$ value due to Mikkelsen *et al.* [64] $= 2275$ a.u.. At the same wavelength but in chloroform and acetone, Stähelin *et al.* [60] obtained $\beta_{vec}(-2\omega;\omega,\omega) = 11666$ a.u. and 17984 a.u., respectively. Obviously, on the basis of very large solvation effects [61,63–65], one cannot compare further our gas-phase theoretical values with the liquid-phase experimental data. On the other hand our SOS/CIS values are in good agreement with more elaborated techniques including the effects of electron correlation.

The SOS/CIS results point out that *p*-NA presents the largest β value among the 3 isomers; *m*-NA and *o*-NA having $\beta_{vec}(0;0,0) = 747$ a.u. and 625 a.u., respectively. If the largest value associated with p-NA does not suffer from any dispute, both the experimental and theoretical results of Oudar and Chemla [6] as well as the recent experimental investigation of Cheng *et al.* [59] propose that o-NA has a larger β value than m-NA; the $\beta(o\text{-NA})/\beta(m\text{-NA})$ ratio ranging with the experimental conditions between 1.3 and 1.8 whereas our estimates for the static and dynamic ($\lambda = 1064$ nm) are 0.84 and 0.91. Moreover, by adopting the SOS/CIS scheme, Dirk *et al.* [57] have calculated $\beta(o\text{-NA})/\beta(m\text{-NA})$ ratios smaller or larger than unity according to the choice of the parameters of their Pariser Parr Pople Hamiltonian. For pursuing the theoretical comparison between o-NA and m-NA, it is necessary to investigate the effects of including higher-order excitations with a larger basis set as well as the magnitude of the solvation effects. Oudar and Chemla have associated this ordering with the larger charge transfer contribution in o-NA than in m-NA; p-NA presenting the largest charge transfer contribution to β.

The second aspect of our nitroaniline investigation concerns the analysis of $\langle\alpha\rangle$ and β_{vec} in terms of the excited states contributions. The N-dependence of $\langle\alpha(0; 0)\rangle$ is very similar for the three compounds and saturates for N = 20 (Fig. 9). Moreover, at most 10 excited states contribute significantly. For $\beta_{\text{vec}}(0; 0, 0)$, the σ_n graphs (Figs. 10–12) exhibit some differences among the isomers. For p-NA, 3 states present a dominant σ_n: state 3 ($\sigma_n = 0.96$), state 6 ($\sigma_n = 0.62$) and state 13 ($\sigma_n = -0.47$). For the two other

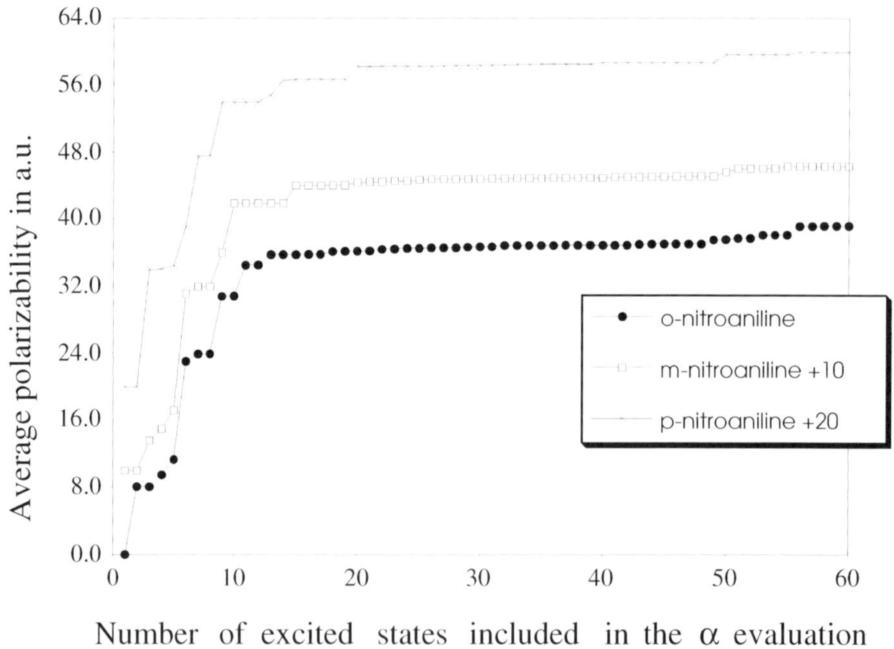

Fig. 9. Evolution with the number of excited states considered in the SOS expressions of the ⟨α(0; 0)⟩ of the three isomers of nitroaniline evaluated at the SOS/CIS/6-31G level of approximation. All the values are given in a.u..

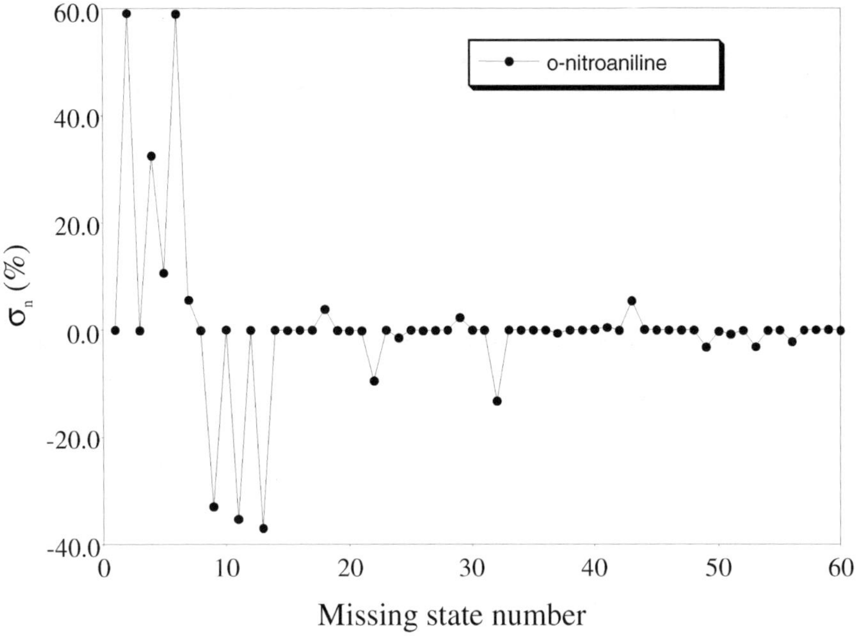

Fig. 10. SOS/CIS/6-31G Missing state analysis for $\beta_{vec}(0; 0, 0)$ of o-NA

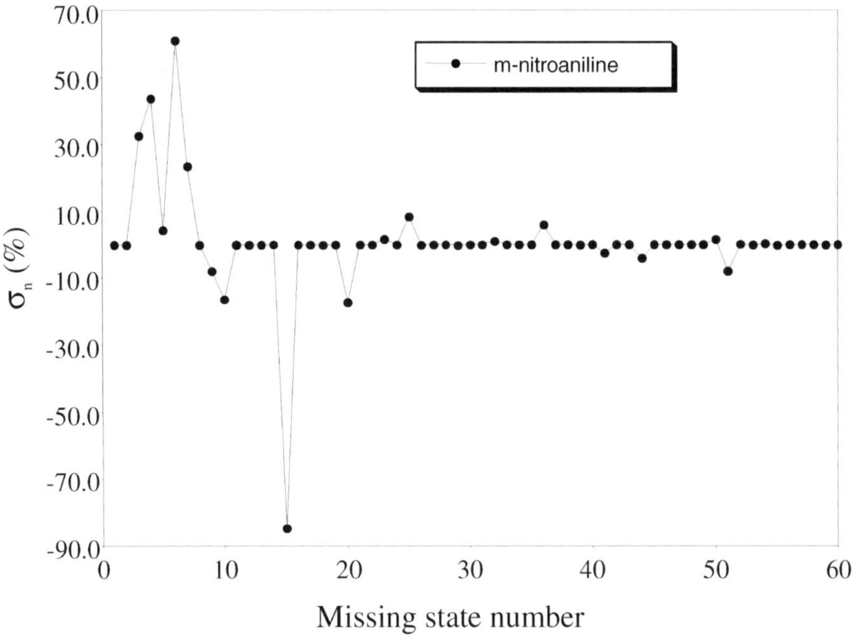

Fig. 11. SOS/CIS/6-31G Missing state analysis for $\beta_{vec}(0; 0, 0)$ of m-NA

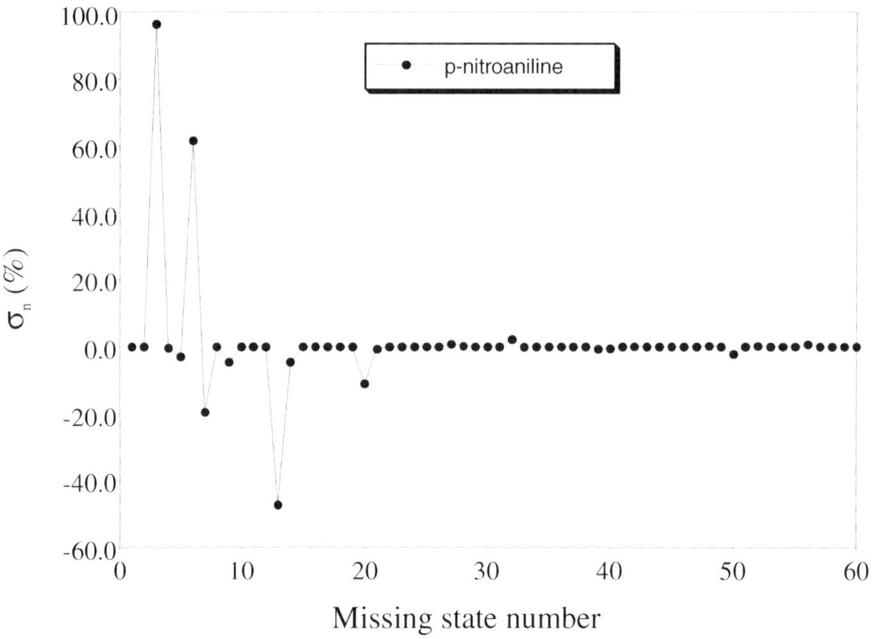

Fig. 12. SOS/CIS/6-31G Missing state analysis for $\beta_{vec}(0; 0, 0)$ of p-NA

isomers, there is a larger number of important excited states with smaller $|\sigma_n|$ values (mainly for *o*-NA) but in any case, only a few excited states present dominant contributions and they are all located in the lower-energy region. Such missing state picture where 3 or more excited states are dominant show also that the 2- or 3-state models have a rather limited domain of application [6].

4. Conclusions and Perspectives

For the first time, systematic *ab initio* SOS/CI calculations of the static and dynamic polarizabilities and first hyperpolarizabilities have been performed. This initial investigation has been carried out at the CI-singles level of approximation on a set of eight small molecules as well as on the three isomers of nitroaniline. By employing the 6-311G** + SPD basis set, shown to provide results close to the basis set limit, we have pointed out for the eight small molecules that i) the average polarizability is larger than the CPHF/TDHF values, smaller than the MP2, CCSD, CCSD(T), ... correlated values and also less than 20% smaller than the experimental values; ii) the vector component of the first hyperpolarizability, $\beta_{//}$, is in most cases overestimated and higher-order electron correlation corrections are required to meet agreement with the experimental quantities; iii) there are a lot of contributing states to both $\langle\alpha\rangle$ and $\beta_{\rm vec}$ but convergence with respect to the number of states is generally achieved within less than 100 configurations. In addition, we have shown that the SOS/CIS/6-31G approach gives for *p*-NA a $\beta_{vec}(0; 0, 0)$ estimate which is bracketted by the MCSCF and FF/MP2 results. Although several improvements are desired, our approach presents some advantages with respect to the CPHF/TDHF schemes but also with respect to various correlated schemes in the sense that frequency-dependent quantities can be obtained with the same ease for all the wavelength domain (large and small λ, below or above the resonances). Moreover, additional frequency-dependent data can be obtained at a very low cost since one has just to run the computationally low-demanding $\alpha\beta$-SOS [19] code. The post Hartree-Fock nature of our CIS/SOS scheme would also enable an easy extension for dealing with solvation effects on the NLO properties [63–66].

The next steps in our SOS/CI approach concern the inclusion of higher-order electron correlation corrections and particularly the inclusion of the double excitations [7–8,67] to describe the ground and excited states. In view of calculating the second hyperpolarizability these double excitations are of crucial importance [68]. Similar approaches have already been followed at various semiempirical levels [69–70] but *ab initio* approaches have the advantage that they can systematically be improved through enlarging the basis set and including electron correlation up to higher orders. Our further developments will tackle the computational aspects of CI matrix resolution and consider suitable schemes such as the frequency-dependent moment method [71].

Note added in proofs

Since the writing of this manuscript a number of papers have appeared where α and β have been evaluated by including electron correlation corrections either at DFT (S.J.A. van Gisbergen, J.G. Snijders, and E.J. Baerends, J. Chem. Phys. **109**, 10657 (1998); A.J. Cohen, N.C. Handy and D.J. Tozer, Chem. Phys. Lett. **303**, 391 (1999)] or conventional correlated [O. Christiansen, C. Hättig and J. Gauss, J. Chem. Phys. **109**, 4745 (1998); H. Larsen, J. Olsen, C. Hättig, P. Jørgensen, O. Christiansen and J. Gauss, J. Chem. Phys. **111**, 1917 (1999); T. Kobayashi, K. Sasagne, F. Aiga, and K. Yamaguchi, J. Chem. Phys **111**, 842 (1999)] levels. A new experimental work has also been reported [P. Kaatz, E.A. Donley, and D.P. Shelton, J. Chem. Phys. **108**, 849 (1998)].

Acknowledgements

The authors are indebted to D. Jacquemin for fruitful discussions. This study is the result of a scientific cooperation established and supported by the Bulgarian Academy of Sciences, the Belgian National Fund for Scientific Research (FNRS) and the Commissariat Général aux Relations Internationales de la Communauté Française de Belgique (CGRI). The calculations have been performed on the IBM RS6000 cluster of the Namur Scientific Computing Facility (Namur-SCF). The authors gratefully acknowledge the financial support of the FNRS-FRFC, the 'Loterie Nationale' for the convention No. 2.4519.97. This work is also financially supported by the Bulgarian Fund for Scientific Research under the project X-412. One of us (MS) is grateful to the Facultés Universitaires Notre-Dame de la Paix, for financial support during her stay in the Laboratoire de Chimie Théorique Appliquée.

References

1. Int. J. Quantum Chem. special issue on Molecular Nonlinear Optics **43** (1) (1992) edited by M.A. Ratner and P.O. Löwdin.
2. Chemical Reviews, thematic issue on Optical Nonlinearities in Chemistry **94** (1) (1994) edited by D.A. Burland and J. Michl.
3. B.J. Orr and J.F. Ward, Mol. Phys. **20**, 513 (1971); D.M. Bishop, J. Chem. Phys. *100*, 6535 (1994).
4. J.L. Oudar, J. Chem. Phys. **67**, 1446 (1977).
5. J.L. Oudar and D.S. Chemla, J. Chem. Phys. **66**, 2664 (1977).
6. D.R. Kanis, M.A. Ratner, and T. Marks, Chem. Rev. **94**, 195 (1994).
7. J.O. Morley, J. Phys. Chem. **99**, 10166 (1995).
8. I.D.L. Albert, T.J. Marks, and M.A. Ratner in *Nonlinear Optical Materials. Theory and Modeling*, edited by S.P. Karna and A.T. Yeates, ACS Symposium Series 628 (ACS, Washington, DC, 1996), 116.
9. T. Tsunekawa and K. Yamaguchi, J. Phys. Chem. **96**, 10268 (1992).
10. M. Tomonari, N. Ookubo, T. Takada, M. W. Feyereisen, and J. Almlöf, Chem. Phys. Lett. **203**, 603–610 (1993).
11. M. Tomonari, N. Ookubo, and T. Takada, Chem. Phys. Lett. **215**, 45 (1993).
12. M. Tomonari, N. Ookubo, and T. Takada, J. Mol. Struct. (Theochem) **117**, 221 (1994).
13. M. Tomonari, N. Ookubo, and T. Takada, Chem. Phys. Lett. **236**, 475 (1995).
14. A. Dalgarno and J.M. McNamee, J. Chem. Phys. **35**, 1517 (1961); P.W. Langhoff, M. Karplus, and R.P. Hurst, J. Chem. Phys. **44**, 505 (1966); T.C. Caves and M. Karplus, J. Chem. Phys. **50**, 3649 (1969).
15. J.M. André, C. Barbier, V.P. Bodart, and J. Delhalle, in *Nonlinear Optical Properties of Organic Molecules and Crystals*, edited by D.S. Chemla and J. Zyss (Academic, San Diego, 1987), Vol. 2, 137; J.G. Fripiat, C. Barbier, V.P. Bodart, and J.M. André, J. Comp. Chem. **7**, 756 (1986).
16. B. Champagne, Chem. Phys. Lett. **261**, 57 (1996); B. Kirtman and B. Champagne, Int. Rev. Phys. Chem. **16**, 389 (1997). D.M. Bishop, Adv. Chem. Phys. **104**, 1 (1998).
17. J. B. Foresman, M. Head-Gordon, J.A. Pople, and M.J. Frisch, J. Phys. Chem. **96**, 135 (1992).
18. M.J. Frisch, G.W. Trucks, H.B. Schlegel, P.M.W. Gill, B.G. Johnson, M.A. Robb, J.R. Cheeseman, T. Keith, G.A. Petersson, J.A. Montgomery, K. Raghavachari, M.A. Al-Laham, V.G. Zakrzewski, J.V. Ortiz, J.B. Foresman, J. Cioslowski, B.B. Stefanov, A. Nanayakkara, M. Challacombe, C.Y. Peng, P.Y. Ayala, W. Chen, M.W. Wong, J.L. Andres, E.S. Replogle, R. Gomperts, R.L. Martin, D.J. Fox, J.S. Binkley, D.J. DeFrees, J. Baber, J.P. Stewart, M. Head-Gordon, C. Gonzalez, and J.A. Pople, (Gaussian Inc., Pittsburgh, PA, 1995).
19. M. Spassova and B. Champagne, $\alpha\beta$-SOS, a Code for Evaluating Static and Dynamic Polarizabilities and Hyperpolarizabilities (Namur & Sofia, 1996).
20. H. Sekino and R.J. Bartlett, J. Chem. Phys. **85**, 976 (1986); S.P. Karna and M. Dupuis, J. Comp. Chem. **12**, 487 (1991).
21. J. Linderberg and Y. Öhrn *Propagators in Quantum Chemistry* (Academic Press, New York, 1973).
22. J. Oddershede, Adv. Quantum Chem. **11**, 257 (1978).
23. M.W. Schmidt, K.K. Baldridge, J.A. Boatz, S.T. Elbert, M.S. Gordon, J.H. Jensen, S. Koseki, N. Matsunaga, K.A. Nguyen, S.J. Su, T.L. Windus, M. Dupuis, and J.A. Montgomery, J. Comput. Chem. **14**, 1347 (1993).
24. W.J. Hehre, R.F. Stewart, and J.A. Pople, J. Chem. Phys. **51**, 2657 (1969); W.J. Hehre, R. Ditchfield, R.F. Stewart, and J.A. Pople, J. Chem. Phys. **52**, 2769 (1970).
25. W.J. Hehre, R. Ditchfield, and J.A. Pople, J. Chem. Phys. **56**, 2257 (1972).
26. P.C. Hariharan and J.A. Pople, Chem. Phys. Letters **66**, 217 (1972).
27. P.C. Hariharan and J.A. Pople, Theor. Chim. Acta **28**, 213 (1973).
28. R. Krishnan, J.S. Binkley, R. Seeger, and J. Pople, J. Chem. Phys. **72**, 650 (1980); A.D. McLean and G.S. Chandler, J. Chem. Phys. **72**, 5639 (1980).
29. G.J.B. Hurst, M. Dupuis, and E. Clementi, J. Chem. Phys. **89**, 385 (1988); J. Perez and M. Dupuis, J. Phys. Chem. **95**, 6525 (1991).
30. A.J. Sadlej, Coll. Czech. Chem. Commun. **53**, 1995 (1988); A.J. Sadlej, Theoret. Chim. Acta **79**, 123 (1991).

31. R.A. Kendall, T.H. Dunning, Jr, and R.J. Harrison, J. Chem. Phys. **96**, 6796 (1992).
32. D.E. Woon and T.H. Dunning, Jr, J. Chem. Phys. **98**, 1358 (1993).
33. *Spectroscopic Data Relative to Diatomic Molecules* (Pergamon, New York, 1970).
34. *Handbook of Chemistry and Physics*, Vol.51, edited by R.C. Weast (CRC Press Inc., Cleveland, Ohio, 1971), F157.
35. W.J. Hehre, L. Radom, P.v.R. Schleyer, and J.A. Pople, *Ab Initio Molecular Orbital Theory* (Wiley, New York, 1986) p.139.
36. M. Spassova, unpublished results.
37. H. Sekino and R.J. Bartlett, J. Chem. Phys. **98**, 3022 (1993).
38. G. Maroulis, J. Chem. Phys. **94**, 1182 (1991).
39. J.E. Rice and N.C. Handy, Int. J. Quant. Chem. **43**, 91 (1992).
40. B.L. Hammond and J.E. Rice, J. Chem. Phys. **97**, 1138 (1992).
41. G.J.M. Velders and D. Feil, Theor. Chim. Acta **86**, 391 (1993).
42. S. Yamada, M. Nakano, I. Shigemoto, S. Kiribayashi, and K. Yamaguchi, Chem. Phys. Lett. **267**, 445 (1997).
43. H.J. Werner and W. Meyer, Mol. Phys. **31**, 855 (1976).
44. R.J. Bartlett and G.D. Purvis III, Phys. Rev. A **20**, 1313 (1979).
45. R.J. Bartlett and G.D. Purvis III, Phys. Rev. A **23**, 1594 (1981).
46. G.H.F. Diercksen and A.J. Sadlej, J. Chem. Phys. **75**, 1253 (1981).
47. M.A. Spackman, J. Phys. Chem. **93**, 7594 (1989).
48. D.P. Chong and S.R. Langhoff, J. Chem. Phys. **93**, 570 (1990).
49. J.E. Rice and N.C. Handy, J. Chem. Phys. **94**, 4959 (1991).
50. D.E. Woon and T.H. Dunning, Jr, J. Chem. Phys. **100**, 2975 (1994).
51. H. Sekino and R.J. Bartlett, Chem. Phys. Lett. **234**, 87 (1995).
52. Y. Luo, H. Ågren, P. Jørgensen, and K.V. Mikkelsen, Adv. Quantum Chem. **26**, 165 (1995).
53. J.G. Guan, P. Duffy, J.T. Carter, D.P. Chong, K.C. Casida, M.E. Casida, and M. Wrinn, J. Chem. Phys. **98**, 4753(1993). A.M. Lee and S.M. Colwell, J. Chem. Phys. **101**, 9704 (1994).
54. B.I. Dunlap and S.P. Karna, in *Nonlinear Optical Materials Theory and Modeling,* edited by S.P. Karna and A.T. Yeates, ACS Symposium Series 628, (ACS, Washington, DC, 1996), p. 164.
55. C.W. Dirk and M.G. Kuzyk, Phys. Rev. A **39**, 1219 (1989).
56. C.C. Teng and A.F. Garito, Phys. Rev. B **28**, 6766 (1983).
57. C.W. Dirk, R.J. Twieg, and G. Wagnière, J. Am. Chem. Soc. **108**, 5387 (1986).
58. S.P. Karna, P.N. Prasad, and M. Dupuis, J. Chem. Phys. **94**, 1171 (1991).
59. L.T. Cheng, W. Tam, S.H. Stevenson, G.R. Meredith, G. Rikken, and S.R. Marder, J. Phys. Chem. **95**, 10631 (1991).
60. M. Stähelin, D.M. Burland, and J.E. Rice, Chem. Phys. Lett. **191**, 245 (1992).
61. A. Willetts, J.E. Rice, D.M. Burland, and D.P. Shelton, J. Chem. Phys. **97**, 7590 (1992).
62. F. Sim, S. Chin, M. Dupuis, and J. E. Rice, J. Phys. Chem. **97,** 1158-1163 (1993).
63. J. Yu and M.C. Zerner, J. Chem. Phys. **100**, 7487 (1994).
64. K.V. Mikkelsen, Y. Luo, H. Ågren, and P. Jørgensen, J. Chem. Phys. **100**, 8240 (1994).
65. C. Dehu, F. Meyers, E. Hendrickx, K. Clays, A. Persoons, S.R. Marder, and J.L. Brédas, J. Am. Chem. Soc. **117**, 10127 (1995).
66. R. Cammi, M. Cossi, B. Mennucci, and J. Tomasi, J. Chem. Phys. **105**, 10556 (1996).
67. I. Kanev, J. Chim. Phys. **79**, 397 (1982).
68. D. Jacquemin, B. Champagne, and J.M. André, Chem. Phys. **197**, 107 (1995).
69. B.M. Pierce, J. Chem. Phys. **89**, 791 (1989).
70. I.D.L. Albert, J.O. Morley, and D. Pugh, J. Chem. Phys. **102**, 237 (1995).
71. S. Iwata, Chem. Phys. Lett. **102**, 544 (1983); T. Inoue and S. Iwata, Chem. Phys. Lett. **167**, 566 (1990); Y. Nomura, S. Miura, M. Fukunaga, S. Narita, and T.I. Shibuya, J. Chem. Phys. **106**, 3243 (1997).

Correlation Energies for Diatomic Molecules: A Re-Evaluation of the Empirical Estimates for the N_2, CO, BF and NO^+ Systems

H.M. Quiney[a], D. Moncrieff[b] and S. Wilson[c]*

[a] *School of Chemistry, University of Melbourne, Parkville, Victoria 3052, Australia*

[b] *Supercomputer Computations Research Institute, Florida State University, Tallahassee, Florida, U.S.A.*

[c] *Rutherford Appleton Laboratory, Chilton, Oxfordshire OX11 0QX, England*

Abstract

Many-body treatments of electron correlation effects in small molecules, when implemented in a high performance computing environment and exploiting large and flexible basis sets, can lead to results of high precision. The results of such calculations are usually assessed by comparison with empirical estimates of the correlation energy and it is, therefore, important that these estimates be reliable. The empirical correlation energies for the N_2, CO and BF molecules and the NO^+ ion are re-evaluated by making use of recent Hartree-Fock results of μhartree accuracy and new estimates of the relativistic energy components which are based on four component atomic and molecular relativistic electronic structure calculations and take account of the non-additivity of correlation and relativistic effects. A definition of the *post*-Hartree-Fock energy is proposed.

1. Introduction

Recent increases in computer power and developments in the software available for quantum chemical calculations have made possible precise calculations of non-relativistic ground-state electronic energies of diatomic molecules in the independent-particle approximation. Finite-difference calculations of molecular electronic structure may be regarded as numerical standards, in much the same way as atomic calculations have served the development of other techniques of approximation, such as the use of finite basis set expansions. Careful numerical analysis of the finite-difference approximation of these model systems yields Hartree-Fock energies with errors less than 1 μhartree. The electronic energy differences which accompany chemical change, however, are a small fraction of the total electronic energy of a system, and comparable in magnitude to the error which is introduced by invoking the independent-particle approximation. In order to remedy the deficiencies of this model, many-body theories have been developed and implemented in quantum chemical studies of electronic structure [1–6].

Following Löwdin [7], we define the electron correlation energy, E_{corr}, to be

$$E_{corr} = E_{NR} - E_{HF} \tag{1}$$

where E_{NR} is the total non-relativistic electronic energy of the system, and E_{HF} is the electronic energy in the Hartree-Fock approximation. Although most work in quantum chemistry has been directed to the evaluation of the three quantities which appear in Eq.

*To whom correspondence should be addressed.

A. Hernández-Laguna et al. (eds.), Quantum Systems in Chemistry and Physics, Vol. 1: Basic Problems and Model Systems, 127–142.

(1), none of them are directly measurable by any experiment, either singly, or in combination with one another. For atoms and diatomic molecules we may calculate E_{HF} to high accuracy, and methods exist for the exact determination of E_{NR} in helium-like systems. In all other cases, we must deduce the value of E_{NR} from experimental data and estimates of the physical effects not included in the spin-independent non-relativistic theory.

The numerical determination of E_{corr} by the use of many-body theory is a formidable task, and estimates of it based on E_{NR} and E_{HF} serve as important benchmarks for the development of methods for calculating electron correlation effects. The purpose of this work is to obtain improved estimates of E_{NR} by combining the leading-order relativistic and many-body effects which have been omitted in Eq. (1) with experimentally determined values of the total electronic energy, and precise values of E_{HF}. We then obtain empirical estimates of E_{corr} for the diatomic species N_2, CO, BF, and NO^+ using E_{NR} and E_{HF} and the definition of E_{corr} in Eq. (1).

The plan of this paper is as follows:- In section 2, the basic experimental data required in the re-evaluation of the empirical correlation energies of the N_2 CO, BF and NO^+ molecules are collected. The essential theoretical ingredients of our re-determination are given in section 3 including new fully relativistic calculations including the frequency independent Breit interaction and electron correlation effects described by second order diagrammatic perturbation theory for the Be-like ions B^+, C^{2+}, N^{3+}, O^{4+} and F^{5+}. Energies derived from the experimental and theoretical data presented in sections 2 and 3 are presented in section 4, beginning with the total molecular ground state energies and then the post-Hartree-Fock energies, the relativistic energies and then the total molecular non-relativistic ground state energies, before giving the empirical correlation energies. A comparison of the empirical correlation energies with recent high precision calculations is made in section 5. Section 6 contains our concluding remarks.

2. Experimental Data

In this section, we collect the experimental data which are required for the determination of the empirical electron correlation energy. In Table 1 the total atomic energies are

Table 1 *Total atomic energies*[1,2]

Atom, X	E_X/eV	E_X/hartree
B	670.98452	24.658214
C	1030.1057	37.85567
N^+	1471.5321	54.07778
N	1486.0662	54.61189
O	2043.8032	75.10823
F	2715.8646	99.80615

[1]Calculated from 'Ionization potentials of atoms and atomic ions' in the CRC Handbook of Chemistry and Physics [8]
[2]Conversion factor: 1 hartree = 27.2114 eV

displayed. These data were derived from the ionization potentials of the relevant atoms and atomic ions given in the CRC Handbook of Chemistry and Physics [8]. The molecular dissociation energies for N_2, CO, BF and NO^+ taken from the compendium of Huber and Herzberg [9] are collected in Table 2 whilst, in Table 3, the corresponding fundamental frequencies of vibration, also taken from the work of Huber and Herzberg [9], are displayed. Anharmonic effects are not significant in the present study in which the most significant experimental errors are associated with the ionization potentials used to determine the total atomic energies given in Table 1.

3. Theoretical Data

Two key pieces of theoretical data are required to obtain an empirical estimate of the correlation energy from the experimental data collected in the preceding section: the total molecular Hartree-Fock energy and the relativistic 'corrections'. It is implicit in the definition of the correlation energy presented in Eq. (1) that the total electronic energy of a given molecule, E_T, may be divided into three constituent parts,

$$E_T = E_{HF} + E_{corr} + E_R, \tag{2}$$

in which E_{HF} is the non-relativistic Hartee-Fock energy, E_{corr} is the correlation energy, and E_R is a residual energy correction usually attributed to the effects of relativity.

The partition (2) is quite arbitrary, since one could with equal validity absorb all one-body relativistic effects of order $(Z\alpha)^2$ into the mean-field energy of the Dirac-Hartree-

Table 2 *Molecular dissociation energies*[1,2]

Molecule	D_0/eV	$D_0/\text{hartree}$
N_2	9.759_4	0.35865_4
CO	11.09_2	0.4076_2
BF	7.8_1	0.2870_1
NO^2	10.850_6	0.39875_2

[1] From Huber and Herzberg [9]
[2] Conversion factor: 1 hartree = 27.2114 eV

Table 3 *Fundamental frequencies of vibration*[1,2]

Molecule	ω_e/cm^{-1}	$\frac{1}{2}\hbar\omega_e/\text{hartree}$
N_2	2358.57	0.01074643
CO	2169.81	0.00988640
BF	1402.1_3	0.00638857
NO^+	2376.42	0.01082776

[1] From Huber and Herzberg [9]
[2] Conversion factor: 1 hartree = 219474.63 cm^{-1}

Fock-Breit model, E_{DHFB}, and define all other energy corrections to this order of relativistic quantum electrodynamics, E_{MB}, as a many-body effects, so that

$$E_T = E_{DHFB} + E_{MB}. \tag{3}$$

This definition of the relativistic many-body energy involves both electrostatic and magnetic interactions between pairs of electrons. Effects due to the Lamb shift and errors associated with the non-additivity of the electronic energy and the energy due to nuclear motion are neglected in the present discussion. The decomposition of the total energy defined by Eq. (3) has the attractive feature that it represents, to order $(Z\alpha)^2$ in the external field potential and to order α^2 in the electron-electron interaction, a precise partition of E_T into one-body and many-body effects. In contrast, Eq. (2) contains the relativistic 'correction', E_R, which includes a chemically significant many-body component, so that the conventional definition of E_{corr} in Eq. (1) is not the entire many-body component of E_T, and E_R is not simply a one-body relativistic energy correction to it.

For systems containing light elements, however, Eq. (2) currently offers the most accurate method for calculating of E_T from first principles and for the estimation of E_{corr} for the calibration of non-relativistic many-body theory. For diatomic molecules high precision accurate numerical methods are available for the calculation of E_{HF}, and, furthermore, E_{corr} is about an order of magnitude larger than E_R for light elements.

3.1. *Computer programs for electronic structure calculations*

A number of electronic structure programs have been used to obtain the data discussed in this paper. Some of this data has been presented previously elsewhere [10,11]. Here a brief description of each of these programs is given in the context of their use in the present investigation.

(i) *BERTHA* [12] is a relativistic molecular structure program which generates four-component Dirac spinors from G-spinor basis functions. It has the capability to construct spinors generated in a molecular self-consistent field including both the Coulomb and Breit interactions, and to calculate second-order many-body corrections in the no-virtual pair Dirac-Coulomb approximation, using direct self-consistent-field algorithms.

(ii) *SWIRLES* [13], the precursor to *BERTHA*, is a relativistic atomic structure program which generates four-component Dirac spinors in an S-spinor basis set, assuming a point-nuclear model. This program is able to include the Breit interaction in the evaluation of many-body corrections.

(iii) *GRASP* [14] is a General Relativistic Atomic Structure Program which solves radially-reduced Dirac equations for atoms using finite-difference algorithms. It contains highly-developed algorithms for investigating the electronic structure of particular complex atomic terms.

(iv) *GAUSSIAN94* [15] is a widely-used, commercially available molecular electronic structure code.

(v) *2dHF* [16,17] is a two-dimensional, finite difference code for diatomic molecules which can solve the Hartree-Fock equations on a prolate spheriodal coordinate system to high precision.

3.2. Molecular Hartree-Fock energies

Until recently, only estimates of the Hartree-Fock limit were available for molecular systems. Now, finite difference [16–24] and finite element [25–28] calculations can yield Hartree-Fock energies for diatomic molecules to at least the 1 μhartree level of accuracy and, furthermore, the ubiquitous finite basis set approach can be developed so as to approach this level of accuracy [29,30] whilst also supporting a representation of the whole one-electron spectrum which is an essential ingredient of subsequent correlation treatments.

In Table 4, finite difference Hartree-Fock energies, E_{fdHF}, for the four diatomic systems considered in this work are displayed together with the nuclear separations at which the calculations were performed.

3.3 Relativistic energy components in atoms

The conventional approach to the estimation of relativistic energy components in molecules involves the calculation of the relativistic and non-relativistic electronic energies of the constituent atoms and ions. In Table 5, numerical data are presented for atoms and ions relevant to this investigation obtained from the finite difference program, *GRASP* [14]. For a given atomic configuration, we find that the term dependence of the relativistic energy component, E_R, to the non-relativistic Hartree-Fock energy, E_{NR}, is of the order 10^{-4} hartree, while the term dependence of the first-order Breit interaction energy, E_B, is an order of magnitude smaller.

To understand this term dependence, we recall that atomic four-component spinors are labelled by the relativistic quantum number set $\{n, \kappa, m, a\}$, rather than the non-relativistic set $\{n, l, m_l, m_s\}$. Of the p-type orbitals in atomic carbon, for example, the $2p_{1/2}$ orbital is lower in energy than the $2p_{3/2}$ orbital, and possesses a non-vanishing electron density close to the nucleus. In the LS-average of configurations constructed from these jj-coupled one-electron orbitals, fine-structure effects are considered explicitly in the model, and care must be taken to obtain relativistic corrections to either a particular term, or to the average of the configuration constructed according to the same prescription in both the relativistic and non-relativistic cases. In this study, we have employed the 'Average Level' model in the *GRASP* code [14], and determined the non-

Table 4 *Molecular total Hartree-Fock energies*[1]

Molecule	E_{fdHF}	r
N_2	−108.993 825 7	2.068
CO	−112.790 907 3	2.132
BF	−124.168 779 2	2.386
NO^+	−128.977 740 7	2.0092

[1]Taken from Moncrieff *et al.* [10]
[2]Energies are in hartree; nuclear separations are in bohr.

Table 5 *Atomic and ionic estimates of relativistic energy corrections. The Dirac-Hartree-Fock energy, E_{DHF}, has been obtained using GRASP [14] using the Average Level (AL) option, including Fermi nuclear model and reduced mass correction, with the non-relativistic limit, E_{NR}, obtained by setting $c = 1.37 \times 10^5$ a.u. The first-order expectation value of the low-frequency Breit operator is labelled E_B; all relativistic corrections to E_{NR} are collected in E_R. The atomic terms are labelled using LS-coupled notation, and the term label Av denotes the configurational average in the GRASP AL approximation. All energies in atomic units.*

		E_{DHF}	E_{NR}	E_B	E_R
B^+	^1S	−24.24516478	−24.23757367	0.00148102	−0.00611009
B	^2P	−24.53661624	−24.52905928	0.00150619	−0.00605077
C	^3P	−37.70485607	−37.68833928	0.00280994	−0.01370686
C	Av	−37.67604073	−37.65969455	0.00278178	−0.01356450
N^+	^3P	−53.92006678	−53.88775588	0.00476217	−0.02758737
N	^4S	−54.43087011	−54.39933354	0.00470445	−0.02713213
N	Av	−54.32772183	−54.29616211	0.00471126	−0.02684850
O	^3P	−74.86541515	−74.80925522	0.00751759	−0.04864233
O	Av	−74.82498609	−74.76918802	0.00750050	−0.04829750
F	^2P	−99.50227579	−99.40932581	0.01144864	−0.08151346
F^-	^1S	−99.55126379	−99.45943039	0.01141142	−0.08042198

relativistic limit of this theory by scaling the speed of light to 1000 times its natural value to extract directly a representative set of relativistic corrections.

At the equilibrium molecular geometry, the atomic dissociation products of the molecule need not provide the most suitable elementary model of the consituent parts of the molecular electronic structure. This is particular true of BF, which is polar and well described by an ionic model in which an electron is transfered from the boron atom to the fluorine atom to form a chemical bond. We have included, therefore, calculations of the relativistic corrections to the the ground-state energies of B^+ and F^- in Table 5. The additional electron in the outer shell and the slight destabilization of the core electrons is sufficient to reduce the relativistic energy component of F^- by 0.001 hartree, compared with the corresponding quantity in neutral atomic fluorine. We conclude that the term dependence of the relativistic energy shifts and the Breit interaction energy in this simple model is too small to influence significantly our investigation of the molecular systems, but the ionic nature of the chemical bond in BF is likely to make a significant contribution to the calculation of total relativistic molecular energy shifts.

3.4. *Relativistic energy components in molecules*

While the calculation of atomic structures provides a useful route to the understanding of qualitative behaviour of electronic relativistic effects, it is desirable that the crude division of a molecule into constituent atomic or ionic sub-systems should be avoided. Some ambiguity has already arisen as to the best choice of these sub-systems in these first row compounds, and the choice can only become more difficult for heavier molecules, in which more electrons may present a greater number of contributing atomic terms to deal with, and a larger values of the relativistic corrections to each such term.

The theoretical apparatus and associated computational algorithms for performing relativistic molecular electronic structure studies by invoking the algebraic approximation is now well established [33, 34]. One of the principal virtues of the kinetic balance prescription, which has been widely adopted in relativistic electronic structure calculations, is that it reproduces exactly the non-relativistic limit of a system in a given basis set. As Dyall, Grant and Wilson [35] demonstrated for the special case of the $\kappa = -1$ atomic symmetry type, the strict one-to-one mapping of the large- and small-component sets implied by kinetic balance ensures that the operator identity

$$(\boldsymbol{\sigma} \cdot \boldsymbol{p})(\boldsymbol{\sigma} \cdot \boldsymbol{p}) = p^2\mathbf{I} \tag{4}$$

is also a matrix identity in a finite-dimensional basis set representation.

The relativistic molecular structure program, *BERTHA* [12], enforces strictly the kinetic balance prescription on the elements of the basis set of Gaussian-type functions. The non-relativistic limit of the electronic energy is obtained exactly by employing Eq. (4) and by retaining only large-component contributions to the matrix elements of electron-nuclear attraction and electron-electron repulsion. One may specify a set of Gaussian exponential parameters and calculate the matrix representations of the Dirac-Hartree-Fock and Hartree-Fock mean-field equations, and the difference between these two calculations is a direct estimate of the molecular relativistic energy shift.

In Table 6, we present a series of calculations of the molecular structures of N_2, CO, BF, and NO^+ using sets of published Gaussian basis set parameters [36]. The relativistic Dirac-Hartree-Fock electronic energies, E_{DHF}, and the non-relativistic Hartree-Fock energies, E_{HF}, exhibit all of the undesirable qualities that one would associate with the use of basis sets of modest dimension. The small increase in the size of the uncontracted atom-centred basis from $10s6p$ to $13s7p$ decreases the electronic energies by about 0.005 hartree in all cases, while the addition of two d-type polarization functions further decreases the energy by roughly 0.1 hartree. In absolute terms, our calculations of E_{DHF} and E_{HF} are rather far from the basis set limit.

However, in this preliminary study, we are working within the partition of energy defined by Eq. (2) and not Eq. (3), so we need only to know E_R to an accuracy of about 10^{-4} a.u., rather than E_{DHFB} to the same accuracy. In this more restricted task, our calculations using small basis sets prove to be more than adequate.

The ability to use precisely the same basis set parameters in the relativistic and non-relativistic calculations means that the basis set truncation error in either calculation cancels, to an excellent approximation, when we calculate the relativistic energy correction by taking the difference. The cancellation is not exact, because the relativistic calculation contains additional symmetry-types in the small component basis set, but the small-component overlap density of molecular spinors involving basis functions whose origin of coordinates are located at different centres is so small as to be negligible. The non-relativistic molecular structure calculation is, for all practical purposes, a precise counterpoise correction to the four-component relativistic molecular structure calculation in which the only surviving residue is the relativistic correction to the energy. So, although the total electronic energies fluctuate by 0.1 a.u. in this study, and the results for the largest basis sets are at least 0.005 hartree above the numerical Hartree-Fock limit in all cases, the fluctuation in the relativistic corrections, E_R, are significantly less than 10^{-4} hartree, which is more than sufficiently accurate for the present study.

Significantly, the first-order expectation value of the Breit energy varies only at the 10^{-6} hartree level for all systems and for all basis sets considered here. The reason for this remarkable stability is that effects due to the Breit interaction are localized to regions in which the magnitudes of the single-particle electronic current densities are the largest, which is always in the neighbourhood of the nuclei. Even basis sets of modest dimension have a sufficient number of large Gaussian exponents that the representation of the small-component amplitudes near the nuclei are saturated, and insensitive to the addition of further s- or p-type basis functions or the polarization of the charge density by d-type functions which only makes a significant contribution to the description of the bonding region of the molecules.

3.5. *Comparison of atomic and molecular estimates of relativistic energy corrections*

To the extent that the molecular estimates of relativistic energy corrections in Table 6 should be regarded as having converged, which is about 10^{-4} hartree, the estimates of the same quantities in the independent particle model using data derived from atomic calculations are revealed to be in good agreement, although some uncertainty exists at the chemically significant level of 0.001 hartree.

Table 6 *Matrix Dirac-Hartree-Fock (E_{DHF}) and Hartree-Fock (E_{HF}) energies calculated using BERTHA. The Gaussian exponential parameters are those of the non-relativistic sets derived by van Duijenveldt and tabulated in Poirier et al. [36]. The first-order molecular Breit energy, E_B, was calculated using methods described in [12]; relativistic corrections to E_{HF} are collected in the column labelled E energies are in atomic units.*

	R	Basis	E_{DHF}	E_{HF}	E_{B}	E_{R}
N_2	2.068	$10s6p$	−108.96670469	−108.90402116	0.00953154	−0.05315199
		$13s7p$	−108.97238690	−108.90957975	0.00953174	−0.05327541
		$13s7p2d$	−109.04777442	−108.98502104	0.00953062	−0.05322276
CO	2.132	$10s6p$	−112.78273329	−112.71110124	0.01038049	−0.06125156
		$13s7p$	−112.78719810	−112.71541241	0.01038193	−0.06140376
		$13s7p2d$	−112.85584115	−112.78411091	0.01037961	−0.06135063
BF	2.386	$10s6p$	−124.21868170	−124.11947338	0.01295886	−0.08624946
		$13s7p$	−124.22458983	−124.12519316	0.01296098	−0.08643569
		$13s7p2d$	−124.26368539	−124.16432936	0.01295608	−0.08639995
NO^+	2.0092	$10s6p$	−128.94073123	−128.85364198	0.01228541	−0.07480384
		$13s7p$	−128.94720334	−128.85993260	0.01228672	−0.07488815
		$13s7p2d$	−129.05194050	−128.96476563	0.01228208	−0.07489279

In the case of BF, the ionic model for the estimation of relativistic energy corrections to E_{HF} proves to be excellent, while the data derived from neutral atomic data overestimates the magnitude of this quantity by about 0.001 hartree. In all other cases, however, estimates of E_R derived from the lowest-energy terms of the ground-state electronic configurations are larger in magnitude than the values derived from molecular calculations, while the use of the average of configuration model appears to improve the agreement with the more directly obtained molecular value.

The apparent superiority of the average of configuration model over estimates based on ground-state term values is actually an incidental observation, however, and not really an explanation of why agreement with the molecular estimates appears to be more precise. The average of configuration model contains some terms which should not be included in the calculation of the relativistic shift to ground-state molecular energy because they cannot be consituents of the molecular wavefunction on symmetry grounds. The significant observation is that the superposition of ground-state atomic terms overestimates the magnitude of the relativistic energy correction because this model assumes implicitly that the ground-state molecular wavefunction is constructed from the superposition of ground-state atomic wavefunctions, each with unit weight. In fact, the formation of a chemical bond may be thought of as introducing into the expansion of the molecular wavefuction excited state atomic configurations, each of which necessarily exhibits a smaller relativistic energy shift than does the ground-state. These excited states are not necessarily those included in the configurational average generated by *GRASP*, and the important point is that the molecular structure calculations do away with this ill-defined attempt to identify atomic configurations contributing to the total molecular electronic environment.

On the other hand, the estimate of the Breit interaction energy is, in all cases, quite satisfactory in any atomic or ionic superposition model. The reason for this has already been discussed; the Breit interaction energy arises due to electron current density in the neighbourhood of the nuclei, which is dominated by the core electrons and is apparently insensitive to the valence electron environment.

This observation regarding the calculation of the Breit interaction energy in molecules has quite important practical implications. Since the expectation value of the molecular Breit interaction energy is very well approximated by the sum of its constituent atomic or ionic components, it follows that in the *ab initio* calculation of the molecular electronic structure using the Dirac-Coulomb-Breit operator, the only interactions of any consequence involve basis functions whose origins coincide at a single centre. In the direct iterative approach to molecular structure calculations adopted in *BERTHA*, we only calculate those contributions to the self-consistent field matrix which make a significant contribution to the change in the matrix from one iteration to the next, and these can only include one-centre contributions. We already make use of this observation in the calculation of some classes of Coulomb interaction integral, but clearly the time has now come to extend the capabilities of *BERTHA* to include special methods for calculating one-centre two-electron matrix elements of the Breit operator using Racah algebra techniques borrowed from atomic physics. This offers a very significant computational saving in any study of electronic structure and bonding based on Eq. (3).

3.6. Relativistic many-body corrections

In previous studies of the argon atom [37] we have shown that relativistic many-body energy corrections are dominated by the effects of virtual transverse photon exchange between core electrons, and that relativistic corrections to pair correlation energies involving valence electrons are negligible, at least for the light systems under investigation here. Consequently, since we are currently unable to undertake the complete many-body calculation defined by Eq. (3) we will estimate relativistic corrections to the electron correlation energy in our target systems by finding the corresponding quantities in Be-like electronic cores with appropriate nuclear charges.

One might be tempted to assume that if relativistic corrections to the mean-field energies of these molecular systems are small quantities, then the relativistic correction to the correlation energy would be dominated by Coulomb correlations in the core, and that these would be negligibly small compared with the dominant one-body effects. That this assumption is false is revealed in Table 8, the contents of which also explain why there is no point in extending blindly our subtraction technique to the calculation of relativistic corrections to molecular correlation energies without the Breit interaction in place in the many-body theory. Relativistic corrections to the correlation energy in the Dirac-Coulomb approximation are smaller than the corresponding contributions in the Dirac-Coulomb-Breit approximation by an order of magnitude, and of opposite sign.

In order to establish the plausibility of the argument that the calculation of relativistic corrections to the second-order correlation energy in Be-like ions may be used to estimate reliably the corresponding quantity in many-electron molecules, we have included a calculation, in Table 8, of the many-body corrections to Ne^{6+}. The relativistic correction to the correlation derived from these calculations is 1.257×10^{-6} hartree, which should be compared with our previous calculations [38] of the same quantity in Ne^{8+}, 1.277×10^{-6} hartree, and with the independent calculations by Drake

Table 7 *Estimates of total relativistic correction,* E_R^{est}, *and the first-order Breit energy correction,* E_B^{est}, *obtained by combining the atomic or ionic contributions indicated by the second column. They may be compared with the values of the total relativistic correction,* E_R, *and the first-order Breit interaction,* E_B, *obtained directly from matrix Dirac-Hartree-Fock and Hartree-Fock calculations of the molecular structure using BERTHA [12]. Only the results of the 13s7p2d atom-centred basis sets for* E_R *and* E_B *are quoted. All energies in atomic units.*

		E_R^{est}	E_R	E_B^{est}	E_B
N_2	$N(^4S) + N(^4S)$	−0.05426426	−0.05327541	0.00940890	0.00953174
	$N(Av) + N(Av)$	−0.05369700		0.00942252	
CO	$C(^3P) + O(^3P)$	−0.06234919	−0.06135063	0.01032753	0.01038193
	$C(Av) + O(Av)$	−0.06186200		0.01028228	
BF	$B^+(^1S) + F^-(^1S)$	−0.08653190	−0.08639995	0.0128924	0.01295608
	$B(^2P) + F(^2P)$	−0.08755211		0.01295483	
NO^+	$N(^4S) + O(^3P)$	−0.07577446	−0.07480384	0.01222204	0.01228541
	$N^+(^3P) + O(^3P)$	−0.07622970		0.01227976	
	$N(Av) + O(Av)$	−0.07514600		0.01221176	

Table 8 *Second-order many-body perturbation theory corrections to beryllium-like ions using non-relativistic (E_2^{NR}), Dirac-Coulomb (E_2^{DC}) and Dirac-Coulomb-Breit (E_2^{DCB}) hamiltonians, obtained using the atomic precursor to BERTHA, known as SWIRLES. Basis sets are even-tempered S-spinors of dimension N = 17, with exponent sets, $\{\lambda_i\}$ generated by $\lambda_i = \alpha\beta^{i-1}$, with $\alpha = 0.413$, and $\beta = 1.376$. Angular momenta in the range $0 \leqslant l \leqslant 6$ have been included in the partial wave expansion of each second-order energy, and the total relativistic correction to E_2^{NR} has been collected as E_2^R. All energies in hartree.*

	E_2^{NR}	E_2^{DC}	E_2^{DCB}	E_2^R
B^+	−0.085787	−0.085759	−0.086090	−0.000303
C^{2+}	−0.095318	−0.095283	−0.095770	−0.000452
N^{3+}	−0.103826	−0.103766	−0.104452	−0.000626
O^{4+}	−0.111702	−0.111323	−0.112525	−0.000823
F^{5+}	−0.119159	−0.119004	−0.120194	−0.001035
Ne^{6+}	−0.126322	−0.125625	−0.127579	−0.001257

[39] using explicitly correlated wavefunctions, who obtained 1.252×10^{-6} hartree for the same quantity. Drake's correlated wavefunction includes, for all practical purposes, wavefunction corrections to all orders in many-body perturbation theory, and his estimate of the relativistic correction to the correlation energy includes the first-order effect of the Breit interaction.

From this, we may deduce that the relativistic correction to the correlation energy is dominated by the contribution from the $1s^2$ electron pair, and that the total relativistic effect involving the exchange of a single transverse Breit photon is obtained to sufficient accuracy for our present purposes at second-order in many-body perturbation theory.

The results presented in Table 8 were obtained in an S-spinor basis set and a point-nuclear model, using *SWIRLES* [13]. The components of the second-order correlation energy were obtained by examining a systematic sequence of basis sets, and each component appears to have converged to about 10^{-5} hartree. The relativistic correction to the correlation energy, obtained by taking the difference between Dirac-Coulomb-Breit and non-relativistic values for each basis set, however, converged to about 10^{-6} hartree using a partial-wave expansion or the intermediate state spectrum limited to $0 \leqslant l \leqslant 6$.

4. Derived Energy Corrections

4.1. *Total molecular electronic energies*

The total molecular ground-state electronic energies of the target molecules, E_T^{XY}, are derived from the total atomic energies of the dissociation fragments, E_T^X and E_T^Y, the dissociation energy, D_0, and the zero-point vibrational energy, $\hbar\omega_e/2$ from the relation

$$E_T^{XY} = E_T^X + E_T^Y - D_0 - \tfrac{1}{2}\hbar\omega_e \tag{5}$$

Table 9 *Total molecular electronic energy, E_T, derived from experimental data, and Eq. (5) All energies in atomic units.*

Molecule	E_T
N_2	−109.587 82
CO	−113.376 57
BF	−124.754 55
NO^+	−129.590 29

Numerical values of E_T^{XY} are presented in Table 9 for the four molecular systems considered in this work.

4.2 post-Hartree-Fock energies

The *post*-Hartree-Fock energy for the system XY may be defined as

$$\Delta E_{XY} = E_{XY} - E_{XY}^{HF} \tag{6}$$

where E_{XY} is the total molecular ground state energy, which is available from experiment, and E_{XY}^{HF} is the exact Hartree-Fock energy, which is available from finite difference and finite element calculations, and from finite basis set studies in which convergence with respect to basis set has been monitored. Values of the *post*-Hartree-Fock energies for the four systems considered in this work are displayed in Table 10. The accuracy with which the *post*-Hartree-Fock energy can be determined depends both on the accuracy of the finite difference Hartree-Fock energy and that of the total molecular energy derived from experiment. Errors in the experimentally derived total molecular energy are the dominant source of error in the *post*-Hartree-Fock energies.

4.3. Relativistic energy components

Relativistic corrections from all sources, including the one-body estimates derived from molecular calculations, E_1^R, and two-body estimates, E_2^R, obtained by summing the *Be*-like ion values for the constituent nuclei, are assembled for each molecule in Table 11.

Table 10 *Values of the post-Hartree-Fock energy, ΔE_{XY}*

Molecule	ΔE_{XY}
N_2	−0.594 00
CO	−0.585 67
BF	−0.585 77
NO^+	−0.612 55

Table 11 *Total relativistic corrections, E_T^R, and the one- and two-body components, E_1^R and E_2^R.*

Molecule	E_1^R	E_2^R	E_T^R
N_2	−0.053 222 76	−0.001 252	−0.054 474 76
CO	−0.061 350 63	−0.001 275	−0.062 602 63
BF	−0.086 399 95	−0.001 338	−0.087 737 95
NO^+	−0.074 803 84	−0.001 449	−0.076 337 15

The uniformity of the value of E_2^R for all systems is noteworthy, as is the expected strong dependence of E_1^R on nuclear charge.

4.4. Non-relativistic ground-state electronic energies

The non-relativistic ground state electronic energies, which are the target of the vast majority of high precision *ab initio* molecular electronic structure calculations are given in Table 12 for the four diatomic systems considered in the present work. Their accuracy is determined by that of the total molecular energies given in Table 9 which, in turn, is determined by the accuracy of the experimental ionization potentials used to calculate the total atomic energies given in Table 1.

4.5. Empirical correlation energy

Our new empirical estimates of the electron correlation energy for the four species considered in this paper are collected in Table 13.

5. Comparison of Empirical Correlation Energies with Recent *ab initio* Results

Recently, we have presented [11] high-precision second-order many-body perturbation theory studies of the electron correlation energy of N_2, CO, BF and NO^+. In Table 14, the second order correlation energy components, E_2, presented in reference are compared with the empirical correlation energy estimates given in Table 13 and the *post*-Hartree-Fock energies given in Table 10. The difference between E_2 and E_{XY}^{corr} is

Table 12 *Non-relativistic ground state electronic energies, E_{NR}*

Molecule	E_{NR}
N_2	−109.533 35
CO	−113.313 97
BF	−124.666 81
NO^+	− 129.513 95

Table 13 *Empirical estimates of the correlation energy, E_{corr}, in atomic units.*

Molecule	E_{corr}
N_2	−0.539 52
CO	−0.523 06
BF	−0.498 03
NO^+	−0.537 65

Table 14 *Comparison of calculated correlation energies with empirical correlation energies and post-Hartree-Fock energies*

Molecule	E_2	E_{XY}^{corr}	ΔE_{XY}	δ^{corr}	δ
N_2	−0.535 11	−0.539 52	−0.594 00	0.004 11	0.058 89
CO	−0.518 72	−0.523 06	−0.585 67	0.004 34	0.066 95
BF	−0.489 75	−0.498 03	−0.585 77	0.008 28	0.096 02
NO^+	−0.573 14	−0.537 65	−0.612 55	−0.035 49	0.039 41

denoted by δ^{corr}, and δ is the difference between E_2 and ΔE_{XY}. For the three neutral systems studied here δ^{corr} is an order of magnitude smaller than δ. For NO^+, δ^{corr} and δ are roughly comparable.

6. Concluding Remarks

In these first row diatomic molecules, we have found that one-body relativistic energy corrections are about an order of magnitude smaller than our empirical estimates of the correlation energy, and that two-body relativistic effects are two-orders smaller in magnitude than E_{corr}. However, the task of *ab initio* quantum chemistry is to determine E_T and not simply E_{NR}, and both classes of relativistic energy correction contribute significantly to E_T. If we choose to estimate one-body relativistic energy corrections by adding atomic energy fragments, we find that the accuracy of this procedure depends on choosing those fragments to reflect the electronic environment corresponding to a particular geometry. The largest relativistic two-body effect is first-order in the Breit interaction, and first-order in relativistic many-body theory, and the energy correction arising from these interaction contributes about 0.001 a.u. to E_T. One might assume that in spectroscopic studies, values of E_R for different states would cancel when energy differences are calculated, but the evidence from this modest study suggests that these relativistic effects may be state dependent, and that they have the potential to make chemically significant contributions to E_T, even for systems as light as we have studied here. Additivity of the fragmentary atomic relativistic effects within a molecule is likely to be satisfied only when the fragments are far apart and weakly interacting. In the case of BF, we have established that E_R is a chemically significant function of the internuclear separation, because it is well approximated by the sum of the ionic constituents B^+ and F^- at the equilibrium geometry, and by definition it comprises the sum of the relativistic energy shifts for the neutral atoms B and F at the dissociation limit.

There have been very significant improvements in our ability to calculate both E_{HF} and E_{DHFB} in recent years, and in the practical implementation of both relativistic and non-relativistic many-body theories. The availability of these complementary theories suggests their use in concert for the study of heavy element systems in which $E_R \gg E_{corr}$, in which the atomic additivity approximation that we have assumed for the two-body relativistic effect is clearly not valid. Molecular properties other than the energy, such as electric and magnetic interactions, exhibit relativistic effects which are more pronounced. Fundamental to the practical application of these methods to molecular systems is the use of finite basis sets. Systematic refinement has reduced the basis set truncation error in diatomic molecule Hartree-Fock studies to a level approaching that achieved in fully numerical studies. For the molecules studied in this work (for which a second-order description of correlation effects is appropriate), the basis set truncation error has been reduced to a level where other components of the *post*-Hartree-Fock energy are an order of magnitude larger than the error in the non-relativistic correlation energy estimate.

Acknowledgements

One of us (DM) acknowledges the support of the Office of Energy Research, Office of Basic Energy Sciences, Division of Chemical Sciences US Department of Energy under grant DE-FG02-97ER-14758. Support by Florida State University through the allocation of supercomputing resources on the SGI-Power Challenge is acknowledged. SW acknowleges the support of the Engineering and Physical Sciences Research Council under Grant GR/L65567. HQ wishes to acknowledge the hospitality of the School of Chemistry at the University of Melbourne and the facilities made available by Prof. F. P. Larkins.

References

1. A.C. Hurley, 1976, *Electron correlation in small molecules*, Academic Press, London.
2. S. Wilson, 1984, *Electron correlation in molecules*, Clarendon Press, Oxford.
3. S. Wilson, ed., 1987, *Electron correlation in atoms and molecules*, Meth. Comput. Chem. **1**, Plenum Press, New York.
4. R. McWeeny, 1989, *Methods of Molecular Quantum Mechanics*, 2nd Edition, Academic Press, London.
5. F.E. Harris, H.J. Monkhorst and D.L. Freeman, *Algebraic and diagrammatic methods in many-fermion theory*, Oxford University Press, New York.
6. S. Wilson and G.H.F. Diercksen, 1992, *Methods in Computational Molecular Physics*, Plenum, New York.
7. P.-O. Löwdin, 1959, Adv. Chem. Phys. **2**, 207.
8. D.R. Lide, Editor-in Chief, 1992, *C.R.C. Handbook of Chemistry and Physics*, 73rd edition, C.R.C. Press, Boca Raton.
9. K.P. Huber and G. Herzberg, 1979, *Molecular spectra and molecular structure. Vol. 4. Constants of diatomic molecules*, Van Nostrand Reinhold.

10. D. Moncrieff, J. Kobus and S. Wilson, 1995, J. Phys. B: At. Mol. Opt. Phys. **28** 4555.
11. D. Moncrieff and S. Wilson, 1998, J. Phys. B: At. Mol. Opt. Phys. **31** 3819.
12. H.M. Quiney, H. Skaane and I.P. Grant, 1998, Adv. Quant. Chem., **32**, 1.
13. H.M. Quiney, 1990, *Supercomputational Science*, (ed. R.G. Evans and S. Wilson), Plenum Press (New York), pp 159–199.
14. K.G. Dyall, I.P. Grant, C.T. Johnson, F.A. Parpia and E.P. Plummer, 1989, Computer Phys. Comm., **55**, 425.
15. Gaussian 94, Revision E.3, M.J. Frisch, G.W. Trucks, H.B. Schlegel, P.M.W. Gill, B.G. Johnson, M.A. Robb, J.R. Cheeseman, T. Keith, G.A. Petersson, J.A. Montgomery, K. Raghavachari, M.A. Al-Laham, V.G. Zakrzewski, J.V. Ortiz, J.B. Foresman, J. Cioslowski, B.B. Stefanov, A. Nanayakkara, M. Challacombe, C.Y. Peng, P.Y. Ayala, W. Chen, M.W. Wong, J.L. Andres, E.S. Replogle, R. Gomperts, R.L. Martin, D.J. Fox, J.S. Binkley, D.J. Defrees, J. Baker, J.P. Stewart, M. Head-Gordon, C. Gonzalez, and J.A. Pople, Gaussian, Inc., Pittsburgh PA, 1995.
16. J. Kobus, 1993, Chem. Phys. Lett. **202**, 7.
17. J. Kobus, 1994, Comput. Phys. Commun. **78**, 247.
18. J. Kobus, D. Moncrieff and S. Wilson, 1994, J. Phys. B: At. Mol. Opt. Phys. **27**, 2867.
19. J. Kobus, D. Moncrieff and S. Wilson, 1994, J. Phys. B: At. Mol. Opt. Phys. **27**, 5139.
20. J. Kobus, D. Moncrieff and S. Wilson, 1995, Molec. Phys. **86**, 1315.
21. D. Moncrieff, J. Kobus and S. Wilson, 1998, Molec. Phys. **93**, 713.
22. J. Kobus, L. Laaksonen and D. Sundholm, 1996, Comput. Phys. Comm. **98**, 346.
23. L. Laaksonen, P. Pyykkö and D. Sundholm, 1983, Intern. J. Quantum Chem. **23**, 309.
24. L. Laaksonen, P. Pyykkö and D. Sundholm, 1983, Intern. J. Quantum Chem. **23**, 319.
25. D. Heinemann, B. Fricke and D. Kolb, Phys. Rev. A **38**, 4994 (1988).
26. D. Heinemann, A. Rosen and B. Fricke, Physica Scripta **42**, 692 (1990).
27. S. Hackel, D. Heinemann, D. Kolb and B. Fricke, Chem. Phys. Lett. **206**, 91 (1993).
28. L. Yang, D. Heinemann and D. Kolb, Phys. Rev. A **48**, 2700 (1993).
29. D. Moncrieff and S. Wilson, 1993, Chem. Phys. Lett. **209**, 423.
30. D. Moncrieff and S. Wilson, 1993, J. Phys. B: At. Mol. Opt. Phys. **26** 1605.
31. D. Moncrieff and S. Wilson, 1996, J. Phys. B: At. Mol. Opt. Phys. **29**, 2425.
32. D. Moncrieff and S. Wilson, 1996, J. Phys. B: At. Mol. Opt. Phys. **29**, 6009.
33. H.M. Quiney, I.P. Grant and S. Wilson, 1989, in *Many-Body Methods in Quantum Chemistry*, ed. U Kaldor, Lecture Notes in Chemistry **52** 307, Springer Verlag, Berlin.
34. S. Wilson, I.P. Grant and B.L. Gyorffy (ed.), 1991, *The Effects of Relativity in Atoms, Molecules and the Solid State*, Plenum, New York.
35. K.G. Dyall, I.P. Grant and S. Wilson, J. Phys. B: At. Mol. Opt. Phys, 1984, **17**, L45, 493, 1201.
36. R. Poirier, R. Kari, and I.G. Csizmadia, 1985, *Handbook of Gaussian Basis Sets* Elsevier, Amsterdam.
37. H. M. Quiney, I.P. Grant and S. Wilson, 1990, J.Phys. B: At. Mol. Opt. Phys., **23**, L271–L278.
38. H. M. Quiney, 1991, in *Relativistic effects in atoms, molecules and the solid state*, (eds. S. Wilson, I.P. Grant and B.F. Gyorffy), Plenum Press (New York) pp 83–123.
39. G.W.H. Drake, 1988, Can. J. Phys., **66**, 586.

Influence of Electron Correlation on the Electronic Structure of Superconducting Y-Ceramics

I.G. Kaplan[a], J. Hernández-Cóbos[b] and J. Soullard[c]

[a]*Instituto de Investigaciones en Materiales, UNAM, Apartado Postal 70-360, 04510 México, D.F., Mexico*

[b]*Laboratorio de Cuernavaca, Instituto de Física, UNAM, Apartado Postal 48-3, Cuernavaca 62251, Morelos, Mexico*

[c]*Instituto de Física, UNAM, Apartado Postal 20-364, 01000 México, D.F., Mexico*

Abstract

The methodology of applying the embedded cluster method in the frame of the Gaussian-94 program for a calculation of the electronic structure of superconducting ceramics at the electron correlation level is described in detail. The elaborated methodology is applied to a ground-state electronic structure calculation of the $YBa_2Cu_3O_7$ ceramics at the ROHF, UHF and MP2 levels. This allows to study the influence of electron correlation effects with a different account of the latter. The obtained results reveal the strong dependence of the charge distribution upon taking into account the electron correlation effects, especially for copper: the MP2 charges on the copper ions decrease in 1.5 times in comparison with the ROHF charges. Allowance for the electron correlation is found to be crucial for determination of the spin location and symmetry of the holes.

1. Introduction

After more than ten years of extensive experimental and theoretical studies of the phenomenon of the high T_c superconductivity (HTSC) [1], we still do not know a microscopic mechanism responsible for this phenomenon. Numerous theories of pairing, which lead to high T_c values, are based on models [2–9] and cannot connect a specific chemical composition of HTSC ceramics with the value of the transition temperature T_c. For creating a quantitative theory of the HTSC phenomenon further comparative studies of the electronic structure and their relative properties of SC and non-SC ceramics are needed. In this paper, we confine ourselves to calculations of the electronic structure of the SC yttrium ceramics.

At present, a great number of electronic structure calculations of HTSC ceramics are performed by variety of methods, see Pickett's review [10] and Refs. [11–31], which range from the band approach to the embedded cluster method (ECM). It is well known that unusually strong localization effects in oxides of transition metals make the task of electronic structure calculations very difficult [32]. As was revealed in spectroscopic experiments using different techniques: X-ray and ultraviolet photo-emission [33], electron energy-loss spectroscopy [34], polarization dependent X-ray absorption spectroscopy [35], in copper-oxide ceramics both copper 3d electron and holes are localized, the latter mainly on the oxygen sites. There is a strong and to a great extent covalent coupling between Cu and O atoms in the lattice. Now there are many evidences that the electron correlation plays an extremely important role in copper-oxide materials, see the review by Brenig [36]. Band structure calculations based on the one-electron

A. Hernández-Laguna et al. (eds.), Quantum Systems in Chemistry and Physics, Vol. 1: Basic Problems and Model Systems, 143–158.

approach, therefore, cannot be a good approximation to study the electronic structure of copper-oxide ceramics. More sophisticated and precise band calculation [24,30,31] can represent many spectroscopic features of the HTSC ceramics. However, to reveal the peculiarities of the electronic structure responsible for the HTSC phenomenon, precise band calculations must be supplemented by calculations of local configurations at the electron correlation level. The most appropriate approach for this is the ECM, see [37–39]. Although, we must keep in mind that the ECM approaches, due to the lack of the periodical conditions, underestimates the metallic character of the systems studied. Note also that in these approaches, as in most of electronic structure calculations, all atoms in the cluster and the background point charges are considered as static hence, no interaction between the electron system and the lattice vibrations is taken into account. Different methodologies of the ECM are distinguished by two general aspects:

i) The preciseness of the computational method used for a quantum mechanical description of cluster.
ii) The embedding scheme used for coupling the cluster to the environment which must represent an infinite crystal.

It is well established that in $YBa_2Cu_3O_7$ ceramics, the small changes of oxygen concentration, which preserve the crystal structure, can completely destroy the superconductivity [40]. So, a complete study of the electronic structure of HTSC should include a study of the effect of crystal stochiometry on the electronic structure. In the case of Y123 ceramics an insight on this problem can be obtained by comparing the electronic structure of the superconducting overdoped ceramics $Y123O_7$ to that of the non superconducting underdoped $Y123O_6$ ceramics. We believe that this can give valuable information on the mechanisms which are supposed for the transition from the superconducting state to the insulator state. In the present paper we report only the results relative to the $Y123O_7$ ceramics, the comparative study of the $Y123O_6$ ceramics will be the object of a future publication.

Below is a brief review of the published calculations of yttrium ceramics based on the ECM approach. In studies by Goodman et al. [20] and Kaplan et al. [25,26], the embedded quantum clusters, representing the $YBa_2Cu_3O_{7-x}$ ceramics (with different x), were calculated by the discrete variation method in the local density approximation (LDA). Although in these studies many interesting results were obtained, it is necessary to keep in mind that the LDA approach has a restricted applicability to cuprate oxides, e.g. it does not describe correctly the magnetic properties [41] and gives an inadequate description of anisotropic effects [42,43]. Therefore, comparative *ab initio* calculations in the frame of the Hartree-Fock approximation are desirable.

Such kind of calculations with a precise self-consistent account of crystal surrounding were performed by, at least, four scientific groups: Baetzold [16], Das with coworkers [22,23], Winter et al. [28] and Ladik with coworkers [29]. Winter et al. [28] performed cluster calculation by restricted Hartree-Fock (RHF) method, so they did not take into account the electron correlation. The others groups used the unrestricted Hartree-Fock (UHF) method for cluster calculations which allows to some extent the electron correlation. The strong covalent C-O bonding in planes and chains was revealed (in accordance with results obtained in Refs. [20,25,26]). For covalent systems, the precision of calculations depends upon the extent of the electron correlation account. Thus, it is important to take into account the electron correlation beyond the UHF method.

The precise quantum cluster calculations of the electronic structure of SC ceramics were performed in Refs. [13,17,21]. Guo et al. [13] used the generalized valence bond method, Martin and Saxe [17] and Yamamoto et al. [21] performed calculations at the configuration interaction level. But in these studies the calculations were carried out for isolated clusters, the second aspect of the ECM scheme, see above, was not fulfilled. The influence of crystal surrounding may considerably change the results obtained.

Thus, to the best of our knowledge, there is a lack of embedding cluster studies on the yttrium ceramics where with a sufficient precision both aspects of the ECM were taken in account. In the study [44], we attempted to fill such a gap and carried out the electronic structure calculations of the $YBa_2Cu_3O_7$ ceramics at the Møller-Plesset level with a self-consistent account of the infinite crystal surrounding to the quantum cluster. The Gaussian basis set employed (6-311G) was larger than those used in previous cluster calculations [16,20,22,29].

In this paper we present the approach [44] in more detail and add to the results obtained in Ref. [44] the results of calculation of the $YBa_2Cu_3O_7$ ceramics at the restricted open shell Hartree-Fock (ROHF) level. The ROHF, UHF and MP2 calculations with the same basis set allow to study the influence of the electron correlation at different levels. As follows from the results obtained, the electron correlation has an essential or in some cases a crucial influence on charge and spin distributions in the superconducting $YBa_2Cu_3O_7$ ceramics.

The plan of the paper is the following. In section 2 we describe the methodology of applying the ECM in the frame of the Gaussian-94 program. We give the results and discussion in section III and present our conclusions in section IV.

2. Methodology

The ECM has been frequently used to study the electronic properties of materials with ionic or partially ionic bonding like in metal oxides. There are many procedures to couple the quantum cluster to its environment, they have been reviewed by Sousa et al. [45]. All these methods have three points in common:

i) The symmetry of the crystal must be preserved.
ii) The Madelung potential of the infinite crystal has to be reproduced on all cluster sites.
iii) The final system including the quantum cluster and the background point charges has to be neutral.

2.1. The embedding scheme

A finite array of charges is built taking into account the symmetry elements of the crystal. The charges of the outermost ions are adjusted in order to provide the correct value of the Madelung potential on each cluster site as well as the electrical neutrality of the final set. This is achieved by the procedure originally proposed by Kelires and Das [22,46]. Thus, to each shell of n_i equal exterior charges q_i which decreates on the site j a potential $V_i(j)$ is associated a charge scaling factor f_i such that the Madelung potential on the site j due to the infinite crystal, $V_M(j)$, is equal to the potential of each shell of scaled charges plus that created by the whole set of charges minus those included in the shells, denoted as $V_q(j)$. In order to determine the scaling factors f_i, the Madelung potential of the infinite lattice is reproduced on the k different sites of the cluster and the f_i are obtained as solutions of the following system of $k+1$ linear equations (similar to that in Ref. [47]):

$$f_1 V_1(1) + f_2 V_2(1) + \ldots + f_{k+1} V_{k+1}(1) + V_q(1) = V_M(1)$$

$$f_1 V_1(2) + f_2 V_2(2) + \ldots + f_{k+1} V_{k+1}(2) + V_q(2) = V_M(2)$$

$$\ldots\ldots\ldots\ldots\ldots\ldots\ldots\ldots\ldots\ldots\ldots\ldots\ldots\ldots$$

$$f_1 V_1(k) + f_2 V_2(k) + \ldots + f_{k+1} V_{k+1}(k) + V_q(k) = V_M(k)$$

$$f_1 n_1 q_1 + f_2 n_2 q_2 + \ldots + f_{k+1} n_{k+1} q_{k+1} + Q = 0 \tag{1}$$

The last equation is associated to the electrical neutrality of the whole set of charges, Q being the net charge of the charge array not included in the scaled shells. It must be noted that the number of cluster sites is generally greater than k (the number of distinct cluster sites) because, a cluster generally possesses several equivalent sites. However, as the procedure preserves the crystal symmetry, the reproduction of the Madelung potential on the distinct cluster sites only, through the solution of the system of equations (1), allows to reproduce the Madelung potential on all cluster sites. This point will be emphasized later in the particular case of the cluster considered to study the electronic structure to the $YBa_2Cu_3O_7$ ceramics. Generally, in order to apply this method, k + 1 shells have to be defined on the external surface of the complete charge array and the Madelung potential on the k distinct cluster sites has to be known.

2.1.1. *Madelung potential calculation*

The calculation method was taken from the ICECAP [38] (or from the program HADES [48]) where the Ewald method [49] is implemented. The Madelung potential, which is the solution of the Poisson equation for gaussian distributions of charges representing the periodic arrangement of positive and negative charges of the crystal, is expressed as [48]:

$$V_M(j) = \sum_{i,l} \Gamma(1/2, \eta^2 R_{il}^2) q_i/\pi^{1/2} R_{il} + (1/\pi V) \sum_k S_k \exp(-g_k^2/\eta^2)/g_k^2 - 2\pi^{-1/2}\eta q_j, \tag{2}$$

where R_{il} is the position vector of the ith ion in the lth cell and q_i is its charge, g_k is a reciprocal lattice vector and the index i runs over all the ions of the unit cell which volume is V, S_k is the structure factor, η is the width of the gaussian distributions and $\Gamma(\frac{1}{2}, \eta^2 R_{il}^2)$ is the incomplete gamma function. The Madelung potential is essentially given by two summations, the contribution of the first one over the direct lattice points being less important. The second summation over the reciprocal lattice points is a fast converging series which value is independent of η.

The Madelung potential on the 13 sites of the unit cell of the $YBa_2Cu_3O_7$ ceramic is given in Table 1. In order to obtain the Madelung potential values with ten significant figures, 3543 ions were considered.

2.1.2. *The quantum cluster and its surroundings*

The cluster studied in this work is shown in Figure 1 with its neighbors in a unit cell based on yttrium ions. With this choice, the ions Cu1, Cu2 and O4 each have all their nearest neighbors included into the cluster. We did not correct the influence of nearest point charges on the boundary ions: O1, O2, and O3, by introducing pseudopotentials, as in the ICECAP program, see [38]. An account of such corrections complicates the calculation scheme used. The real influence of this effect will be checked in future studies. The interionic distances were taken from Jorgensen et al. [50].

The complete charge array is built by the juxtaposition of this cell in three dimensions so that to obtain a block of $3 \times 3 \times 3$ cells, the cluster being located in the central cell. In that case the cluster is well centered in an array of 475 ions. Practically and for computational purposes, the basic symmetry elements of the space group Pmmm (3 mirror planes perpendicular to 3 rotation axes of order 2 as well as the translations of the primitive orthorhombic Bravais lattice) are applied to a group of ions which corresponds to $1/8$ of the unit cell. The procedure ensures that the crystalline symmetry is preserved.

In our case there are six different sites in the considered cluster. So, to apply the method described previously (system of equations (1)), we need seven shells to reproduce the Madelung potential on these six sites. There are various possibilities to define seven shells by considering various grouping of the 218 charges located on the exterior surface of the complete array. As an example consider the following definition of shells which we used in our calculation: S1 contains 32 yttrium ions located on the upper and lower faces of the array; S2 contains 24 yttrium ions on the lateral faces; S3

Table 1 *Madelung potential on the sites of the $YBa_2Cu_3O_7$ unite cell obtained with Gupta-Gupta [14] charges, in a.u. (infinite crystal)*

Site charge		Potential
Ba	1.81	−0.6458040017
Y	2.58	−0.8074713101
Cu1	1.55	−0.8862608839
Cu2	1.64	−0.7041223128
O1	−1.61	0.4686010081
O2	−1.51	0.7092282530
O3	−1.51	0.7201305019
O4	−1.65	0.5411489261

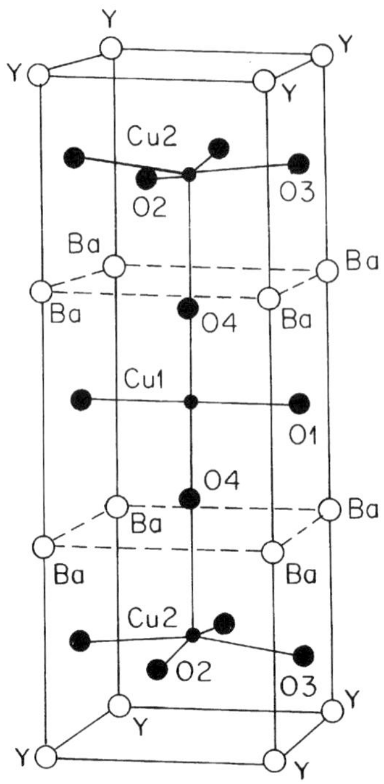

Fig. 1. The quantum cluster Cu_3O_{12} (black circles) embedded into a unit cell.

contains 36 oxygen O2 located on 2 opposite lateral faces; S4 contains 36 oxygens O3 on the other 2 opposite faces; S5 contains 18 oxygens O1; S6 contains 24 barium ions located on the 4 edges of the cluster; S7 and S8 contain 24 barium ions on the faces perpendicular to the a and b axis respectively. In that particular case, eight shells have been defined but, 7 shells are only needed to adjust the array. Thus, the charges of one shell must be left unchanged or 2 shells must be combined. Several attempts with various groups of seven shells were realized in order to obtain the scaling factors associated with a background self-energy (electrostatic energy of background point charges) as small as possible.

In Table 2 we present the shells which correspond to our best choice as well as the values of the scaling factors which result from the solution of the system of equations (1). The Madelung potential values on each cluster site calculated with the finite adjusted array are given in Table 3. When these values are compared to that of the Table

Table 2 *Shells and associated scaling factors for the adjusted charge array*

Shells	S1	S2	S3	S4	S4	S6	S7
Scaling factor	0.424	0.391	0.388	0.280	0.683	−1.35	1.063

Table 3 *Madelung potential on the cluster Cu_3O_{12} sites obtained with the Gupta-Gupta [14] charges, in a.u. (finite crystal)*

Site charge		Potential
Cu1	1.55	−0.8862608839
Cu2	1.64	−0.7041223128
Cu2	1.64	−0.7041223127
O1	−1.61	0.4686010081
O1	−1.61	0.4686010081
O4	−1.65	0.5411489261
O4	−1.65	0.5411489261
O2	−1.51	0.7092282530
O2	−1.51	0.7092282530
O3	−1.55	0.7201305019
O3	−1.55	0.7201305019
O2	−1.51	0.7092282530
O2	−1.51	0.7092282530
O3	−1.55	0.7201305019
O3	−1.55	0.7201305019

1, we conclude that the Madelung potential values of the infinite crystal are reproduced on the 15 sites of the cluster with the desired precision. It must be noted however, that only 6 sites were considered to adjust the cluster. The values of scaling factors and background self-energy obtained with larger arrays were found to be larger than the present ones and less suitable for our purpose.

Generally, the charges of cluster ions obtained after a calculation at the MP2 level differ from that of the background ions and in the case where the cluster does not coincide with the unit cell this causes that the unit cell associated with new ionic charges is not neutral. In order to get consistency between cluster and background charges, a series of calculations has been performed where the charges obtained as a result of a calculation are taken as background charges for the next calculation; at each step, the ionic charges are modified in order to assure the electrical neutrality of the unit cell. To realize this modification, two parameters: the cation charge variation (Δq_c) and the anion charge variation (Δq_a) are determined from a set of 2 equations expressing the unit cell neutrality and the fact that the cluster charge must be a definite integer q_0, the value of which is determined by the initial charge distribution. So, at each step of iterations we preserve the cluster charge and the unit cell neutrality. In our case these equations are the following:

$$3\Delta q_c + 7\Delta q_a = -a_0,$$
$$3\Delta q_c + 12\Delta q_a = q_0, \tag{3}$$

where a_0 represent the departure from neutrality. The flow chart corresponding to the complete embedding procedure is shown in Fig. 2.

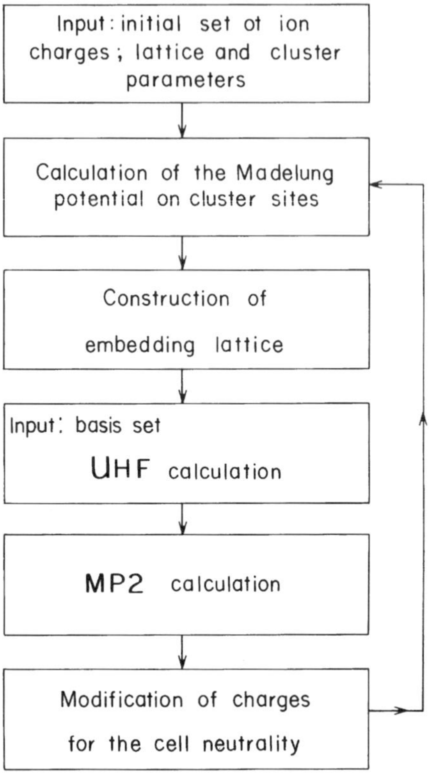

Fig. 2. Interaction cycle in the self-consistent charge calculation.

2.2. *The quantum cluster calculations*

The *ab initio* calculations of the cluster were done using the Gaussian-94 program. The finite array of background charges (see subsection 2.1) was introduced as solvent charges using the CHARGE keyword of the program. The triply split valence 6-311G basis set, provided with the program, was used in the ROHF *y* UHF calculations while the more precise account of the electron correlation was performed by the Møller Plesset method up to the second order using the UHF results as initial. To calculate charge distribution we used the Mulliken population analysis along with the Natural Bond Orbital (NBO) option provided by the program. We realize that the Mulliken population analysis is based on some conventions and often can lead to unaccurate charge values. However, as we are mainly interested in the comparative study of the electron correlation effects on ionic charges, the Mulliken procedure is sufficient. On the other hand, the NBO analysis gives more realistic values of charges on atoms and bonds.

It is clear that the quality of an *ab initio* calculation depends critically on the size and quality of the basis set used. However, the cost of the calculation also increases with the size of the basis set, so it is necessary to arrive to a compromise that allows to obtain reliable results within reasonable CPU times. In previous study [52], we found that the 6-311G basis set yields results comparable to those obtained with larger basis sets, so we are confident that, even if the exact values can change, the observed behavior will not change with the use of a larger basis set.

We found many convergence problems for this system. In particular, it is very difficult to obtain good results trying to do directly the UHF calculation starting from the standard initial guess calculated by the program. In this case there are at least two important problems that has to be noted. The first one is the very slow convergence showed. It was necessary to increase the maximum number of iterations for the SCF procedure to more than three hundreds in some cases and frequently we did not get convergence at all for the SCF procedures. The second important problem is that the solution obtained with this procedure converges to a wave functions with a very high value of the S^2 operator, indicating a large spin contamination of the solution by excited states. Many of the charge distributions obtained with this highly spin contaminated wave functions are unsymmetrical, and all of them show unrealistic values for the net charge of the atoms in the cluster.

We make several attempts to solve this problem changing the initial guess for the SCF procedure. In particular we try to start from the solution of the isolated cluster with different charges (including neutral) and with and without background charges. With this procedure we found that the spin contamination changes depending on the initial guess used and consequently the charge distribution obtained. We also try using different backgrounds, all of them reproducing the Madelung potential at cluster sites, but with different self-energy. Backgrounds with self-energy much higher than the cluster energy were avoided in order to get a realistic total energy.

The results obtained up to this point show that the amount of spin contamination was crucial to obtain good results for this system. We found that even small spin contamination leads, in some cases, to disparate charge distributions. Thus, we used the restricted open shell procedure (ROHF) to calculate an initial guess for the UHF calculation. Since the ROHF solution has the correct value of the spin, it was found that it provides a better starting point for the UHF procedure. Unfortunately it is not a final solution to this problem because we found some cases where even this procedure leads to a very large spin contamination.

So, the final procedure can be described as follows: we calculate an initial guess performing an ROHF calculation of the isolated and neutral cluster. Then, we add both the cluster charge and the background set of charges. If the value of the S^2 operator differs from 0.75 only at the third digit, we take this solution as acceptable and proceed to the MP2 calculation and population analysis. If the spin contamination is too large, we try to use different initial solutions until an acceptable solution is reached.

3. Results and Discussion

Below, we present the results of applying the methodology of the ECM, described in section 2, to the calculation of the superconducting ground state of the $YBa_2Cu_3O_7$

ceramics. We begin from an analysis of the electron correlation account on charge and spin distributions.

3.1. Influence of the electron correlation and the Madelung field on charge and spin distributions

In the self-consistent charge calculation (see block–schema in Fig. 2), the final charge distribution must not depend on the initial charge set. We performed such calculation using the Gupta–Gupta [14] initial charge set. Only 5 iterations were enough for obtaining the self-consistent charge set with an error less than 0.03 e, see Table 4, where, as in Tables 5 and 6, charges are calculated by the Mulliken procedure.

After obtaining the self-consistent charges in lattice, we can evaluate the influence of the Madelung potential field created by these charges on the charge and spin distributions in the quantum cluster. In Table 5 we present the self-consistent charge and spin distributions calculated at the MP2 level for the isolated cluster $[Cu_3O_{12}]^{-14}$ (the charge -14 corresponds to the Gupta–Gupta [14] charge set) and the same cluster in the crystal $YBa_2Cu_3O_7$. The influence of the crystal field on the spin distribution is very

Table 4 *Iterative self-consistent charge calculation at the MP2 level starting from the Gupta-Gupta [14] charge set with fixed charges on Ba and Y ions.*

	Cu1	Cu2	O1	O2	O3	O4	Ba	Y
Gupta-Gupta charges (tight binding method)	1.55	1.64	−1.61	−1.51	−1.55	−1.65	1.81	2.58
$MP2_1$	1.20	1.17	−1.12	−1.55	−1.57	−1.39		
$MP2_2$	0.95	1.11	−1.44	−1.51	−1.50	−1.13		
$MP2_3$	1.07	1.10	−1.28	−1.52	−1.53	−1.25		
$MP2_4$	1.02	1.10	−1.37	−1.51	−1.52	−1.18		
$MP2_5$	1.04	1.10	−1.33	−1.52	−1.52	−1.21		
Modified set providing cell neutrality	1.07	1.13	−1.30	−1.49	−1.49	−1.18	1.84	2.61

Table 5 *The influence of background on the charge and spin distributions in the cluster $[Cu_3O_{12}]^{-14}$, calculations at the MP2 level, in a.u.*

	Isolated cluster		Cluster in cristal		Madelung potential
	charge	spin	charge	spin	on cluster sites
Cu1	0.96	0.81	1.04	0.80	−0.508
Cu2	0.68	0.05	1.10	0.00	−0.646
O1	−1.52	0.03	−1.33	0.05	0.515
O2	−1.50	0.00	−1.52	0.00	0.537
O3	−1.48	0.00	−1.52	0.00	0.538
O4	−0.68	0.02	−1.21	0.05	0.471

Table 6 *Self–consistent charge and spin distributions at different electron correlation levels*

	Charges on atoms			Valence orbital population			Spin distribution		
	ROHF	UHF	MP2	ROHF	UHF	MP2	ROHF	UHF	MP2
Cu1	1.53	1.42	1.04	$4s^{0.41}\ 3d^{8.81}$	$4s^{0.33}\ 3d^{9.04}$	$4s^{0.56}\ 3d^{9.05}$	0.00	0.89	0.80
Cu2	1.73	1.69	1.10	$4s^{0.35}\ 3d^{8.69}$	$4s^{0.35}\ 3d^{8.74}$	$4s^{0.38}\ 3d^{9.33}$	0.00	0.00	0.00
O1	−1.63	−1.36	−1.33	$2s^{1.97}\ 2p^{5.71}$	$2s^{1.98}\ 2p^{5.41}$	$2s^{1.95}\ 2p^{5.34}$	0.00	0.02	0.05
O2	−1.57	−1.67	−1.52	$2s^{1.97}\ 2p^{5.64}$	$2s^{1.97}\ 2p^{5.75}$	$2s^{1.95}\ 2p^{5.52}$	0.13	0.00	0.00
O3	−1.58	−1.67	−1.52	$2s^{1.97}\ 2p^{5.65}$	$2s^{1.97}\ 2p^{5.75}$	$2s^{1.96}\ 2p^{5.52}$	0.11	0.00	0.00
O4	−1.58	−1.36	−1.21	$2s^{1.95}\ 2p^{5.69}$	$2s^{1.96}\ 2p^{5.43}$	$2s^{1.94}\ 2p^{5.24}$	0.00	0.04	0.05

small in contrast with that on the charge distribution. The large charge changes take place for the Cu2, O4 and O1 ions in the quantum cluster. On the other hand, the charges on Cu1, O2 and O3 undergo small changes.

Next, we assess the role of the electron correlation. We performed the electron correlation calculations at two levels: the UHF and MP2. The comparison of these results with the calculation at the ROHF level allows to study the influence of the electron correlation with the different allowance of the latter.

The self-consistent spin and charge distributions calculated at the ROHF, UHF and MP2 levels are presented in Table 6. Already at the ROHF level, we have not the formal valence values for an ionic crystal: +2 for copper and −2 for oxygen. This is the well–known result: the bonding in copper oxides has a mixed covalent–ionic character. The calculation at the UHF level leads to a reduction of the absolute values of charges on copper and oxygen ions, except the O2 and O3 ions where the negative charge is increased. The calculation at the MP2 level leads to a further reduction of the absolute values of charges on copper and all oxygen ions. This decrease is the consequence of a more precise account of the electron correlation in the MP2 approach. The influence of the total account of the electron correlation (ROHF → MP2) is more important for copper charges than for oxygen ones. Although for O4 and O1 it is also large and equals 23.4% and 18.4%, respectively.

For copper ions, the account of the electron correlation at the MP2 level decreases the charges about 1.5 times or on 32 % for the Cu1 and on 36.4% for the Cu2. The change of copper charges for the transition in calculation methods ROHF → UHF is not such essential as for the transition UHF → MP2. The account of the electron correlation causes also the essential change of the occupation orbital numbers of the 4s and 3d copper shells. The 4s copper electrons are involved in the covalent bonding with the 2p oxygen electrons.

3.2. *Nature of bonding in the $YBa_2Cu_3O_7$ lattice*

The average charge on ions in Table 6 is calculated in frame of the Mulliken population analysis according to which the charges on atom–atom bond (or on lines connected different atoms pairs) are divided in two equal parts and added to the diagonal values of the charge matrix. The diagonal values are also presented in Table 6 and Fig. 3. The charges on bonds are exhibited in Fig. 3, they are given in fractions of the electron charge, so their values are negative. The largest values of the electron density are indicated on the Cu–O bonds in planes and on the Cu1–O4 bond. The value of the electron density on the Cu1–O1 bond is smaller. The latter reflects the relative weakness of this bond. As follows from experimental data, the deoxygenization occurs in chains: in the $YBa_2Cu_3O_6$ ceramics the oxygen O1 is absent.

Thus, the results obtained give a convincing evidence of the covalent bonding which is stronger in planes and between the chain copper, Cu1, and the apex oxygen, O4. The small value of the electron density on the line Cu2–O4, 0.0006 e, points out that the coupling between chains and planes is essentially ionic. The suggestion [53] that the covalent bonding can also occur through O–O interactions does not agree with our calculation because the charge on the line O4–O1 is very small, 0.0017 e, and becomes even smaller on the lines O4–O2 and O4–O3, 0.0005 e and 0.0004 e, respectively.

3.3. Spin localization, symmetry of holes

At the ROHF level, the unpaired spin is located in the CuO_2 planes on oxygens, O2 and O3. The picture is drastically changed when we include the electron correlation. According to Table 6, in both at the UHF level as well as at the MP2 level, the unpaired spin is located in chains on the Cu1. In previous ECM calculations [22,29] at the UHF level, the unpaired spin was indicated in planes on Cu2. The reasons for this contradiction are not clear, because in our calculations the location of unpaired spin at the UHF level is the same as at the MP2 one. Note that the location of the unpaired spin is found to be very sensitive to the spin contamination in the UHF calculation. In our case,

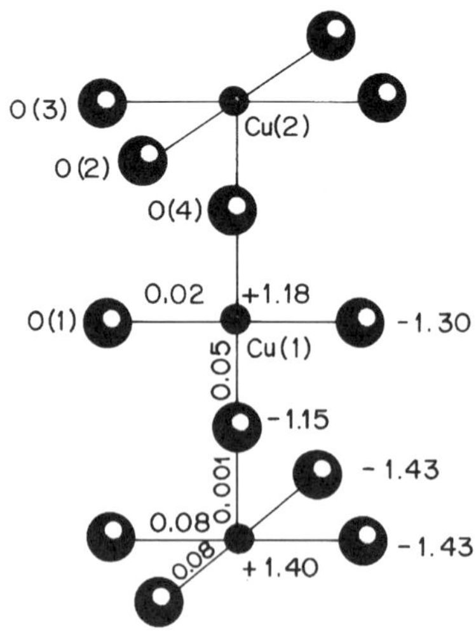

Fig. 3. Charge distribution obtained at the MP2 level.

we found that the value of S^2 in cluster may differ from 0.75 only in the third digit for a physically correct result, even a value of 0.76 for S^2 leads to the location of the unpaired spin elsewhere in the cluster.

The calculations at the electron correlation level do not reveal the unpaired spin in the CuO_2 planes indicating the formation of the singlet spin state. This supports the surmise [54] that due to a strong Cu–O hybridization, all spins in the CuO_2 planes are coupled (the, so-called, Zhang-Rice singlet).

The electron paramagnetic resonance experiments on the yttrium ceramics, on the other hand, are ambiguous. In the study by Murrieta et al. [55], the EPR signal of $YBa_2Cu_3O_7$ sample was interpreted as a superposition of two different lines attributed to the Cu1 and Cu2 sites. In some other studies of yttrium ceramics, the EPR signal was not detected or was attributed to an impurity phase [56]. Thus, further more refined EPR experiments are needed to confirm the location of the unpaired spin in the cluster.

Finally, we turn to the symmetry of the holes on oxygen and copper ions. This problem has a long history. In the early band calculations [11,12], the 2pσ symmetry of the holes on oxygen ions was obtained. While, in the more precise generalized bond calculations [13], but without taking into account a crystal surrounding, the 2pπ symmetry of the holes on oxygen ions was reported. In the Xα ECM calculations of $YBa_2Cu_3O_7$ ceramics [25,26], the 2pσ symmetry of the holes on oxygen ions in planes was obtained. In the present more precise calculation at the MP2 level, we confirmed this result: the holes on plane oxygen ions have the 2pσ symmetry, see Fig. 4. This is in

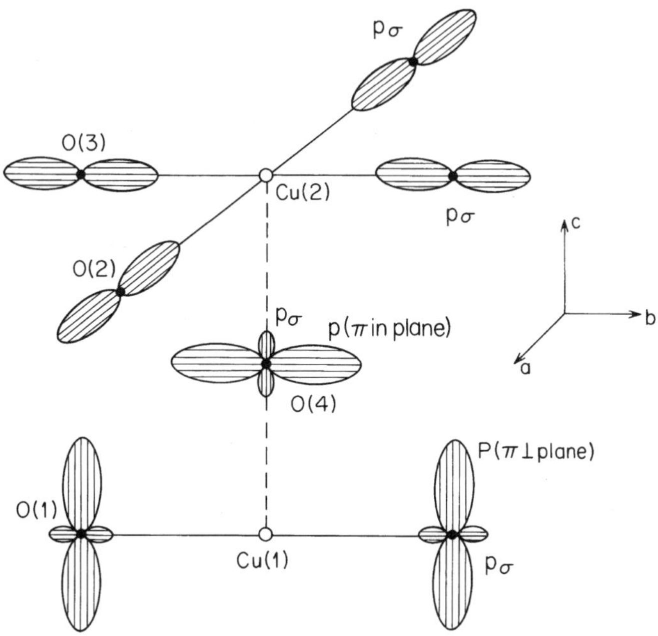

Fig. 4. Symmetry of holes on oxygens (the NBO analysis).

agreement with NMR experiments [57,58] and recent X-ray absorption spectroscopy [35] studies. We also revealed that the symmetry of the holes on the O1 and O4 has predominantly the 2pπ character; or more precisely (at the MP2 level): 0.45 2pπ + 0.14 2pσ on O1 and 0.56 2pπ + 0.21 2pσ on O4. On the Cu1 ion there is a partially unoccupied state associated to the orbitals $3d_z{}^2{}_{-y}{}^2$ which is favourable to the covalent bonding with the O4 and O1 atoms. For the Cu2 ion the symmetry of unoccupied orbitals has a mixed character: 0.17 $3d_z{}^2{}_{-y}{}^2$ + 0.46 $3d_x{}^2$.

Note, that an account of the electron correlation has a great influence on the symmetry of holes. For example, at the ROHF level the holes on the O2 and O3 ions, instead of the 2pσ symmetry, have a mixed symmetry: 0.22 2pσ + 0.13 2pπ although on the O1 and O4 ions, the symmetry of the uncorrelated holes is a pure 2pσ. Thus, as in the case of the spin location, the allowance of the electron correlation is crucial for determining the symmetry of holes.

4. Conclusions

As follows from our study, both aspects of the embedded quantum cluster approach are found to be important for calculations of the ground-state electronic structure of superconducting Y-ceramics. Analysis of the charge distribution in the quantum cluster reveals that Madelung field has a great influence on the Cu2, O4 and O1 ions and much less on the Cu1, O2 and O3 ions. We revealed the strong influence of the electron correlation effects on the charge and spin distributions. The allowance of the electron correlation at the MP2 level decreases the charges on copper ions in 1.5 times. The electron correlation is crucial for a determination of the spin location and for revealing the symmetry of the holes. As was expected, the chemical bond in the $YBa_2Cu_3O_7$ crystal has a mixed covalent-ionic nature. The covalent bonding is more predominant in the CuO_2 planes and between Cu1 and O4 than in the chains while the coupling between the CuO_2 planes and chains has essentially an ionic character. The calculated symmetry of the holes on planar oxygens is 2pσ in accordance with recent experiment results. Our results confirm the existence in the CuO_2 planes of the, so-called, Zhang-Rice singlet [54] and predict the location of the unpaired spin on the Cu1 in chains.

Acknowledgements

We are grateful to R.C. Baetzold, J. Ladik and R. Pandey for helpful discussions and reprints and to S. Shrinivas for sending some unpublished material of her Thesis. This work was supported by DGAPA-UNAM Project IN 108697.

References

1. E. Kaldis et al. (eds): *High – Tc superconductivity 1996: Ten Years after the Discovery*, Klumer Acad. Publ., the Netherlands, (1997)
2. P.W. Anderson: Science **235**, 1196 (1987)
3. V.J. Emory: Nature **333**, 14 (1987)
4. J.R. Schrieffer, X.-G. Wen and S.-C. Xa Zhang : Phys. Rev. Lett. **60**, 944 (1988)

5. V.J. Emory, S.A. Kielson and H.-Q. Lin: Phys. Rev. Lett. **64**, 475 (1990)
6. P. Monthoux, A.V. Balatsky and D. Pines: Phys. Rev. Lett. **67**, 3448 (1991)
7. A.S. Alexandrov and N.F. Mott: *High Tc Superconductivity and Other Superfluids*, Taylor, London 1994; *Polarons and Bipolarons*, World Scientific, Singapore, (1995)
8. P.W. Anderson: *The Theory of Superconductivity in High Tc Cuprates*, Princeton U.P., Princeton, (1997)
9. S.C. Zhang: Science, **275**, 1089 (1997)
10. W.E. Pickett: Revs. Mod. Phys., **61**, 433 (1989)
11. L.F. Mattheiss: Phys. Rev. Lett., **58**, 1028 (1987)
12. J. Yu, A.F. Freeman and J.H. Hu: Phys. Rev. Lett. **58**, 1035 (1987)
13. Y. Guo, Y.M. Langlois and W.A. Goddard III: Science, **239**, 896 (1998)
14. R.P. Gupta and M. Gupta, Solid. St. Comm: **67**, 129, (1988)
15. H. Chen, J. Callaway and P.K. Misra: Phys. Rev. **38**, 195 (1988)
16. R.C. Baetzold: Phys. Rev. B38, 11304 (1988); J. Phys. Chem. Solid **54**, 793 (1993)
17. R.L. Martin and P.W. Saxe : Intern. J. Quant. Chem. **S22**, 237 (1988)
18. H. Adachi and M. Tokano : Physica **C157**, 169 (1989)
19. H. Wang, F. Liu and C. Wang: J. Phys. **C1**, 1983 (1989)
20. G.C. Goodman, D.E. Ellis, E.E. Alp and L. Soderholm: J. Chem. Phys. **91**, 2983 (1989)
21. S. Yamamoto, K. Yamaguchi and K. Nasu: Phys. Rev. **B44**, 7028 (1991)
22. N. Sahoo, S. Markert, T.P. Das and K. Nagamine: Phys. Rev. **B41**, 220 (1990)
23. S.B. Sulaiman, N. Sahoo, T.P. Das, O. Donzelli, E. Torikai and K. Nagamine: Phys. Rev. **B44**, 7028 (1991)
24. I.I. Mazin, S.N. Rashkeev, A.I. Lichtenstein and O.K. Andersen: Phys. Rev. **B46**, 11232 (1992)
25. I.G. Kaplan, B.S. Vaisberg and V.N. Smutny: Sov. Phys. Dokl. Akad. Nauk **322**, 324 (1992)
26. B.S. Vaisberg, I.G. Kaplan and V.N. Smutny: Supeconductivity **5**, 975 (1992)
27. R.L. Martin and P.J. Hay: J. Chem. Phys. **98**, 8680 (1993)
28. N.W. Winter, C.I. Mezbacher and C.E. Violet: Appl. Spectrosc. Revs. **28**, 123 (1993)
29. J. Li, H.L. Liu, and J. Ladik: Chem. Phys. Lett. **230**, 414 (1994)
30. H. Kim and J. Ihm: Phys. Rev. **B51**, 3886 (1995)
31. D. Wechsler and J. Ladik: Phys. Rev. **B55**, 8544 (1997); **B56**, 13750 (1997)
32. N.F. Mott: Proc. Phys. Soc. **A62**, 416 (1956)
33. A. Fujimori, E. Takayama – Muromachi, Y. Uchida and B. Okai: Phys. Rev. **B35**, 8814 (1987)
34. N. Nücker, H. Romberg, X.X. Xi, J. Fink, B. Gegenheimer and Z.X. Zhao: Phys. Rev. **B39**, 6619 (1989)
35. N. Nücker, E. Pellegrin, P. Schweiss, J. Fink, S.L. Molodtsov, C.T. Simmons, G. Kaindl, W. Frentrup, A. Erb and G. Muller–Vogt: Phys. Rev. **B51**, 8529 (1995)
36. W. Brenig, Phys. Rep. **251**, 153 (1995)
37. C.R.A. Catlow and W.C. Mackrodt, (eds): Computer Simulation of Solids, Springer–Verlag, Berlin (1982)
38. J. Vail, R. Pandey, and A.B. Kunz, Revs. Solid State Sci. **5**, 181 (1991)
39. A.L. Shluger, A.H. Harker, V.E. Puchin, N. Itoh and C.R.A. Catlow, Modelling Simul. Mater. Sci. Eng. **1**, 673 (1993)
40. R.J. Cava, Science **247**, 556 (1990)
41. B. Szpunar and V.H. Smith: Jr., Intern. J. Quant. Chem. **S22**, 33 (1988)
42. K. Schwarz, C. Ambrosch – Draxl and P. Blaha: Phys. Rev. **B42**, 2051 (1990)
43. D.J. Singh, K. Shwarz and P. Blaha: Phys. Rev. **B46**, 5849 (1992)
44. I.G. Kaplan, J. Soullard, J. Hernadez – Cobos and R. Pandey: J. Phys: Condensed Matter (in Press)
45. C. Sousas, J. Casanovas, J. Rubio and F. Illas: J. Comp. Chem. **14**, 680 (1993)
46. P.C. Kelires and T.P. Das: Hyperfine Interact. **34**, 285 (1987)
47. S. Shrinivas: Ph. D. Thesis, Suny, Albany, (1995) (unpublished)
48. M.J. Norgett: Harwell Rep. No AERE-R. 7650, (1974)
49. P.P. Ewald: Ann. Phys. **64**, 253 (1921)
50. J.D. Jorgensen, B.W. Veal, A.P. Paulicas, L.J. Nowicki, G.W. Crabtree, H. Claus and W.K. Kwok: Phys. Rev. **B41**, 1863 (1990)
51. Gaussian-94, Revision C. 3, M.J. Frisch, G.W. Trucks, H.B. Schlegel, P.M.W. Gill, B.G. Johnson, M.A.

Robb, J.R. Cheeseman, T. Keith, G.A. Petersson, J.A. Montgomery, K. Raghavachari, M.A. Al-Laham, V.G. Zakrzewski, J.V. Ortiz, J.B. Foresman, J. Cioslowski, B.B. Stefanov, A. Nanayakkara, M. Challacombe, C.Y. Peng, P.Y. Ayala, W. Chen, M.W. Wong, J.L. Andres, E.S. Replogle, R. Gomperts, R.L. Martin, D.J. Fox, J.S. Binkley, D.J. Defrees, J. Baker, J.P. Stewart, M. Head-Gordon, C. Gonzales and J.A. Pople, Gaussian, Inc., Pittsburgh PA, (1995)

52. I.G. Kaplan, J. Hernandez-Cobos, I. Ortega-Blake y O. Novaro: Phys. Rev. **A53**, 2493 (1996)
53. P. Monthoux: Preprint DMR 88-17613
54. F.S. Zhang and T.M. Rice: Phys. Rev. **B37**, 3759 (1988)
55. H. Murrieta, G. Aguilar, J. Ramirez, T. Akachi, R.A. Barrio, R. Escudero and J. Rubio: J. Phys. C: Solid State Phys. **21**, 4999 (1988)
56. J. Albino, O. De Aguiar, J. van der Berg, H.B. Brom, G.J. Nieuwenhuys, J.A. Mydosh, J.A. Mydosh, F.P.F. van Berkel and H.W. Zandbergen: Physica **C156**, 571 (1988)
57. Alloul, T. Ohno and P. Mendels: Phys. Rev. Lett. **63**, 1700 (1989)
58. M. Takigawa, P.C. Hammel, R.H. Heffner, Z. Fisk, J.L. Smith and R.B. Schwarz: Phys. Rev. **B39**, 300 (1989)

Part III

Relativistic Formulations

Energies and Other Properties of Heavy Atoms and Molecules

Uzi Kaldor and Ephraim Eliav
School of Chemistry, Tel Aviv University, 69978 Tel Aviv, Israel

Abstract

The relativistic coupled cluster method starts from the four-component solutions of the Dirac-Fock or Dirac-Fock-Breit equations, and correlates them by the coupled-cluster approach. The Fock-space coupled-cluster method yields atomic transition energies in good agreement (usually better than 0.1 eV) with known experimental values. This is demonstrated here by the electron affinities of group-13 atoms. Properties of superheavy atoms which are not known experimentally can be predicted. Here we show that the rare gas eka-radon (element 118) will have a positive electron affinity. One-, two-, and four-components methods are described and applied to several states of CdH and its ions. Methods for calculating properties other than energy are discussed, and the electric field gradients of Cl, Br, and I, required to extract nuclear quadrupoles from experimental data, are calculated.

1. Introduction

Heavy atoms exhibit large relativistic effects, often too large to be treated perturbatively. The Schrödinger equation must be supplanted by an appropriate relativistic wave equation such as Dirac-Coulomb or Dirac-Coulomb-Breit. Approximate one-electron solutions to these equations may be obtained by the self-consistent-field procedure. The resulting Dirac-Fock or Dirac-Fock-Breit functions are conceptually similar to the familiar Hartree-Fock functions; the Hartree-Fock orbitals are replaced, however, by four-component spinors. Correlation is no less important in the relativistic regime than it is for the lighter elements, and may be included in a similar manner.

The present chapter describes methodology for high-accuracy calculations of systems with heavy and super-heavy elements. The no-virtual-pair Dirac-Coulomb-Breit Hamiltonian, which is correct to second order in the fine-structure constant α, provides the framework of the method. Correlation is treated by the coupled cluster (CC) approach. The method is described in section 2, with the calculation of properties other than energies discussed in 2.4. Several recent applications are reviewed in section 3. The main properties of interest for the heavy atoms are transition energies (ionization potentials, electron affinities, excitation energies), which are obtained with high accuracy (usually 0.1 eV or better); here we show the electron affinities of group-13 atoms, some of which have experimental error bounds of 50–100% and are best determined by calculation. The interesting questions regarding super-heavy elements concern the very large relativistic effects, which may change the basic chemical properties of the atom. Eka-radon (element 118), a rare gas with positive electron affinity, is shown as an example. Molecules including heavy elements exhibit significant relativistic effects too; one-, two-, and four-component calculations for CdH are

A. Hernández-Laguna et al. (eds.), Quantum Systems in Chemistry and Physics, Vol. 1: Basic Problems and Model Systems, 161–176.

described. Finally, preliminary calculations are presented for the electric field gradients at the Cl, Br, and I nuclei, from which the nuclear quadrupole moments may be derived.

2. Methodology

2.2. The Relativistic Hamiltonian

The relativistic many-electron Hamiltonian cannot be written in closed form; it may be derived perturbatively from quantum electrodynamics [1]. The simplest form is the Dirac-Coulomb (DC) Hamiltonian, where the nonrelativistic one-electron terms in the Schrödinger equation are replaced by the one-electron Dirac operator h_D,

$$H_{\mathrm{DC}} = \sum_i h_D(i) + \sum_{i<j} 1/r_{ij}, \tag{1}$$

with

$$h_D = c\boldsymbol{\alpha} \cdot \boldsymbol{p} + \beta c^2 + V_{\mathrm{nuc}}. \tag{2}$$

$\boldsymbol{\alpha}$ and β are the four-dimensional Dirac matrices, and V_{nuc} is the nuclear attraction operator, with the nucleus modeled as a point or finite-size charge. Only the one-electron terms in the DC Hamiltonian include relativistic effects, and the two-electron repulsion remains in the nonrelativistic form. The lowest-order correction to the two-electron repulsion is the Breit [2] operator

$$B_{12} = -\frac{1}{2}[\boldsymbol{\alpha}_1 \cdot \boldsymbol{\alpha}_2 + (\boldsymbol{\alpha}_1 \cdot \boldsymbol{r}_{12}) \cdot (\boldsymbol{\alpha}_2 \cdot \boldsymbol{r}_{12})/r_{12}^2]/r_{12}, \tag{3}$$

yielding the Dirac-Coulomb-Breit (DCB) Hamiltonian

$$H_{\mathrm{DCB}} = \sum_i h_D(i) + \sum_{i<j} (1/r_{ij} + B_{ij}). \tag{4}$$

All equations are in atomic units.

The DC or DCB Hamiltonians may lead to the admixture of negative-energy eigenstates of the Dirac Hamiltonian in an erroneous way [3,4]. The no-virtual-pair approximation [5,6] is invoked to correct this problem: the negative-energy states are eliminated by the projection operator Λ^+, leading to the projected Hamiltonians

$$H_{\mathrm{DC}}^+ = \Lambda^+ H_{\mathrm{DC}} \Lambda^+ \tag{5}$$

or

$$H_{\mathrm{DCB}}^+ = \Lambda^+ H_{\mathrm{DCB}} \Lambda^+. \tag{6}$$

H_{DCB}^+ is correct to second order in the fine-structure constant α, and is expected to be highly accurate for all neutral and weakly-ionized atoms [7]. Higher quantum electrodynamic (QED) terms are required for strongly-ionized species; these are outside the scope of this chapter. A comprehensive discussion of higher QED effects and other aspects of relativistic atomic physics may be found in the proceedings of the 1988 Santa Barbara program [8].

2.3. *The one-electron equation*

The no-pair DCB Hamiltonian (6) is used as a starting point for variational or many-body relativistic calculations [9]. The procedure is similar to the nonrelativistic case, with the Hartree-Fock orbitals replaced by the four-component Dirac-Fock-Breit (DFB) functions. The spherical symmetry of atoms leads to the separation of the one-electron equation into radial and spin-angular parts [10]. The radial four-spinor has the so-called large component $P_{n\kappa}$ in the upper two places and the small component $Q_{n\kappa}$ in the lower two. The quantum number κ (with $|\kappa| = j + 1/2$) comes from the spin-angular equation, and n is the principal quantum number, which counts the solutions of the radial equation with the same κ. Defining

$$\phi_{n\kappa} = \begin{pmatrix} P_{n\kappa}(r) \\ Q_{n\kappa}(r) \end{pmatrix}, \tag{7}$$

the DFB equation has the form

$$F_\kappa \phi_{n\kappa} = \varepsilon_{n\kappa} \phi_{n\kappa}, \tag{8}$$

where the one-electron DFB operator F_κ is [11–15]

$$F_\kappa = \begin{pmatrix} V_{\text{nuc}} + U^{LL} & c\Pi_\kappa + U^{LS} \\ c\Pi_\kappa^{+} + U^{SL} & V_{\text{nuc}} + U^{SS} - 2c^2 \end{pmatrix}, \tag{9}$$

with

$$\Pi_\kappa = -d/dr + \kappa/r \tag{10}$$

and

$$\Pi_\kappa^{+} = d/dr + \kappa/r. \tag{11}$$

V_{nuc} is the nuclear attraction potential. In the uniform charge distribution model used here, the charge of a nucleus of atomic mass A is distributed uniformly over a sphere with radius $R = 2.2677 \times 10^{-5} A^{-1/3}$. The nuclear potential for a nucleus with charge Z is then

$$V_{\text{nuc}} = \begin{cases} -Z/r & \text{for } r > R \\ -(Z/2R)(3 - r^2/R^2) & \text{for } r \leqslant R \end{cases}. \tag{12}$$

The terms U^{LL} etc. represent the one-body mean-field potential, which approximates the two-electron interaction in the Hamiltonian, as is the practice in SCF schemes. In the DFB equations this interaction includes the Breit term (3) in addition to the electron repulsion $1/r_{ij}$.

The radial functions $P_{n\kappa}(r)$ and $Q_{n\kappa}(r)$ may be obtained by numerical integration [16,17] or by expansion in a basis (for recent reviews see [18,19]). Since the Dirac Hamiltonian is not bound from below, failure to observe correct boundary conditions leads to 'variational collapse' [20,21], where admixture of negative-energy solutions may yield energies much below experimental. To avoid this failure, the basis sets used for expanding the large and small components must maintain 'kinetic balance' [22,23].

In the nonrelativistic limit ($c \to \infty$), the small component is related to the large component by [20]

$$Q_{n\kappa}(r) = (2c)^{-1}\Pi_\kappa^+ P_{n\kappa}(r), \tag{13}$$

where Π_κ^+ is defined in (11). The simplest way to obtain kinetic balance is to derive the small-component basis functions from those used to span the large component by

$$\chi_{\kappa j}^S = \Pi_\kappa^+ \chi_{\kappa j}^L. \tag{14}$$

Ishikawa and coworkers [15,24] have shown that G-spinors, with orbitals spanned in Gaussian-type functions (GTF) chosen according to (14), satisfy kinetic balance for finite c values if the nucleus is modeled as a uniformly-charged sphere.

2.4. *The Fock-space coupled-cluster method*

The coupled-cluster method is well-known by now, and only a brief account of aspects relevant to our applications is given here.

The Dirac-Coulomb-Breit Hamiltonian H_{DCB}^+ may be rewritten in second-quantized form [5,15] in terms of normal-ordered products of spinor creation and annihilation operators $\{r^+s\}$ and $\{r^+s^+ut\}$,

$$H = H_{\mathrm{DCB}}^+ - \langle 0|H_{\mathrm{DCB}}^+|0\rangle = \sum_{rs} f_{rs}\{r^+s\} + \frac{1}{4}\sum_{rstu}\langle rs||tu\rangle\{r^+s^+ut\}, \tag{15}$$

where

$$\langle rs||tu\rangle = \langle rs|tu\rangle - \langle rs|ut\rangle \tag{16}$$

and

$$\langle rs|tu\rangle = \int d\mathbf{x}_1 d\mathbf{x}_2 \Psi_r^*(\mathbf{x}_1)\Psi_s^*(\mathbf{x}_2)(r_{12}^{-1} + B_{12})\Psi_t(\mathbf{x}_1)\Psi_u(\mathbf{x}_2). \tag{17}$$

Here f_{rs} and $\langle rs||tu\rangle$ are, respectively, elements of one-electron Dirac-Fock and antisymmetrized two-electron Coulomb-Breit interaction matrices over Dirac four-component spinors. The effect of the projection operators Λ^+ is now taken over by the normal ordering, denoted by the curly braces in (15), which requires annihilation operators to be moved to the right of creation operators as if all anticommutation relations vanish. The Fermi level is set at the top of the highest occupied positive-energy state, and the negative-energy states are ignored.

By adopting the no-pair approximation, a natural and straightforward extension of the nonrelativistic open-shell CC theory emerges. The multireference valence-universal Fock-space coupled-cluster approach is employed [25], which defines and calculates an effective Hamiltonian in a low-dimensional model (or P) space, with eigenvalues approximating some desirable eigenvalues of the physical Hamiltonian. The effective Hamiltonian has the form [26]

$$H_{\mathrm{eff}} = PH\Omega P \tag{18}$$

where Ω is the normal-ordered wave operator,

$$\Omega = \{\exp(S)\}. \tag{19}$$

The Fock-space approach starts from a reference state (closed-shell in our applications, but other single-determinant functions may also be used), correlates it, then adds and/or removes electrons one at a time, recorrelating the whole system at each stage. The sector (m, n) of the Fock space includes all states obtained from the reference determinant by removing m electrons from designated occupied orbitals, called valence holes, and adding n electrons in designated virtual orbitals, called valence particles. The practical limit is $m + n \leqslant 2$, although higher sectors have also been tried [27]. The excitation operator is partitioned into sector operators

$$S = \sum_{m\geqslant 0}\sum_{n\geqslant 0} S^{(m,n)}. \tag{20}$$

This partitioning allows for partial decoupling of the open-shell CC equations. The equations for the (m, n) sector involve only S elements from sectors (k, l) with $k \leqslant m$ and $l \leqslant n$, so that the very large system of coupled nonlinear equations is separated into smaller subsystems, which are solved consecutively: first, the equations for $S^{(0,0)}$ are iterated to convergence; the $S^{(1,0)}$ (or $S^{(0,1)}$) equations are then solved using the known $S^{(0,0)}$, and so on. This separation, which does not involve any approximation, reduces the computational effort significantly. The effective Hamiltonian (18) is also partitioned by sectors. An important advantage of the method is the simultaneous calculation of a large number of states.

Each sector excitation operator is, in the usual way, a sum of virtual excitations of one, two, . . ., electrons,

$$S^{(m,n)} = \sum_{l} S_l^{(m,n)}, \tag{21}$$

with l going, in principle, to the total number of electrons. In practice, l has to be truncated. The level of truncation reflects the quality of the approximation, i.e., the extent to which the complementary Q space is taken into account in the evaluation of the effective Hamiltonian. In the applications described below the series (21) is truncated at $l = 2$. The resulting CCSD (coupled cluster with single and double excitations) scheme involves the fully self-consistent, iterative calculation of all one- and two-body virtual excitation amplitudes and sums all diagrams with these excitations to infinite order. As negative-energy states are excluded from the Q space, the diagrammatic summations in the CC equations are carried out only within the subspace of the positive-energy branch of the DF spectrum.

2.5. *Properties other than energy*

Much information of interest for atomic and molecular systems involves properties other than energy, usually observed via the energy shifts generated by coupling to some external field. The desired property is then the derivative of the energy with respect to the external field, which may be obtained by two different approaches. The finite-field method solves the Schrödinger equation in the presence of the external field, yielding the energy as a (numerical) function of the field, from which the derivative is obtained. Alternatively, the derivatives may be calculated analytically by the Z-vector or Λ-operator method [28]. The latter method is much preferred if many properties are desired; it is particularly useful for optimizing molecular geometries and mapping potential surfaces, where one requires energy derivatives with respect to all internal coordinates. The finite-field method, on the other hand, has the advantage of being very easy to program and add to an existing computer code, and may be useful if one particular property is needed. An early example is the Stark effect of the hyperfine structure of atomic Li [29]. The spin density at the Li nucleus was calculated with finite values of the electric field, and the derivative obtained numerically. Section 3.5 below describes a similar approach to the electric field gradients (EFG) at the Cl, Br, and I nuclei, needed to obtain the nuclear quadrupole moments Q: the Hamiltonian is modified to include a finite nuclear quadrupole, and the EFG is calculated from the derivative of the total energy with respect to Q.

Conflicting considerations govern the size of the external field added to the Hamiltonian. On one hand, the effect must be large enough not to disappear in the precision of the calculation. On the other hand, a large perturbation may go beyond the linear regime, introducing errors in the derivative. Linearity is easily checked by monitoring the energy change as function of the field. Electric fields of 10^3–10^4 kV/cm 0were used in the Li example [29], and nuclear quadrupoles of 50–500 barn were taken in the halogen atoms case. These values are unphysically large, but linearity was maintained in both cases.

3. Applications

3.1. Atoms

Different ways of implementing the relativistic coupled cluster (RCC) method are known. A numerical procedure for solving the pair equation has been developed by Lindgren and coworkers [30] and applied to two-electron atomic systems [31]. Other approaches use discrete basis sets of local or global functions. This makes the application of the projection operators onto the positive-energy space much easier than in the numerical scheme; one simply ignores the negative-energy branch of the one-electron spectrum. A technique based on local splines was developed by Blundell et al. [32], while the Göteborg group introduced another type of local basis, obtained by discretizing the radial space [33]. The first relativistic coupled cluster calculation in a global basis [34] appeared in 1990, but was limited to *s* orbitals only, both in the occupied and virtual space. A more general and sustained implementation started two years later, with pilot calculations for light atoms in closed-shell [35] and open-shell [36] states. The method has since been applied to many heavy atoms, where relativistic effects are crucial to the correct description of atomic structure. Calculated properties include ionization potentials, excitation energies, electron affinities, fine-structure splittings, and for super-heavy elements — the nature of the ground state. The additivity of relativistic and correlation effects was also studied. Systems investigated include the gold atom [37], few-electron ions [38], the alkali-metal atoms Li–Fr [39], the Xe atom

[40], the f^2 shells of $Pr3^+$ and U^{4+} [41], the ytterbium [42], lutetium [42], mercury [43], barium [44], radium [44], thallium [45], and bismuth [46] atoms, and the superheavy elements lawrencium [42], rutherfordium [47], 111 [48], 112 [43], 113 [45], 115 [46], and 118 [49]. Representative applications are described below.

The spherical symmetry of atoms, which leads to angular decomposition of the wave function and coupled-cluster equations, is used at both the Dirac-Fock-Breit [15] and RCC [37,39] stages of the calculation. The energy integrals and CC amplitudes which appear in the Goldstone-type diagrams defining the CC equations are decomposed in terms of vector-coupling coefficients, expressed by angular-momentum diagrams, and reduced Coulomb-Breit or S matrix elements, respectively. The reduced equations for single and double excitation amplitudes are derived using the Jucys-Levinson-Vanagas theorem [26] and solved iteratively. This technique makes possible the use of large basis sets with high l values, as a basis orbital gives rise to two functions at most, with $j = l \pm 1/2$, whereas in Cartesian coordinates the number of functions increases rapidly with l. Typically we go up to h ($l = 5$) or i ($l = 6$) orbitals. To account for core-polarization effects, which may be important for many systems, we correlate at least the two outer shells, usually 20–40 electrons. Finally, uncontracted Gaussians are used, since contraction leads to problems in satisfying kinetic balance and correctly representing the small components. On the other hand, it has been found that high-energy virtual orbitals have little effect on the transition energies we calculate, since these orbitals have nodes in the inner regions of the atom and correlate mostly the inner-shell electrons, which we do not correlate anyway. These virtual orbitals, with energies above 80 or 100 hartree, are therefore eliminated from the RCC calculation.

3.2. *Electron affinities of group-13 elements*

Of the five group-13 elements, only B and Al have experimentally well characterized electron affinities. Lists of recommended EAs [50,51] show errors ranging from 50% to 100% for Ga, In, and Tl. Very few calculations have appeared for the latter atoms. These include the multireference configuration interaction (MRCI) of Arnau *et al.* using pseudopotentials [52], our relativistic coupled cluster work on Tl [45], and the multiconfiguration Dirac-Fock (MCDF) computation of Wijesundera [53].

Comparison with experimental values are meaningful only for boron and aluminum. The MRCI values for Al (0.45 eV) and the MCDF results for B (0.26 eV) and Al (0.43 eV) are in good agreement with the Hotop and Lineberger values (0.28 and 0.44 eV, respectively). The MRCI and MCDF EAs for the other atoms agree with each other (0.29 and 0.30 eV for Ga, 0.38 and 0.39 eV for In, 0.27 and 0.29 eV for Tl). The RCC EA of Tl is much higher at 0.40(5) eV. A major difference between the RCC and the other two methods lies in the number of electrons correlated. While [52] and [53] correlate valence electrons only, three for the neutral atom and four for the anion, we correlated 35 electrons in Tl and 36 in Tl^-. A RCC study of all five elements was undertaken [54], with the aim of determining all five EAs and, in particular, the effect of inner-shell correlation and virtual space used on the calculated values.

Large basis sets of Gaussian-type orbitals (up to $35s27p21d9g6h4i$) were used, taken from the universal basis set of Malli *et al.* [55]. Many electrons were correlated (5 in B,

11 in Al, 27 in Ga, 21 in In, and 35 in Tl), to account for core polarization. Triple excitations proved important for B but not for the heavier elements (see [54] for details). The calculated EAs are shown in Table 1 and compared with experiment [50] and with the MCDF results [53]. Very good agreement with experimental values is obtained when the latter are known (B and Al). Our results are close to corresponding MCDF values for all atoms except Tl. The main difference between the two calculations is in the number of electrons correlated, three in the MCDF work vs. 35 in RCC. A careful study of the contribution of the different core shells to the electron affinity was therefore undertaken. Table 2 shows that correlation of the $4f$, $5s$, and $5p$ shells has small but not insignificant effect (0.015 eV); the $5d$ shell, on the other hand, has a substantial contribution (0.08 eV), accounting for most of the difference between RCC and MCDF values. Additional contributions to this difference come from the truncation of the MCDF space at $l = 3$; the RCC value with this truncation and correlating only the $6s$ and $6p$ electrons is very close to the MCDF results, confirming the source of the difference. Neglect of dynamic correlation in MCDF calculations has affected energies in other cases: the RCC values for excitation energies in Pr^{3+} [41] have one fourth the error of corresponding MCDF values [56]; and dynamic correlation changes the order of the two low level of rutherfordium (element 104) [47]. A similar study carried out

Table 1 *Group 13 electron affinities (eV).*

Atom	Expt. [50,51]	RCCSD [54]	MCDF [53]
B	0.277(10)	0.279[a]	0.260
Al	$0.44094(^{+66}_{-48})$	0.427[a]	0.433
Ga	0.30(15)	0.301	0.305
In	0.30(20)	0.419	0.393
Tl	0.20(20)	0.40(5)	0.291

[a]Including triple excitations

Table 2 *Dependence of Tl EA on amount of correlation (eV)*

Basis	Correlated Tl electrons	EA [54]
$l \leqslant 6$	$4f^{14}5s^25p^65d^{10}6s^26p$	0.417
	$4f^{14}5p^65d^{10}6s^26p$	0.416
	$5p^65d^{10}6s^26p$	0.409
	$5d^{10}6s^26p$	0.402
	$6s^26p$	0.319
$l \leqslant 5$	$6s^26p$	0.318
$l \leqslant 4$	$6s^26p$	0.315
$l \leqslant 3$	$6s^26p$	0.304
MCDF [53]		
$l \leqslant 3$	$6s^26p$	0.291

for In (Table 3) shows dynamic correlation effects one-third those of Tl (~0.04 eV), explaining the good agreement between MCDF and RCC. Finally, calculated ionization potentials of group-13 elements (Table 4) show excellent agreement with experiment.

3.3. *Element 118 — a rare gas with electron affinity*

One of the most dramatic effects of relativity is the contraction and concomitant stabilization of s orbitals. An intriguing question is whether the $8s$ orbital of element 118, the next rare gas, would be stabilized sufficiently to give the atom a positive electron affinity. Using the neutral atom Dirac-Fock orbitals as a starting point raises a problem, since the $8s$ orbital has positive energy and tends to 'escape' to the most diffuse basis functions. This may be avoided by calculating the unoccupied orbitals in an artificial field, obtained by assigning partial charges to some of the occupied shells. The unphysical fields are compensated by including an appropriate correction in the perturbation operator. A series of calculations with a variety of fields gave electron affinities differing by a few wave numbers. Thus, assigning a charge of 0.8e to the $7p_{3/2}$ electrons gave an EA of 454 cm^{-1}; a charge of 0.75e on the $7s$ electrons yielded 449 cm^{-1}; and putting 0.9e on all $7s$ and $7p$ electrons yielded an electron affinity of 437 cm^{-1}. Several other tests of the stability and reliability of the calculations were carried out. The basis set convergence was checked by calculating several l limits (with

Table 3 *Dependence of In EA on amount of correlation (eV)*

Basis	Correlated In electrons	EA [54]
$l \leq 6$	$4s^2 4p^6 4d^{10} 5s^2 5p$	0.419
	$4p^6 4d^{10} 5s^2 5p$	0.417
	$4d^{10} 5s^2 5p$	0.421
	$5s^2 5p$	0.387
$l \leq 5$	$5s^2 5p$	0.386
$l \leq 4$	$5s^2 5p$	0.384
$l \leq 3$	$5s^2 5p$	0.377
MCDF [53]		
$l \leq 3$	$5s^2 5p$	0.393

Table 4 *Ionization potentials of group 13 elements (eV)*

Atom	Expt. [70]	RCCSD [54]
B	8.298	8.287
Al	5.986	5.938
Ga	5.998	5.984
In	5.786	5.798
Tl	6.108	6.110

a charge of $0.8e$ on the $7p_{3/2}$ electrons). The f, g, and h limits came out as 427, 447, and 452 cm^{-1}, respectively. The correlated relativistic electron affinity is put at 0.056 eV, with an estimated error of 0.01 eV. It should be emphasized that correlated nonrelativistic or relativistic uncorrelated calculations yield no electron affinity. The Rn atom does not show a bound state of the anion even at the RCC level. Further details may be found in [49].

3.4. Molecules: CdH

Molecules are more difficult to treat accurately than atoms, because of the reduced symmetry. An additional complication arises in relativistic calculations: the Dirac-Fock-(-Breit) orbitals will in general be complex. One way to circumvent this difficulty is by the Douglas-Kroll-Hess transformation [57], which yields a one-component function with computational effort essentially equal to that of a nonrelativistic calculation. Spin-orbit interaction may then be added as a perturbation. Implementation to AuH and Au_2 has been reported [58]. Progress has also been made in the four-component formulation [59], and the MOLFDIR package [60] has been extended to include the CC method. Application to SnH_4 has been described [61]; here we present a recent calculation of several states of CdH and its ions [62], with one-, two-, and four-component methods.

The present calculation uses the Fock-space scheme

$$\mathrm{CdH}^+(0,0) \rightarrow \mathrm{CdH}(0,1) \rightarrow \mathrm{CDH}^-(0,2) \tag{22}$$

to calculate molecular data including bond length, adiabatic and vertical ionization potentials, transition energies, vibrational and anharmonic parameters and dissociation energies. Starting with the CdH^+ closed-shell state, an all-electron SCF function is first calculated. This function may be nonrelativistic Hartree–Fock, relativistic four-component Dirac–Fock, or relativistic one-component Douglas–Kroll. The external 18 electrons are correlated at the CCSD level, so that core-valence polarization is included explicitly. The two-component scheme involves adding the spin-orbit coupling terms as a perturbation to the one-component DK SCF function. The five lowest states of CdH are obtained by adding an electron in the 5σ, $5\pi_{1/2}$, $5\pi_{3/2}$, 6σ and 7σ valence orbitals (the 7σ was dropped in two- and four-component calculations because of convergence problems in the CC procedure). The ground $5\sigma^2$ state of CdH^- is then obtained by adding a second electron to the lowest $^2\Sigma$ state of CdH.

The molecular orbitals in the nonrelativistic and one-component calculations and the large component in the Dirac–Fock functions were spanned in the Cd $(18s14p9d)/[9s7p6d]$ basis of [63] and the H $(5s2p)/[3s1p]$ set [61]. Contraction coefficients were taken from corresponding atomic SCF calculations. The basis for the small components in the Dirac–Fock calculations is derived by the MOLFDIR program from the large-component basis. The basis set superposition error is corrected by the counterpoise method [64]. The Breit interaction was found to have a very small effect and is therefore not included in the results.

Table 5 shows the RCC spectroscopic parameters of CdH^+ and compares our results with experiment [65] and with some previous calculations, including nonrelativistic

Table 5 *Spectroscopic parameters of CdH^+ $X^1\Sigma^+$ state*

Method	Ref.	R_e()	ω_e(cm^{-1})	$\omega_e x_e$(cm^{-1})	D_e(eV)
R-SCF	[66]	1.762	1728		0.67
NR-GVB+CI	[63]	1.709	1696		1.82
NR-SCF	[67]	1.73	1810		0.78
R-SCF	[67]	1.68	1880		0.87
R-CAS	[67]	1.72	1660		1.74
R-CI	[67]	1.70	1690		1.86
NRCC	[62]	1.758	1625	42.4	1.812
RCC-1	[62]	1.708	1661	34.5	1.893
RCC-2	[62]	1.708	1666	35.5	1.899
RCC-4	[62]	1.709	1672	35.3	1.908
Expt.	[65]	1.667	1771	35.4	2.081

generalized valence bond (GVB) plus configuration interaction (CI) [63], relativistic Hartree–Fock (RHF) [66], and relativistic Hartree–Fock, complete active space (CAS), and CI, with inner electrons represented by effective core potentials 67. The comparison of relativistic with nonrelativistic coupled-cluster data reveals significant bond strengthening, with a bond contraction of 0.05, a dissociation energy higher by 0.1 eV, and ω_e larger by 47 cm^{-1}. The one- and two-component functions account for most of the relativistic correction and yield results close to four-component values. The agreement of the RCC numbers with experiment is very good; the largest source of remaining errors is probably basis set incompleteness.

The only calculation we found for CdH is the work of Balasubramanian [68], using CI with relativistic effective core potentials. The coupled-cluster results are presened in Table 6. Calculated values for R_e, ω_e and D_e agree very well with experiment. Relativity contracts the bond by 0.04 and *reduces* the binding energy by 0.16 eV. The one- and two-component DK method reproduce the relativistic effects closely. Similar trends are observed for the excited states (Tables 7–9). Comparison with experiment is difficult for these states, since many of the experimental values are based on incomplete or uncertain data [65]. Calculated results for the CdH^- anion are shown in Table 10. The

Table 6 *Spectroscopic parameters of* CdH $X^2\Sigma^+$ *ground state*

Method	R_e()	ω_e(cm^{-1})	$\omega_e x_e$(cm^{-1})	IP(cm^{-1})	D_e(eV)
CI [68]	1.794	1298			0.48
NRCC [62]	1.820	1386	30.3	57114	0.867
RCC-1 [62]	1.784	1377	36.5	58619	0.709
RCC-2 [62]	1.780	1366	39.3	58722	0.718
RCC-4 [62]	1.778	1370	37.2	58660	0.703
Expt. [65]	1.781	1337			0.68

Table 7 *Spectroscopic parameters of CdH A $^2\Pi$ states (when two rows appear, they show values for $^2\Pi_{1/2}$ and $^2\Pi_{3/2}$, respectively)*

Method	R_e()	$\omega_e(cm^{-1})$	$\omega_e x_e(cm^{-1})$	$T_e(cm^{-1})$
NRCC [62]	1.765	1571	41.2	20590
RCC-1 [62]	1.711	1640	35.3	21998
RCC-2 [62]	1.708	1637	36.1	21644
	1.707	1652	37.3	22650
RCC-4 [62]	1.702	1652	38.9	21680
	1.701	1659	39.2	22670
d-FOCI [68]	1.72	1592		23027
	1.722	1590		23919
Expt. [65]	1.669	1677		22117
	1.657	1758	38.6	23116

Table 8 *Spectroscopic parameters of CdH B $^2\Sigma^+$ state*

Method	R_e()	$\omega_e(cm^{-1})$	$\omega_e x_e(cm^{-1})$	$T_e(cm^{-1})$
NRCC [62]	2.715	843	14.7	26701
RCC-1 [62]	2.617	894	10.3	27940
d-FOCI [68]	2.433	967		24494
Expt. [65]	2.39	1000	17	24961

Table 9 *Spectroscopic parameters of CdH C $^2\Sigma^+$ state*

Method	R_e()	$\omega_e(cm^{-1})$	$\omega_e x_e(cm^{-1})$	$T_e(cm^{-1})$
NRCC [62]	1.763	1560	31.6	37501
RCC-1 [62]	1.716	1678	49.7	38440
RCC-2 [62]	1.710	1650	50.1	38600
RCC-4 [62]	1.709	1654	48.1	38556
d-FOCI [68]	1.734			42794
Expt. [65]	1.68	1567	50	40202

Table 10 *Spectroscopic parameters of CdH^- X $^1\Sigma^+$ state [62]*

Method	R_e()	$\omega_e(cm^{-1})$	$\omega_e x_e(cm^{-1})$	$EA(cm^{-1})$
NRCC	1.847	1442	35.7	3680
RCC-1	1.808	1390	38.9	3561
RCC-2	1.806	1389	40.5	3575
RCC-4	1.806	1389	42.0	3567

predicted adiabatic electron affinity is 0.44 eV, while the vertical EA at the equilibrium separation of the neutral molecule is 0.43 eV. We are not aware of an experimental measurement.

3.5. Properties: electric field gradients and nuclear quadrupoles

One method of determining nuclear quadrupole moment Q is by measuring the quadrupole coupling constant, given by eqQ/h, where e is the charge of the electron and q the electric field gradient due to the electrons at the atomic nucleus. The extraction of Q depends on an accurately calculated q. As a test of our finite-field relativistic coupled cluster approach, preliminary results for Cl, Br, and I are presented.

The closed-shell states of the anions serve as reference, from which the neutral atoms are reached by removing one electron. A $32s24p18d$ basis is used. The external 16 electrons of Cl^- and 26 electrons of Br^- and I^- are correlated. Interaction with a finite Q is included explicitly in the perturbation part of the Hamiltonian, together with the correlation perturbation. As explained above, Q has to be chosen large enough to eliminate precision problems, and small enough to retain linearity. A typical test of linearity is shown in Table 11. Table 12 summarizes the calculated moments. Large correlation effects appear for all three atoms. Relativistic effects are small ($\sim$2%) for Br, but increase to 7% for iodine. The Breit term has little effect on the results. The RCC results are in very good agreement with the best available estimates [69]. These are not our final values, since convergence with respect to the basis is currently being tested.

Table 11 *Test of linearity for Br: the quadrupole coupling constant Δ (in 10^{-8} au) vs. quadrupole moment Q (barn)*

Q	Δ	Δ/Q
50	881	17.62
100	1762	17.62
150	2642	17.61

Table 12 *Calculated nuclear quadrupoles (barn)*

	Nonrel		Rel			'Best' [69]
	HF	CC	DF	CC	+Breit	
^{35}Cl			−0.1015	−0.0824		−0.0825
^{79}Br	0.400	0.341	0.392	0.332	0.333	0.331
^{127}I	−0.956	−0.834	−0.893	−0.774	−0.778	−0.789

4. Summary and Conclusion

The relativistic coupled-cluster method includes simultaneously relativistic terms through second order in the fine-structure constant α and correlation effects summed to all orders of the one- and two-electron excitations. In atomic systems, where spherical symmetry allows the use of large basis sets, the method makes possible calculation of large numbers of heavy-atom states with unprecedented accuracy, and gives reliable predictions for superheavy elements. The largest remaining source of error is probably the omission of triple virtual excitations. Molecules present a more difficult challenge, and few four-component RCC applications have been reported to date. The one-component Douglas-Kroll-Hess transformation provides a cheap, accurate alternative when spin-orbit interaction may be neglected. This interaction may be added as a perturbation in a two-component approach. Properties other than energy may also be calculated with high accuracy.

Acknowledgements

Large parts of the work described in this chapter have been done in collaboration with Yasuyuki Ishikawa of the University of Puerto Rico, Bernd Hess of Bonn University, and Pekka Pyykkö of Helsinki. Support was provided by the Israel Science Foundation, the Ministry of Science, the U.S.-Israel Binational Science Foundation and the German-Israeli Foundation for Scientific Research and Development.

References

1. See, e.g., J. Sucher, in *Relativistic, Quantum Electrodynamic, and Weak Interaction Effects in Atoms*, ed. W. Johnson, P. Mohr, and J. Sucher (American Institute of Physics, New York, 1989), p. 28.
2. G. Breit, *Phys. Rev.* **B34**, 553 (1929); *ibid.* **36**, 383 (1930); *ibid.* **39**, 616 (1932).
3. G.E. Brown and D.G. Ravenhall, *Proc. Roy. Soc.* A **208**, 552 (1951).
4. H.A. Bethe and E.E. Salpeter, *Quantum Mechanics of One- and Two-Electron Atoms* (Springer-verlag, Berlin, 1957).
5. J. Sucher, *Phys. Rev.* A **22**, 348 (1980); *Phys. Scr.* **36**, 271 (1987).
6. W. Buchmüller and K. Dietz, *Z. Phys.* C **5**, 45 (1980).
7. I. Lindgren, in *Many-Body Methods in Quantum Chemistry*, ed. U. Kaldor, Lecture Notes in Chemistry Vol. 52 (Springer-Verlag, Heidelberg, 1989) p. 293; *Nucl. Instrum. Methods* **B31**, 102 (1988).
8. *Relativistic, Quantum Electrodynamic, and Weak Interaction Effects in Atoms*, ed. W. Johnson, P. Mohr, and J. Sucher (American Institute of Physics, New York, 1989).
9. M. Mittleman, *Phys. Rev.* A **4**, 893 (1971); *ibid.* **5**, 2395 (1972); *ibid.* **24**, 1167 (1981).
10. A.S. Davydov, *Quantum Mechanics* (NEO Press, Peaks Island, Maine, 1966) chap. VIII.
11. Y.-K. Kim, *Phys. Rev.* **154**, 17 (1967).
12. T. Kagawa, *Phys. Rev.* A **12**, 2245 (1975); *ibid.* **22**, 2340 (1980).
13. W.R. Johnson and J. Sapirstein, *Phys. Rev. Lett.* **57**, 1126 (1986); W.R. Johnson, M. Idrees, and J. Sapirstein, *Phys. Rev.* A **35**, 3218 (1987); S.A. Blundell, W.R. Johnson, Z.W. Liu, and J. Sapirstein, *Phys. Rev.* A **39**, 3768 (1989); W.R. Johnson, S.A. Blundell, and J.W. Sapirstein, *Phys. Rev.* A **37**, 307 (1988); *ibid.* **37**, 2764 (1988); *ibid.* **41**, 1689 (1990).
14. H.M. Quiney, I.P. Grant, and S. Wilson, *J. Phys.* B **20**, 1413 (1987); *Phys. Scr.* **36**, 460 (1987); in *Many-Body Methods in Quantum Chemistry*, ed. U. Kaldor, Lecture Notes in Chemistry Vol. 52 (Springer-Verlag, Heidelberg, 1989); *J. Phys.* B **23**, L271 (1990); I.P. Grant and H.M. Quiney, *Adv. Atom. Molec.*

Phys. **23**, 37 (1988).

15. Y. Ishikawa, R.C. Binning, and H. Sekino, *Chem. Phys. Lett.* **160**, 206 (1989); Y. Ishikawa, *Phys. Rev.* A **42**, 1142 (1990); *Chem. Phys. Lett.* **166**, 321 (1990); Y. Ishikawa and H.M. Quiney, *Phys. Rev.* A **47**, 1732 (1993); Y. Ishikawa and K. Koc, *Phys. Rev.* A **50**, 4733 (1994).
16. J.-P. Desclaux, *Comput. Phys. Commun.* **9**, 31 (1975).
17. I.P. Grant, B.J. McKenzie, P.H. Norrington, D.F. Mayers, and N.C. Pyper, *Comput. Phys. Commun.* **21**, 207 (1980).
18. Y. Ishikawa and U. Kaldor, in *Computational Chemistry: Review of Current Trends*, ed. J. Leszczynski, (World Scientific, Singapore, 1996), vol. I, p. 1.
19. U. Kaldor and E. Eliav, *Adv. Quantum Chem.* **31**, 313 (1998).
20. Y.-S. Lee and A.D. McLean, *J. Chem. Phys.* **76**, 735 (1982).
21. P.J.C. Aerts and W.C. Nieuwpoort, *Chem. Phys. Lett.* **113**, 165 (1985); *ibid.* **125**, 83 (1986).
22. R. E. Stanton and S. Havriliak, *J. Chem. Phys.* **81**, 1910 (1984).
23. Y. Ishikawa, R. Baretty, and R.C. Binning, *Intern. J. Quantum Chem. Symp.* **19**, 285 (1985); Y. Ishikawa and H. Sekino, *Chem. Phys. Lett.* **165**, 243 (1990).
24. Y. Ishikawa, R.C. Binning, and K.M. Sando, *Chem. Phys. Lett.* **101**, 111 (1983); *ibid.* **105**, 189 (1984); *ibid.* **117**, 444 (1985); Y. Ishikawa, R. Baretty, and R.C. Binning, *Chem. Phys. Lett.* **121**, 130 (1985); Y. Ishikawa and H.M. Quiney, *Intern. J. Quantum Chem. Symp.* **21**, 523 (1987).
25. For a discussion of Fock-space coupled cluster see U. Kaldor, *Theor. Chim. Acta* **80**, 427 (1991) and references therein.
26. I. Lindgren and J. Morrison, *Atomic Many-Body Theory*, 2nd ed. (Springer Verlag, Berlin, 1986).
27. S.R. Hughes and U. Kaldor, *Chem. Phys. Lett.* **194**, 99 (1992); *ibid.* **204**, 339 (1993); *Phys. Rev.* A **47**, 4705 (1993); *J. Chem. Phys.* **99**, 6773 (1993); *Intern. J. Quantum Chem.* **55**, 127 (1995).
28. J. D. Goddard, N. C. Handy, and H. F. Schaefer, *J. Chem. Phys.* **71**, 1259 (1979); E. A. Salter, G. W. Trucks, and R. J. Bartlett, *J. Chem. Phys.* **90**, 1752 (1989).
29. U. Kaldor, *J. Phys.* B **6**, 71 (1973).
30. S. Salomonson, I. Lindgren, and A.-M. Mårtensson, *Phys. Scr.* **21**, 351 (1980).
31. E. Lindroth, *Phys. Rev.* A **37**, 316 (1988).
32. S.A. Blundell, W.R. Johnson, Z.W. Liu, and J. Sapirstein, *Phys. Rev.* A **39**, 3768 (1989); *ibid.* **40**, 2233 (1989); S.A. Blundell, W.R. Johnson, and J. Sapirstein, *Phys. Rev. Lett.* **65**, 1411 (1990); *Phys. Rev.* A **43**, 3407 (1991); Z.W. Liu and H.P. Kelly, *ibid.* **43**, 3305 (1991).
33. S. Salomonson and P. Öster, *Phys. Rev.* A **40**, 5548 (1989); E. Lindroth and S. Salomonson, *ibid.* **41**, 4659 (1990); A.C. Hartly, E. Lindroth and A-M. Mårtensson-Pendrill, *J. Phys.* B **23**, 1990 (1990); A.C. Hartly and A-M. Mårtensson-Pendrill, *Z. Phys.* D **15**, 309 (1990); *J. Phys.* B **24**, 1193 (1991); A.-M. Mårtensson-Pendrill, S.A. Alexander, L. Adamowicz, N. Oliphant, J. Olsen, P. Öster, H.M. Quiney, S. Salomonson, and D. Sundholm, *Phys. Rev.* A **43**, 3355 (1991); E. Lindroth, H. Persson, S. Salomonson and A-M. Mårtensson-Pendrill, *Phys. Rev.* A **45**, 1493 (1992); E. Lindroth and J. Hvarfner, *ibid.* **45**, 2771 (1992).
34. H. Sekino and R.J. Bartlett, *Intern. J. Quantum Chem. Symp.* **24**, 241 (1990).
35. E. Ilyabaev and U. Kaldor, *Chem. Phys. Lett.* **194**, 95 (1992).
36. E. Ilyabaev and U. Kaldor, *J. Chem. Phys.* **97**, 8455 (1994); *Phys. Rev.* A **47**, 137 (1993).
37. E. Eliav, U. Kaldor, and Y. Ishikawa, *Phys. Rev.* A **49**, 1724 (1994).
38. E. Eliav (Ilyabaev), U. Kaldor, and Y. Ishikawa, *Chem. Phys. Lett.* **222**, 82 (1994).
39. E. Eliav, U. Kaldor, and Y. Ishikawa, *Phys. Rev.* A **50**, 1121 (1994).
40. E. Eliav, U. Kaldor, and Y. Ishikawa, *Intern. J. Quantum Chem. Symp.* **28**, 205 (1994).
41. E. Eliav, U. Kaldor, and Y. Ishikawa, *Phys. Rev.* A **51**, 225 (1995).
42. E. Eliav, U. Kaldor, and Y. Ishikawa, *Phys. Rev.* A **52**, 291 (1995).
43. E. Eliav, U. Kaldor, and Y. Ishikawa, *Phys. Rev.* A **52**, 2765 (1995).
44. E. Eliav, U. Kaldor, and Y. Ishikawa, *Phys. Rev.* A **53**, 3050 (1996).
45. E. Eliav, U. Kaldor, Y. Ishikawa, M. Seth, and P. Pyykkö, *Phys. Rev.* A **53**, 3926 (1996).
46. E. Eliav, U. Kaldor, and Y. Ishikawa, *Mol. Phys.* **94**, 181 (1998).
47. E. Eliav, U. Kaldor, and Y. Ishikawa, *Phys. Rev. Lett.* **74**, 1079 (1995).
48. E. Eliav, U. Kaldor, P. Schwerdtfeger, B.A. Heß, and Y. Ishikawa, *Phys. Rev. Lett.* **73**, 3203 (1994).

49. E. Eliav, U. Kaldor, Y. Ishikawa, and P. Pyykkö, *Phys. Rev. Lett.* **77**, 5350 (1996).
50. H. Hotop and W. C. Lineberger, *J. Phys. Chem. Ref. Data* **4**, 539 (1975); *ibid.* **14**, 731 (1985).
51. D. Calabrese, A. M. Covington, J. S. Thompson, R. W. Marawar, and J. W. Farley, *Phys. Rev.* A **54**, 2797 (1996).
52. F. Arnau, F. Mota, and J. J. Novoa, *Chem. Phys.* **166**, 77 (1992).
53. W. P. Wijesundera, *Phys. Rev.* A **55**, 1785 (1997).
54. E. Eliav, Y. Ishikawa, P. Pyykkö, and U. Kaldor, *Phys. Rev.* A **56**, 4532 (1997).
55. G. L. Malli, A. B. F. Da Silva, and Y. Ishikawa, *Phys. Rev.* A **47**, 143 (1993).
56. Z. Cai, V. Meiser Umar, and C. Froese Fischer, *Phys. Rev. Lett.* **68**, 297 (1992).
57. M. Douglas and N. M. Kroll, *Ann. Phys. (NY)* **82**, 89 (1974); B.A. Heß, *Phys. Rev.* A **32**, 756 (1982); *ibid.* **33**, 3742 (1986); G. Jansen and B.A. Heß, *Phys. Rev.* A **39**, 6016 (1989); R. Samzow, B.A. Heß, and G. Jansen, *J. Chem. Phys.* **96**, 1227 (1992); B.A. Heß, R. J. Buenker, and P. Chandra, *Intern. J. Quantum Chem. Symp.* **29**, 737 (1986); R. Samzow and B.A. Heß, *Chem. Phys. Lett.* **184**, 491 (1991).
58. U. Kaldor and B.A. Heß, *Chem. Phys. Lett.* **230**, 1 (1994).
59. K.G. Dyall, *J. Chem. Phys.* **100**, 2118 (1994); L. Visscher, K.G. Dyall and T.J. Lee, *Intern. J. Quantum Chem. Symp.* **29**, 411 (1995).
60. P.J.C. Aerts, O. Visser, L. Visscher, H. Merenga, W.A. de Jong, and W.C. Nieuwpoort, MOLFDIR, University of Groningen, The Netherlands; L. Visscher, T.J. Lee and K.G. Dyall, RELCCSD, NASA Ames Research Center, Moffett Field, California.
61. E. Eliav and U. Kaldor, *Chem. Phys. Lett.* **248**, 405 (1996).
62. E. Eliav, U. Kaldor, and B. A. Heß, *J. Chem. Phys.* **108**, 3409 (1998).
63. J. B. Schilling, W. A. Goddard III, and J. L. Beauchamp, *J. Amer. Chem. Soc.* **109**, 5565 (1987).
64. S. F. Boys and F. Bernardi, *Mol. Phys.* **19**, 553 (1970).
65. K.P. Huber and G. Herzberg, *Molecular Spectra and Molecular Structure Constants of Diatomic Molecules*, (Van Nostrand, New York, 1979).
66. G. Malli, in *Relativistic and Electron Correlation Effects in Molecules and Solids*, Nato ASI Series B Vol. 318, ed. G. Malli (Plenum, New York, 1994).
67. D. Strömberg, O. Gropen and U. Wahlgren, *Chem. Phys. Lett.* **133**, 207 (1989).
68. K. Balasubramanian, *J. Chem. Phys.* **93**, 8061 (1990).
69. I. Mills, T. Cvitaš, K. Homann, N. Kallay, and K. Kuchitsu, *Quantities, Units and Symbols in Physical Chemistry*, 2nd edition, Blackwell, Oxford, 1993; P. Pyykkö, private communication.
70. C.E. Moore, *Atomic Energy Levels*, Natl. Bur. of Stand. (U.S.) Circ. No. 467 (U.S. GPO, Washington, DC, 1948).

Variational Principle in the Dirac Theory: Theorems, Examples and Counterexamples

Jacek Karwowski, Grzegorz Pestka and Monika Stanke
Instytut Fizyki, Uniwersytetu Nikołaja Kopernika, ul. Grudziądzka 5, 87-100 Toruń, Poland

Abstract

The variational Dirac-Coulomb and the corresponding Lévy-Leblond problems, in which the large and the small components are treated independently, are analyzed. Close similarities between these two variational problems are emphasized. Several examples in which the so called *strong minimax principle* is violated are discussed.

1. Introduction

An application of the variational principle to an unbounded from below Dirac-Coulomb eigenvalue problem, requires imposing upon the trial function certain conditions. Among these the most important are the symmetry properties, the asymptotic behaviour and the relations between the large and the small components of the wavefunction related to the so called *kinetic balance* [1,2,3]. In practical calculations an exact fulfilment of these conditions may be difficult or even impossible. Therefore a number of minimax principles [4–7] have been formulated in order to allow for some less restricted choice of the trial functions. There exist in the literature many either purely intuitive or derived from computational experience, rules which are commonly used as a guidance in generating basis sets for variational relativistic calculations.

In this paper we give a number of examples based on simple models showing the influence of the non-exact fulfilment of either the boundary conditions or the relationship between the large and the small components by the trial functions on the results of variational calculations. In particular, we show that there exist bound variational solutions degenerate with the exact energy but with incorrect wavefunctions. Also several examples of violation of the 'strong minimax principle' [7] are given. The results obtained for the Dirac equation are compared with those of the nonrelativistic Schrödinger equation in the Lévy-Leblond [8] form. Similarities and differences between variational solutions of the Dirac and the Lévy-Leblond equations are discussed. The behaviour of the variational solutions that are a consequence of the unboundedness of the Dirac Hamiltonian are separated from those that are due to the two-component structure of the orbitals.

Atomic units are used in this paper. In some cases, in order to make equations more transparent, the electron mass m is written explicitly.

2. Dirac and Lévy-Leblond Equations

In the standard (Dirac-Pauli) representation, the Dirac equation for an electron in the field of a stationary potential V reads

A. Hernández-Laguna et al. (eds.), Quantum Systems in Chemistry and Physics, Vol. 1: Basic Problems and Model Systems, 177–193.

$$\begin{pmatrix} V - E & c(\boldsymbol{\sigma} \cdot \boldsymbol{p}) \\ c(\boldsymbol{\sigma} \cdot \boldsymbol{p}) & V - E - 2mc^2 \end{pmatrix} \begin{pmatrix} \psi_D^L \\ \psi_D^S \end{pmatrix} = 0, \tag{1}$$

where $\boldsymbol{\sigma}$ are the Pauli spin matrices, ψ_D^L and ψ_D^S are traditionally called the large and the small components of the wavefunction and E is the energy relative to mc^2.

It has been noticed by Lévy-Leblond [8] that the non-relativistic Schrödinger equation may be written as

$$\begin{pmatrix} V - E & c(\boldsymbol{\sigma} \cdot \boldsymbol{p}) \\ c(\boldsymbol{\sigma} \cdot \boldsymbol{p}) & -2mc^2 \end{pmatrix} \begin{pmatrix} \psi_{\mathcal{L}}^L \\ \psi_{\mathcal{L}}^S \end{pmatrix} = 0. \tag{2}$$

Elimination of $\psi_{\mathcal{L}}^S$ gives

$$\left[\frac{(\boldsymbol{\sigma} \cdot \boldsymbol{p})(\boldsymbol{\sigma} \cdot \boldsymbol{p})}{2m} + V - E\right] \psi_{\mathcal{L}}^L = 0, \tag{3}$$

i.e. the Schrödinger equation. Eq. (2) may be obtained from the Dirac equation (1) if $V - E$ is neglected relative to mc^2. It may be interpreted as a representation of the Schrödinger equation in the Hilbert space augmented by the spin-space.

The Dirac and the Lévy-Leblond equations establish relationships between the large and the small components of the wavefunctions. If these relationships are to be fulfilled by the functions derived from a variational procedure, the basis sets for the large and for the small components have to be constructed accordingly. In particular, the relation

$$\psi^S \propto \left(1 + \frac{E - V}{2mc^2}\right)^{-1} (\boldsymbol{\sigma} \cdot \boldsymbol{p}) \psi^L \tag{4}$$

is referred to as the *atomic balance condition* [3]. A simpler relation

$$\psi^S \propto (\boldsymbol{\sigma} \cdot \boldsymbol{p}) \psi^L, \tag{5}$$

is called the *kinetic balance condition* [2,3]. A variational basis set constructed so that condition (5) is fulfilled is called kinetically balanced. A non-fulfilment of the kinetic balance condition may lead to the variational collapse not only in the Dirac but also in the Lévy-Leblond case — Eqs. (2) and (3) are equivalent only if the kinetic balance condition (5) is fulfilled.

If the kinetic balance condition (5) is fulfilled then the spectrum of the Lévy-Leblond (and Schrödinger) equation is bounded from below. Then, in each case there exists the lowest value of E referred to as the ground state. In effect, this equation may be solved using the variational principle without any restrictions. On the contrary, the spectrum of the Dirac equation is unbounded from below. It contains the negative ('positronic') continuum. Therefore the variational principle applied unconditionally would lead to the so called 'variational collapse' [2,3,7]. The variational collapse may be avoided by properly selecting the trial functions so that they fulfil the boundary conditions specific for the bound-state solutions [1].

3. Minimax Principle

It is highly desirable to formulate a variational principle valid for the Dirac Hamiltonian. The first attempts are due to Drake and Goldman [9], Wood *et al.* [10], Talman [4] and Datta and Deviah [5]. Very recently the subject has been discussed in detail by Griesemer and Siedentop [6] and also by Kutzelnigg [7] and by Quiney *et al.* [11].

3.1. One-electron problems

A relativistic variational principle has to be formulated as a recipe for reaching the saddle point on the energy hypersurface in the space of variational parameters. According to refs. [4,5,6], the ground state of a Dirac electron is given by

$$E = \min_{\{l\}} \left[\max_{\{s\}} \frac{\langle \Psi | H | \Psi \rangle}{\langle \Psi | \Psi \rangle} \right], \tag{6}$$

where $\{l\}$ and $\{s\}$ refer, respectively, to the spaces in which large and small components of the wavefunction are represented. Eq. (6) has been named by Talman the *minimax principle*. Kutzelnigg [7] has called this equation a *weak minimax principle*, to distinguish from a *strong minimax principle*

$$E = \min_{\{l\}} \max_{\{s\}} \frac{\langle \Psi | H | \Psi \rangle}{\langle \Psi | \Psi \rangle}. \tag{7}$$

In the weak minimax principle the optimization of the Rayleigh quotient consists of two steps. In the first step it is subject to a maximization with respect to *all* small components for *each* large component kept fixed. In the second step the saddle point (i.e. the ground state energy) is obtained as the minimum with respect to the large component of the numbers obtained in the first step. Hence, if ψ^L is the exact large component then the maximum over ψ^S is obtained for the exact ψ^S while if ψ^L is arbitrary, the maximum over ψ^S yields an upper bound to the exact E. In the strong minimax principle both steps may be performed simultaneously or even in the reversed order. According to the strong minimax principle, also the following sentence should be true: If ψ^S is the exact small component then the minimum over ψ^L is obtained for the exact ψ^L while if ψ^S is arbitrary, the minimum over ψ^L yields a lower bound to the exact E. If the space of the small components is restricted (for example by selecting and keeping fixed the exact small component), then conditions of Eq. (6) are not fulfilled. In such a case a minimization of the Rayleigh quotient with respect to the parameters of the large component corresponds to an application of the strong minimax principle as it is formulated in Eq. (7). Unfortunately, as it has been pointed out already by Kutzelnigg [7] and will be demonstrated in several examples in the next section, the strong minimax principle is, in general, false and may be considered, at most, as a computational recipe. Nevertheless, effectively, it constitutes a base for many relativistic calculations.

Let us select the trial function in the form

$$\Psi = \begin{pmatrix} \psi^L \\ \psi^S \end{pmatrix} = \frac{1}{r} \begin{pmatrix} A^L \Phi^L(r) \\ A^S \Phi^S(r) \end{pmatrix}, \tag{8}$$

where A^L and A^S contain the exact angular and the spin parts, while Φ^L and Φ^S are radial functions which depend upon some non-linear parameters. Combining Eqs. (1) and (6) we get

$$E_{\mathcal{D}} = \langle V_+ \rangle - mc^2 + |\langle V_- \rangle + mc^2| \sqrt{1 + \frac{2\langle T \rangle}{mc^2} \left(1 + \frac{\langle V_- \rangle}{mc^2}\right)^{-2}}, \tag{9}$$

where $E_{\mathcal{D}}$ is the Dirac energy corresponding to Φ^L and Φ^S,

$$\langle V_\pm \rangle = \frac{1}{2} \left(\frac{\langle \psi^L | V | \psi^L \rangle}{\langle \psi^L | \psi^L \rangle} \pm \frac{\langle \psi^S | V | \psi^S \rangle}{\langle \psi^S | \psi^S \rangle} \right), \tag{10}$$

and

$$\langle T \rangle = \frac{1}{2m} \frac{\langle \psi^L | \boldsymbol{\sigma} \cdot \boldsymbol{p} | \psi^S \rangle \langle \psi^S | \boldsymbol{\sigma} \cdot \boldsymbol{p} | \psi^L \rangle}{\langle \psi^L | \psi^L \rangle \langle \psi^S | \psi^S \rangle}. \tag{11}$$

Using Eq. (2) instead, we obtain the analogous expression for the Lévy-Leblond energy:

$$E_{\mathcal{L}} = \langle T \rangle + \langle V \rangle, \tag{12}$$

where $\langle V \rangle = \langle V_+ \rangle + \langle V_- \rangle$. The same expression may easily be obtained at the limit of $c \to \infty$ from Eq. (9). As one can easily check, if and only if Φ^S and Φ^L fulfil the kinetic balance condition (5), then

$$\langle T \rangle = \left\langle \psi^L \middle| \frac{p^2}{2m} \middle| \psi^L \right\rangle \langle \psi^L | \psi^L \rangle^{-1},$$

and Eq. (12) becomes

$$E_{\mathcal{L}} = \left\langle \psi^L \middle| \frac{p^2}{2m} + V \middle| \psi^L \right\rangle \langle \psi^L | \psi^L \rangle^{-1}, \tag{13}$$

being the correct non-relativistic limit.

A surprising result is obtained if in Eq. (8) we set $\Phi^L = \Phi^S = \Phi$. In such a case $\langle V_- \rangle = 0$. Assuming that $\int_0^\infty |\Phi(r)|^2 dr = 1$ we have

$$\langle V_+ \rangle = \langle V \rangle = \int_0^\infty \Phi^*(r) V(r) \Phi(r) dr \tag{14}$$

and

$$\langle T \rangle = \frac{1}{2m} \left| \int_0^\infty \Phi^*(r) \left(-\frac{d}{dr} + \frac{k}{r} \right) \Phi(r) dr \right|^2, \tag{15}$$

where $k = \pm 1, \pm 2, \ldots$ is the relativistic angular momentum quantum number. If Φ is real, the matrix element

$$\int_0^\infty \Phi \frac{d\Phi}{dr} dr = \frac{1}{2}[\Phi(\infty)^2 - \Phi(0)^2] \tag{16}$$

vanishes for all Φ such that $\Phi(0) = \Phi(\infty) = 0$. In such a case, for $V = -(Z/r)$, we have

$$\langle V \rangle = -ZX, \text{ and } \langle T \rangle = \frac{k^2}{2} X^2, \tag{17}$$

where

$$X = \int_0^\infty \frac{\Phi^2}{r} dr. \tag{18}$$

A substitution to Eqs. (9) and (12) gives, respectively,

$$E_{\mathcal{D}}(X) = c\sqrt{c^2 + k^2 X^2} - ZX - c^2, \tag{19}$$

and

$$E_{\mathcal{L}}(X) = \frac{1}{2} k^2 X^2 - ZX. \tag{20}$$

Each of these functions has one extremum: a minimum. In the Dirac case it corresponds to

$$X_0^{\mathcal{D}} = \frac{Zc}{|k|\sqrt{k^2 c^2 - Z^2}} \tag{21}$$

and in the Lévy-Leblond case to

$$X_0^{\mathcal{L}} = \frac{Z}{k^2}. \tag{22}$$

The energies corresponding to these minima are equal to

$$E_{\mathcal{D}}(X_0) = \frac{c}{|k|}\sqrt{c^2 k^2 - Z^2} - c^2, \tag{23}$$

and

$$E_{\mathcal{L}}(X_0) = -\frac{Z^2}{2k^2}, \tag{24}$$

i.e. to the *exact* Dirac and Schrödinger energies, respectively.

If we select

$$\Phi(r) \propto r^L \exp(-\alpha r^d), \tag{25}$$

then

$$X = (2\alpha)^{\frac{1}{d}} \Gamma\left(\frac{2L}{d}\right) \left[\Gamma\left(\frac{2L+1}{d}\right)\right]^{-1}. \tag{26}$$

In particular, for $d = 1$, i.e. for the Slater-type functions,

$$X_{Slater} = \frac{\alpha}{L} \tag{27}$$

and for $d = 2$, i.e. for the Gaussian functions,

$$X_{Gausss} = 2^{2L-1}\sqrt{\frac{2\alpha}{\pi}}\frac{\Gamma(L)^2}{\Gamma(2L)}. \tag{28}$$

As one can see, for each set of parameters Z, k, L and d, one can select α in such a way that Eqs. (21) and (22) are fulfilled. Then, the resulting expectation values correspond to the variational minima and are equal to the appropriate *exact* eigenvalues of either Dirac or Schrödinger (or rather Lévy-Leblond) Hamiltonian. However the corresponding functions *do not* fulfil the pertinent eigenvalue equations: they are *not* eigenfunctions of these Hamiltonians. This example demonstrates that the value of the variational energy cannot be taken as a measure of the quality of the wavefunction, unless the appropriate relation between the components of the wavefunction is fulfilled [2].

3.2. Two-electron problems

A two-electron Dirac-Coulomb Hamiltonian is defined as

$$H(1,2) = H(1) \otimes I(2) + I(1) \otimes H(2) + h(1,2), \tag{29}$$

where $I(1)$, $I(2)$ stand for the identity operators, $H(1)$, $H(2)$ are one-electron Dirac Hamiltonians and $h(1,2)$ describes the electrostatic interaction. The domain of this operator is defined as the antisymmetric part of the space spanned by the direct products of one-electron Dirac spinors:

$$\Psi(1,2) = \mathcal{A}[\Psi(1) \otimes \Psi(2)], \tag{30}$$

where $\mathcal{A}$ is the antisymmetrizer. Then, $\Psi(1,2)$ has four components, denoted ψ^{LL}, ψ^{LS}, ψ^{SL}, and ψ^{SS}, each of them being a four-component spinor composed of two large (LL), large and small (LS and SL) or two small (SS) components of the one-electron spinors.

A natural generalization of Eq. (6) would be to choose the parameters in all one-electron small components of the two-electron wavefunction (30) to maximize E and then to choose the parameters in all one-electron large components to minimize E. However, in order to solve variationally the eigenvalue problem of the Dirac-Coulomb Hamiltonian, Kołakowska *et al.* [12] advocated, on the basis of rather intuitive arguments, the following rule:

$$E = \min_{\{ll\}}\left[\max_{\{ls\},\{sl\},\{ss\}} \frac{\langle\Psi|H|\Psi\rangle}{\langle\Psi|\Psi\rangle}\right], \tag{31}$$

where $\{ll\}$, $\{ls\}$, $\{sl\}$, and $\{ss\}$ denote spaces in which the appropriate types of spinors are represented. An example showing that Eq. (31) is incorrect is given in the next section.

4. Results and Discussion

In this section several examples of the application of the minimax principle and several counterexamples, which demonstrate the invalidity of its strong form, are discussed. For the one-electron case the ground state of a hydrogen-like atom has been calculated using a trial wavefunction as given in Eq. (8). For the two-electron case calculations have been performed for the ground state of a two-electron atom described by the Hamiltonian defined in Eq. (29) with $h(1, 2) \equiv 0$ and a trial wavefunction given by Eq. (30). In order to emphasize that most of the computational problems associated with the variational solving of the Dirac equation result from the independent variation of the different components of the multi-component wavefunction rather than from the unboundedness from below of the Dirac Hamiltonian, in all cases the calculations are also performed using the Lévy-Leblond equation.

4.1. Hydrogen-like atom in a hydrogenic basis

Let $V = -(Z/r)$ and

$$\Phi^L \propto r^L e^{-\alpha r}, \qquad \Phi^S \propto r^S e^{-\beta r}, \tag{32}$$

where L and S are not necessarily integers. Then

$$\langle V_\pm \rangle = -\frac{Z}{2}\left(\frac{\alpha}{L} \pm \frac{\beta}{S}\right) \tag{33}$$

and

$$\langle T \rangle = \frac{1}{2m}\left(\frac{L\beta - S\alpha}{\alpha+\beta} - 1\right)^2 \frac{\alpha\beta}{LS}\left(\frac{2\alpha}{\alpha+\beta}\right)^{2L}\left(\frac{2\beta}{\alpha+\beta}\right)^{2S}\frac{\Gamma(L+S)^2}{\Gamma(2L)\Gamma(2S)}. \tag{34}$$

In this case the ground state energy in both the Dirac and Lévy-Leblond case depends upon four nonlinear parameters:

$$E = E(\alpha, \beta, L, S). \tag{35}$$

The exact solution of the Dirac equation is given by

$$E^0_{\mathcal{D}} = E(Z, Z, \gamma, \gamma) = c^2\left(\sqrt{1 - \frac{Z^2}{c^2}} - 1\right), \tag{36}$$

where $\gamma = \sqrt{1 - Z^2/c^2}$. In the Lévy-Leblond case the exact ground-state energy is

$$E^0_{\mathcal{L}} = E(Z, Z, 1, 1) = -\frac{Z^2}{2}. \tag{37}$$

The shapes of the surfaces $E = E_{\mathcal{D}}(\alpha, \beta, \gamma, \gamma)$, $E = E_{\mathcal{L}}(\alpha, \beta, 1, 1)$, $E = E_{\mathcal{D}}(Z, Z, L, S)$ and $E = E_{\mathcal{L}}(Z, Z, L, S)$ for $Z = 90$ are shown in figure 1 ($\mathcal{D}1$, $\mathcal{L}1$, $\mathcal{D}2$ and $\mathcal{L}2$ respectively). In the first pair of the diagrams ($\mathcal{D}1$ and $\mathcal{L}1$) E is given as a function of α and β; in the second pair – as a function of L and S. In spite of a rather large value of Z (the relativistic correction to the energy is equal to 618 hartree, i.e. 12% of the

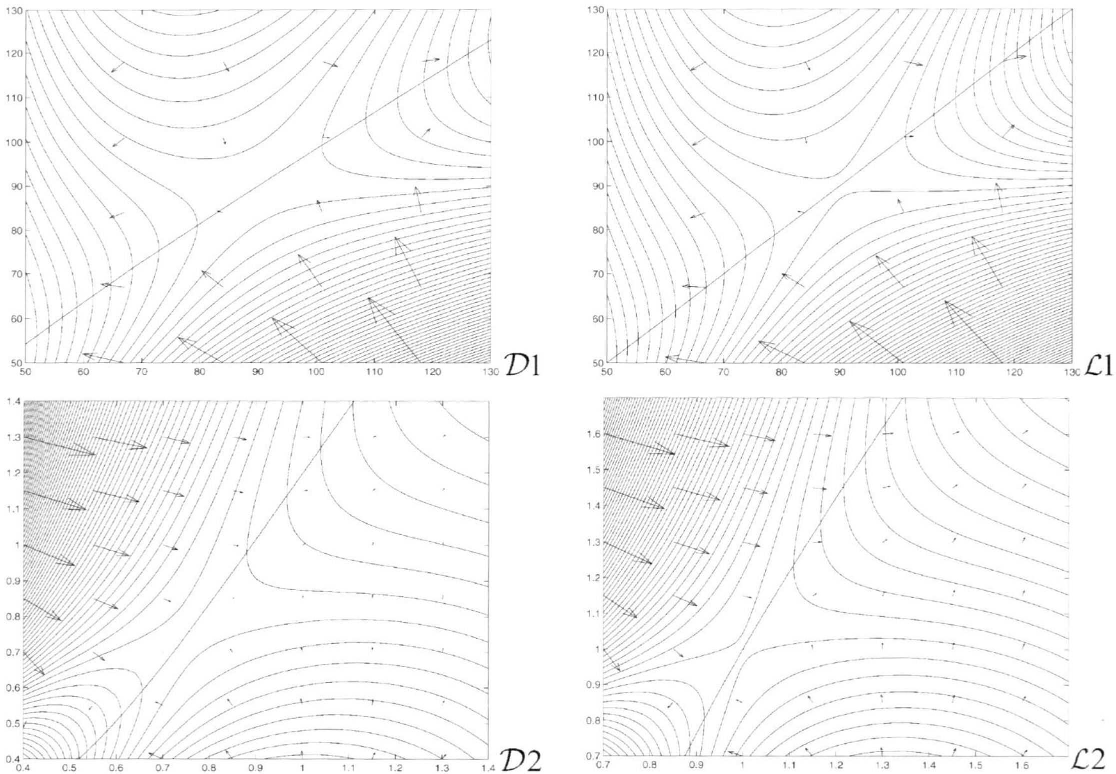

Fig. 1. *The ground state energies of a $Z = 90$ hydrogen-like atom obtained from the Dirac ($\mathcal{D}$) and from the Lévy-Leblond ($\mathcal{L}$) equations as functions of the nonlinear parameters. In the upper-row figures ($\mathcal{D}1$ and $\mathcal{L}1$) the α (abscissa) and β (ordinate) dependence of E is displayed when L and S are set equal to the values corresponding to the exact solutions. In the lower-row figures ($\mathcal{D}2$ and $\mathcal{L}2$) the L (abscissa) and S (ordinate) dependence of E is displayed when α and β are set equal to the exact value. The arrows show directions of the gradient; their length is proportional to the value of the gradient. The solid line crossing the saddle corresponds to the functions $\beta = \beta_{\max}(\alpha)$ (in $\mathcal{D}1$ and $\mathcal{L}1$) or $S = S_{\max}(L)$ $\mathcal{D}2$ and $\mathcal{L}2$. For the definitions of these functions see text.*

total energy), the Dirac and the Lévy-Leblond surfaces are very similar to each other. The exact eigenvalue corresponds to the saddle point. The solid line crossing the saddle corresponds to the function $\beta = \beta_{\text{max}}(\alpha)$ (in the case of $\mathcal{D}1$ and $\mathcal{L}1$) or $S = S_{\text{max}}(L)$ (in the case of $\mathcal{D}2$ and $\mathcal{L}2$) defined as these values of the small component parameters at which E is maximum while the large component parameters are fixed.

More information about the saddle point and about the minimax principle may be derived from figure 2. It also corresponds to the ground state of $Z = 90$ one-electron atom. The first pair of diagrams ($\mathcal{D}1$ and $\mathcal{L}1$) gives E as a function of the small component parameters (β and S) while the large component is exact. In the second pair of diagrams ($\mathcal{D}2$ and $\mathcal{L}2$) E is displayed as a function of the large component parameters while the small component is exact. In the first case ($\mathcal{D}1$, $\mathcal{L}1$) the exact energy corresponds to the maximum of the surface, in agreement with both weak and strong minimax principles. In the second case ($\mathcal{D}2$, $\mathcal{L}2$), the exact energy is a saddle point while, according to the strong minimax principle, it should be a minimum. However the last result is not contradictory to the weak minimax principle.

The dependence of the energy hypersurface on a single parameter while all the remaining parameters assume their optimum values (i.e. are the same as in the exact wavefunction) is shown in figure 3: in diagrams $\mathcal{D}1$ and $\mathcal{L}1$ the dependence on α and β is shown while in $\mathcal{D}2$ and $\mathcal{L}2$ – the dependence on L and S. The corresponding curves in the relativistic (Dirac) and in the non-relativistic (Lévy-Leblond) models are nearly identical (except for the shift by 618 hartree along the energy axis). The exact energy always corresponds to an extremum. As a function of α, β and S it behaves in agreement with both strong and weak minimax principles. The behaviour of E as a function of L violates the strong minimax principle (the exact energy corresponds here to a maximum rather than to a minimum). The minimum with respect to L taken over the manifold of the wavefunctions with the exact small component and $\alpha = Z$ is located well below the exact ground state energy. The origin of this behaviour of the energy surface may be understood after a careful inspection of figure 2 and analyzing the 'geometry' of the saddle. However the weak minimax principle is always fulfilled: it is illustrated by the solid lines which in $\mathcal{D}1$ and $\mathcal{L}1$ represent the energy as a function of α when $\beta = \beta_{\text{max}}(\alpha)$ and in $\mathcal{D}2$ and $\mathcal{L}2$ – as a function of L when $S = S_{\text{max}}(L)$. In all these cases the exact energy corresponds to the global minimum of the curve.

In the cases shown in figures 1–3 the trial wavefunctions fulfil exactly the boundary conditions at $r \to 0$ and at $r \to \infty$ and for the nonlinear parameters corresponding to the saddle point we obtain the exact solution of the pertinent equation. The energy surfaces as functions of α and β in a case when the boundary conditions at the nucleus are not fulfilled are shown in figure 4. Here $L = S = 1$ (rather than $L = S = \gamma$) have been taken for the Dirac equation and $L = S = \gamma$ (rather than $L = S = 1$) for the Lévy-Leblond equation. As one should expect, with these values of the parameters, the saddle points are located above the exact energies. The strong minimax principle, valid in this case for the Lévy-Leblond equation (c.f. $\mathcal{L}2$), is not fulfilled for the Dirac equation (c.f. $\mathcal{D}2$). As it is seen in $\mathcal{D}2$, the saddle point corresponds here to an inflexion point (rather than to a minimum) with respect to the large component. The weak minimax principle is fulfilled in both cases – the energy taken as a function of α for $\beta = \beta_{\text{max}}(\alpha)$ has a minimum at the saddle point. It is interesting to note, that if we set $\alpha = \beta$ then,

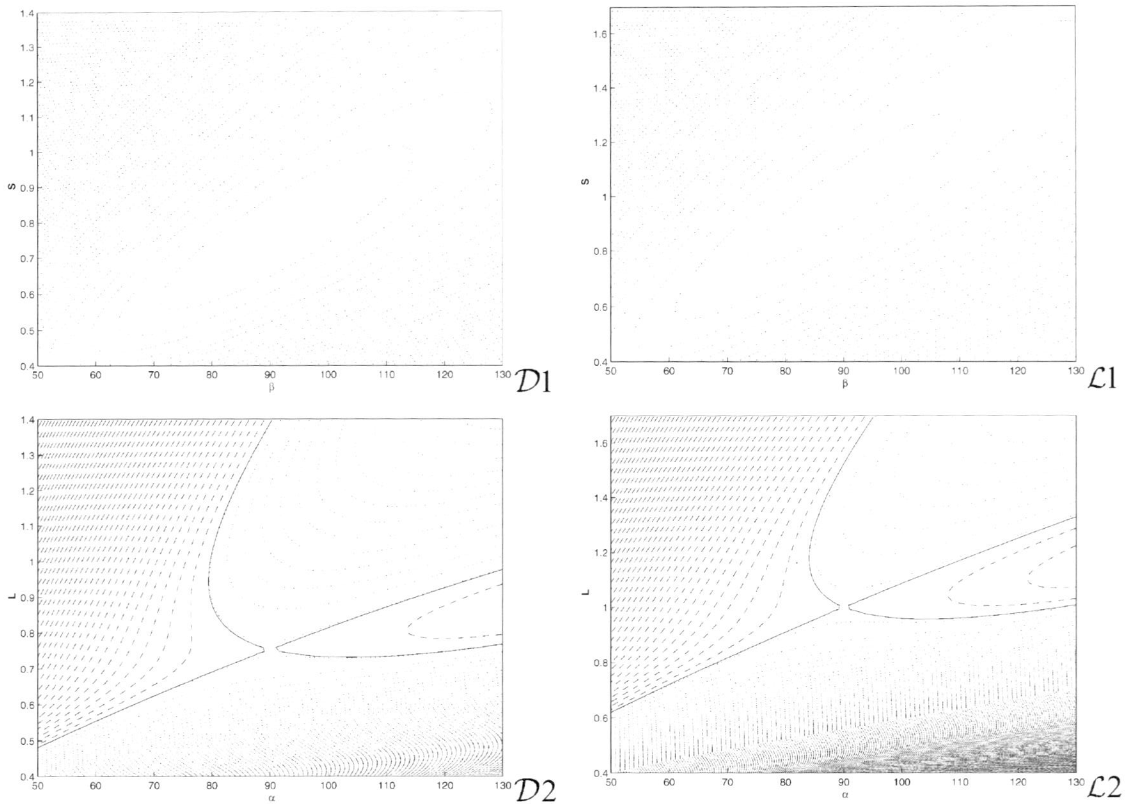

Fig. 2. *The upper-row figures show the dependence of E upon the small component parameters (β-abscissa, S-ordinate) when the large component is exact. The lower-row figures show the dependence of E upon the large component parameters (α-abscissa, L-ordinate) when the small component is exact. The dotted and the broken lines correspond to the energies which are, respectively, lower and higher than the exact one. In figures $\mathcal{D}1$ and $\mathcal{L}1$ the exact energy corresponds to the maximum of the surface. In figures $\mathcal{D}2$ and $\mathcal{L}2$ the exact energy is represented by the solid lines.*

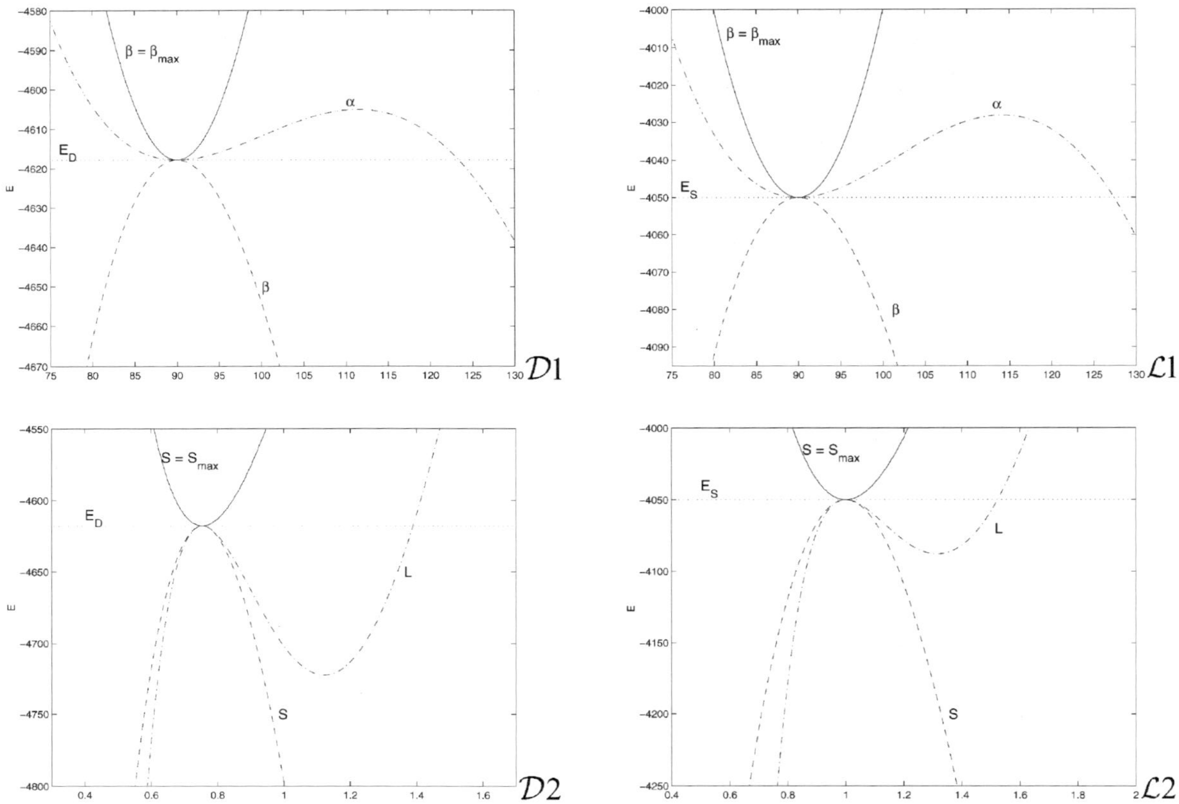

Fig. 3. *The solid lines illustrate the weak minimax principle. They describe the dependence of E on α when $\beta = \beta_{\max}(\alpha)$ (in $\mathcal{D}1$ and $\mathcal{L}1$) and the dependence of E on L when $S = S_{\max}(L)$ (in $\mathcal{D}2$ and $\mathcal{L}2$). The broken lines and the broken lines with dots show the dependence of the energy hypersurface on a single parameter (indicated at the curve) when the values of the remaining parameters are the same as in the exact wavefunctions. The broken lines describe the behaviour of the small components and the broken lines with dots – of the large ones. The horizontal dotted lines show the exact values of the ground state energies.*

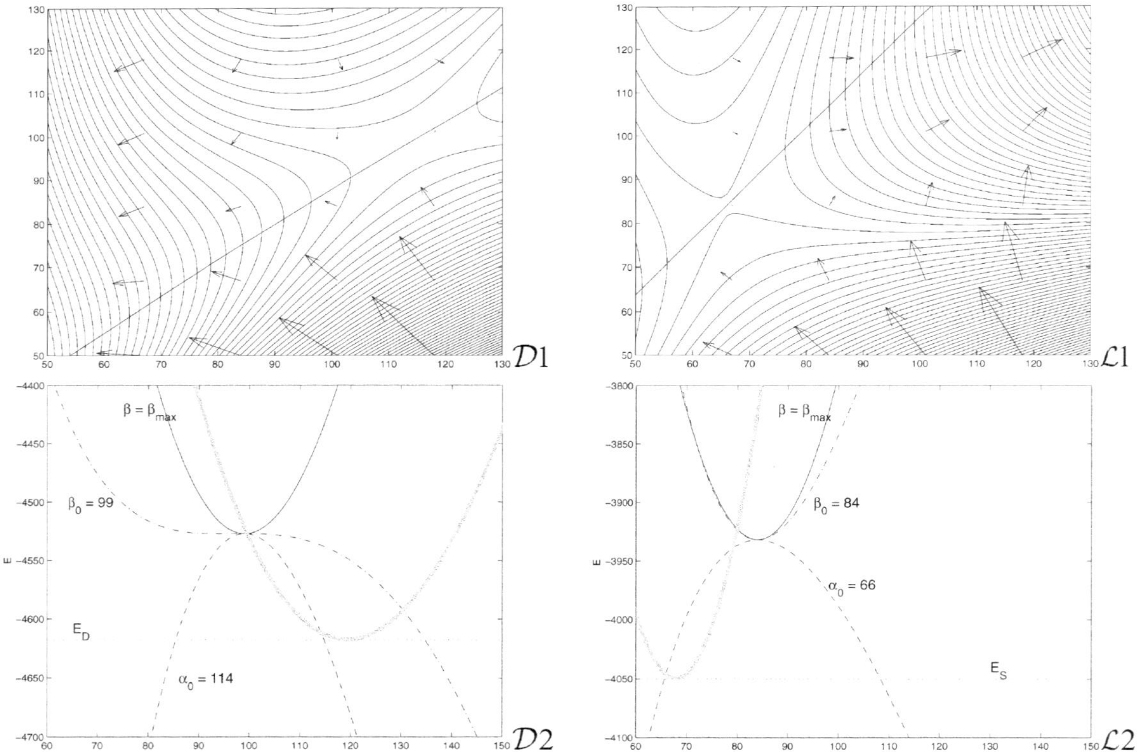

Fig. 4. The same as in figures 1 and 3, but the values of L and S are not equal to the ones of the exact solutions. In the case of the Dirac equation $L = S = 1$ and in the case of the Lévy-Leblond equation $L = S = \gamma$ have been taken. The saddle point coordinates (α_0, β_0) are equal to (114,99) in the Dirac case and (66,84) in the Lévy-Leblond case. The cross-sections of the energy surface by the planes $\alpha = \beta$ (broad solid line), and $\alpha = \alpha_0$ (broken line) are plotted versus β, while the cross-sections by the plane $\beta = \beta_0$ (broken line with dots) and by the surface $\beta = \beta_{\max}(\alpha)$ (thin solid line) are plotted versus α, in $\mathcal{D}2$ (Dirac) and in $\mathcal{L}2$ (Lévy-Leblond). The scale of β is shown in the horizontal axes. The scale of α has been chosen so that the curve for which $\beta = \beta_0$ and the one for which $\alpha = \alpha_0$ match at the saddle point. The scale of α may be obtained by adding $\alpha_0 - \beta_0$ to the values of β displayed in the axes of $\mathcal{D}2$ and $\mathcal{L}2$.

according to Eqs. (19) – (24) and (27), the energy is bounded from below and the minimum is equal to the *exact* value in both Dirac and Lévy-Leblond cases. The corresponding plots are shown in diagrams $\mathcal{D}2$ and $\mathcal{L}2$.

4.2. Hydrogen-like atom in a Gaussian basis

The Gaussian orbitals are very important in practical applications. In spite of their wrong asymptotic behaviour at both $r \to 0$ and $r \to \infty$, nearly all molecular electronic structure calculation programs have been constructed using Gaussian sets of one-electron functions. In this example the Gaussian basis has been selected as

$$\Phi^L \propto re^{-\alpha r^2}, \qquad \Phi^S \propto re^{-\beta r^2}. \tag{38}$$

Then

$$\langle V_\pm \rangle = -Z\sqrt{\frac{2}{\pi}}(\sqrt{\alpha} \pm \sqrt{\beta}) \tag{39}$$

and

$$\langle T \rangle = \frac{1}{\pi m}\left(\frac{4\sqrt{\alpha\beta}}{\alpha+\beta}\right)^3 \frac{\alpha^2}{\alpha+\beta}. \tag{40}$$

The area of the saddle point at the energy surface of a hydrogen-like atom with $Z = 30$ is plotted versus α and β in figure 5. The cross-sections of the energy surfaces by the planes $\alpha = \alpha_0$ and $\beta = \beta_0$ are shown in diagrams $\mathcal{D}2$ (Dirac) and $\mathcal{L}2$ (Lévy-Leblond) of figure 5. Also here the saddle point corresponds to a double maximum rather than to a minimax, showing a violation of the strong minimax principle. The weak minimax principle is fulfilled since E as a function of α and $\beta_{\max}(\alpha)$ has a minimum at the saddle point. The energy corresponding to $\alpha = \beta$ is bounded from below. Similarly as in the case presented in figure 4, the minima are located below the saddle points and are equal to the *exact* Dirac and Schrödinger energies in the respective cases.

4.3. Two-electron atom in a hydrogen-like basis

The variational procedure in a many-electron space may be considered as several consecutively executed variational procedures in one-electron spaces (the procedure may be iterative if a self-consistency is required). This means that fulfilment of the one-electron minimax principle is a necessary (but, in general, not sufficient) condition for the fulfilment of a similar principle in a many-electron case. Therefore one should not expect a many-electron generalization of Eq. (7) being valid when, say, parameters L and S are varied.

We assume for simplicity that the two-electron atom is described by a Hamiltonian (29) in which $H(1)$ and $H(2)$ are the hydrogen-like Dirac Hamiltonians and $h(1, 2) \equiv 0$. Apparently, after this simplification the problem is trivial since it becomes separable to two one-electron problems. Nevertheless we present this example because it sheds some light upon formulations of the minimax principle in a many-electron case.

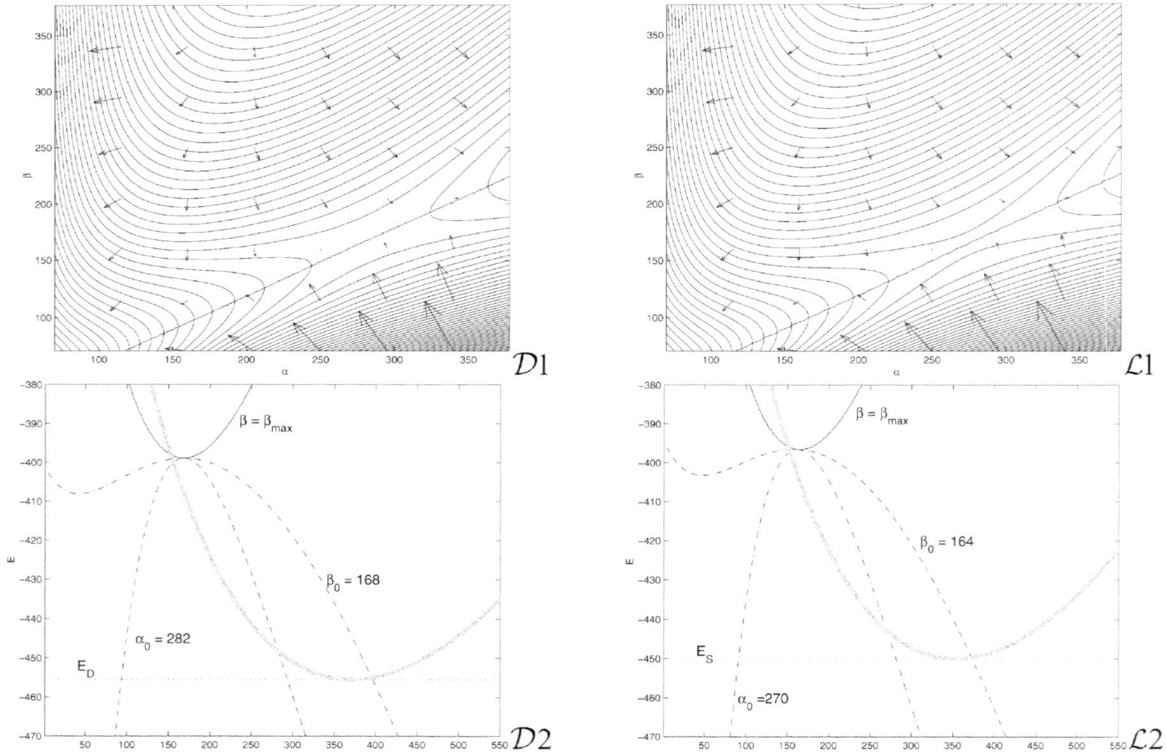

Fig. 5. The ground state energies of a $Z = 30$ hydrogen-like atom in the Gaussian basis obtained from the Dirac ($\mathcal{D}$) and from the Lévy-Leblond ($\mathcal{L}$) equations as functions of α and β. The saddle point coordinates (α_0, β_0) are equal to (282,168) in the Dirac case and (271,164) in the Lévy-Leblond case. The cross-sections of the energy surface by the planes $\alpha = \beta$ and $\alpha = \alpha_0$ are plotted versus β, while the cross-sections by the plane $\beta = \beta_0$ and by the surface $\beta = \beta_{\max}(\alpha)$ – versus α, in $\mathcal{D}2$ (Dirac) and in $\mathcal{L}2$ (Lévy-Leblond). The scale of β is shown in the abscissas. The scale of α has been chosen so that the curve for which $\beta = \beta_0$ and the one for which $\alpha = \alpha_0$ match at the saddle point. The scale of α may be obtained by adding $\alpha_0 - \beta_0$ to the values of β displayed in the abscissas of $\mathcal{D}2$ and $\mathcal{L}2$. All conventions concerning the notation are the same as in figures 1 and 4.

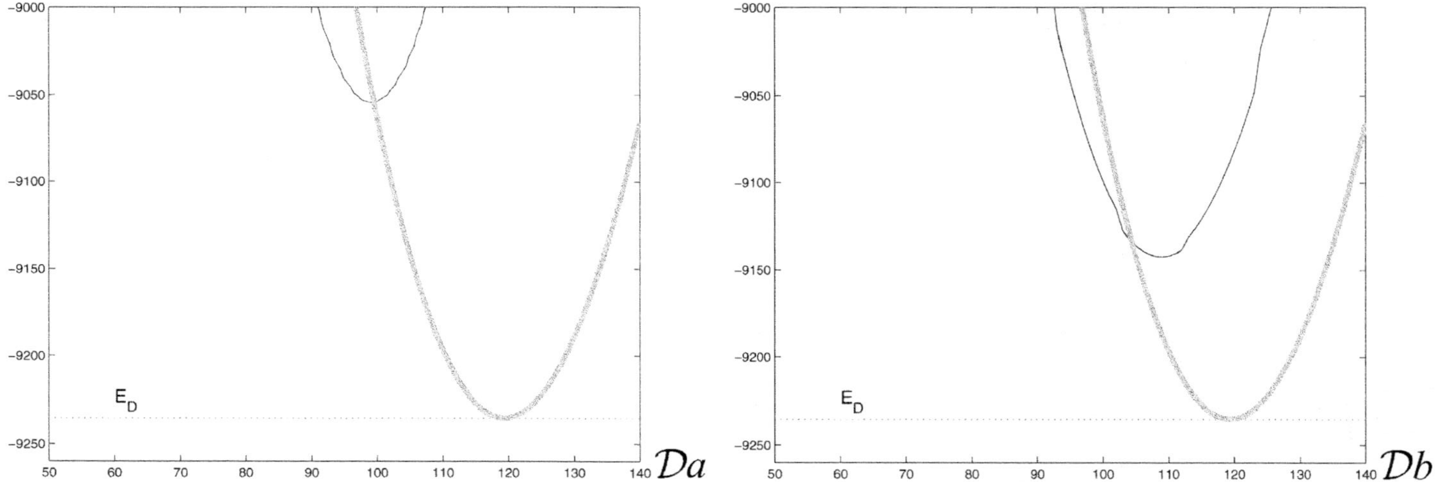

Fig. 6. *The application of the weak minimax principle to the ground state of a* $Z = 90$ *two-electron atom described by the simplified two-electron Dirac Hamiltonian using the hydrogen-like basis with* $L = S = 1$. *The thin solid lines represent the energy as a function of* α *when* $\beta = \beta_{\max}(\alpha)$. *The broad solid lines give the energy as a function of* α *when* $\beta = \alpha$. *The horizontal dotted lines show the exact value of the ground state energy. In diagram* $\mathcal{D}a$ *the results of the weak minimax principle formulated according to Eq. (6) (all one-electron small components are associated with the parameter* β *and all one-electron large components – with the parameter* α*) are displayed. In diagram* $\mathcal{D}b$ *the results of the minimax principle as it is defined in Eq. (31) are shown.*

The calculations have been performed for $Z = 90$ using the hydrogenic trial functions with $L = S = 1$, i.e. with an incorrect behaviour at $r \to 0$.

The results are displayed in figure 6. In the plots shown in diagram $\mathcal{D}a$, the exponent β has been assigned to all one-electron small components and the exponent α to all one-electron large components. Then a weak minimax principle has been applied. In this procedure $E[\alpha, \beta_{\max}(\alpha)]$ has been obtained through a maximalization of the Rayleigh quotient with respect to β for all fixed values of α. The minimum of $E[\alpha, \beta_{\max}(\alpha)]$, i.e. the variational energy of this system, is equal to -9060 a.u. It is twice larger than the corresponding one-electron energy obtained using the same trial function (c.f. figure 3). An application of the weak minimax principle formulated according to Eq. (31) is presented in diagram $\mathcal{D}b$. Here the resulting minimum is equal to -9140 a.u., i.e. it is by 80 a.u. too low. We conclude that Eq. (31) leads to a deeper, but incorrect, energy minimum.

5. Conclusions

Using several simple examples we have demonstrated that the minimax principle, if applied in its *strong* form (as it is done in a majority of calculations), may lead to erroneous results. Its results depend upon the way in which the saddle point of the energy surface in the space of the variational parameters is seen when a given parameter is varied, i.e. they depend upon the 'geometry' of the saddle. Easy to use, strictly correct and general variational prescription most likely cannot be formulated. The formally correct weak minimax principle [4,5,6] or the alternative formulations which require establishing some relations between the large and small components of the wavefunction and imposing appropriate boundary conditions [1,7,11] are rather difficult to implement.

It should be emphasized that in a variational calculation establishing correct relations between the components of the wavefunction is, in general, more difficult than eliminating the negative-energy solutions. This may be seen by comparing the Dirac and Lévy-Leblond energy surfaces: for the types of parameters studied in this work their qualitative behaviour is identical. Incorrect relations between the two components may produce quite unexpected results, as e.g. the exact energies with completely wrong wavefunctions. This feature of the equations, when misunderstood, may lead to serious mistakes; however, when consciously used, may lead to new simple methods for the estimation of the energy values.

Acknowledgements

This work has been supported by the Polish KBN, projects No. 2 P03B 126 14 and 2 P03B 023 12 and by the European Science Foundation within the REHE programme. The authors are indebted to L. Wolniewicz for a penetrating remark. Helpful comments of a referee are gratefully acknowledged.

References

1. I.P. Grant, in *Atomic, Molecular & Optical Physics Handbook*, ed. G.W.F. Drake, Chapt. 22, p. 287. AIP Press, Woodbury, New York, USA, 1996.
2. W. Kutzelnigg, Int. J. Quantum Chem. **15**, 107 (1984).
3. L. Visscher, P.J.C. Aerts, O. Visser and W.C. Nieuwpoort, Int. J. Quantum Chem. **S25**, 131 (1991).
4. J.D. Talman, Phys. Rev. Letters **57**, 1091 (1986).
5. S.N. Datta and G. Deviah, Pramana, **30**, 387 (1988).
6. M. Griesemer and H. Siedentop, J. London Math. Soc. **xx**, xxx (1999).
7. W. Kutzelnigg, Chem. Phys. **225**, 203 (1997).
8. J.M. Lévy-Leblond, Commun. Math. Phys. **6**, 286 (1967).
9. G.W.F. Drake and S.P. Goldman, Phys. Rev. A **23**, 2093 (1981).
10. J. Wood, I.P. Grant and S. Wilson, J. Phys. B: At. Mol. Phys. **18**, 3027 (1985).

11 H.M. Quiney, H. Skaane and I.P. Grant, Advan. Quantum Chem. **32**, xxx (1998).

12 A. Kołakowska, J.D. Talman and K. Aashamar, Phys. Rev. A **53**, 168 (1996).

13 A. Kołakowska, J. Phys. B: At. Mol. Phys. **29**, 4515 (1996).

Perspectives in Relativistic Thomas-Fermi Calculations for Atomic Systems

I. Porras and A. Moya
Departamento de Física Moderna, Facultad de Ciencias, Universidad de Granada, E-18071, Granada, Spain

Abstract

After reviewing the different approaches in the literature for the formulation of a relativistic Thomas-Fermi procedure for the study of complex electron systems, we will make contact between quantum mechanics with first-order relativistic corrections and the weak relativistic limit of quantum electrodynamics for finding explicit energy functionals that will be studied. In addition to this the possibility of using alternative near-nuclear corrections instead of gradient ones is discussed.

1. Introduction

The Thomas-Fermi (TF) and related methods such as the Thomas-Fermi-Dirac (TFD) have played an important role in the study of complex fermionic systems due to their simplicity and statistical nature [1]. For atomic systems, they are able to provide some knowledge about general features such as the behaviour with the atomic number Z of different ground state properties [2,3].

With the development of powerful computers, these methods have been restricted in practice to the application in specific problems where no other calculations are available. For non-relativistic atomic systems, they have been replaced by quantum mechanical calculations like Monte Carlo or multiconfigurational Hartree-Fock ones. Nevertheless, Thomas-Fermi estimates can be easily evaluated by non-specialists in theoretical calculations and in some problems they provide a starting point for more sophisticated procedures. Moreover, they are interesting for theoretical purposes such as finding relationships among different average quantities [4].

One of the reasons why TF approaches have lost attention is the problem of the systematic improvement of the results, in spite of the well established theoretical basis. In the non-relativistic framework, they are approximations to the Hartree-Fock (HF) method, asymptotically exact in the large Z-limit. However, for real atoms, these approximations are not able to match precisely the HF results, the main reason being the wrong contribution of electrons near the nucleus where the system differs most from a local Fermi gas. The gradient expansion of the energy functional, the most popular technique utilized for the improvement of the original TF(D) theory, has some problems in its application: sixth order terms diverge, fourth order ones make calculations very involved for a simple procedure and second order calculations make contact to HF values when a parameter obtained from theoretical grounds as $1/9$ is replaced by an effective value of $1/5$ [5].

The relativistic problem of many electron systems, described by quantum electrodynamics (QED), is nowadays an interesting topic of research. The high complexity of

A. Hernández-Laguna et al. (eds.), Quantum Systems in Chemistry and Physics, Vol. 1: Basic Problems and Model Systems, 195–213.

this approach for these systems makes it still an open problem, where different techniques are applied, such as relativistic density functional theory with the use of orbital dependent exchange and correlation functionals [6], relativistic coupled cluster methods [7] or multiconfigurational Dirac-Fock calculations [8]. In this field TF models encounter more difficulties for their application, but it is where they might be more interesting for performing simple estimates for a very complex problem.

Moreover, as the Thomas-Fermi model for atomic systems becomes more accurate when increasing Z (asymptotically exact in the large Z-limit) with respect to the non-relativistic solution of Schrödinger equation, but relativistic effects increases with Z does, the inclusion of these is demanded for its application.

We will start from a brief review of the relativistic extensions of TF approaches, the problems which arise and the alternatives for avoiding them.

2. Review of Relativistic Extensions of Thomas-Fermi Theory

2.1. Basis of Thomas-Fermi theory

The Thomas-Fermi approach is based on the minimization of an energy functional of the particle density

$$E[\rho(\mathbf{r})] = \int d\mathbf{r}\epsilon(\rho) \tag{1}$$

subject to the normalization of ρ to the number of particles N.

In non-relativistic Thomas-Fermi theory the functional is given by

$$E[\rho] = T[\rho] + U[\rho] \tag{2}$$

where

$$T[\rho] = C_T \int d\mathbf{r}\rho^{5/3}(\mathbf{r}) \qquad C_T = \frac{3}{10}(3\pi^2)^{2/3}\frac{\hbar^2}{m} \tag{3}$$

and where for atoms

$$U[\rho] = -Ze^2 \int d\mathbf{r}\frac{1}{r}\rho(\mathbf{r}) + \frac{e^2}{2}\int d\mathbf{r}\int d\mathbf{r}'\frac{\rho(\mathbf{r})\rho(\mathbf{r}')}{|\mathbf{r}-\mathbf{r}'|} \tag{4}$$

The minimization of $E[\rho]$ leads to a second-order differential equation in an auxiliar screening function $\phi(x)$, which determines the mean potential $V(r)$ and the density $\rho(r)$ [1,2]; where $x = br$, and b is a scale factor.

Coulomb exchange effects are commonly introduced by means of the Dirac-Slater expression for the exchange energy of a electron gas:

$$K[\rho] = \int d\mathbf{r}k(\rho) \qquad k(\rho) = -e^2\frac{3}{4}\left(\frac{3}{\pi}\right)^{1/3}\rho^{4/3} \tag{5}$$

which, by insertion in the energy functional leads to the so-called Thomas-Fermi-Dirac method.

Gradient corrections to the energy density for improving the Thomas-Fermi-(Dirac) method are usually included by the addition to the energy functional of the so-called Weizsäcker term [1]:

$$T_W[\rho] = \frac{1}{9}\frac{\hbar^2}{8m}\int d\mathbf{r}\frac{|\nabla\rho|^2}{\rho} \tag{6}$$

(in the original work by von Weizsäcker the factor 1/9 was absent)

2.2. *First approaches*

The relativistic formulation of Thomas-Fermi theory started at the same time as the original non-relativistic one, the first work being of Vallarta and Rosen [9] in 1932. The result they arrived at can be found by replacing the kinetic energy functional by the result of the integration of the relativistic kinetic energy in terms of the momentum p times the number of electrons with a given momentum p from $p = 0$ to the Fermi momentum $p = p_F$:

$$T_R[\rho] = \int d\mathbf{r}\tau_R(\rho) \tag{7}$$

$$\begin{aligned}\tau_R(\rho) &= \int_0^{p_F}\frac{p^2 dp}{\pi^2\hbar^3}\left[\sqrt{p^2c^2+m^2c^4} - mc^2\right]\\ &= \frac{1}{8\pi^2\hbar^3}\left[\frac{p_F}{c^2}\left(\sqrt{p_F^2c^2+m^2c^4}\right)^3 + p_F^3\sqrt{p_F^2c^2+m^2c^4} - m^4c^5\,\mathrm{arcsinh}\left(\frac{p_F}{mc}\right)\right]\\ &\quad -\frac{mc^2p_F^3}{3\pi^2\hbar^3}\end{aligned} \tag{8}$$

$$p_F = \hbar(3\pi^2\rho)^{1/3} \tag{9}$$

The minimization of the energy functional using this kinetic energy term leads to a relativistic Thomas-Fermi differential equation

$$\frac{d^2\phi}{dx^2} = \frac{\phi^{3/2}}{x^{1/2}}\left(1+\gamma\frac{\phi}{x}\right)^{3/2} \tag{10}$$

where

$$\gamma = \left(\frac{4}{3\pi}\right)^{2/3}\alpha^2 Z^{4/3} \tag{11}$$

and $\alpha = e^2/(\hbar c)$ is the fine structure constant.

This equation is singular at $x = 0$, which implies a divergent behaviour for the density obtained from ϕ by:

$$\rho(r) = \frac{Z}{4\pi b^{3/2}}\frac{[\phi(r/b)]^{3/2}}{r^{3/2}}\left[1+\gamma b\frac{\phi(r/b)}{r}\right]^{3/2} \sim \frac{1}{r^3} \qquad \text{when } r\to 0 \tag{12}$$

in a way that ρ is not even normalizable.

This divergent behaviour at the origin can be avoided by considering instead of a point-like nucleus a uniform charged sphere of radius R [10,11,12]. Then the density is forced to drop to zero at the center of the nucleus, which makes it normalizable, and the energy is finite. However, this quantity as well as ρ near the nucleus are highly overstimated, and for example the relativistic correction to the energy

$$\Delta E = E_R - E_{NR} \tag{13}$$

is overstimated by one or two orders of magnitude. The problem can be understood from the non-relativistic Thomas-Fermi theory, which fails in the description of the density near the nucleus, where the relativistic effects are most important, with the result that the density becomes too large for small values of r but greater than the nuclear radius.

This problem can be illustrated by estimating, for a weak relativistic atom, the first correction to the kinetic energy. Using the usual expansion

$$\sqrt{p_F^2 c^2 + m^2 c^4} - mc^2 \sim \frac{p^2}{2m} - \frac{p^4}{8m^3 c^2} \tag{14}$$

in Eq. (8) we obtain

$$T_R[\rho] = C_T \int d\mathbf{r}\rho^{5/3} - C_{T2}\alpha^2 \int d\mathbf{r}\rho^{7/3} + O(\alpha^4) \tag{15}$$

where α is the fine structure constant.

The first correction cannot be evaluated perturbatively from the non-relativistic TF density because $\rho \sim 1/(r^{3/2})$ at short distances and the integral of $\rho^{7/3}$ diverges.

For performing energy estimations within this framework, corrections for strongly bound electrons (replacing their TF contribution for the energy by the values obtained from the Sommerfeld formula for a bare nuclear field) are necessary. This has been done, using the non-relativistic density by Schwinger [13] and with the relativistic one (with the nuclear finite extent) by Hill et al [12]. For including further relativistic corrections, contributions from Darwin and Breit terms were evaluated *a posteriori.* [14]

2.3. *Near-nuclear corrections*

With the purpose of evaluate not only the energy but also the electron density itself, Ashby and Holzman [15] performed calculations in which the relativistic TF density was replaced at short distancies from the nucleus from the one obtained for the $1s$ Dirac orbital for an hydrogenic atom, matched continuously to the semiclassical density at a switching radius r_0 where the kinetic energy density of both descriptions also match.

Therefore the main assumptions of this approach, provided that such matching radius r_0 exists, can be summarized as: i) for $r > r_0$ the system is adequately described as a local relativistic Fermi gas, ii) for $r < r_0$ the main contribution is due to the $1s$ single-particle state, and iii) the potential near the nucleus is approximated to $-\beta/r$. In addition to this, exchange effects were not considered.

This procedure was applied to single ionized atoms (results for the energy in Table

Table 1 *Energies (in KeV) of single positive ions evaluated with: (AH) a full relativistic kinetic energy functional without exchange [15]; the c^{-2}-order semi-relativistic functional (Eq. 46) without (1) and with (2) the relativistic exchange correction (ρ^2-term), all using near-nuclear corrections, compared to Dirac-Fock (DF) values.*

Z	$-E$(AH)	$-E$(1)	$-E$(2)	$-E$(DF)
29	40.17	42.08	42.06	44.87
37	73.26	76.19	76.16	80.90
47	132.2	136.8	136.7	144.7
55	195.0	201.4	201.3	211.6
79	482.5	501.4	500.8	517.6
89	616.8	645.4	644.5	661.1

1), with a large improvement with respect to previous RTF calculations. However, as the relativistic effects are small for most atoms, the deviations from Dirac-Fock energies are greater than the differences among these and non-relativistic Hartree-Fock ones. Nevertheless, this method provides fair values of ΔE.

2.4. *Gradient corrections to the kinetic energy*

As in the non-relativistic framework, the expansion of the energy functional in gradients of the density was explored as a theoretically well based way to correct the pure local Fermi gas description accounting for the variations of the potential and the density.

The different techniques utilized in the non-relativistic case were applied to this problem, becoming more involved (the presence of negative energy states is one of the reasons). The most popular procedures employed are the Kirznits operator conmutator expansion [16,17], or the $\hbar$ expansion of the Wigner-Kirkwood density matrix [18], which is performed starting from the Dirac hamiltonian for a mean field and does not include exchange. By means of these procedures the relativistic kinetic energy density results:

$$\tau_{RG}(\rho) = \tau_R(\rho) + \frac{1}{9}\frac{\hbar^2 c^2}{8(p_F^2 c^2 + m^2 c^4)}\left[\sqrt{p_F^2 c^2 + m^2 c^4} + 2 p_F c \operatorname{arcsinh}\left(\frac{p_F}{mc}\right)\right]\frac{|\nabla\rho|^2}{\rho} \tag{16}$$

2.5. *Relativistic exchange corrections*

Before the progress with the relativistic gradient expansion of the kinetic energy took place, and due to a growing interest of applying the Kohn-Sham scheme of density functional theory [19] in the relativistic framework, an explicit functional for the exchange energy of a relativistic electron gas was found [20,21]:

$$k_R(\rho) = \frac{e^2}{8\pi^3\hbar^4}\left[-2p_F^4 + 3m^4c^4\left(\frac{p_F}{m^2c^3}\sqrt{p_F^2c^2 + m^2c^4} - \arcsin\frac{p_F}{mc}\right)^2\right] \tag{17}$$

2.6. A relativistic Thomas-Fermi-Dirac-Weizsäcker approach

As we said in the introduction, the only consistent framework for a relativistic many-electron system is QED. By means of the Hartree-Fock limit of this theory, after renormalization, and using gradient techniques, Engel and Dreizler [22] found a complete energy functional where both terms of the two previous sections appear naturally.

$$E[\rho] = \int d\mathbf{r}\tau_{RG}(\rho) + \int d\mathbf{r}k_R(\rho) + U[\rho] \tag{18}$$

The minimization of this functional, which includes second order gradient corrections leads to the relativistic analogous of the Thomas-Fermi-Dirac-Weizsäcker model and constitutes the state of the art in relativistic semiclassical approaches for many-electron systems.

The practical implementation of this method leads to an integro-differential equation which has been solved for atoms [23] leading to the most consistent relativistic Thomas-Fermi estimations known up to now. The energy values for some atoms are displayed in Table 2 compared to Dirac-Fock ones.

Some conclusions of this work are that adequate estimates of the energy are found, similar to the precision of the TFDW method in the non-relativistic case. As a difference, no effective parameter replacing the theoretical prediction $1/9$ of the factor leading the gradient term gives general accuracy as $1/5$ does in the non-relativistic case.

Also, the fourth order terms of the gradient expanssion of the kinetic energy have been evaluated [24], leading to more involved expressions. This is one of the problems of the methods based on the gradient expansion: the systematic improvement of the results by adding higher orders is not possible because of the asymptotic nature, and

Table 2 *Values of relativistic energies (E) and differences among relativistic and non-relativistic energies (ΔE) for neutral atoms in atomic units with the present approach using the functional given by Eq. (46) not including (1) or including (2) the ρ^2 term, compared to the results of Engel and Dreizler (ED) [23] using the relativistic Thomas-Fermi-Dirac-Weizsäcker approach described in Section 2.6, and to Dirac-Fock values (DF) [30].*

Z	$-E(1)$	$-\Delta E(1)$	$-E(2)$	$-\Delta E(2)$	$-E$(ED)	$-\Delta E$ (ED)	$-E$(DF)	$-\Delta E$(DF)
20	616.97	2.505	616.78	2.310	718.59	−2.281	679.50	2.744
40	3364.8	54.82	3362.5	52.50	3738.2	20.92	3594.8	55.84
60	9149.9	354.9	9139.6	344.6	9963.6	267.3	9615.9	332.2
80	19001.	1455.	18961.	1415.	20351.	1216.	19624.	1215.
100	34499.	4560.	34398.	4460.	36247.	3838.	34806.	3524.

makes the expressions too involved for a procedure which is intended to provide simple estimates of quantum mechanical calculations for complex systems.

3. Semi-relativistic Functionals from the Breit Hamiltonian

One of the purposes of this work is to make contact between relativistic corrections in quantum mechanics and the weakly relativistic limit of QED for this problem. In particular, we will check how performing plane-wave expectation values of the Breit hamiltonian in the Pauli approximation (only terms depending on c^{-2} in atomic units) we obtain the proper semi-relativistic functional consistent in order $p_F/(mc^2)$, with the possibility of analyzing the separate contributions of terms with different physical meaning. Also the role of these terms compared to next order ones will be studied.

Neglecting spin-orbit contributions (smaller than other relativistic corrections for the ground state of atoms, and zero for closed-shell ones), the Breit hamiltonian in the Pauli approximation [25] (weak relativistic systems) can be written for a many electron system as:

$$H = H_0 + H_{MV} + H_{CRC} + H_D + H_{SS} \tag{19}$$

where H_0 stands for the non-relativistic hamiltonian:

$$H_0 = \frac{1}{2m}\sum_j \mathbf{p}_j^2 + U, \tag{20}$$

H_{MV} for the mass variation leading term:

$$H_{MV} = -\frac{1}{8m^2c^2}\sum_j \mathbf{p}_j^4, \tag{21}$$

H_{CRC} represents the relativistic correction of Coulomb interaction due to the retardation of the electromagnetic field:

$$H_{CRC} = -\frac{e^2}{2m^2c^2}\sum_{j<k}\frac{1}{r_{jk}}\left[\mathbf{p}_j\cdot\mathbf{p}_k + \frac{\mathbf{r}_{jk}\cdot(\mathbf{r}_{jk}\cdot\mathbf{p}_j)\mathbf{p}_k}{r_{jk}^3}\right], \tag{22}$$

H_{SS} denotes the spin-spin interaction:

$$H_{SS} = \frac{e^2\hbar^2}{m^2c^2}\sum_{j<k}\left\{-\frac{8\pi}{3}(\mathbf{s}_j\cdot\mathbf{s}_k)\delta(\mathbf{r}_{jk}) + \frac{1}{r_{jk}^3}\left[\mathbf{s}_j\cdot\mathbf{s}_k - 3\frac{(\mathbf{s}_j\cdot\mathbf{r}_{jk})(\mathbf{s}_k\cdot\mathbf{r}_{jk})}{r_{jk}^2}\right]\right\}; \tag{23}$$

finally, H_D accounts for the Darwin one- and two-body terms:

$$H_D = \frac{i\hbar}{(2mc)^2}\sum_j \mathbf{p}_j\cdot\nabla_j U, \tag{24}$$

and U stands for the total potential energy:

$$U \equiv U(\mathbf{r}_j, r_{jk}) = \sum_j v_{ext}(r_j) + \sum_{j<k}\frac{e^2}{r_{jk}}; \tag{25}$$

where v_{ext} is the external potential, which for the case of atoms it is just the potential energy operator for the nuclear field:

$$v_{ext}(r) = \frac{-Ze^2}{r}. \tag{26}$$

Atomic units will be used throughout. The explicit density functionals representing the different contributions to the energy from the different terms of the hamiltonian are found performing expectation values taking Slater determinants of local plane waves as in the standard Fermi gas model. Those representing the first relativistic corrections are calculated in the Appendix.

Two types of corrections to the Thomas-Fermi-Dirac non-relativistic energy density appear. The first is the correction to the kinetic energy given by the mass-variation term:

$$\epsilon_{MV} = -\frac{3^{7/3}\pi^{8/3}}{56c^2}\rho^{7/3} \tag{27}$$

and all the rest here found, depending on ρ^2 are relativistic coulomb exchange corrections

$$\epsilon_{exc} = \left(\epsilon_{EE}^{R}\right)_{exc} + (\epsilon_{CRC})_{exc} + (\epsilon_{SS})_{exc} = \frac{3\pi}{2c^2}\rho^2 \tag{28}$$

Up to now we have omitted the contribution of the Darwin terms which needs more discussion which will be given below.

The expressions (27–28) give the first correction given by the fully relativistic functionals, respectively for the kinetic and exchange energy densities (Eqs. 8 and 17) when performing a weak relativistic limit (p_F/c small):

$$\tau_R[\rho] = \tau_{NR} - \frac{3^{7/3}\pi^{8/3}}{56c^2}\rho^{7/3} + \frac{3\pi}{16c^4}\rho^3 + O(c^{-6}) \tag{29}$$

$$k_R[\rho] = K_{NR} + \frac{3\pi}{2c^2}\rho^2 - \frac{3^{8/3}\pi^{7/3}}{10c^4}\rho^{8/3} + O(c^{-6}) \tag{30}$$

where NR stands for the non-relativistic values and we have included the next order terms (in c^{-4}).

We have reproduced the c^{-2} terms by means of the analysis of the different terms of the Breit hamiltonian with the exception of Darwin ones. These terms, denoted by H_D and given by Eq. (24) are not hermitian in general. The adjoint is given by

$$H_D^{\dagger} = -\frac{i\hbar}{(2mc)^2}\sum_j \mathbf{p}_j \cdot \nabla_j U \tag{31}$$

H_D is hermitian when restricted to Slater determinants of single particle wavefunctions of the type

$$\psi_j(\mathbf{r}) = R_j(r)Y_{l_j}^{m_j}(\theta, \phi) \tag{32}$$

where $R_j(r)$ are general real and bounded functions. We have proved this statement with a lenghty but straighforward algebra using multipole expansions of r_{jk}^{-1}.

But these results do not hold for plane waves. Therefore, for our purposes we should replace H_D for the hermitian combination H'_D:

$$H'_D = \frac{1}{2}(H_D + H_D^\dagger) = \frac{\hbar^2}{8(mc)^2}\sum_j(\nabla_j^2 U) \tag{33}$$

Nevertheless, these terms should not be considered for a pure Thomas-Fermi formulation, because they depend on second derivatives of the potential operator, which are neglected in this formulation. Using one of the techniques for obtaining Thomas-Fermi plus gradient corrections these terms appears as a $\hbar^2$ contribution and should be neglected in consistency with the neglect of the density derivatives. Although only the complete $\hbar$ contribution can be assumed to be small and not necessary a particular term, we will notice in the results of Section 5 that its effect in the energy values is not very significant.

However, if we perform the expectation value of H'_D as in the previous cases we would arrive to (not considering local values, to which the one-body term contributes):

$$(\epsilon_D)_{dir} = -\frac{\pi}{2c^2}\rho^2 \tag{34}$$

$$(\epsilon_D)_{exc} = \frac{\pi}{4c^2}\rho^2 \tag{35}$$

Therefore, it appears that the Darwin corrections contribute similarly as the previous exchange relativistic corrections, but this result can be considered as spurious. It does not match with the weak relativistic limit of the fully relativistic functionals from QED but has been considered (the direct part) in previous work [26] and could be considered in the present approach just replacing the coefficient of ρ^2 (1.5 times π/c^2) by 1.25 times π/c^2.

The main problem of the application of these simple functionals that we will denote from now on by

$$T^{(2)}[\rho] = -C_{T2}c^{-2}\int d\mathbf{r}\rho^{7/3} \tag{36}$$

$$K^{(2)}[\rho] = C_{K2}c^{-2}\int d\mathbf{r}\rho^2 \tag{37}$$

where

$$C_{T2} = \frac{3^{7/3}\pi^{8/3}}{56} \tag{38}$$

$$C_{K2} = \frac{3\pi}{2} \tag{39}$$

for weakly relativistic systems (or for testing this feature) is that they diverge when evaluated by the density obtained with the standard TF method, and therefore incorporate a strong divergence of the density at the nucleus when introducing them directly in the energy functional to be minimized, as a consequence of the non-validity of the expanssion in powers of c^{-2} of the energy functional near divergencies of the potential.

However, they can be suitable for being used when correcting the near nuclear region as in the Ashby-Holzman method [15] mentioned previously. The contribution of the outer region ($r > 1/Z$) to this functionals can be easily evaluated by using the Thomas-Fermi density, which provides the asymptotic (large Z) estimates for neutral atoms:

$$\int_{1/Z}^{\infty} d\mathbf{r}\rho^t = \frac{2^{3+3t/2}}{3^{t+1}\pi^{2t-1}(t-2)} Z^{3(t-1)} \qquad \text{for } t > 2 \tag{40}$$

and

$$\int_{1/Z}^{\infty} d\mathbf{r}\rho^2 = \left(\frac{4Z}{3\pi}\right)^3 \ln Z \tag{41}$$

which allow us to compare second and fourth order (c^{-4}) terms:

$$T^{(2)}[\rho] = 3.93\,10^{-5} Z^4 \tag{42}$$

$$K^{(2)}[\rho] = 1.93\,10^{-5} Z^3 \ln Z \tag{43}$$

(appreciably smaller than the former correction),

$$T^{(4)}[\rho] = \frac{3\pi}{16c^4} \int d\mathbf{r}\rho^3 = 1.07\,10^{-10} Z^6 \tag{44}$$

which is less than 3% of $T^{(2)}[\rho]$ even for $Z = 100$, and

$$|K^{(4)}[\rho]| = \frac{3^{8/3}\pi^{7/3}}{10c^4}\rho^{8/3} \ll T^{(4)}[\rho] \tag{45}$$

for common values of Z. These estimations justify the treatment of most of the atoms as weakly relativistic systems with this procedure with the exception of the near-nuclear region. An application follows.

4. Application of the Semi-relativistic Functionals with Near-Nuclear Corrections

Here we will present a semi-relativistic Thomas-Fermi-Dirac approach for the evaluation of ground state atomic properties for not too large Z specifically designed for being used with near-nuclear corrections as those mentioned in a previous section, which becomes an alternative approach of gradient corrections procedures.

We will consider from now on the c^{-2} corrections as small, and solve consistently up to this order, first minimizing the energy functional:

$$E_{RTFD} = C_T \int \rho^{5/3} d^3r - C_{T2}c^{-2} \int \rho^{7/3} d^3r - C_K \int \rho^{4/3} d^3r + C_{K2}c^{-2} \int \rho^2 d^3r + U[\rho] \tag{46}$$

where

$$C_K = \frac{3}{4}\left(\frac{3}{\pi}\right)^{1/3} \tag{47}$$

This functional has some limitations because of the truncation of the c^{-2} expansion, which has been shown not to be accurate enough for $Z \geqslant 90$ when using gradient corrections, so in principle its application should be restricted to a smaller range of Z-values as mentioned above.

The energy given by Eq. (46) becomes stationary for a ρ which holds:

$$\frac{5}{3}\rho^{2/3} - \frac{7}{3}C_{T2}c^{-2}\rho^{4/3} - \frac{4}{3}C_K\rho^{1/3} + C_{K2}c^{-2}\rho = \epsilon_F - V \tag{48}$$

From this equation, and following the lines of standard TF theory, an explicit expression for ρ in terms of V can be found in order c^{-2} by substituting $\rho = \rho_0 + c^{-2}\rho_2$, where ρ_0 is the solution of the previous equation for $c^{-2} = 0$ (TFD method). This result can be expressed in terms of the screening function $\phi(x)$ defined as in the TFD approach by

$$a^2 - V + \epsilon_F = \frac{Z}{r}\phi(x) \tag{49}$$

where $a^2 = 1/(2\pi^2)$, $r = bx$, being b the scale factor of TF theory

$$b = \frac{1}{2}\left(\frac{3\pi}{4}\right)^{2/3} Z^{-1/3} \tag{50}$$

with the result

$$\rho = \frac{2^{3/2}}{3\pi^2}\left[a + \sqrt{\frac{Z}{b}\frac{\phi(r/b)}{x}}\right]^3 + \frac{c^{-2}}{\sqrt{2\pi}}\sqrt{\frac{r}{Z\phi(r/b)}}\left[a + \sqrt{\frac{Z}{b}\frac{\phi(r/b)}{x}}\right]^6 - c^{-2}\frac{8C_{K2}}{3\pi^4}\sqrt{\frac{r}{Z\phi(r/b)}}\left[a + \sqrt{\frac{Z}{b}\frac{\phi(r/b)}{x}}\right]^5 \tag{51}$$

Using $\nabla^2 V = -4\pi\rho$ in terms of ϕ we arrive at the second-order differential equation

$$\frac{d^2\phi}{dx^2} = x\left(\beta + \sqrt{\frac{\phi}{x}}\right)^3 + c^{-2}\gamma\sqrt{\frac{x}{\phi}}x\left(\beta + \sqrt{\frac{\phi}{x}}\right)^6 + c^{-2}\sigma\sqrt{\frac{x}{\phi}}x\left(\beta + \sqrt{\frac{\phi}{x}}\right)^5 \tag{52}$$

where

$$\beta = \frac{3}{2(6\pi Z)^{2/3}} \tag{53}$$

$$\gamma = \left(\frac{6}{\pi^2}\right)^{1/3} Z^{4/3} \tag{54}$$

and

$$\sigma = C_{K2}\frac{32}{3\sqrt{2}\pi^3}\left(\frac{3\pi}{4}\right)^{1/3} Z^{2/3} \tag{55}$$

The main interest of this approximation, which is justified when removing from the statistical treatment the near-nuclear region, becomes clear from this expression, because the technique to be employed is absolutely similar to the implementation of the TFD method.

The screened potential V and the electron density ρ are obtained from the numerical values of ϕ by means of Eqs. (49) and (51).

As in the TFD method, this equation has to be solved with the boundary conditions $\phi(0) = 1$ and $x_c\phi'(x_c) = \phi(x_c)$ where $r_c = bx_c$ is the cutoff point where the pressure of the electron gas becomes zero. The value of $\phi(x_c)/x_c$ necessary for this differs from the non-relativistic case and can be found from similar grounds [27], just by making zero the pressure of a homogeneous electron gas, given by

$$p = \rho^2 \frac{\delta}{\delta\rho}\frac{\epsilon}{\rho} = \frac{2}{3}C_T\rho^{5/3} - \frac{4}{3}C_{T2}c^{-2}\rho^{7/3} - \frac{1}{3}C_K\rho^{4/3} + C_{K2}c^{-2}\rho^2 = 0 \qquad (56)$$

which again can be solved in order c^{-2}. Substituting the value ρ obtained in this way in Eq. (51) and solving for ϕ/x we find that

$$\frac{\phi(x_c)}{x_c} = y_c(1 - c^{-2}\delta) \qquad (57)$$

where $y_c = \beta^2/16$ is the TFD-value and

$$\delta = \frac{5^4 3}{2^6 7\pi^2} - \frac{5^3 C_{K2}}{6\pi^3} \qquad (58)$$

The Fermi energy ϵ_F is found by imposing that at the cutoff radius the potential should be equal to

$$V(r_c) = -\frac{Z - N}{r_c} \qquad (59)$$

Note that replacing c^{-2} by zero in this procedure, we find all the TFD expresions, and if we make $C_{K2} = 0$ we supress the relativistic exchange corrections. When we include these in our calculations we have always chosen the value of $3\pi/2$, the one which matches the weak relativistic limit of the fully relativistic energy functional.

We should remember that this simplified method cannot be applied when $r \to 0$, because the density, as it can be seen in Eq. (51) diverges as $r^{-5/2}$ and also because the expansion in powers of c^{-2} of the energy functional is not valid at short distancies from the nucleus. We have applied this method using the near-nuclear correction of Ashby and Holzman [15], which consists of a replacement of the energy and electron densities for r below a given r_0 by the relativistic quantum mechanical expresions for the K-shell spinor provided by these authors. This corresponds to solving Dirac equation for an effective mean potential equal to $-Z^*/r$, where Z^* is an effective charge (denoted β in ref. [15]). We determine this parameter, as well as the normalization constant of the spinor by matching ρ and its derivative to the semiclassical density at a r_0 where the quantum mechanical kinetic energy density equals the one in Eq. (46).

The only weak point of this procedure is the existence of the matching point determined by this *ad-hoc* argument. In fact, no solution for the original Ashby-

Holzmann procedure was found beyond $Z \sim 90$ as pointed out by Gross and Dreizler [28].

The values of the total energy of atomic systems is calculated then integrating the quantum mechanical energy density for $r < r_0$ and the semiclassical one for $r_c > r > r_0$. Our first calculation was performed for single positive ions, neglecting all exchange effects (even the non-relativistic ones) in order to compare our procedure to the results of Ref. [15] where they were not considered, as a test of the validity of the mass variation correction in c^{-2}. The differences are about 1% for $Z = 55$, 2% for $Z = 79$ and less than 3% for $Z = 89$.

Then we have applied the present method, including all terms of Eq. (46) for evaluating the energy of single positive ions. In Table 1 a comparison to Ashby-Holzman results as well as to Dirac-Fock ones [29] can be found. Calculations without the ρ^2 term are also displayed. It can be observed how the inclusion of non-relativistic exchange improves appreciably the results, accounting for about half of the gap between previous estimates and Dirac-Fock values. Also, the relativistic correction to exchange (ρ^2 terms) is found to be few significant for most of the atoms.

We have also performed calculations of the energy for neutral atoms and compared to those obtained from the fully relativistic functional of Engel and Dreizler (Eq. 18) [23] including gradient corrections (the best and most consistent values known up to now) as well as to Dirac-Fock values [30]. The values of the energy and of the differences among relativistic and non-relativistic values ΔE are shown for the three approaches in Table 2. We see how the accuracy of the present method is similar to that of the gradient expansion for Z not very large, so it can constitute an alternative procedure. Moreover, the differences ΔE are estimated more precisely except for large Z, which suggests that the remaining errors in the present semiclassical approach with the near nuclear corrections above mentioned are similar in the non-relativistic and in the relativistic case.

It is worth to mention how the present method provides fair results for $Z = 100$, what was not expected, because the Ashby-Holzmann original procedure was not applicable for such a Z-value. There are differences among their procedure and ours, from the inclusion of exchange effects with the additional fact that different boundary conditions have been utilized, and from the truncation of the energy functional. All these facts alter the values of the matching point found with a self-contained procedure where ρ, ρ' and τ become continuous at the same point.

We want to mention here that the application of the near-nuclear corrections could have been performed with the complete relativistic functional, and we have utilized the semi-relativistic expressions just for simplicity and for testing them. For not large Z, the remaining errors above mentioned should be addressed to limitations of the semiclassical approach and of the procedure utilized for the near-nuclear corrections, rather than to the truncation of the expansion in powers of c^{-2}.

Within the context of the gradient expansion, and for taking into account greater orders from the Foldy-Wouthuysen representation, Gross and Dreizler used the full relativistic kinetic energy part [16] instead of the mass variation term. Using this in the present approach might improve the results for large Z, but greater orders of the exchange energy should then also be taken into account, by the use of Eq. (17).

Nevertheless, when we include the near nuclear corrections (where the fully relativistic kinetic energy is used), the truncation of the energy functional only in the outer region up to c^{-2} order both in the kinetic and exchange energies turns out to be an adequate approximation.

5. Conclusions and Suggestions for Future Work

Semiclassical methods from quantum mechanics with first-order relativistic corrections obtained from the Foldy-Wouthuysen transformation match with the weak relativistic limit of functionals obtained from quantum electrodynamics, neglecting the (spurious) Darwin terms.

For describing electron and energy densities, the near-nuclear region has to be described fully relativistically. Outside this region, electrons are weakly relativistic, and the most important relativistic effect in the energy values is the mass-variation one.

Results of similar accuracy as relativistic TFDW are found with a simple procedure based on near-nuclear correction which leave space for further improvements. For the reasons mentioned at the end of previous section the direct way to improve the present approach seems to be the refinement of the near nuclear corrections, a problem that we have just tackled with success in the non-relativistic framework [31,32]. The aim was to describe the near-nuclear region accurately by means of using the quantum mechanical exact asymptotic expression up to r^3 of the different ns eigenstates of Schödinger equation with a fit of the semiclassical potential at short distancies to the exact asymptotic behaviour (with four terms) of the potential near the nucleus. The result is that the density below r_0 becomes very close to Hartree-Fock values and the improvement of the energy values is large (as an example, the energy of Cs^+ is improved from the Ashby-Holzman result of -189.5 keV up to -205.6, very close to the HF value of -204.6 keV). This result makes us expect that a similar procedure in the relativistic framework may provide results comparable to Dirac-Fock ones.

Acknowledgements

The authors acknowledge useful discussions with Dr. E. Engel. This work has been partially supported by the Spanish Comisión Interministerial de Ciencia y Tecnología (CICYT) under contract PB95-1211 and by the Junta de Andalucía.

Appendix: Obtention of Explicit Functionals for the Relativistic Corrections to the Energy

In this appendix, explicit functionals of the density representing the different relativistic corrections are found by performing expectation values taking Slater determinants of local plane waves as in the standard Fermi gas model. For a one-body operator we use:

$$A[\rho] \equiv \langle\Psi|\sum_{j=1}^{N} A(\mathbf{r}_j, \sigma_j)|\Psi\rangle = \sum_s \int_{p\leq p_F} d\mathbf{p} \int d\mathbf{r} \phi^+_{\mathbf{p},s}(\mathbf{r}) \cdot A(\mathbf{r}, \sigma)\phi_{\mathbf{p},s}(\mathbf{r}) \tag{60}$$

(s denoting the third spin component), and for a two body operator we will perform separately the direct and exchange terms:

$$B[\rho] = \langle\Psi| \sum_{j<k=1}^{N} B(\mathbf{r}_j, \sigma_j; \mathbf{r}_k, \sigma_k)|\Psi\rangle = (B[\rho])_{dir} + (B[\rho])_{exc} \tag{61}$$

$$(B[\rho])_{dir} = \frac{1}{2}\sum_{s}\sum_{s'}\int_{p\leqslant p_F} d\mathbf{p}\int_{p'\leqslant p_F} d\mathbf{p}'\int d\mathbf{r}\int d\mathbf{r}'\phi^{+}_{\mathbf{p},s}(\mathbf{r})\phi^{+}_{\mathbf{p}',s'}(\mathbf{r}')\cdot$$

$$B(\mathbf{r}, \sigma; \mathbf{r}', \sigma')\phi_{\mathbf{p},s}(\mathbf{r})\phi_{\mathbf{p}',s'}(\mathbf{r}') \tag{62}$$

$$(B[\rho])_{exc} = -\frac{1}{2}\sum_{s}\sum_{s'}\int_{p\leqslant p_F} d\mathbf{p}\int_{p'\leqslant p_F} d\mathbf{p}'\int d\mathbf{r}\int d\mathbf{r}'\phi^{+}_{\mathbf{p},s}(\mathbf{r})\phi^{+}_{\mathbf{p}',s'}(\mathbf{r}')\cdot$$

$$B(\mathbf{r}, \sigma; \mathbf{r}', \sigma')\phi_{\mathbf{p}',s'}(\mathbf{r})\phi_{\mathbf{p},s}(\mathbf{r}'), \tag{63}$$

The results will be expressed in terms of energy densities. For obtaining the first-order relativistic corrections (in $\alpha^2 = c^{-2}$ in a.u.) it is enough to utilize relativistic plane waves for the non-relativistic hamiltonian H_0 and non-relativistic ones for the other terms depending on c^{-2}.

Relativistic correction to the expectation value of H_0

For the non-relativistic hamiltonian H_0, the relativistic correction is found by using for the evaluation of the expectation value relativistic plane waves for electrons, given by:

$$\phi_{\mathbf{p},s}(\mathbf{r}) = \frac{1}{(2\pi)^{3/2}}\sqrt{\frac{c^2}{E_p}}u_s(\mathbf{p})e^{-i\mathbf{pr}} \tag{64}$$

where $E_p = \sqrt{p^2c^2 + c^4}$ and $u_s(\mathbf{p})$ is a Dirac spinor for a free electron of momentum $\mathbf{p}$ and spin s.

From all the terms of H_0 the only correction in c^{-2} is found for the exchange term of the electron-electron interaction:

$$(\epsilon_{EE})_{exc} = -\frac{1}{2(2\pi)^6}\int_{p\leqslant p_F} d\mathbf{p}\int_{p'\leqslant p_F} d\mathbf{p}'\int d\mathbf{r}'\frac{c^2}{E_p}\frac{c^2}{E'_p}\sum_{s}\sum_{s'}(u^{+}_{s}(\mathbf{p})\cdot u_{s'}(\mathbf{p}'))$$

$$(u^{+}_{s'}(\mathbf{p}')\cdot u_s(\mathbf{p}))\frac{1}{|\mathbf{r}-\mathbf{r}'|}e^{-i(\mathbf{p}-\mathbf{p}')(\mathbf{r}-\mathbf{r}')}. \tag{65}$$

Using $\mathbf{t} = \mathbf{r} - \mathbf{r}'$ and

$$\frac{c^2}{E_p}\frac{c^2}{E'_p}\sum_{s}\sum_{s'}(u^{+}_{s}(\mathbf{p})\cdot u_{s'}(\mathbf{p}'))(u^{+}_{s'}(\mathbf{p}')\cdot u_s(\mathbf{p})) = \frac{(E_p + c^2)(E_{p'} + c^2)}{2E_pE_{p'}}$$

$$\left[1\frac{\mathbf{p}\cdot\mathbf{p}'}{2c^2} + O(c^{-4})\right] = 2\left(1 - \frac{|\mathbf{p}-\mathbf{p}'|^2}{4c^2} + O(c^{-4})\right), \tag{66}$$

a result that can be easily found from the standard algebra of Dirac spinors and neglecting any terms of higher order than c^{-2}, we find that

$$(\epsilon_{EE})_{exc} = -\frac{1}{(2\pi)^6}\int d\mathbf{t}\int_{p\leq p_F} d\mathbf{p}\int_{p'\leq p_F} d\mathbf{p}'\left[1-\frac{|\mathbf{p}-\mathbf{p}'|^2}{4c^2}\right]\frac{1}{t}e^{-i(\mathbf{p}-\mathbf{p}')\mathbf{t}} \tag{67}$$

The integral of the first term in square brackets gives the non-relativistic Dirac-Slater exchange energy, the second giving the relativistic correction:

$$(\epsilon_{EE}^{R})_{exc} = \frac{p_F^6}{32\pi^3c^2} = \frac{\pi}{4c^2}\rho^2 \tag{68}$$

All the rest of the terms of H_0 just provide the usual TF energy functional (Eq. 2).

Mass variation correction

As said above we will restrict ourselves from now on to the use of non-relativistic plane waves:

$$\phi_{\mathbf{p},s}(\mathbf{r}) = \frac{1}{(2\pi)^{3/2}}e^{-i\mathbf{pr}}\chi_s \tag{69}$$

where χ_s is a two-component spin state.

We evaluate now the contribution of H_{MV}, which provides the first order relativistic correction to the kinetic energy:

$$\epsilon_{MV} = -\frac{1}{8c^2}\sum_s\int_{p\leq p_F} d\mathbf{p}\phi^+_{\mathbf{p},s}(\mathbf{r})p^4\phi_{\mathbf{p},s}(\mathbf{r}) = -\frac{1}{4c^2}\int_{p\leq p_F} d\mathbf{p}p^4 = -\frac{1}{56\pi^2c^2}p_F^7$$
$$= -\frac{3^{7/3}\pi^{8/3}}{56c^2}\rho^{7/3} \tag{70}$$

Coulomb relativistic correction

We separate two terms from $H_{CRC} = H^{(1)}_{CRC} + H^{(2)}_{CRC}$:

$$H^{(1)}_{CRC} = -\frac{1}{2c^2}\sum_{j<k}\frac{1}{r_{jk}}\mathbf{p}_j\mathbf{p}_k \tag{71}$$

$$H^{(2)}_{CRC} = -\frac{1}{2c^2}\sum_{j<k}\frac{1}{r^3_{jk}}\mathbf{r}_{jk}\cdot(\mathbf{r}_{jk}\cdot\mathbf{p}_j)\mathbf{p}_j \tag{72}$$

The direct terms of both two-body operators are zero with a single integration in one of the moments. The exchange contribution of the first one is given by (using conventions from the previous section):

$$(\epsilon^{(1)}_{CRC})_{exc} = \frac{1}{2c^2}\frac{1}{(2\pi)^6}\int d\mathbf{t}\int_{p\leq p_F} d\mathbf{p}\int_{p'\leq p_F} d\mathbf{p}'\frac{\mathbf{p}\cdot\mathbf{p}'}{t}e^{-i(\mathbf{p}-\mathbf{p}')\mathbf{t}} \tag{73}$$

We use now $4\mathbf{p}\cdot\mathbf{p}' = (\mathbf{p}+\mathbf{p})^2 - (\mathbf{p}-\mathbf{p}')^2$ and integrating in $\mathbf{t}$ we can write

$$(\epsilon^{(1)}_{CRC})_{exc} = \frac{1}{(2\pi)^5 c^2} \int_{p \leqslant p_F} d\mathbf{p} \int_{p' \leqslant p_F} d\mathbf{p}' \left[\frac{(\mathbf{p}+\mathbf{p})^2}{4(\mathbf{p}-\mathbf{p}')^2} - \frac{1}{4} \right] \tag{74}$$

The first integral can be performed using the change of variables

$$\mathbf{q} = \mathbf{p} - \mathbf{p}' \qquad \mathbf{u} = \mathbf{p} + \mathbf{p}' \tag{75}$$

which gives

$$\int_{p \leqslant p_F} d\mathbf{p} \int_{p' \leqslant p_F} d\mathbf{p}' \frac{(\mathbf{p}+\mathbf{p})^2}{4(\mathbf{p}-\mathbf{p}')^2} = \int \frac{d\mathbf{q}}{q^2} \int u^2 d\mathbf{u} \theta(p_F - |\mathbf{u} + \mathbf{q}/2|)\theta(p_F - |\mathbf{u} - \mathbf{q}/2|) \tag{76}$$

Integrals of functions of u, q restricted to the region $p_F - |\mathbf{u} + \mathbf{q}/2|$ and $p_F - |\mathbf{u} - \mathbf{q}/2|$ ($\mathbf{u}$ lying in the intersection of two spheres of centers located at $q/2$ and $-q/2$ of radius p_F) are easily evaluated with the conventions

$$y = u \cos(\mathbf{u}, \mathbf{q}) \qquad x = u \sin(\mathbf{u}, \mathbf{q}) \tag{77}$$

which lead to:

$$\int d\mathbf{q} \int d\mathbf{u} \theta(p_F - |\mathbf{u} + \mathbf{q}/2|)\theta(p_F - |\mathbf{u} - \mathbf{q}/2|) f(u, q)$$

$$= 16\pi^2 \int_0^{2p_F} dq \int_0^{p_F - q/2} dy \int_0^{\sqrt{p_F^2 - (z+q/2)^2}} dx f(q, \sqrt{x^2 + y^2}) \tag{78}$$

Using this and evaluating the second integral straightforwardly we obtain

$$(\epsilon^{(1)}_{CRC})_{exc} = \frac{1}{24\pi^3 c^2} p_F^6 = \frac{3\pi}{8c^2} \rho^2 \tag{79}$$

The expectation value of Eq. (72) is a bit more involved. The direct term is found to be zero, and the exchange one can be written as

$$(\epsilon^{(2)}_{CRC})_{exc} = \frac{1}{2(2\pi)^6 c^2} \int d\mathbf{u} \int d\mathbf{q} \int d\mathbf{t} e^{-i\mathbf{q}\mathbf{t}} \left[\frac{(\mathbf{t} \cdot \mathbf{u})^2}{t^3} - \frac{(\mathbf{t} \cdot \mathbf{q})^2}{4t^3} \right]$$

$$\times \theta(p_F - |\mathbf{u} + \mathbf{q}/2|)\theta(p_F - |\mathbf{u} - \mathbf{q}/2|) \tag{80}$$

We will perform the integration in $\mathbf{t}$ choosing the z axis in the direction of $\mathbf{q}$. The second integral is quite easy and equal to:

$$\int d\mathbf{t} e^{-i\mathbf{q}\mathbf{t}} \frac{(\mathbf{t} \cdot \mathbf{q})^2}{4t^3} = -\pi \tag{81}$$

and the first one, decomposed as

$$\int d\mathbf{t} e^{-i\mathbf{q}\mathbf{t}} \frac{(\mathbf{t} \cdot \mathbf{u})^2}{t^3} = \sum_{j=1}^{3} \sum_{k=1}^{3} u_j u_k I_{jk} \tag{82}$$

where

$$I_{jk} = \int d\mathbf{t} e^{-iqt\cos\theta} \frac{t_i t_j}{t^3} \tag{83}$$

is given by (only diagonal terms I_{jj} are non-zero)

$$\int d\mathbf{t} e^{-i\mathbf{q}\mathbf{t}} \frac{(\mathbf{t}\cdot\mathbf{u})^2}{t^3} = 4\pi\left(\frac{u^2}{q^2} - 2\frac{u^2}{q^2}\, cos(\mathbf{u}, \mathbf{q})\right) \tag{84}$$

which applied to Eq. (80) leads to the result:

$$(\epsilon^{(2)}_{CRC})_{exc} = \frac{1}{24\pi^3 c^2} p_F^6 = \frac{3\pi}{8c^2}\rho^2 \tag{85}$$

exactly the same as $(\epsilon^{(1)}_{CRC})_{exc}$. Therefore

$$\epsilon_{CRC} = (\epsilon^{(1)}_{CRC})_{exc} + (\epsilon^{(2)}_{CRC})_{exc} = \frac{3\pi}{4c^2}\rho^2 \tag{86}$$

Corrections for spin-spin interactions

The energy density from the first term of H_{SS} (Eq. 23) can be written as

$$(\epsilon^{(1)}_{SS})_{dir} = -\frac{\pi}{3c^2}\frac{1}{(2\pi)^6}\int_{p\leq p_F} d\mathbf{p}\int_{p'\leq p_F} d\mathbf{p}'\int d\mathbf{t}\delta(\mathbf{t})\sum_s\sum_{s'} \mathbf{A}(s, s)\cdot\mathbf{A}(s', s') \tag{87}$$

for the direct contribution, and

$$(\epsilon^{(1)}_{SS})_{exc} = \frac{\pi}{3c^2}\frac{1}{(2\pi)^6}\int_{p\leq p_F} d\mathbf{p}\int_{p'\leq p_F} d\mathbf{p}'\int d\mathbf{t}\delta(\mathbf{t}) e^{-i(\mathbf{p}-\mathbf{p}')\mathbf{t}}\sum_s\sum_{s'} \mathbf{A}(s, s')\cdot\mathbf{A}(s', s) \tag{88}$$

where

$$\mathbf{A}(s, s') = \langle\chi_s|\sigma|\chi_{s'}\rangle \tag{89}$$

being $\sigma = (\sigma_x, \sigma_y, \sigma_z)$ the Pauli matrices. For the second term we write

$$(\epsilon^{(2)}_{SS})_{dir} = \frac{1}{8c^2}\frac{1}{(2\pi)^6}\int_{p\leq p_F} d\mathbf{p}\int_{p'\leq p_F} d\mathbf{p}'\int d\mathbf{t}\frac{1}{t^3}\left[\sum_s\sum_{s'} \mathbf{A}(s, s)\mathbf{A}(s', s') - \frac{1}{t^2}{}_s\sum_{s'}(\mathbf{A}(s, s)\cdot\mathbf{t})(\mathbf{A}(s', s')\cdot\mathbf{t})\right] \tag{90}$$

and

$$(\epsilon^{(2)}_{SS})_{exc} = \frac{1}{8c^2}\frac{1}{(2\pi)^6}\int_{p\leq p_F} d\mathbf{p}\int_{p'\leq p_F} d\mathbf{p}'\int d\mathbf{t}\frac{1}{t^3}\left[\sum_s\sum_{s'} \mathbf{A}(s, s')\mathbf{A}(s', s) - \frac{1}{t^2}\sum_s\sum_{s'}(\mathbf{A}(s, s')\cdot\mathbf{t})(\mathbf{A}(s', s)\cdot\mathbf{t})\right] \tag{91}$$

Using that $\mathbf{A}(1/2, 1/2) = (0, 0, 1)$, $\mathbf{A}(1/2, -1/2) = (1, i, 0)$, $\mathbf{A}(-1/2, 1/2) = (1, -i, 0)$ and $\mathbf{A}(-1/2, -1/2) = (0, 0, -1)$ it is easy to find that

$$\epsilon_{SS} = (\epsilon_{SS}^{(1)})_{exc} = \frac{\pi}{2c^2}\rho^2 \tag{92}$$

because $(\epsilon_{SS}^{(1)})_{dir} = (\epsilon_{SS}^{(2)})_{dir} = (\epsilon_{SS}^{(2)})_{exc} = 0$, which completes the plane-wave evaluation of the Breit-Pauli Hamiltonian with the exception of Darwin terms, discussed in Section 3.

References

1. R.G. Parr and W. Yang: *Density Functional Theory of Atoms and Molecules*, Oxford University Press, New York (1989).
2. N.H. March: *Self-Consistent Fields in Atoms*, Pergamon, Oxford (1975).
3. I.K. Dmitrieva and G.I. Plindov: Z. Phys. A305, 103 (1982).
4. J.S. Dehesa, F.J. Gálvez and I. Porras: Phys. Rev. A40, 35 (1989).
5. Y. Tomishima and K. Yonei: J. Phys. Soc. Jpn. 21, 142 (1966).
6. E. Engel, S. Keller, A. Facco Bonetti, H. Müller and R.M. Dreizler: Phys. Rev. A52, 2750 (1995).
7. E. Ilyabaev and U. Kaldor: J. Chem. Phys. 97, 8455 (1992).
8. J. Hata and I. P. Grant: J. Phys. B 16, 523 (1983).
9. M.S. Vallarta and N. Rosen: Phys. Rev. 41, 708 (1932).
10. H. Jensen: Z. Phys. 82, 794 (1933).
11. J. Ferreirinho, R. Ruffini and L. Stella: Phys. Lett. 91B, 314 (1980).
12. S.H. Hill, P.J. Grout and N.H. March: J. Phys. B 17, 4819 (1984).
13. J. Schwinger: Phys. Rev. A 22, 1827 (1980)
14. S.H. Hill, P.J. Grout and N.H. March: J. Phys. B, 20, 11 (1987).
15. N. Ashby and M.A. Holzman: Phys. Rev. A 1, 764 (1970).
16. E.K.U. Gross and R.M. Dreizler: Phys. Lett. 81 A, 447 (1981).
17. P. Malzacher and R.M. Dreizler: Z. Phys. D 2, 37 (1986).
18. M. Centelles, X. Viñas, M. Barranco and P. Schuck: Ann. Phys. 221, 165 (1993).
19. W. Kohn and L.J. Sham: Phys. Rev. 140, A1133 (1965).
20. A.K. Rajagopal: J. Phys. C 11, L943 (1978).
21. A.H. MacDonald and S.H. Vosko: J. Phys. C 12, 2977 (1979).
22. E. Engel and R.M. Dreizler: Phys. Rev. A 35 3607 (1987).
23. E. Engel and R.M. Dreizler: Phys. Rev. A 38 3909 (1988).
24. W.F. Pohlner and R.M. Dreizler: Phys. Rev. A 44, 7165 (1991).
25. H.A. Bethe and E.E. Salpeter: *Quantum Mechanics of One- and Two-Electron Atoms*, Plenum-Rosetta, New York (1977).
26. B.G. Englert: *Semiclassical Theory of Atoms*, Lect. Notes in Phys. **300**, Springer Verlag, Heidelberg (1988).
27. N.H. March: Adv. Phys. 6, 1 (1957).
28. E.K.U. Gross and R.M. Dreizler: Density Functional approach to time-dependent and to relativistic systems, in R.M. Dreizler and J. da Providência (Eds.), *Density Functional Methods in Physics*, NATO ASI Series B Vol.123, Plenum Press, New York, 81 (1985).
29. J.F. Barnes and R.D. Cowan: Phys. Rev. 132, 236 (1963).
30. J.P. Desclaux: At. Data Nucl. Data Tables 12, 311 (1973).
31. I. Porras and A. Moya: Phys. Rev. A 59, 1859 (1999).
32. A. Moya and I. Porras: Expectation values for ground state atoms from a modified Thomas-Fermi-Dirac approach, in A. Hernández-Laguna et al. (eds.), Quantum Systems in Chemistry and Physics, Vol. I, Kluwer Academic Publishers (2000).

Expectation Values for Ground-State Atoms from a Modified Thomas-Fermi-Dirac Approach

A. Moya and I. Porras
Departamento de Física Moderna, Facultad de Ciencias, Universidad de Granada, E-18071 Granada, Spain

Abstract

By means of a modification of the *TFD* method in the near nuclear region for the electron and energy densities, which introduces exact asymptotic properties, radial expectation values and the atomic density at the nucleus are evaluated, comparing fairly closely to the HF results, with a large improvement of the TF estimates. In addition to this, momentum expectation values can be estimated from semiclassical relations.

1. Introduction

Among the average properties which play a special role in the study of quantum fermionic systems are the radial expectation value $\langle r^\alpha \rangle$, the momentum expectation value $\langle p^\alpha \rangle$ and the atomic density at the nucleus $\rho(0) = \langle \delta(\mathbf{r}) \rangle$. These density-dependent quantities are defined by

$$\langle r^\alpha \rangle = \langle \Psi | \sum_{i=1}^{N} r_i^\alpha | \Psi \rangle = \int d\vec{r} r^\alpha \rho(\vec{r})$$

$$\langle p^\alpha \rangle = \langle \Psi | \sum_{i=1}^{N} (-i\hbar \vec{\nabla}_i)^\alpha | \Psi \rangle = \int d\vec{p} p^\alpha \gamma(\vec{p})$$

where

$$\rho(\vec{r}) = \langle \Psi | \sum_{i=1}^{N} \delta(\vec{r} - \vec{r}_i) | \Psi \rangle$$

and

$$\gamma(\vec{p}) = \langle \tilde{\Psi} | \sum_{i=1}^{N} \delta(\vec{p} - \vec{p}_i) | \tilde{\Psi} \rangle$$

where $\tilde{\Psi}(\vec{p}_1, \ldots, \vec{p}_N)$ is the 3N-Fourier transform of $\Psi(\vec{r}_1, \ldots, \vec{r}_N)$.

$$\rho(0) = \int \delta(\vec{r}) \rho(\vec{r}) d\vec{r}$$

Some of these quantities are experimentally measurable from scattering experiments [1,2] or relate to physically interesting quantities which appear in different problems. For example: $\langle p^2 \rangle$ gives the kinetic energy of any nonrelativistic system, $\langle r^2 \rangle$ is related to the diamagnetic susceptibility, $\langle r^{-1} \rangle$ relates to the electron-nucleus potential energy,

A. Hernández-Laguna et al. (eds.), Quantum Systems in Chemistry and Physics, Vol. 1: Basic Problems and Model Systems, 215–225.

$\langle p^{-1}\rangle$ is twice the peak height of the atomic Compton profile and $\rho(0)$ plays a role in the asymptotic behaviour of momentum-space properties [3].

When dealing with a model wavefunction or density, these quantities are usually evaluated for comparison to accurate calculations, because they constitute a test of the description of the system in different regions of space for different orders α.

In addition to this, average properties like $\langle r^{\alpha}\rangle$ or $\langle p^{\alpha}\rangle$ play a special role in the formulation of bounds or approximations to different properties like the kinetic energy [4,5], the average of the radial and momentum densities [6,7] and $\rho(0)$ itself [8,9,10]; they also are the basic information required for the application of bounds to the radial electron density $\rho(\mathbf{r})$, the momentum one density $\gamma(\mathbf{p})$, the form factor and related functions [11,12,13]. Moreover they are required as input in some applications of the Maximum-entropy principle to modelize the electron radial and momentum densities [14,15].

Although numerical values of $\langle r^{\alpha}\rangle$ and $\langle p^{\alpha}\rangle$, as well as $\rho(0)$ can be easily obtained for near Hartree-Fock wavefunctions [16,17], some effort has been done in order to study the general features of these quantities as a function of the atomic number [18,19,20]. The easiest way to estimate these comes from the *TF* model [21,22], and has the important property of providing the exact asymptotic behaviour at large Z.

In addition to the energy value, it is interesting to evaluate the expectation values mentioned above by means of *TF* and it extensions as tests of the validity of the semiclassical method in different regions of space for the electron distribution.

The original *TF* model has some limitations for the evaluation of $\langle r^{\alpha}\rangle$, specially addressed in the description of the electron density near the nucleus, where the electron cloud differs most from a local Fermi gas, and at large distances. This limits the values of α within $(-3/2, 3)$ [18], otherwise they diverges. The use of a density cutoff, as it is used in the *TF*D method (including exchange effects) [23] removes the upper limit, although higher orders are not expected to be described accurately. But the main problem lies on the lowest-order ones, $\rho(0)$ and $\langle r^{-2}\rangle$ diverge both in *TF* and *TFD*, and $\langle r^{-1}\rangle$ is not accurate for real atoms. Nevertheless, the *TF* estimates provide the asymptotic values for neutral atoms [20]:

$$\langle r^{-1}\rangle = 1.7936 \cdot Z^{4/3}$$

$$\langle r\rangle = 3.1859 \cdot Z^{2/3}$$

$$\langle r^{2}\rangle = 41.9625 \cdot Z^{1/3}$$

These asymptotic expansions are not reached in the case of real atoms in nature, for which Z is not large enough, as it can be seen in Figures 3, 4 and 5.

The most popular approach for the improvement of *TF*, the Thomas-Fermi- Dirac-Weizsäcker approach, based on the gradient expansion of the energy functional, although providing reasonable energy estimates, is not enough for describing accurately the density $\rho(\mathrm{r})$ near the nucleus, even including fourth order terms [24].

In this work we propose an improvement of the alternative method of Ashby-Holzman [25], which is based on matching the *TF* density at short distances with the density corresponding to the 1s-hydrogenic orbital at a point where the kinetic energy equals the *TF* one. We include the electron exchange (TF-Dirac) and a different treatment of the short-range density from the asymptotically exact small-r limit of the s-orbitals and of the potential in this region.

2. Modified TFD Approach

The standard *TFD* is based on the minimization of the energy functional (atomic units used throughout the work)

$$E[\rho] = \int d\vec{r}\varepsilon_{TFD} \tag{1}$$

$$\varepsilon_{TFD} = C_k \rho_{TFD}^{5/3} + V(\vec{r})\rho_{TFD} - C_e \rho_{TFD}^{4/3} \tag{2}$$

$$C_k = \frac{3}{10}(3\pi^2)^{2/3}; \qquad C_e = -\frac{4}{3}\left(\frac{3}{\pi}\right)^{1/3}$$

where

$$V(\vec{r}) = -\frac{Z}{r} + \int \frac{\rho_{TFD}(\vec{r}')}{|\vec{r} - \vec{r}'|} d\vec{r}' \tag{3}$$

which leads to the electron density

$$\rho_{TFD}(r) = \frac{2^{3/2}}{3\pi}\left[\frac{1}{\sqrt{2}\pi} + \sqrt{\frac{1}{2\pi^2} - V(r) + \varepsilon_F}\right]^3 \tag{4}$$

All the quantities can be expressed in terms of the screening function defined by

$$V(r) - \varepsilon_F - \frac{1}{2\pi^2} = -\frac{Z}{r}\phi(x) \tag{5}$$

that satisfy the *TFD* differential equation

$$\frac{d^2\phi}{dx^2} = x\left[\beta + \left(\frac{\phi}{x}\right)^{1/2}\right]^3 \tag{6}$$

with

$$\beta = \frac{3}{2(6\pi Z)^{2/3}}; \qquad r = bx; \qquad b = \frac{1}{2}\left(\frac{3\pi}{4}\right)^{-1/3} = 0.885341 \cdot Z^{-1/3}$$

Eq. (6) is solved with the boundary conditions $\phi(0) = 1$, which provides the proper small-r limit for $V(r)$, and $(\phi(x_l)/x_l) = (\beta^2/16)$ for null pressure of the electron cloud at a limiting atomic radius $r_l = bx_l$. At this point the density is cutoff.

We will adopt this model for ε_{TFD} and ρ_{TFD} from an inner radius $r = r_0$ to the cutoff radius $r = r_l = bx_l$. In the inner region they will be replaced by quantum mechanical values than can be found from the solution of Schrödinger equation with the asymptotic expansion of the potential

$$V(r) = -\frac{Z}{r} + V_0 + V_1 r + O(r^2) \tag{7}$$

where V_0 and V_1 are found from a fitting of the *TFD* potential (5) and leads to an expansion of the electron density of

$$\rho_I = A\left[1 - 2Zr + (2C + Z^2)r^2 + \frac{1}{3}(Z^3 - 10ZC + V_1)r^3\right] \tag{8}$$

and of the energy density

$$\varepsilon_I = A(Z^2 + V_0 - 3C)\left[1 - 2Zr + (2C + Z^2)r^2 + \frac{1}{3}(Z^3 - 10ZC + V_1)r^3\right] \tag{9}$$

where A and C are the constants from the resolution of Schrödinger equation up to order r^3. They are determined from a matching of ρ and ρ' to the *TFD* values at a switching radius r_0, found by continuity of the energy density in a self-consistent procedure which converges very fast [26].

We will evaluate the density as

$$\rho(r) = \begin{cases} \rho_I(r) & \textit{if} \quad r < r_0 \\ \rho_{TFD}(r) & \textit{if} \quad r \geqslant r_0 \end{cases}$$

and the energy as

$$E = 4\pi\left[\int_0^{r_0} r^2 \varepsilon_I(r)\, dr + \int_{r_0}^{r_1} r^2 \varepsilon_{TFD}\, dr\right].$$

The radial expectation values are given by

$$\langle r^k \rangle = 4\pi \int_0^{r_1} r^{k+2} \rho(r)\, dr$$

As a test of this procedure, we can obtain the energy which improves greatly the *TF* estimates and compares fairly closely to *HF* results. As an example we show those for the Krypton atom. The energy obtained by the Hartree-Fock method is $E_{HF} = 2752.06$, whereas *TF* gives $E_{TF} = 3252.27$, that is, a difference of 18.18%. In the present work we obtain an energy for the Krypton of $E_{PW} = 2719.37$, that is a 1.19% deviation. This is a general behaviour for all the atoms. Even for the atoms with few electrons we obtain the same difference with *HF*, which is remarkable for a semiclasical model that employs average shell effects. For example for the Neon atom, we obtained $E_{PW} = 125.893$, while $E_{HF} = 128.547$ (a difference of 2%).

One of the facts we found when we solved all the atoms is that the switching point r_0 we have defined before fits very close to the expresion $r_0 = 1/(2Z)$ as it can be seen in Table 1.

Table 1 *Values of the atomic electron density at the nucleus, $\rho(0)$ evaluated with the present modified* TFD *method compared to* HF *values by means of the percent deviation (%). Also, the values of $2 \cdot Z \cdot r_0$ are displayed where r_0 is the switching point among the quantum mechanical and the semiclassical description (see text).*

Z	$\rho(0)_{PW}$	$\rho(0)_{HF}$	%	$2 \cdot Z \cdot r_0$
3	15.3	13.8	10.8	0.9954
6	135	128	5.6	1.0032
10	653	620	5.1	1.00546
18	3988	3840	3.7	1.00476
27	13881	13371	3.7	1.01498
36	33359	32228	3.4	1.00152
45	65852	63716	3.2	1.00055
54	114973	111164	3.3	1.00062
86	463200	457800	1.2	0.98258

3. Results

3.1. *Density at the origin ($\rho(0)$)*

The *TF* and modified methods based on average shell effects does not reproduce fairly closely local properties like $\rho(0)$. It diverges with *TF* and *TFD* and only after introducing gradient corrections, can we obtain at least a finite value. In the present work we have obtain results quite close to *HF* values (Table 1). As an example, in Table 2 we present the evolution of this value through the different theories in the case of Krypton. The improvement by the present approach is found to be large.

We see how the difference among *HF* values and ours is going down when Z increases. This is because *TF* increases accuracy when Z grows. Therefore we have studied the asymptotic behaviour of $\rho(0)$ with this method. It is well known that

Table 2 *Values of $\rho(0)$ for Krypton provided by different approaches compared to the HF ones*

Method	$\rho(0)$
TF,TFD	∞
TFDW(1/9) [24]	312170
TFDW(1/5) [24]	126620
TFDW + Hodges [24]	68199
Present Work	33358
HF [16]	32228

$\rho(0) \propto Z^3$, but the proportionality constant has not been provided by a semiclassical method. In Figure 1 we represent $\rho(0)/Z^3$ for the neutral atoms from $Z = 10$ to 120.

We see how for bigger Z the values goes to a constant. If we fit the values for the atoms form 80 to 120 we obtain the line in the graph, which corresponds to

$$\rho(0)/Z^3 = (0.7320 \pm 0.0021)$$

3.2. *Radial expectation values $\langle r^\alpha \rangle$*

As we said previously, the radial expectation values constitute a test of the description of the density in different regions depending on α. For α smaller than -1 the most important region is the near nuclear one. Now we show in the Table 3 the values from $\alpha = -2$ to $\alpha = 2$ obtained with the present method and with *HF*. In Figures 2 and 3 are illustrated $\alpha = -2$ and $\alpha = -1$ compared with the *HF* values.

In the Table it is remarkable that the results for $\alpha = -2$ and $\alpha = -1$, (in the cases the region we have corrected has the biggest importance) comparing with the other methods where $\alpha = -2$ diverges and $\alpha = -1$ gives no realistic values. For positive α the method gives the average behaviour of *HF* (this one gives a shell structure), like *TF*. The difference between *TF* and the present method for $\alpha = 1$ and $\alpha = 2$ is ilustrated in

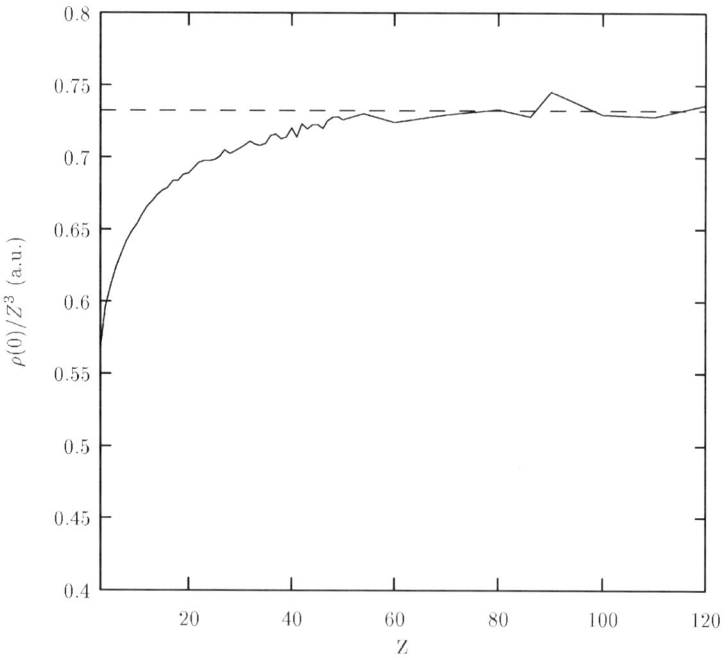

Fig. 1. Plot of $\rho(0)/Z^3$ for neutral atoms using the approach of present work.

Table 3 *Some radial expectation values $\langle r^{\alpha}\rangle$ evaluated in the present work (PW) compared to the HF ones for some neutral atoms.*

Z	$\langle r^{-2}\rangle_{PW}$	$\langle r^{-2}\rangle_{HF}$	$\langle r^{-1}\rangle_{PW}$	$\langle r^{-1}\rangle_{HF}$	$\langle r\rangle_{PW}$	$\langle r\rangle_{HF}$	$\langle r^{2}\rangle_{PW}$	$\langle r^{2}\rangle_{HF}$
3	30.34	30.22	5.40	5.72	3.5	5.02	6.5	18.6
6	138.35	138.77	14.62	14.69	6.5	7.1	11.9	13.8
10	414.59	414.90	30.0	31.1	10.0	7.9	17.9	9.4
18	1451	1465	68.5	69.7	16.04	16.07	27.7	26.0
27	3428	3456	126.8	122.1	23.3	21.9	36.8	39.4
36	6271	6331	186.8	182.9	27.6	26.2	44.6	39.5
45	10009	10114	244.9	248.1	32.67	32.56	51.4	52.7
54	14669	14818	315.1	317.9	37.5	39.1	57.7	62.7
86	38330	39027	597.7	604.4	53.03	53.65	76.5	81.2

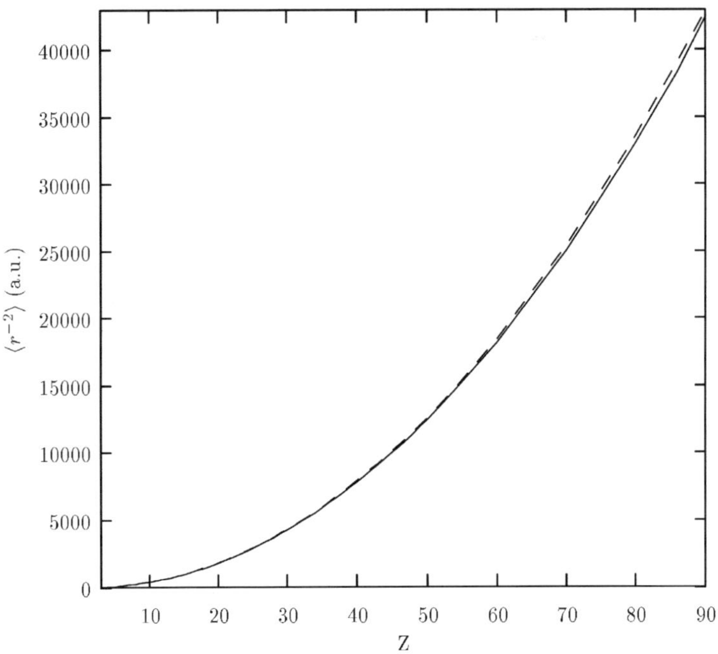

Fig. 2. Comparison of $\langle r^{-2}\rangle$ values (solid line) using the present approach to HF ones (dashed line).

Figures IV and V and is mainly assigned to exchange effects (standard *TFD* values are quite close to ours for these two quantities).

In [18] Dimitrieva and Plindov found the asymptotic behaviour for the $\langle r^{\alpha}\rangle$ expectation values. Now we check this behaviour and the coefficients they give. For $\langle r^{-2}\rangle$

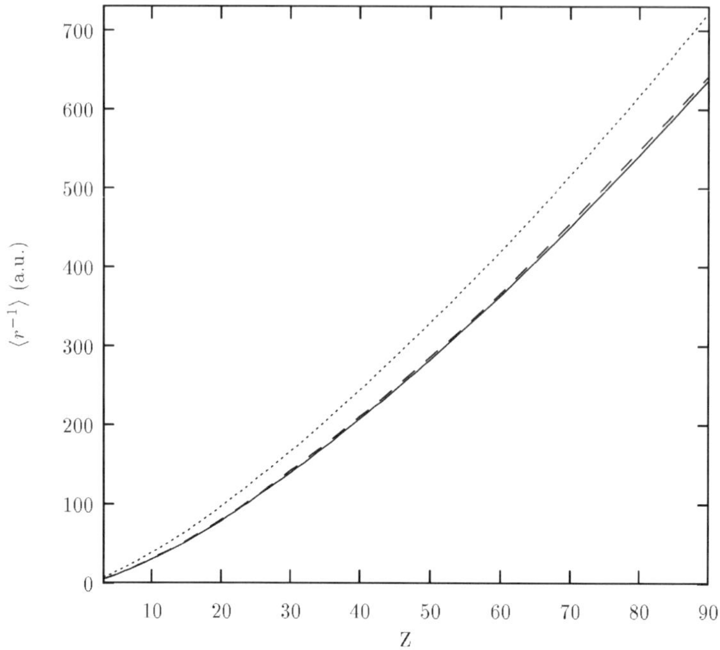

Fig. 3. Comparison of $\langle r^{-1}\rangle$ values using the present approach (solid line) and standard TF (dotted line) to HF results (dashed line).

which also diverges in *TF* and *TFD* approaches, we test a relation $\langle r^{-2}\rangle \sim \mathrm{C}\cdot \mathrm{Z}^2 + \mathrm{C}'\cdot \mathrm{Z}^{5/3}$. A fit from $Z = 80$ to 120 gives

$$\langle r^{-2}\rangle = (6.1 \pm 0.5)\cdot \mathrm{Z}^2 - (3.9 \pm 2.4)\cdot \mathrm{Z}^{5/3}$$

This overlaps with the theoretical prediction of [18] $\langle r^{-2}\rangle = 6.58\cdot \mathrm{Z}^2 - 6.022\cdot \mathrm{Z}^{5/3}$.

3.3. Momentum expectation values $\langle p^{\alpha}\rangle$

The momentum expectation values $\langle p^{\alpha}\rangle$ are not directly related to the electron density, but to the wave function via its Fourier transform, the momentum density. However we can make use of a semiclassical relation for a local Fermi gas for estimating these values:

$$\langle p^{\alpha}\rangle = \frac{3}{\alpha+3}(3\pi^2)^{\alpha/3}\int \rho^{1+(\alpha/3)}\, d\vec{r}$$

In Table 4 we compare the most physical momentum expectation values $\alpha = 2$ and $\alpha = -1$ with the *HF* ones. For α=2 (related with the kinetic energy) we compare with the value -2E obtained by our method in order to check how close the aproximations utilized verify the virial theorem.

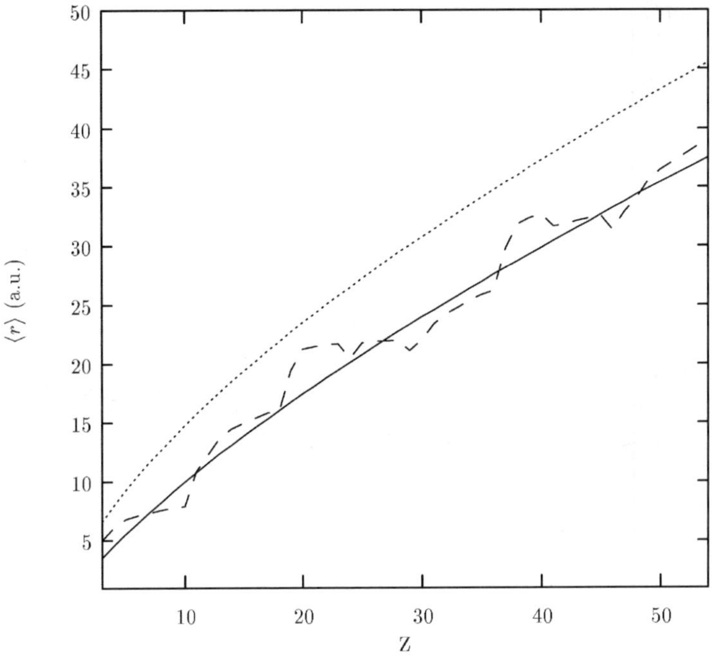

Fig. 4. Comparison of $\langle r\rangle$ values using the present approach (solid line) and standard TF (dotted line) to HF results (dashed line).

We see how, though we have use two aproximations, the estimates of $\langle p^{\alpha}\rangle$ are adequate except for low Z and the deviation for the virial relation is not large.

4. Conclusions

We can conclude that the present method of correcting *TF* calculations provides adequate estimations of expectation values for ground state atoms taking into account the simplicity of the model and it self-consistent nature, where no empirical parameters are used. It provides information about the asymptotic behaviour of quantities such as $\rho(0)$ and $\langle r^{-2}\rangle$ that cannot be evaluated with the standard semiclassical approach and allow us to estimate momentum expectation values which are not directly related to the density in an exact way.

Acknowledgements

This work has been supported by the Spanish Comision Interministerial de Ciencia y Tecnologia (CICYT) under contract PB$-$1211 and from the Junta de Andalucia.

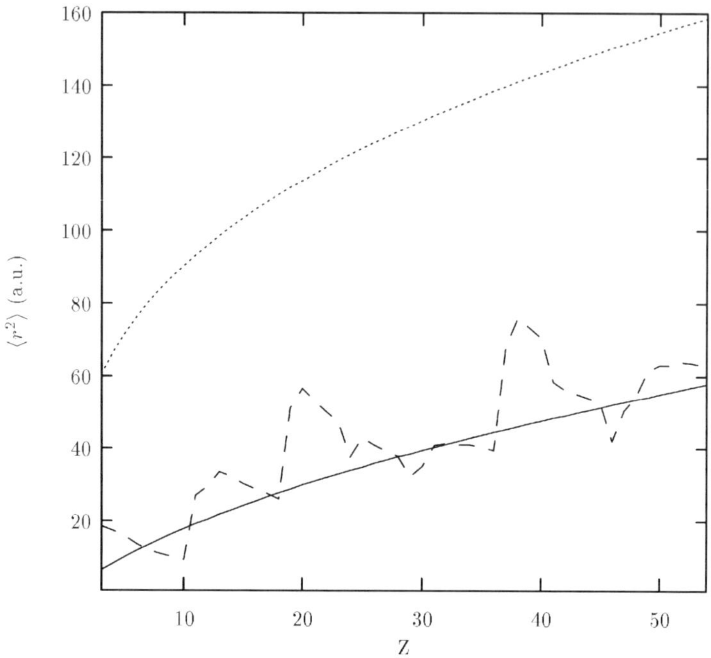

Fig. 5. Comparison of $\langle r^2 \rangle$ values using the present approach (solid line) and standard TF (dotted line) to HF results (dashed line).

Table 4 *Some momentum expectation values $\langle p^{\alpha} \rangle$ evaluated in the present work (PW) compared to the HF ones for some neutral atoms*

Z	$\langle p^{-1} \rangle_{HF}$	$\langle p^{-1} \rangle_{PW}$	$\langle p^2 \rangle_{HF}$	$\langle p^2 \rangle_{PW}$	$-2E_{PW}$
10	5.4557	8.201	257.09	216.51	251.79
18	10.128	11.369	1053.6	914.5	1036.0
27	13.97	14.1	2762.8	2459.3	2725.2
36	14.474	16.35	5504.0	4946.3	5438.7
45	16.693	18.29	9371.1	8492.9	9273.6
54	19.755	20.02	14463	13199	14315

References

1. R. Pucci and N.H. March: J. Chem. Phys. 76, 4089 (1982)
2. I.R. Epstein: Phys. Rev. A 8, 160 (1973)
3. A.J. Thakkar and V.H. Smith: J. Phys. B 11, 3803 (1978)
4. W. Thirring: *A Course on Mathematical Physics*, Vol. 3, Springer, New York (1981)
5. F.J. Gálvez and I. Porras: J. Phys. B 24, 3343 (1991)
6. S.R. Gadre and S.J. Chakravorty: Chem. Phys. Lett. 132, 535 (1986)

7. J.S. Dehesa, F.J. Gálvez and I. Porras: Phys. Rev. A 39, 494 (1989)
8. M. Hoffmann-Ostenhof, T. Hoffmann-Ostenhof and W. Thirring: J. Phys. B 11, L571 (1978)
9. F.W. King: J. Chem. Phys. 80, 4317 (1984)
10. F.J. Gálvez, I. Porras, J.C. Angulo and L.S. Dehesa: J. Phys. B 21, L271 (1988)
11. F.W. King: J. Chem. Phys. 78, 2459 (1983); 78, 3091 (1983)
12. F.W. King: J. Math. Phys. 24, 1891 (1983)
13. F.J. Gálvez and I. Porras: Phys. Rev. A 51, 2857 (1995)
14. A. Zarzo, J.C. Angulo and J. Antolín: J. Phys. B 26, 4663 (1993)
15. A. Zarzo, J.C. Angulo, J. Antolín and R.J. Yáñez: Int. J. Quantum Chem. 56, 747 (1994)
16. E. Clementi and C. Roetti: At. Data Nucl. Data Tables 14, 177 (1974)
17. A.D. Mclean and R.S. Mclean: At. Data Nucl. Data Tables 26, 199 (1981)
18. I.K. Dimitrieva and G.I. Plindov: Z. Phys. A 305, 103 (1982)
19. Y. Tal and M. Levy: J. Chem. Phys. 72, 4009 (1980)
20. J. Katriel and M.R. Nyden: J. Chem. Phys. 74, 1221 (1981)
21. L.H. Thomas: Proc. Camb. Phil. Soc. 23, 542 (1927)
22. E. Fermi: Z. Phys. 48, 73 (1928)
23. R.G. Parr and W. Yang: *Density Functional Theory of Atoms and Molecules*, Oxford U.P., New York (1989)
24. E. Engel and R.M. Dreizler: J. Phys. B 22, 1901 (1989)
25. N. Ashby and M.A. Holzman: Phys. Rev. A 1, 764 (1970)
26. I. Porras and A. Moya: Phys. Rev. A 59, 1859 (1999)

Correlated Effective Single-Particle Theory: Relativistic Optimized-Potential Method

E. Engel and A. Facco-Bonetti
Institut für Theoretische Physik, Universität Frankfurt, Robert-Mayer-Straße 8–10, D-60054 Frankfurt/Main, Germany

Abstract

We analyze the relativistic exchange-correlation (xc) energy functional E_{xc} of density functional theory (DFT), extending the conventional approach to E_{xc}, explicitly based on the density, by the concept of orbital-dependent representations. We introduce two different schemes for the derivation of xc-functionals depending on Kohn-Sham orbitals. On the one hand, a perturbation expansion, using the Kohn-Sham Hamiltonian as noninteracting reference system, allows to derive a perturbation expansion of E_{xc} in powers of e^2. Alternatively, the adiabatic connection formalism directly yields an RPA-like xc-functional. The xc-potential corresponding to these orbital-dependent functionals can be evaluated via the optimized-potential-method (OPM), which is also presented. The atomic exchange-only ground state energies resulting from the OPM are compared to Hartree-Fock and conventional DFT energies. Finally, we give some first OPM results for the correlation energies of the neon isoelectronic series.

1. Introduction

Density functional theory (DFT) is a well established approach for the description of many electron systems, comprising atoms, molecules, clusters and solids [1–4]. Its main advantage, as compared to other correlated many-body methods, like the configuration interaction or the multi-configuration Hartree-Fock (HF) scheme, is its particular computational efficiency. This is achieved by representing the nonclassical components of the electron-electron interaction as a functional of the density n, the so-called exchange-correlation (xc) energy functional $E_{xc}[n]$. However, the exact functional dependence of E_{xc} on n is unknown to date, so that one has to resort to suitable approximations. While for a long time explicitly density-dependent variants of $E_{xc}[n]$, like the local density approximation (LDA) or the generalized gradient approximation (GGA) [5–8], were exclusively applied, now a new trend is emerging: In analogy to the kinetic energy functional T_s [9], also E_{xc} can be represented in terms of the Kohn-Sham (KS) orbitals, leading to the concept of orbital-dependent functionals, i.e. implicit density functionals [10–19].

An orbital-dependent representation has first been used for the exchange (x-only) energy functional E_x [20,5]. For E_x the Fock expression, written in terms of KS orbitals, represents the natural definition, as it exactly eliminates the self-interaction energy inherent in the classical Coulomb (Hartree) energy. Although the density dependence of this $E_x[n]$ is unknown, it is nevertheless possible to obtain the corresponding multiplicative exchange potential $v_x = \delta E_x/\delta n$ via the optimized potential method (OPM) [21]. Together with this representation of E_x, which is the simplest and most frequently applied orbital-dependent functional, various forms of the correlation energy

A. Hernández-Laguna et al. (eds.), Quantum Systems in Chemistry and Physics, Vol. 1: Basic Problems and Model Systems, 227–250.

$E_c = E_{xc} - E_x$ have been employed. On the one hand, LDA and GGA type correlation functionals have been used [14]. However, the success of the LDA (and, to a lesser extent, also the GGA) partially depends on an error cancellation between the exchange and correlation contributions, which is lost as soon as the exact E_x is used. On the other hand, the semiempirical orbital-dependent Colle-Salvetti functional [22] has been investigated [15]. Although the corresponding atomic correlation energies compare well [15] with the exact data extracted from experiment [23], the Colle-Salvetti correlation potential deviates substantially from the exact $v_c = \delta E_c/\delta n$ [24] in the case of closed subshell atoms [25].

Based on a perturbation expansion using the KS Hamiltonian [26,27], recently a new systematic scheme for the derivation of orbital-dependent E_c has been proposed [12]. While this representation is exact in principle, an explicit evaluation requires the solution of a highly nonlinear equation, coupling E_{xc} and the corresponding v_{xc} [19]. For a rigorous treatment of this E_{xc} one thus has to resort to an expansion in powers of e^2, which allows to establish a recursive procedure for the evaluation of E_{xc} and the accompanying v_{xc}.

A fully relativistic extension of the scheme put forward in [12] has been introduced in [19], including the transverse electron-electron interaction (Breit + ...) and vacuum corrections. Restricting the discussion to the no-pair approximation [28] for simplicity, we here compare this perturbative approach to orbital-dependent E_{xc} to the relativistic variant of the adiabatic connection formalism [29], demonstrating that the latter allows for a direct extraction of an RPA-like orbital-dependent functional for E_{xc}. In addition, we provide some first numerical results for atomic E_c.

The paper is structured as follows. After a brief summary of relativistic DFT in Section 2.1, the nonlinear equation which determines the exact E_{xc} as a functional of the KS orbitals and eigenvalues is derived via KS perturbation theory in Sections 2.2, 2.3. It is explicitly shown that an expansion in powers of e^2 allows to solve this nonlinear relation in a recursive manner. In Section 2.4 the adiabatic connection scheme is extended to inhomogeneous relativistic systems, extracting an RPA-like form for E_{xc}. The relativistic OPM (ROPM), which allows to evaluate the v_{xc} corresponding to orbital and eigenvalue dependent functionals is reviewed in Section 3. While the ROPM can be used straightforwardly with the recursive scheme for E_{xc}, its application to nonlinear forms of E_{xc} requires additional discussion. An approximate version of the ROPM, which allows to deal with such nonlinear forms is indicated. Finally, in Section 4 the orbital-dependent E_{xc} is specialized to the pure Coulomb interaction in order to facilitate comparison with standard many-body methods. The Coulomb form of the functionals also allows for a direct extraction of the nonrelativistic limit. A brief comparison of x-only ROPM results with relativistic HF (RHF) data for closed subshell atoms demonstrates the physical equivalence of the x-only ROPM with the RHF approach. Moreover, the corresponding LDA and GGA results illustrate the progress achieved by using the orbital-dependent exchange. First results for the correlation energies of the neon isoelectronic series (exact to the order e^4) are also given in Section 4, and compared with conventional relativistic many-body perturbation theory (RMBPT) data [30]. It is found that for neutral neon the ROPM and RMBPT correlation energies differ by roughly 20%, while this disagreement vanishes with increasing Z.

The asymptotic $-1/r$-behaviour of the KS x-only potential, which leads to a Rydberg series of excited levels even for neutral atoms, is identified as the main source of this difference.

2. Systematic Approaches to the Exchange-Correlation Energy Functional

2.1. Relativistic density functional theory

The main feature of the relativistic generalization of the Hohenberg-Kohn theorem [31–35], as compared to its nonrelativistic counterpart [36], is the transition from the ground state density n as the basic DFT variable to the ground state four current $j^\mu = (n, \boldsymbol{j})$. Its crucial statement is: The ground state $|\Phi\rangle$ of any stationary N-electron system is uniquely determined (up to gauge transformations) by the corresponding ground state four current

$$j^\mu(\boldsymbol{x}) = \langle\Phi|\hat{j}^\mu(x)|\Phi\rangle. \tag{2.1}$$

Consequently, $|\Phi\rangle$ may be understood as a functional $|\Phi[j^\mu]\rangle$ of the ground state four current. As an immediate result, j^μ also determines all ground state expectation values, in particular the ground state energy $E = E[j^\mu]$.

Both E and j^μ can in principle be simultaneously calculated using the variational property of the energy, $\delta E[j]/\delta j = 0$ (the no-pair approximation is implied throughout this paper — the implications of an extension of the Hohenberg-Kohn theorem to the full QED problem including the negative energy states have been discussed in [34,35]). The energy minimization is most easily implemented by representing j^μ via auxiliary single particle spinors,

$$j^\mu(\boldsymbol{x}) = \sum_{-mc^2<\epsilon_k\leq\epsilon_F} \overline{\phi}_k(\boldsymbol{x})\gamma^\mu\phi_k(\boldsymbol{x}) \tag{2.2}$$

(where it has been assumed that a noninteracting system with an external potential v^μ_{KS} exists whose ground state four current is identical with the current of the actually interesting interacting system). The total energy E can then be decomposed as

$$E = T_s + \int d^3x j_\mu(\boldsymbol{x})v^\mu_{ext}(\boldsymbol{x}) + E_H + E_{xc}, \tag{2.3}$$

where T_s represents the kinetic energy of the 'auxiliary particles',

$$T_s = \int d^3x \sum_{-mc^2<\epsilon_k\leq\epsilon_F} \overline{\phi}_k(\boldsymbol{x})[-ic\boldsymbol{\gamma}\cdot\nabla + mc^2]\phi_k(\boldsymbol{x}), \tag{2.4}$$

and v^μ_{ext} denotes the given external potential (nuclei, applied fields). Furthermore, E_H is the 'covariant' Hartree energy,

$$E_H = \frac{1}{2}\int d^3x \int d^4y D^0_{\mu\nu}(\boldsymbol{x}-\boldsymbol{y},\, y^0) j^\mu(\boldsymbol{x}) j^\nu(\boldsymbol{y})$$

$$= \frac{e^2}{2} d^3x \int d^3y \frac{j_\nu(\boldsymbol{x}) j^\nu(\boldsymbol{y})}{|\boldsymbol{x}-\boldsymbol{y}|}, \tag{2.5}$$

with $D^0_{\mu\nu}$ being the noninteracting photon propagator (given precisely in (2.19)) and E_{xc} is the xc-energy, which is defined via Eq. (2.3). The energy minimization is now performed with respect to the orbitals, which leads to the single particle equations of relativistic DFT, introduced by Rajagopal, MacDonald and Vosko [32,33],

$$\{-ic\boldsymbol{\alpha}\cdot\nabla + \beta mc^2 + \alpha_\mu v^\mu_{KS}(\boldsymbol{x})\}\phi_k(\boldsymbol{x}) = \epsilon_k \phi_k(\boldsymbol{x}) \tag{2.6}$$

($\alpha^\mu = \gamma^0\gamma^\mu$). The corresponding effective single particle potential v^μ_{KS} reflects the decomposition of the total energy,

$$v^\mu_{KS}(\boldsymbol{x}) = v^\mu_{ext}(\boldsymbol{x}) + v^\mu_H(\boldsymbol{x}) + v^\mu_{xc}(\boldsymbol{x}) \tag{2.7}$$

$$v^\mu_H(\boldsymbol{x}) = e^2 \int d^3y \frac{j^\mu(\boldsymbol{y})}{|\boldsymbol{x}-\boldsymbol{y}|} \tag{2.8}$$

$$v^\mu_{xc}(\boldsymbol{x}) = \frac{\delta E_{xc}[j]}{\delta j_\mu(\boldsymbol{x})}, \tag{2.9}$$

with v^μ_H being the Hartree potential and v^μ_{xc} the xc-potential. Eqs. (2.2, 2.6–2.9) have to be solved selfconsistently, using some suitable approximation for the functional $E_{xc}[j]$.

While it is common practice to apply purely density-dependent approximations in (2.9), we here want to review the concept of orbital-dependent E_{xc}. Two different approaches to orbital-dependent E_{xc} have been introduced in the nonrelativistic context, both providing an exact representation of E_{xc}. To develop these approaches we now establish a connection between relativistic DFT and QED, which provides the most general framework for the discussion of the Coulomb many-body problem.

2.2. *Field theoretical background*

Starting from the standard QED Lagrangian, the Hamiltonian characterizing a system of interacting electrons in a static external potential $v^\mu_{ext}(\boldsymbol{x})$ is readily obtained as the 00-component of the energy-momentum tensor (see e.g. [35]),

$$\hat{H} = \hat{H}_e(x^0) + \hat{H}_\gamma(x^0) + \hat{H}_{int}(x^0). \tag{2.10}$$

The individual components, the electronic Hamiltonian $\hat{H}_e$, the free photon Hamiltonian $\hat{H}_\gamma$ and the electron-photon coupling Hamiltonian $\hat{H}_{int}$ are given by [37]

$$\hat{H}_e(x^0) = \int d^3x \hat{\bar{\psi}}(x)(-ic\gamma \cdot \nabla + mc^2 + \gamma_\mu v^\mu_{ext}(\boldsymbol{x}))\hat{\psi}(x) \tag{2.11}$$

$$\hat{H}_\gamma(x^0) = -\frac{1}{8\pi}\int d^3x \{\partial^0 \hat{A}_\nu(x)\partial^0 \hat{A}^\nu(x) + \nabla \hat{A}_\nu(x) \cdot \nabla \hat{A}^\nu(x)\} \tag{2.12}$$

$$\hat{H}_{int}(x^0) = e\int d^3x \hat{j}^\mu(x)\hat{A}_\mu(x). \tag{2.13}$$

Besides the nuclear attraction, $v^\mu_{ext}(\boldsymbol{x})$ could also include additional external fields, if present. $\hat{\psi}(x)$ denotes the fermion field operator of the interacting, inhomogeneous system characterized by $\hat{H}$, $\hat{j}^\mu(x)$ is the corresponding fermion four current operator,

$$\hat{j}^\mu(x) = \hat{\bar{\psi}}(x)\gamma^\mu \hat{\psi}(x) \tag{2.14}$$

and $\hat{A}_\mu(x)$ represents the field operator of the photons for which the covariant quantization scheme (Feynman gauge) has been used.

Following the common approach in relativistic field theory, which aims at a manifestly covariant representation of the dynamics inherent in the field operators, so far all quantities have been introduced in the Heisenberg picture. To develop the framework of relativistic DFT, however, it is common practice to transform to the Schrödinger picture, so that the relativistic theory can be formulated in close analogy to its nonrelativistic limit. As usual we choose the two pictures to coincide at $x^0 = 0$. Once the field operators in the Schrödinger-picture have been identified via $\hat{\psi}_S(\boldsymbol{x}) = \hat{\psi}(\boldsymbol{x}, x^0 = 0)$, etc, the Hamiltonians $\hat{H}_{e,S}$, $\hat{H}_{\gamma,S}$ and $\hat{H}_{int,S}$ are immediately obtained in terms of the Schrödinger-picture field operators.

Let us for the moment assume that the KS-potential (2.7) is known. This potential then allows to define a noninteracting KS Hamiltonian

$$\hat{H}_{KS} = \int d^3x \hat{\bar{\psi}}_S(\boldsymbol{x})(-ic\gamma \cdot \nabla + mc^2)\hat{\psi}_S(\boldsymbol{x}) + \int d^3x \hat{j}_{S,\mu}(\boldsymbol{x}) v^\mu_{KS}(\boldsymbol{x}) + \hat{H}_{\gamma,S}, \tag{2.15}$$

which provides a decomposition of $\hat{H}$,

$$\hat{H} = \hat{H}_{KS} + \hat{H}_{1,S}, \tag{2.16}$$

which is particularly suitable for our present purpose. We will later come back to the point how to calculate the selfconsistent v^μ_{KS}. As electrons and photons do not interact in $\hat{H}_{KS}$, the ground state $|\Phi_{KS}\rangle$ of $\hat{H}_{KS}$,

$$\hat{H}_{KS}|\Phi_{KS}\rangle = E_{KS}|\Phi_{KS}\rangle, \tag{2.17}$$

can be factorized into a product of the photon vacuum $|0_\gamma\rangle$ (no free photons are present in the ground state) and an electronic ground state $|\Phi_{KS,e}\rangle$,

$$|\Phi_{KS}\rangle = |\Phi_{KS,e}\rangle \times |0_\gamma\rangle; \; \langle 0_\gamma|\hat{A}^\mu_0(x)|0_\gamma\rangle = 0, \tag{2.18}$$

where $\hat{A}^\mu_0(x)$ denotes the free photon field operator. As a consequence, standard vacuum QED results can be used for the photon sector of the noninteracting problem defined by $\hat{H}_{KS}$, as e.g. the free photon propagator (Feynman gauge)

$$D^{0,\mu\nu}(x-y) = -i\frac{e^2}{c}\langle 0_\gamma | T\hat{A}_0^\mu(x)\hat{A}_0^\nu(y)|0_\gamma\rangle$$

$$= g^{\mu\nu}\int\frac{d^4q}{(2\pi)^4}e^{-iq(x-y)}\frac{-4\pi e^2}{q^2+i\eta}. \tag{2.19}$$

In the electronic sector the presence of the potential v_{KS}^μ leads to an inhomogeneous reference system. Within the no-pair approximation,

$$\hat{\psi}_{S,np}(\boldsymbol{x}) = \mathcal{L}_+\hat{\psi}_S(\boldsymbol{x})\mathcal{L}_+, \tag{2.20}$$

where $\mathcal{L}_+$ is the projection operator onto the space spanned by the positive energy eigenstates ϕ_k of $\hat{H}_{KS}$ [28, 38], the noninteracting field operators are given by

$$\hat{\psi}_{0,np}(x) = e^{i\hat{H}_{KS}x^0/c}\hat{\psi}_{S,np}(\boldsymbol{x})e^{-i\hat{H}_{KS}x^0/c} = \sum_{-mc^2<\epsilon_k}\hat{b}_k\phi_k(\boldsymbol{x})e^{-i\epsilon_k x^0/c} \tag{2.21}$$

$$\hat{\psi}^\dagger_{0,np}(x) = \sum_{-mc^2<\epsilon_k}\hat{b}_k^\dagger\phi_k^\dagger(\boldsymbol{x})e^{i\epsilon_k x^0/c} \tag{2.22}$$

(As all quantities discussed in this publication are understood within the no-pair approximation, we will omit the index np in the following for brevity). In Eqs. (2.21, 2.22) $\hat{b}_k$ and $\hat{b}_k^\dagger$ are the annihilation and creation operators for positive energy KS states, which allow to write the electronic ground state as

$$|\Phi_{KS,e}\rangle = \prod_{-mc^2<\epsilon_k\leq\epsilon_F}\hat{b}_k^\dagger|0_e\rangle, \tag{2.23}$$

where the electronic single particle states have been assumed to be filled up to the Fermi level ϵ_F (both the $\hat{\psi}_0^{(\dagger)}$ and the $\hat{b}_k^{(\dagger)}$ satisfy the standard commutation relations, of course). The corresponding electron propagator is given by

$$G^0(x,y) \equiv G^0(\boldsymbol{x},\boldsymbol{y},x^0-y^0) = -i\langle\Phi_{KS,e}|T\hat{\psi}_0(x)\hat{\bar{\psi}}_0(y)|\Phi_{KS,e}\rangle \tag{2.24}$$

$$= -\,i\Theta(x^0-y^0)\sum_{\epsilon_F<\epsilon_k}\phi_k(\boldsymbol{x})\overline{\phi}_k(\boldsymbol{y})e^{-i\epsilon_k(x^0-y^0)/c}$$

$$+\,i\Theta(y^0-x^0)\sum_{-mc^2<\epsilon_k\leq\epsilon_F}\phi_k(\boldsymbol{x})\overline{\phi}_k(\boldsymbol{y})e^{-i\epsilon_k(x^0-y^0)/c}. \tag{2.25}$$

Using (2.14, 2.21–2.23) the ground state four current

$$j^\mu(\boldsymbol{x}) = \langle\Phi_{KS}|\hat{j}^\mu(x)|\Phi_{KS}\rangle, \tag{2.26}$$

is obtained in the form introduced in Eq. (2.2) and the ground state energy reads

$$E_{KS} = T_s + \int d^3x j_\mu(\boldsymbol{x})v_{KS}^\mu(\boldsymbol{x}). \tag{2.27}$$

In fact, the field theoretical formulation of the KS problem via Eqs. (2.15–2.27) in the no-pair approximation is the actual origin of the forms (2.2, 2.4) for the KS four current and T_s.

In the following the KS Hamiltonian is used to analyze the total energy E of the ground state $|\Phi\rangle$ of $\hat{H}$,

$$\hat{H}|\Phi\rangle = E|\Phi\rangle, \tag{2.28}$$

using a perturbation expansion in terms of $\hat{H} - \hat{H}_{KS}$. For expansions of this type different techniques have been used in the literature. On the one hand, in the framework of QED usually the level shift formula of Gell-Mann and Low [39] in the symmetric form of Sucher [40] is applied (see e.g. [41–44]). In the context of nonrelativistic DFT, on the other hand, two versions of a coupling constant integration have been utilized for such purposes. While the so-called adiabatic connection [29,45] has been particularly useful for the analysis of E_{xc} [46–48] in the context of explicit density functionals, the second scheme [26,27] allows a systematic extraction of orbital-dependent forms for E_{xc}, i.e implicit density functionals, so that we first discuss the extension of this latter variant to the relativistic domain.

2.3. Perturbation theory on Kohn-Sham basis

In order to obtain a formula for the energy difference between the complete ground state energy E and the noninteracting energy E_{KS}, and thus for E_{xc}, the interaction Hamiltonian $\hat{H}_1 = \hat{H} - \hat{H}_{KS}$ is supplemented by a dimensionless coupling strength parameter g in such a way,

$$\hat{H}(g) = \hat{H}_{KS} + \hat{H}_{1,S}(g) \tag{2.29}$$

$$\hat{H}_1(x^0) = \int d^3x \hat{j}_\mu(x)[g^{1/2} e\hat{A}^\mu(x) - g v^\mu_{Hxc}(\boldsymbol{x})] \tag{2.30}$$

$$v^\mu_{Hxc}(\boldsymbol{x}) \equiv v^\mu_{KS}(\boldsymbol{x}) - v^\mu_{ext}(\boldsymbol{x}) = v^\mu_{H}(\boldsymbol{x}) + v^\mu_{xc}(\boldsymbol{x}), \tag{2.31}$$

that the scaling of the two contributions to $\hat{H}_1$ with g reflects their dependence on the actual coupling constant e^2 (at least to lowest order of v_{Hxc} — see below). The original Hamiltonian (2.10) is then obtained from (2.29) for $g = 1$. The desired expression for the energy shift induced by the interaction is obtained by first differentiating the g-dependent ground state energy

$$E(g) = \langle\Phi(g)|\hat{H}(g)|\Phi(g)\rangle \tag{2.32}$$

with respect to the coupling parameter. One finds

$$\frac{\partial}{\partial g} E(g) = \int d^4x \delta(x^0) \langle\Phi(g)|\hat{j}_\mu(x)\left[\frac{e}{2g^{1/2}}\hat{A}^\mu(x) - v^\mu_{Hxc}(\boldsymbol{x})\right]|\Phi(g)\rangle, \tag{2.33}$$

where the normalization of $|\Phi(g)\rangle$,

$$\langle\Phi(g)|\Phi(g)\rangle = 1, \tag{2.34}$$

has been used. In the second step integration over g leads to

$$E(1) - E(0) = E - E_{KS} \equiv E_1$$
$$= \int_0^1 dg \int d^4x \delta(x^0) \langle \Phi(g) | \hat{j}_\mu(x) \left[\frac{e}{2g^{1/2}} \hat{A}^\mu(x) - v^\mu_{Hxc}(\boldsymbol{x}) \right] | \Phi(g) \rangle . \quad (2.35)$$

Further evaluation of Eq. (2.35) requires an expression connecting $|\Phi(g)\rangle$ (assumed to be nondegenerate) with $|\Phi_{KS}\rangle$ (also assumed to be nondegenerate). This link is established via the interaction-picture time-evolution operator $\hat{U}_{I,\epsilon}$, i.e. by an adiabatic switching of $\hat{H}_1$,

$$|\Phi\rangle = A \lim_{\epsilon \to 0} \frac{\hat{U}_{I,\epsilon}(0, \mp\infty)|\Phi_{KS}\rangle}{\langle \Phi_{KS}|\hat{U}_{I,\epsilon}(0, \mp\infty)|\Phi_{KS}\rangle} \quad (2.36)$$

$$A = \lim_{\epsilon_1, \epsilon_2 \to 0} \left[\frac{\langle \Phi_{KS}|\hat{U}_{I,\epsilon_1}(+\infty, 0)|\Phi_{KS}\rangle \langle \Phi_{KS}|\hat{U}_{I,\epsilon_2}(0, -\infty)|\Phi_{KS}\rangle}{\langle \Phi_{KS}|\hat{U}_{I,\epsilon_1}(+\infty, 0)\hat{U}_{I,\epsilon_1}(0, -\infty)|\Phi_{KS}\rangle} \right]^{1/2} \quad (2.37)$$

$$\hat{U}_{I,\epsilon}(t, t') = \sum_{n=0}^{\infty} \frac{(-i)^n}{n!} \int_{t'}^{t} dt_1 \cdots \int_{t'}^{t} dt_n \exp[-\epsilon(|t_1| + \cdots + |t_n|)] T[\hat{H}_{1,I}(t_1) \cdots \hat{H}_{1,I}(t_n)] \quad (2.38)$$

$$\hat{H}_{1,I}(x^0) = e^{i\hat{H}_{KS}x^0/c} \hat{H}_{1,s} e^{-i\hat{H}_{KS}x^0/c} = \int d^3x \hat{j}_{0,\mu}(x) [g^{1/2} e \hat{A}^\mu_0(x) - g v^\mu_{Hxc}(\boldsymbol{x})] \quad (2.36)$$

(for brevity, the g-dependence of the various operators and states involved is no longer noted explicitly). Together with the normalization of $\langle \Phi_{KS}|\Phi_{KS}\rangle = 1$ the factor (2.37) ensures the validity of (2.34) for all g. Insertion of (2.36) into (2.35) and use of the additivity of the time-evolution operator gives

$$E_1 = \lim_{\epsilon \to 0} \int_0^1 dg \int d^4x \delta(x^0) \sum_{n=0}^{\infty} \frac{(-i)^n}{n!} \int_{-\infty}^{\infty} dt_1 \cdots \int_{-\infty}^{\infty} dt_n e^{-\epsilon(|t_1| + \cdots + |t_n|)}$$
$$\times \frac{\langle \Phi_{KS}|T \hat{j}_{0,\mu}(x)[(e/2g^{1/2})\hat{A}^\mu(x) - v^\mu_{Hxc}(\boldsymbol{x})] \hat{H}_{1,I}(t_1) \cdots \hat{H}_{1,I}(t_n)|\Phi_{KS}\rangle}{\langle \Phi_{KS}|\hat{U}_{I,\epsilon}(+\infty, -\infty)|\Phi_{KS}\rangle} . \quad (2.40)$$

Eq. (2.40) directly leads to a perturbative expansion of E_1 in powers of e,

$$E_1 = \lim_{\epsilon \to 0} \int_0^1 dg \sum_{n=0}^{\infty} \frac{(-i)^n}{n! c^n} \sum_{k=0}^{n} \binom{n}{k} g^{n-k/2} \int d^4x \delta(x^0) \int d^4x_1 \cdots \int d^4x_n e^{-\epsilon(|x_1^0| + \cdots + |x_n^0|)}$$
$$\times \langle \Phi_{KS,e}|T \hat{j}^\mu_0(x) \hat{j}^{\mu_1}_0(x_1) \cdots \hat{j}^{\mu_n}_0(x_n)|\Phi_{KS,e}\rangle_c (-1)^{n-k} v_{Hxc,\mu_{k+1}}(\boldsymbol{x}_{k+1}) \cdots v_{Hxc,\mu_n}(\boldsymbol{x}_n)$$
$$\times e^k \left\{ \frac{e}{2g^{1/2}} \langle 0_\gamma|T \hat{A}_{0,\mu}(x) \hat{A}_{0,\mu_1}(x_1) \ldots \hat{A}_{0,\mu_k}(x_k)|0_\gamma\rangle \right.$$
$$\left. - v_{Hxc,\mu}(\boldsymbol{x}) \langle 0_\gamma|T \hat{A}_{0,\mu_1}(x_1) \ldots \hat{A}_{0,\mu_k}(x_k)|0_\gamma\rangle \right\}, \quad (2.41)$$

where the index c indicates that only those diagrammatic contributions are to be included in which all vertices $x_1 \cdots x_n$ are connected to the vertex x (the remaining terms are exactly cancelled by the denominator of (2.40)). Eliminating all photon operators in favor of the photon propagator (2.19), the expansion (2.41) can be directly ordered in powers of e^2. Recalling that due to (2.18) the vacuum expectation value of an odd number of $\hat{A}_0$ vanishes, only terms with an even number of photon operators contribute. In the latter case all possible contractions, i.e. permutations of the $\hat{A}_0$, are obtained by

$$\langle 0_\gamma | T\hat{A}_0^{\mu_1}(x_1) \cdots \hat{A}_0^{\mu_{2n}}(x_{2n}) | 0_\gamma \rangle$$

$$= \sum_{k=2}^{2n} \langle 0_\gamma | T\hat{A}_0^{\mu_1}(x_1)\hat{A}_0^{\mu_k}(x_k)|0_\gamma\rangle \langle 0_\gamma | T\hat{A}_0^{\mu_2}(x_2) \cdots \hat{A}\!\!/_0^{\mu_k}(x_k) \cdots \hat{A}_0^{\mu_{2n}}(x_{2n})|0_\gamma\rangle \tag{2.42}$$

(here $\hat{A}\!\!/_0^{\mu_k}(x_k)$ indicates that this operator has to be dropped from the series $k = 2, \ldots, 2n$). From Eq. (2.18) it therefore follows that k must be odd in the first term inside the curly brackets in (2.41) and even in the second. The interchange of the summation order of k and n now allows to separate the contributions with even from those with odd k. Subsequently the photon vacuum expectation values are evaluated using (2.42). Taking into account the multiplicity originating from the number of possible contractions of the $\hat{A}_0$ (Wick's theorem) this leads to

$$\begin{aligned} E_1 = \lim_{\epsilon \to 0} \int_0^1 dg \sum_{l=0}^{\infty} \frac{(ig)^l}{l! c^l} \sum_{k=0}^{l} \binom{l}{k} \left(-\frac{1}{2}\right)^k \int d^4x \delta(x^0) \int d^4x_1 \cdots \int d^4x_{l+k} e^{-\epsilon(|x_1^0| + \cdots + |x_{l+k}^0|)} \\ \times D^0_{\mu_1\mu_2}(x_1 - x_2) \cdots D^0_{\mu_{2k-1}\mu_{2k}}(x_{2k-1} - x_{2k}) v_{Hxc,\mu_{2k+1}}(\boldsymbol{x}_{2k+1}) \cdots v_{Hxc,\mu_{l+k}}(\boldsymbol{x}_{l+k}) \\ \times \left\{ \frac{1}{2} \int d^4y\, e^{-\epsilon|y^0|} D^0_{\mu\nu}(x - y) \langle \Phi_{KS,e} | T\hat{j}_0^\mu(x)\hat{j}_0^\nu(y)\hat{j}_0^{\mu_1}(x_1) \cdots \hat{j}_0^{\mu_{l+k}}(x_{l+k}) | \Phi_{KS,e}\rangle_c \right. \\ \left. - v_{Hxc,\mu}(\boldsymbol{x}) \langle \Phi_{KS,e} | T\hat{j}_0^\mu(x)\hat{j}_0^{\mu_1}(x_1) \cdots \hat{j}_0^{\mu_{l+k}}(x_{l+k}) | \Phi_{KS,e}\rangle_c \right\}. \end{aligned} \tag{2.43}$$

Finally, this result can be given a more compact form, if the binomial relation is used to eliminate the k-summation. The xc-contribution is extracted from E_1 via (2.3, 2.27, 2.35). Introducing the interaction operator

$$\begin{aligned} \hat{W}(x^0) = \frac{1}{2} \int d^3x \int d^4y\, e^{-\epsilon(|x^0| + |y^0|)} \hat{j}_0^\mu(x) D^0_{\mu\nu}(x - y) \hat{j}_0^\nu(y) \\ - \int d^3x\, e^{-\epsilon|x^0|} \{ v_{H,\mu}(\boldsymbol{x}) + v_{xc,\mu}(\boldsymbol{x}) \} \hat{j}_0^\mu(x), \end{aligned} \tag{2.44}$$

the final result for E_{xc} can be written as

$$E_{xc} = \frac{1}{2}\int d^4x\delta(x^0)\int d^4y\, e^{-\epsilon|y^0|} D^0_{\mu\nu}(x-y)\{\langle\Phi_{KS}|T\hat{j}^\mu_0(x)\hat{j}^\nu_0(y)|\Phi_{KS}\rangle - j^\mu(\boldsymbol{x})j^\nu(\boldsymbol{y})\}$$

$$+\lim_{\epsilon\to 0}\sum_{n=1}^{\infty}\frac{(-i)^n}{(n+1)!}\int_{-\infty}^{\infty}dt_1\cdots\int_{-\infty}^{\infty}dt_n\langle\Phi_{KS}|T\hat{W}(0)\hat{W}(t_1)\cdots\hat{W}(t_n)|\Phi_{KS}\rangle_c, \tag{2.45}$$

with the additional convention that, in analogy to the time-evolution operator, the time-ordering in (2.45) also applies inside (2.44), i.e. before performing the time integrations in the individual $\hat{W}$ the overall time-ordering of (2.45) has to be established.

Without going into details we just remark that Eq. (2.45) can be further simplified by applying Wick's theorem to the electronic sector, utilizing the KS propagator (2.25). Taking into account the explicit form (2.8) for v^μ_H it is then possible to eliminate a further class of diagrammatic contributions (the interested reader is referred to [19] for details). Eq. (2.45), which provides an exact representation of E_{xc} in terms of the KS orbitals, the KS eigenvalues and the xc-potential, $E_{xc}[\phi^{(\dagger)}_k, \epsilon_k, v^\mu_{xc}]$, is the central result of this Section. Note that the $\phi^{(\dagger)}_k$, ϵ_k and v^μ_{xc} are implicit functionals of j^μ, so that the same is true for E_{xc}.

However, Eq. (2.45) is not suitable for a direct evaluation of the xc-energy. In fact, as $v^\mu_{xc} = \delta E_{xc}/\delta j_\mu$ the xc-energy appears on both sides of Eq. (2.45), i.e. Eq. (2.45) represents a highly nonlinear relation for E_{xc}. This highly nonlinear character can be resolved by an expansion in powers of e^2,

$$E_{xc} = \sum_{i=1}^{\infty} e^{2i}E^{(i)}_{xc}; \qquad v^\mu_{xc} = \sum_{i=1}^{\infty} e^{2i}v^{\mu,(i)}_{xc}. \tag{2.46}$$

After insertion of Eq. (2.46) into (2.45) one notices that the lowest order contribution in e^2, the exchange energy $E_x = e^2E^{(1)}_{xc}$, does not depend on v^μ_{xc}. Explicitly one obtains

$$E_x = -\frac{e^2}{2}\int d^3r\int d^3r' \sum_{-mc^2<\epsilon_k,\epsilon_l\leq\epsilon_F}\frac{\cos(\omega_{kl}|\boldsymbol{r}-\boldsymbol{r}'|)}{|\boldsymbol{r}-\boldsymbol{r}'|}\phi^\dagger_k(\boldsymbol{r})\alpha_\mu\phi_l(\boldsymbol{r})\phi^\dagger_l(\boldsymbol{r}')\alpha^\mu\phi_k(\boldsymbol{r}'), \tag{2.47}$$

where ω_{kl} represents the KS single particle transition frequencies

$$\omega_{kl} = |\epsilon_k - \epsilon_l|/c. \tag{2.48}$$

In (2.47) we have again chosen to work in Feynman gauge, which is technically simplest to handle. This choice does not introduce any gauge dependence, as demonstrated in [19]. Thus E_x is a well-defined functional of the $\phi^{(\dagger)}_k$ and ϵ_k. The corresponding exchange potential v^μ_x can be evaluated using the ROPM in the order e^2 (see Section 3).

This then unambiguously defines the e^4-contribution to E_{xc}, as only $v^\mu_x = e^2v^{\mu,(1)}_{xc}$ enters in (2.45) in this order,

$$
\begin{aligned}
E_c^{(2)} = & \frac{i}{2c} \lim_{\epsilon\to 0} \int d^4x\delta(x^0) \int d^4y\, e^{-\epsilon|y^0|} \\
& \times \Big\{ - v_{x,\mu}(\boldsymbol{x}) v_{x,\nu}(\boldsymbol{y}) tr[\gamma^\mu G^0(x, y)\gamma^\nu G_0(y, x)]. \\
& + i \int d^4z\, e^{-\epsilon|z^0|} [D^0_{\mu\rho}(x-z) v_{x,\nu}(\boldsymbol{y}) + D^0_{\nu\rho}(y-z) v_{x,\mu}(\boldsymbol{x})] \\
& \times tr[\gamma^\mu G_0(x, y)\gamma^\nu G_0(y, z)\gamma^\rho G^0(z, x)] \\
& - \frac{1}{2} \int d^4z \int d^4u\, e^{-\epsilon(|z^0|+|u^0|)} D^0_{\mu\nu}(x-z) D^0_{\rho\lambda}(y-u) \\
& \times \ \{tr[\gamma^\mu G^0(x, y)\gamma^\rho G^0(y, x)] tr[\gamma^\nu G^0(z, u)\gamma^\lambda G^0(u, z)] \\
& - 2 tr[\gamma^\mu G^0(x, z)\gamma^\nu G^0(z, u)\gamma^\lambda G^0(u, y)\gamma^\rho G^0(y, x)] \\
& - tr[\gamma^\mu G^0(x, y)\gamma^\rho G^0(y, z)\gamma^\nu G^0(z, u)\gamma^\lambda G^0(u, x)]\} \Big\}, \quad (2.49)
\end{aligned}
$$

so that the corresponding potential $v_c^{(2)}$ can again be calculated by the ROPM. Generalizing this procedure to all orders, it is obvious, that $E_{xc}^{(n)}$ only depends on those $v_{xc}^{\mu,(i)}$ with $i < n$. Therefore a recursive definition of E_{xc} in terms of the $\phi_k^{(\dagger)}$ and ϵ_k is established.

We have thus found a systematic perturbative approach to orbital-dependent representations of E_{xc}. In many physical situations, however, the resummation of certain classes of diagrammatic contributions is required, or at least very helpful. The most simple resummation of this type, the RPA, can be derived most easily within the framework of the adiabatic connection scheme, which is extended to inhomogeneous relativistic systems in the next Section.

2.4. *Adiabatic connection formula*

The relativistic adiabatic connection formula is based on a modified Hamiltonian $\hat{H}(g)$ in which not only the electron-photon coupling strength is multiplied by the dimensionless scaling parameter g but also a g-dependent, multiplicative, external potential is introduced,

$$\hat{H}(g) = \hat{H}_{0,S}(g) + g\hat{H}_{int,S} \tag{2.50}$$

$$
\begin{aligned}
\hat{H}_{0,S}(g) = & \int d^3x \hat{\bar{\psi}}_S(\boldsymbol{x})(-ic\boldsymbol{\gamma}\cdot\nabla + mc^2)\hat{\psi}_S(\boldsymbol{x}) \\
& + \int d^3x \hat{j}_{S,\mu}(\boldsymbol{x}) v_g^\mu(\boldsymbol{x}) + \hat{H}_\gamma(x^0 = 0).
\end{aligned} \tag{2.51}
$$

The physically relevant limit (2.10) is obtained from Eq. (2.50) for $g = 1$, for which v_g^μ coincides with the actual external potential,

$$v^{\mu}_{g=1}(\boldsymbol{x}) \equiv v^{\mu}_{ext}(\boldsymbol{x}). \tag{2.52}$$

Moreover, assuming noninteracting v-representability [2], v^{μ}_{g} is chosen so, that for all $g<1$ the corresponding ground state $|\Phi(g)\rangle$ yields the exact interacting ground state four current

$$j^{\mu}(\boldsymbol{x}) = \langle\Phi(g)|\hat{j}^{\mu}(x)|\Phi(g)\rangle, \qquad \forall g \leqslant 1. \tag{2.53}$$

Consequently, while for vanishing g the noninteracting Hamiltonian $\hat{H}_{0,S}(g)$ is identical with the KS Hamiltonian,

$$v^{\mu}_{g=0}(\boldsymbol{x}) = v^{\mu}_{KS}(\boldsymbol{x}), \tag{2.54}$$

the noninteracting ground state $|\Phi_0\rangle$ depends on g,

$$\hat{H}_{0,S}(g)|\Phi_0(g)\rangle = E_0(g)|\Phi_0(g)\rangle, \tag{2.55}$$

in contrast to the scheme of Section 2.3.

With this construction of the noninteracting reference system one finds for the g-dependent interacting ground state energy $E(g)$,

$$\frac{\partial}{\partial g}E(g) = \langle\Phi(g)|\hat{H}_{int,S}|\Phi(g)\rangle + \int d^3x j_{\mu}(\boldsymbol{x})\frac{\partial}{\partial g}v^{\mu}_{g}(\boldsymbol{x}), \tag{2.56}$$

and thus by integration over g and subsequent insertion of (2.27, 2.52, 2.54)

$$E = T_s + \int d^3x j_{\mu}(\boldsymbol{x}) v^{\mu}_{ext}(\boldsymbol{x}) + E_{int}, \tag{2.57}$$

$$E_{int} = \int_0^1 dg \langle\Phi(g)|\hat{H}_{int,S}|\Phi(g)\rangle. \tag{2.58}$$

At this point one can apply the standard techniques of many-body theory introduced in the previous subsection to obtain (2.40) with $\hat{H}_1$ and $|\Phi_{KS}\rangle$ being replaced by $g\hat{H}_{int}$ and $|\Phi_0\rangle$, respectively. Using (2.42) one can rewrite E_{int} as

$$\begin{aligned} E_{int} = \lim_{\epsilon\to 0} \int_0^1 dg\, g \int d^4x\, \delta(x^0) \int d^4y\, e^{-\epsilon|y^0|} D^0_{\mu\nu}(x-y) \\ \times \sum_{n=1}^{\infty} \frac{(-ig)^{n-1}}{(n-1)!} \int_{-\infty}^{\infty} dt_1 \cdots \int_{-\infty}^{\infty} dt_{n-1} e^{-\epsilon(|t_1|+\cdots+|t_{n-1}|)} \\ \times \frac{\langle\Phi_0|T\hat{j}^{\mu}_0(x)\hat{j}^{\nu}_0(y)\hat{H}_{int,I}(t_1)\cdots\hat{H}_{int,I}(t_{n-1})|\Phi_0\rangle}{\langle\Phi_0|\hat{U}_{I,\epsilon}(+\infty,-\infty)|\Phi_0\rangle} \end{aligned} \tag{2.59}$$

(the relation (2.42) leads to n identical contributions which can be added up after relabelling the integration variables). Distinguishing the two possible time-orderings of x^0 and y^0 one can reorder (2.59) as

$$E_{int} = \lim_{\epsilon\to 0}\int_0^1 dgg\int d^4x\delta(x^0)\int d^4y\, e^{-\epsilon|u^0|}D^0_{\mu\nu}(x-y)$$

$$\times\left\{\Theta(x^0-y^0)\frac{\langle\Phi_0|\hat{U}_{I,\epsilon}(+\infty,\,x^0)\hat{j}^\mu_0(x)\hat{U}_{I,\epsilon}(x^0,\,y^0)\hat{j}^\nu_0(y)\hat{U}_{I,\epsilon}(y^0,\,-\infty)|\Phi_0\rangle}{\langle\Phi_0|\hat{U}_{I,\epsilon}(+\infty,\,-\infty)|\Phi_0\rangle}\right.$$

$$\left.+\,\Theta(y^0-x^0)\frac{\langle\Phi_0|\hat{U}_{I,\epsilon}(+\infty,\,y^0)\hat{j}^\mu_0(y)\hat{U}_{I,\epsilon}(y^0,\,x^0)\hat{j}^\nu_0(x)\hat{U}_{I,\epsilon}(x^0,\,-\infty)|\Phi_0\rangle}{\langle\Phi_0|\hat{U}_{I,\epsilon}(+\infty,\,-\infty)|\Phi_0\rangle}\right\}.$$

Reintroducing the interacting current operator one then obtains

$$E_{int} = \int_0^1 dgg\int d^4x\delta(x^0)\int d^4yD^0_{\mu\nu}(x-y)\langle\Phi(g)|T\hat{j}^\mu(x)\hat{j}^\nu(y)|\Phi(g)\rangle, \tag{2.60}$$

where it has been assumed that the limit $\epsilon\to 0$ is only relevant for the construction of the Heisenberg ground state $|\Phi(g)\rangle$. With the standard definition of the time-ordered current-current response function of the interacting system,

$$\chi^{\mu\nu}_g(x,\,y)=\chi^{\mu\nu}_g(\boldsymbol{x},\,\boldsymbol{y},\,x^0-y^0)=-i[\langle\Phi(g)|T\hat{j}^\mu(x)\hat{j}^\nu(y)|\Phi(g)\rangle-j^\mu(\boldsymbol{x})j^\nu(\boldsymbol{y})], \tag{2.61}$$

one finally obtains

$$E_{xc}=E_{int}-E_H=\frac{i}{2}\int_0^1 dg^2\int d^3x\int d^4yD^0_{\mu\nu}(x-y)\chi^{\mu\nu}_g(x,\,y). \tag{2.62}$$

Eq. (2.62) is the relativistic extension of the adiabatic connection formula introduced by Gunnarsson and Lundqvist [49] and Langreth and Perdew [50]. The latter has proven to be very useful both for the analysis of E_{xc} [46–48] and for the derivation of approximations in the context of explicit density functionals [5]. This is due to the fact that the construction of explicitly density-dependent forms for E_{xc} is usually based on the homogeneous electron gas (HEG), either directly in the case of the LDA, or indirectly via the response of the HEG to a weak perturbation for gradient corrected functionals. While in the former situation Eq. (2.62) directly provides a representation of E_{xc} in terms of the linear HEG response function, in the latter case an expansion of $\chi^{\mu\nu}_g$ in powers of v^μ_g is required, resulting in an expression for E_{xc} which, in general, also depends on the nonlinear response functions of the HEG. Using standard many-body techniques for the HEG response functions, these ingredients of (2.62) only depend on the density via the Fermi momentum $k_F=(3\pi^2n_0)^{1/3}$ of the corresponding noninteracting HEG, which is independent of g. The coupling constant integration in (2.62) can thus be performed either directly in the case of the LDA or after a suitable discussion of the g-dependence of v^μ_g [5].

In the case of arbitrary, inhomogeneous systems the situation is somewhat more complicated, so that for simplicity we restrict the discussion to the RPA, defined via

$$\chi^{\mu\nu}_{g,RPA}(x,\,y)=\chi^{\mu\nu}_0(x,\,y)+\frac{g^2}{c}\int d^4zd^4u\chi^{\mu\rho}_0(x,\,z)D^0_{\rho\kappa}(z-u)\chi^{\kappa\nu}_{g,RPA}(u,\,y), \tag{2.63}$$

as usual. Its basic ingredient is the noninteracting response function $\chi_0^{\mu\nu}$ obtained from (2.61) in the limit $g = 0$, i.e. the KS response function

$$\begin{aligned}\chi_0^{\mu\nu}(x,y) &= -i[\langle\Phi_{KS}|T\hat{j}_0^{\mu}(x)\hat{j}_0^{\nu}(y)|\Phi_{KS}\rangle - j^{\mu}(\boldsymbol{x})j^{\nu}(\boldsymbol{y})] \\ &= -i\mathrm{Tr}[\gamma^{\mu}G^0(x,y)\gamma^{\nu}G^0(y,x)] \\ &= \chi_0^{\mu\nu}(\boldsymbol{x},\boldsymbol{y},x^0-y^0).\end{aligned} \tag{2.64}$$

For all g, however, the g-independence of the corresponding j^{μ} (defining v_g^{μ}) implies that the KS system corresponding to the interacting system with coupling strength $g < 1$ is identical with the KS system corresponding to the fully interacting system with $g = 1$. As a consequence the KS single particle orbitals and thus $\chi_0^{\mu\nu}$ are independent of g. Taking into account the g-dependence of the electron-electron interaction one finds

$$\begin{aligned}E_{xc}^{RPA} = \frac{i}{2}\sum_{n=0}^{\infty}\frac{1}{n}\int d^4x_1\int d^4y_1 \cdots \int d^4x_n\int d^4y_n\delta x_1^0 \\ \times D^0_{\mu_1\nu_1}(x_1-y_1)\chi_0^{\nu_1\mu_2}(y_1,x_2)\cdots D^0_{\mu_n\nu_n}(x_n-y_n)\chi_0^{\nu_n\mu_1}(y_n,x_1).\end{aligned} \tag{2.65}$$

Of course, Eq. (2.65) reduces to the standard result in the case of the homogeneous electron gas [51–53]. Eq. (2.65) thus provides an alternative to Eq. (2.49) for all systems without a gap at the Fermi surface. In practice, a combination of the RPA with the second order functional (2.49) suggests itself as a rather universal form for E_{xc}.

3. Relativistic Optimized-Potential-Method

In Section 2 we have presented two systematic schemes, that allow the extraction of orbital- and eigenvalue-dependent representations of E_{xc}. As already emphasized, however, we are still dealing with current functionals, albeit implicit ones: In fact, the Hohenberg-Kohn theorem for noninteracting particles guarantees that the ϕ_k and ϵ_k are uniquely determined by j^{μ} (up to gauge transformations), as long as they are solutions of single particle equations with a multiplicative potential. The functional dependences $\phi_k([j];\boldsymbol{r})$ and $\epsilon_k[j]$ are established implicitly via the solution of the single particle equations. Consequently, we can follow the standard scheme for the derivation of the single particle equations (2.6) as in the case of explicitly j-dependent E_{xc}.

Both for the proper definition of the functional (2.45) and for selfconsistent calculations with orbital- and eigenvalue-dependent forms for E_{xc}, one then has to provide a method for the evaluation of the corresponding xc-potential $\delta E_{xc}/\delta j^{\mu}$. The basic approach to $\delta E_{xc}/\delta j^{\mu}$ is provided by the OPM, which has first been introduced in the context of nonrelativistic DFT [21,20,5,27,54] and recently been extended to the relativistic domain [16,17,19,55]. The derivation of the crucial OPM integral equation determining the xc-potential can proceed along two different, but equivalent, lines [12,21]: In the original approach [21] the ground state energy E is explicitly minimized with respect to the total KS potential v_{KS}^{μ}, which, due to the uniqueness of v_{KS}^{μ}, indirectly yields the xc-component of v_{KS}^{μ}. The second derivation [12] starts directly from Eq. (2.45), replacing the functional derivative with respect to j^{μ} by derivatives with respect to $\phi_k^{(\dagger)}$ and ϵ_k via the chain rule for functional differentiation. We here follow the latter scheme.

Bearing in mind the inherently recursive character of our first construction scheme for E_{xc}, Eqs. (2.44–2.49), one can again choose between two alternative approaches. On the one hand, it is possible to start with an expansion of E_{xc} in powers of e^2, as outlined in Section 2.3, and subsequently extract the xc-potential order by order. In this case it is important to note, that the lowest order contribution to E_{xc}, i.e. the exchange energy, only depends on the $\phi_k^{(\dagger)}$ and ϵ_k. This allows the calculation of v_x^μ as a functional of the $\phi_k^{(\dagger)}$ and ϵ_k via the ROPM integral equation for the exchange functional. As a consequence, one can eliminate v_x^μ from all higher orders terms in E_{xc} in favor of the $\phi_k^{(\dagger)}$ and ϵ_k. The second order contribution to E_{xc} then becomes a functional of the $\phi_k^{(\dagger)}$ and ϵ_k only, allowing to repeat the procedure. On the other hand, this recursive scheme for the coupled determination of v_{xc} and E_{xc} is not required from the very outset for establishing the ROPM integral equation. We thus first discuss the ROPM integral equation for the general nonlinear form (2.45), indicating a possible shortcut which resolves the coupling of v_{xc}^μ and E_{xc}. Subsequently, the general result will be expanded in powers of e^2.

3.1. *Complete nonlinear ROPM integral equations*

Treating v_{xc}^μ as an independent variable, the chain rule for functional differentiation gives

$$\frac{\delta E_{xc}[\phi_k^{(\dagger)}, \epsilon_k, v_{xc}^\mu]}{\delta j^\nu(\boldsymbol{r})} = \int d^3r' \frac{\delta v_{KS}^\rho(\boldsymbol{r}')}{\delta j^\nu(\boldsymbol{r})} \sum_k \Bigg\{ \int d^3r'' \left[\frac{\delta \phi_k^\dagger(\boldsymbol{r}'')}{\delta v_{KS}^\rho(\boldsymbol{r}')} \frac{\delta E_{xc}}{\delta \phi_k^\dagger(\boldsymbol{r}'')}\bigg|_{expl.} + c.c. \right]$$

$$+ \frac{\delta \epsilon_k}{\delta v_{KS}^\rho(\boldsymbol{r}')} \frac{\partial E_{xc}}{\partial \epsilon_k}\bigg|_{expl.} \Bigg\} + \int d^3r' \frac{\delta v_{xc}^\rho(\boldsymbol{r}')}{\delta j^\nu(\boldsymbol{r})} \frac{\delta E_{xc}}{\delta v_{xc}^\rho(\boldsymbol{r}')}\bigg|_{expl.}, \tag{3.1}$$

where both the unique correspondence between j^μ and v_{KS}^μ and the unique correspondence between v_{KS}^μ and the $\phi_k^{(\dagger)}$, ϵ_k have been used. In the following the derivatives of E_{xc} will always be understood with respect to the explicit dependence on $\phi_k^{(\dagger)}$, ϵ_k and v_{xc}^μ, so that the index *expl.* is dropped from now on. The linear response of the $\phi_k^{(\dagger)}$ and ϵ_k to a variation of v_{KS}^ρ, which appears in (3.1), can be directly obtained from first order perturbation theory,

$$\frac{\delta \phi_k^\dagger(\boldsymbol{r})}{\delta v_{KS}^\rho(\boldsymbol{r}')} = -\phi_k^\dagger(\boldsymbol{r}')\alpha_\rho G_k(\boldsymbol{r}', \boldsymbol{r}) \tag{3.2}$$

$$G_k(\boldsymbol{r}, \boldsymbol{r}') = \sum_{l \neq k} \frac{\phi_l(\boldsymbol{r})\phi_l^\dagger(\boldsymbol{r}')}{\epsilon_l - \epsilon_k} \tag{3.3}$$

$$\frac{\delta \epsilon_k}{\delta v_{KS}^\rho(\boldsymbol{r})} = \phi_k^\dagger(\boldsymbol{r})\alpha_\rho \phi_k(\boldsymbol{r}). \tag{3.4}$$

Multiplying Eq. (3.1) with the static limit of the KS response function (2.64),

$$\chi_0^{\mu\nu}(\boldsymbol{r},\boldsymbol{r}') = \frac{\delta j^\mu(\boldsymbol{r})}{\delta v_{KS,\nu}(\boldsymbol{r}')} = \int_{-\infty}^{\infty} d\tau \chi_0^{\mu\nu}(\boldsymbol{r},\boldsymbol{r}',\tau)$$

$$= - \sum_{-mc^2<\epsilon_k\leq\epsilon_F} \phi_k^\dagger(\boldsymbol{r})\alpha^\mu G_k(\boldsymbol{r},\boldsymbol{r}')\alpha^\nu \phi_k(\boldsymbol{r}') + c.c., \tag{3.5}$$

and integrating over $\boldsymbol{r}$ leads to the ROPM integral equations for the xc-potential,

$$\int d^3r' \chi_0^{\mu\nu}(\boldsymbol{r},\boldsymbol{r}') v_{xc,\nu}(\boldsymbol{r}') = \Lambda_{xc}^\mu(\boldsymbol{r}), \tag{3.6}$$

where the inhomogeneity is given by

$$\Lambda_{xc}^\mu(\boldsymbol{r}) = -\sum_k \int d^3r' \left[\phi_k^\dagger(\boldsymbol{r})\alpha^\mu G_k(\boldsymbol{r},\boldsymbol{r}') \frac{\delta E_{xc}}{\delta \phi_k^\dagger(\boldsymbol{r}')} + c.c. \right] + \sum_k \phi_k^\dagger(\boldsymbol{r})\alpha^\mu \phi_k(\boldsymbol{r}) \frac{\partial E_{xc}}{\partial \epsilon_k}$$

$$+ \int d^3r' \int d^3r'' \chi_0^{\mu\nu}(\boldsymbol{r},\boldsymbol{r}') \frac{\delta v_{xc}^\rho(\boldsymbol{r}'')}{\delta j^\nu(\boldsymbol{r}')} \frac{\delta E_{xc}}{\delta v_{xc}^\rho(\boldsymbol{r}'')}. \tag{3.7}$$

The ROPM integral equation (3.6) has to be solved selfconsistently together with the single particle equations (2.6), i.e. (3.6) replaces the explicit evaluation of $\delta E_{xc}/\delta j^\mu$ of the conventional KS procedure.

The most critical functional derivative in (3.7) is the static xc-kernel

$$\frac{\delta v_{xc}^\mu(\boldsymbol{r})}{\delta j_\nu(\boldsymbol{r}')} = f_{xc}^{\mu\nu}(\boldsymbol{r},\boldsymbol{r}'),$$

which makes (3.6) a highly nonlinear equation for v_{xc}^μ. $f_{xc}^{\mu\nu}$ has to be evaluated in accordance with the actual approximation used for E_{xc}. Apart from the systematic recursive procedure discussed in Section 3.2 an approximation strategy is conceivable: For the second order functional (2.49) one can explicitly evaluate the v_x^μ-dependent component, using the known ROPM exchange potential. For atoms and molecules one finds that this energy component essentially cancels with certain other contributions to $E_c^{(2)}$ (see Section 4). This fosters the hope that those terms which almost cancel each other can be neglected in Λ_{xc}^μ, so that $f_{xc}^{\mu\nu}$ is no longer required for the solution of the ROPM integral equation. Alternatively, one could apply some approximation for $f_{xc}^{\mu\nu}$, as e.g. the LDA. This would immediately allow to deal with nonperturbative classes of diagrammatic contributions to (2.45), as e.g. the combination of the RPA functional (2.65) with (2.49).

3.2. *Second order ROPM equations*

In this Section we discuss the ROPM integral equations for the recursive scheme introduced in Sec. 2.2, restricting the analysis to the order e^4. An extension to higher order can proceed along the same lines. Expanding the energy and the potential into a power series with respect to e^2,

$$E_{xc} \to E_x + E_{c,2} + \cdots \tag{3.8}$$

$$v_{xc} \to v_x + v_{c,2} + \cdots, \tag{3.9}$$

also the ROPM equations (3.6) can be expanded, leading to one integral equation for each order in e^2,

$$\int d^3 r_2 \chi_0^{\mu\nu}(\boldsymbol{r}_1, \boldsymbol{r}_2) v_{x,\nu}(\boldsymbol{r}_2) = \Lambda_x^{\mu}(\boldsymbol{r}_1) \tag{3.10}$$

$$\int d^3 r_2 \chi_0^{\mu\nu}(\boldsymbol{r}_1, \boldsymbol{r}_2) v_{c,\nu}^{(2)}(\boldsymbol{r}_2) = \Lambda_c^{(2),\mu}(\boldsymbol{r}_1), \tag{3.11}$$

where the inhomogeneities are given by

$$\begin{aligned} \Lambda_x^{\mu}(\boldsymbol{r}_1) = &- \sum_k \int d^3 r_2 \left[\phi_k^{\dagger}(\boldsymbol{r}_1) \alpha^{\mu} G_k(\boldsymbol{r}_1, \boldsymbol{r}_2) \frac{\delta E_x}{\delta \phi_k^{\dagger}(\boldsymbol{r}_2)} + c.c. \right] \\ &+ \sum_k \phi_k^{\dagger}(\boldsymbol{r}_1) \alpha^{\mu} \phi_k(\boldsymbol{r}_1) \frac{\partial E_x}{\partial \epsilon_k} \end{aligned} \tag{3.12}$$

$$\begin{aligned} \Lambda_c^{(2),\mu}(\boldsymbol{r}_1) = &- \sum_k \int d^3 r_2 \left[\phi_k^{\dagger}(\boldsymbol{r}_1) \alpha^{\mu} G_k(\boldsymbol{r}_1, \boldsymbol{r}_2) \frac{\delta E_c^{(2)}}{\delta \phi_k^{\dagger}(\boldsymbol{r}_2)} + c.c. \right] \\ &+ \sum_k \phi_k^{\dagger}(\boldsymbol{r}_1) \alpha^{\mu} \phi_k(\boldsymbol{r}_1) \frac{\partial E_c^{(2)}}{\partial \epsilon_k} + \int d^3 r_2 \frac{\delta v_x^{\nu}(\boldsymbol{r}_2)}{\delta v_{KS,\mu}(\boldsymbol{r}_1)} \frac{\delta E_c^{(2)}}{\delta v_x^{\nu}(\boldsymbol{r}_2)} \end{aligned} \tag{3.13}$$

(the unique correspondence between j^{μ} and v_{KS}^{μ} has been used to simplify the last term in Eq. (3.13)).

The evaluation of the functional derivatives of E_x and $E_c^{(2)}$ can be performed directly using Eqs. (2.47, 2.49). The only nontrivial term appearing in the order e^4 is $\delta v_x^{\nu}/\delta v_{KS,\mu}$. For its calculation we rely on the first order ROPM equation (3.10): Taking its functional derivative with respect to v_{KS}^{μ}, one finds after reordering

$$\frac{\delta v_x^{\nu}(\boldsymbol{r}_2)}{\delta v_{KS}^{\mu}(\boldsymbol{r}_1)} = \int d^3 r_3 \chi_0^{-1,\nu\lambda}(\boldsymbol{r}_2, \boldsymbol{r}_3) \left\{ \frac{\delta \Lambda_{x,\lambda}(\boldsymbol{r}_3)}{\delta v_{KS}^{\mu}(\boldsymbol{r}_1)} - \int d^3 r_4 \frac{\delta \chi_{0,\lambda\rho}(\boldsymbol{r}_3, \boldsymbol{r}_4)}{\delta v_{KS}^{\mu}(\boldsymbol{r}_1)} v_x^{\rho}(\boldsymbol{r}_4) \right\}, \tag{3.14}$$

with $\chi_0^{-1,\nu\lambda}$ denoting the inverse response function,

$$\int d^3 r' \chi_0^{-1,\mu\nu}(\boldsymbol{r}, \boldsymbol{r}') \chi_{0,\nu\rho}(\boldsymbol{r}', \boldsymbol{r}'') = \delta^{(3)}(\boldsymbol{r} - \boldsymbol{r}'') \delta_{\nu}^{\mu}.$$

The standard ROPM replacement of functional derivatives (again using the various unique equivalences) directly yields an expression for the first contribution on the right hand side of (3.14) in terms of the orbitals and eigenvalues,

$$\frac{\delta\Lambda_x^\lambda(\boldsymbol{r}_2)}{\delta v_{KS}^\mu(\boldsymbol{r}_1)} = -\sum_k \int d^3r_3 \left[\phi_k^\dagger(\boldsymbol{r}_1)\alpha_\mu G_k(\boldsymbol{r}_1, \boldsymbol{r}_3)\frac{\delta\Lambda_x^\lambda(\boldsymbol{r}_2)}{\delta\phi_k^\dagger(\boldsymbol{r}_3)} + c.c.\right]$$
$$+ \sum_k \phi_k^\dagger(\boldsymbol{r}_1)\alpha_\mu\phi_k(\boldsymbol{r}_1)\frac{\partial\Lambda_x^\lambda(\boldsymbol{r}_2)}{\partial\epsilon_k}. \tag{3.15}$$

The second contribution in (3.14) contains the quadratic response function,

$$\frac{\delta\chi_{0,\mu_1\mu_2}(\boldsymbol{r}_1, \boldsymbol{r}_2)}{\delta v_{KS}^{\mu_3}(\boldsymbol{r}_3)} = \frac{\delta^2 j_{\mu_1}(\boldsymbol{r}_1)}{\delta v_{KS}^{\mu_3}(\boldsymbol{r}_3)\delta v_{KS}^{\mu_2}(\boldsymbol{r}_2)}$$
$$= \sum_{-mc^2<\epsilon_k\leq\epsilon_F} H_{k,\mu_1\mu_2\mu_3}(\boldsymbol{r}_1, \boldsymbol{r}_2, \boldsymbol{r}_3)$$
$$+ \text{all permutations of } \boldsymbol{r}_1\mu_1 \leftrightarrow \boldsymbol{r}_2\mu_2 \leftrightarrow \boldsymbol{r}_3\mu_3, \tag{3.16}$$

with

$$H_{k,\mu_1\mu_2\mu_3}(\boldsymbol{r}_1, \boldsymbol{r}_2, \boldsymbol{r}_3) = \phi_k^\dagger(\boldsymbol{r}_1)\alpha_{\mu_1}G_k(\boldsymbol{r}_1, \boldsymbol{r}_2)\alpha_{\mu_2}G_k(\boldsymbol{r}_2, \boldsymbol{r}_3)\alpha_{\mu_3}\phi_k(\boldsymbol{r}_3)$$
$$- \phi_k^\dagger(\boldsymbol{r}_1)\alpha_{\mu_1}\phi_k(\boldsymbol{r}_1)\phi_k^\dagger(\boldsymbol{r}_2)\alpha_{\mu_2}\int d^3r_4 G_k(\boldsymbol{r}_2, \boldsymbol{r}_4)G_k(\boldsymbol{r}_4, \boldsymbol{r}_3)\alpha_{\mu_3}\phi_k(\boldsymbol{r}_3). \tag{3.17}$$

Eqs. (3.10–3.17) together with (2.47, 2.49, 3.3, 3.5) provide all the necessary ingredients for the determination of v_x and $v_c^{(2)}$. However, it is immediately clear from these relations that a higher order perturbative treatment of E_{xc}, requiring still higher order response functions, becomes prohibitive on the selfconsistent level.

Finally, we note that Eqs. (3.5–3.7) can also be used to extract the behavior of the *xc*-potential in the asymptotic regime of finite systems. Restricting the analysis to spherically averaged systems, a somewhat tedious analysis (the interested reader is referred to [19]) leads to the relativistic form of the Krieger-Li-Iafrate identity for the highest occupied orbital ϕ_h [56],

$$\int d^3r\phi_h^\dagger(\boldsymbol{r})\left\{\phi_h(\boldsymbol{r})v_x^0(\boldsymbol{r}) - \frac{\delta E_x}{\delta\phi_h^\dagger(\boldsymbol{r})}\right\} + c.c. = 0, \tag{3.18}$$

and the asymptotic behavior of v_x^0,

$$v_x^0(r) \underset{r\to\infty}{\longrightarrow} -\frac{e^2}{r}, \tag{3.19}$$

which one would have expected from the analogous nonrelativistic relation [21].

4. First Results for the Coulomb Limit

In order to facilitate a direct comparison with other relativistic many-body methods as the RHF approach and the Møller-Plesset perturbation expansion [30] we now reduce the main results of Section 2 to the limit of a pure Coulomb interaction between the electrons (longitudinal limit). The resulting expressions also allow for the straightforward extraction of the nonrelativistic limit, simply by replacing the four spinors ϕ_k by the corresponding Pauli spinors. The Coulomb limit is obtained by substituting the photon propagator $D^0_{\mu\nu}$ by the instantaneous Coulomb interaction,

$$D^0_{\mu\nu}(x-y) \rightarrow \frac{e^2 g_{\mu 0} g_{\nu 0}}{|\boldsymbol{x}-\boldsymbol{y}|}\delta(x^0-y^0), \tag{4.1}$$

and appropriately reordering the field operators in the interaction Hamiltonian,

$$\hat{j}^0_0(x)\hat{j}^0_0(y)|_{x^0=y^0=ct} \rightarrow \hat{\psi}^\dagger_0(\boldsymbol{x}t)\hat{\psi}^\dagger_0(\boldsymbol{y}t)\hat{\psi}_0(\boldsymbol{y}t)\hat{\psi}_0(\boldsymbol{x}t). \tag{4.2}$$

In this way one ends up with the familiar forms of the Hartree and exchange energies,

$$E^C_H = \frac{e^2}{2}\int d^3r \int d^3r' \frac{n(\boldsymbol{r})n(\boldsymbol{r}')}{|\boldsymbol{r}-\boldsymbol{r}'|} \tag{4.3}$$

$$E^C_x = -\frac{e^2}{2}\int d^3r \int d^3r' \sum_{-mc^2<\epsilon_k,\epsilon_l\leq\epsilon_F} \frac{\phi^\dagger_k(\boldsymbol{r})\phi_l(\boldsymbol{r})\phi^\dagger_l(\boldsymbol{r}')\phi_k(\boldsymbol{r}')}{|\boldsymbol{r}-\boldsymbol{r}'|}, \tag{4.4}$$

while the complete E_c is given by (2.45) evaluated with the Coulomb form of $\hat{W}$,

$$\hat{W}^C(t) = \frac{e^2}{2}\int d^3x \int d^3y \frac{\hat{\psi}^\dagger_0(\boldsymbol{x}t)\hat{\psi}^\dagger_0(\boldsymbol{y}t)\hat{\psi}_0(\boldsymbol{y}t)\hat{\psi}_0(\boldsymbol{x}t)}{|\boldsymbol{x}-\boldsymbol{y}|}$$
$$- \int d^3x \{v_{H,0}(\boldsymbol{x})g_{0\mu} + v_{xc,\mu}(\boldsymbol{x})\}\hat{\psi}^\dagger_0(\boldsymbol{x}t)\alpha^\mu\hat{\psi}_0(\boldsymbol{x}t). \tag{4.5}$$

In particular, the Coulomb limit of the lowest order correlation contribution $E^{(2)}_c$ reads

$$E^{(2),C}_c = E_{MP2} + E_{\Delta HF} \tag{4.6}$$

$$E_{MP2} = \frac{e^4}{2} \sum_{-mc^2<\epsilon_i,\epsilon_j\leq\epsilon_F<\epsilon_k,\epsilon_l} \frac{1}{\epsilon_i+\epsilon_j-\epsilon_k-\epsilon_l} \int d^3r_1 \int d^3r_2 \frac{\phi^\dagger_i(\boldsymbol{r}_1)\phi_k(\boldsymbol{r}_1)\phi^\dagger_j(\boldsymbol{r}_2)\phi_l(\boldsymbol{r}_2)}{|\boldsymbol{r}_1-\boldsymbol{r}_2|}$$
$$\times \int d^3r_3 \int d^3r_4 \left\{ \frac{\phi^\dagger_k(\boldsymbol{r}_3)\phi_i(\boldsymbol{r}_3)\phi^\dagger_l(\boldsymbol{r}_4)\phi_j(\boldsymbol{r}_4)}{|\boldsymbol{r}_3-\boldsymbol{r}_4|} - \frac{\phi^\dagger_k(\boldsymbol{r}_3)\phi_j(\boldsymbol{r}_3)\phi^\dagger_l(\boldsymbol{r}_4)\phi_i(\boldsymbol{r}_4)}{|\boldsymbol{r}_3-\boldsymbol{r}_4|} \right\} \tag{4.7}$$

$$E_{\Delta HF} = \sum_{-mc^2<\epsilon_k\leq\epsilon_F<\epsilon_l} \frac{1}{\epsilon_k-\epsilon_l} \Bigg| \int d^3r_1 \phi^\dagger_k(\boldsymbol{r}_1)\alpha^\mu\phi_l(\boldsymbol{r}_1)v^C_{x,\mu}(\boldsymbol{r}_1)$$
$$+ e^2 \sum_{-mc^2<\epsilon_j\leq\epsilon_F} \int d^3r_1 \int d^3r_2 \frac{\phi^\dagger_k(\boldsymbol{r}_1)\phi_j(\boldsymbol{r}_1)\phi^\dagger_j(\boldsymbol{r}_2)\phi_l(\boldsymbol{r}_2)}{|\boldsymbol{r}_1-\boldsymbol{r}_2|}\Bigg|^2. \tag{4.8}$$

Its first term, E_{MP2}, has exactly the same functional form as the conventional second order Møller-Plesset correlation contribution (MP2). However, in Eqs. (4.7, 4.8) the $\phi^{(\dagger)}_k$ represent KS orbitals, which experience the multiplicative KS potential (2.7), rather than the nonlocal HF potential. The difference between both potentials is most obvious for neutral atoms, for which they produce a completely different spectrum of virtual excitations: While the HF potential is short-ranged and therefore generates no bound states beyond the N occupied orbitals [57], the asymptotic $-e^2/r$ tail of v_x^0, Eq. (3.19), leads to a complete Rydberg series in the case of v_{KS}. The second term, $E_{\Delta HF}$, reflects the conceptual difference between the x-only ROPM and the RHF approach, i.e. it represents the lowest order perturbative contribution to the difference ΔE between the RHF and the x-only ROPM ground state energies,

$$\Delta E = E^{RHF} - E^{x\text{-}only\ ROPM}. \tag{4.9}$$

We just remark that $E_c^{(2),\mathrm{C}}$ also includes the leading contribution to the van-der-Waals interaction between two atoms [19].

To illustrate the relation between the x-only ROPM and the RHF scheme we list the resulting ground state energies of closed subshell atoms in Table 1. All results in Table 1 have been calculated fully numerically on a discretized radial grid of 400–800 mesh points, using standard finite differences techniques for the solution of the radial Dirac equation and for integration. As is immediately obvious from Table 1 the ROPM values are extremely close to the RHF energies, their difference reflecting the slightly reduced variational freedom of the ROPM: While the ground state energy expressions of the x-

Table 1 *X-only Coulomb ground state energies: Selfconsistent ROPM [16], RHF [60], RLDA and PW91-GGA [8] results for neutral atoms with closed subshells. Also given is $E_{\Delta HF}$, Eq. (4.8) (all energies in mhartree).*

Atom	E^{C}	$E^{\mathrm{C}} - E^{\mathrm{C,ROPM}}$			$E_{\Delta HF}$
	ROPM	RLDA	PW91	RHF	
He	−2862	138	6	0	0
Be	−14575	350	18	−1	−1
Ne	−128690	1062	−26	−2	−2
Mg	−199932	1376	−5	−3	−3
Ar	−528678	2341	21	−5	−6
Ca	−679704	2656	−4	−6	−7
Zn	−1794598	4140	−388	−14	−16
Kr	−2788848	5565	−265	−13	−13
Sr	−3178067	5996	−305	−13	−13
Pd	−5044384	7707	−666	−16	−17
Cd	−5593299	8213	−734	−20	−22
Xe	−7446876	9800	−1003	−19	−20
Ba	−8135625	10289	−1188	−19	−20
Yb	−14067621	13272	−3789	−48	−57
Hg	−19648826	17204	−5132	−39	−41
Rn	−23601969	19677	−6530	−35	−38
Ra	−25028027	20460	−7186	−34	−37
No	−36740625	25787	−14645	−57	−67

only ROPM and RHF are identical, the restriction to a multiplicative exchange potential in the case of the ROPM leads to marginally higher energies. This difference is essentially given by $E_{\Delta HF}$, Eq. (4.8), as can also be seen from Table 1.

In order to demonstrate the qualitative progress accompanying the use of the orbital-dependent E_x we also list the ground state energies obtained with the relativistic x-only LDA [32,33] and the exchange component of the Perdew-Wang (PW91) GGA [8]. While the relativistic LDA (RLDA) misrepresents E^{C} throughout the periodic table, the GGA is clearly more accurate for light atoms. The fact that the deviation of the GGA energies from the ROPM values strongly increases with Z indicates the need for the inclusion of relativistic corrections in GGAs. While this problem can be resolved by introducing appropriate relativistic correction factors [58], the performance of the resulting relativistic GGAs is still not completely satisfying. This is most obvious from the fact that the GGA exchange potential for finite systems decays much faster than $-e^2/r$ for large r, leading e.g. to a rather poor description of negative ions.

In Table 2 we show the Coulomb correlation energies of the neon isoelectronic series obtained with the functional (4.6) by insertion of selfconsistent x-only ROPM orbitals, in comparison with MP2 results [30] (the neon isoelectronic series appears to be the only systematic set of data beyond the helium series for which accurate relativistic MP2 data are available in the literature, so that it is ideally suited for this comparison). All necessary r-integrations have been performed on a radial grid of 1600 mesh points, using the numerical spinors resulting from x-only ROPM calculations, both for the bound part of the spectrum and the positive continuum. The contribution of higher Rydberg states have been treated by an analytical formula derived by Kelly [59]. This resummation rests on the observation that the ratio of two expectation values, which differ only in the principal quantum number n of one state involved approaches a simple form in the limit of large n_A, n'_A,

Table 2 *Correlation energies of the neon isoelectronic series: Comparison of* $E^{\mathrm{C}}_{c,2}$, *Eq. (4.6), with the corresponding second order Møller-Plesset (MP2) [30], RLDA [16] and PW91-GGA [8] results. In addition,* $E_{\Delta HF}$, *Eq. (4.8), is compared with the difference* ΔE, *Eq. (4.9), between RHF [60] and x-only ROPM [16] total energies (all energies in mhartree).*

Atom	$-E_c^{MP2}$	$-E_c^{(2),\mathrm{C}}$	$-E_c^{LDA}$	$-E_c^{PW91}$	$-E_{\Delta HF}$	$-\Delta E$
Ne	383.2	476.5	747	382	1.7	1.7
Ca^{10+}	395.4	430.4	1010	474	2.1	2.0
Zn^{20+}	406.4	430.8	1147	509	2.2	2.2
Zr^{30+}	414.1	434.1	1243	529	2.4	2.4
Sn^{40+}	421.6	439.3	1317	544	2.8	2.7
Nd^{50+}	430.2	446.8	1380	554	3.4	3.2
Yb^{60+}	441.1	457.6	1434	562	4.4	4.0
Hg^{70+}	455.5	472.8	1484	568	6.0	5.2
Th^{80+}	475.4	494.9	1530	571	8.6	7.2
Fm^{90+}	504.4	528.2	1575	573	13.0	11.0

$$\frac{\langle n_A, l_A, j_A; B|\hat{O}|C; D\rangle}{\langle n'_A, l_A, j_A; B|\hat{O}|C; D\rangle} \sim \left(\frac{n'_A}{n_A}\right)^{3/2} \tag{4.10}$$

(the orbitals B, C, D are unchanged, as are the angular momenta l_A and j_A). For discrete excited states with $l = 0$, 1 Eq. (4.10) was used to resum the contributions of shells with principal quantum number $n \geqslant 15$. In contrast, for orbitals with $l = 2$, 3, 4 Eq. (4.10) was already utilized for $n \geqslant 6$. The integration over the continuum states has also been done via a Newton-Cotes algorithm, checking carefully that the number of energy mesh points ($\sim$ 50–100) is sufficient to achieve the desired accuracy of 0.1 *mhartree*. All angular momenta up to $l = 10$ have been included for the continuum states, and the corresponding energy cutoff has been chosen to be the absolute value of 400 times the innermost eigenvalue of the ion under investigation.

Focussing on the total correlation energies reported in the first two columns of Table 2, one notices a marked difference of 20% between the conventional MP2 and the corresponding DFT correlation energy (4.6) of neon. This deviation, however, diminishes rapidly to approximately 5% for neonlike zinc. For still higher Z the relative difference remains almost constant. The particular discrepancy for neon can in large part be attributed to the excitations into the Rydberg states, which give a contribution of -43.7 mhartrees. On the other hand, as Z increases along the isoelectronic series, the asymptotic behavior of the HF and KS potentials becomes more and more similar, and thus also the excitation spectra approach each other. It remains to be investigated how energy differences as ionization potentials or electron affinities obtained with the two schemes compare with each other.

Table 2 also shows the well-known incorrect Z-scaling of the LDA correlation energies for isoelectronic series: The error of the LDA increases from almost a factor of 2 to more than a factor of 3 from neon to fermium. On the other hand, the nonrelativistic PW91-GGA [8] agrees closely with the MP2 result for neon. However, while the GGA improves substantially over the LDA also for higher Z, the data nevertheless indicate that its scaling with Z is not completely satisfactory. The fact that the deviation of the GGA is as large for Ca^{10+} as it is for Fm^{10+} demonstrates that this error cannot be explained by the missing relativistic corrections in the functional form of the GGA. In any case, apart from neon, $E_c^{(2),C}$ gives clearly more accurate correlation energies than the GGA. In particular, $E_c^{(2),C}$ shows the correct Z-scaling (this is also true in the nonrelativistic limit).

Finally, in Table 2 one again recognizes that $E_{\Delta HF}$ gives a good account of the difference ΔE between the x-only ROPM and the RHF ground state energies. Although no corresponding estimate of the quantitative impact of $E_{\Delta HF}$ on the selfconsistent $v_c^{(2)}$, Eq. (3.13), is presently available, the small size of $E_{\Delta HF}$, as compared to E_{MP2}, suggests that neglect of $E_{\Delta HF}$ could be a reasonable approximation to (4.6). This would eliminate the v_x-dependence of $E_c^{(2),C}$, thus considerably simplifying the ROPM procedure for this functional.

Acknowledgements

We would like to thank R.M. Dreizler, R.N. Schmid and A. Höck for helpful discussions. Financial support by the Deutsche Forschungsgemeinschaft (project Dr 113/20-2) is gratefully acknowledged.

References

1. R.O. Jones and O. Gunnarsson, Rev. Mod. Phys. **61**, 689 (1989).
2. R.M. Dreizler and E.K.U. Gross, *Density Functional Theory* (Springer, Berlin, 1990).
3. B.G. Johnson, P.M.W. Gill and J.A. Pople, J. Chem. Phys. **98**, 5612 (1993).
4. *Density Functional Theory*, edited by E.K.U. Gross and R.M. Dreizler, Vol. 337 of *NATO Advanced Study Institute, Series B: Physics* (Plenum, New York, 1995).
5. D.C. Langreth and M.J. Mehl, Phys. Rev. B **28**, 1809 (1983).
6. A.D. Becke, Phys. Rev. A **38**, 3098 (1988).
7. C. Lee, W. Yang, and R.G. Parr, Phys. Rev. B **37**, 785 (1988).
8. J.P. Perdew, in *Electronic Structure of Solids 1991*, edited by P. Ziesche and H. Eschrig (Akademie Verlag, Berlin, 1991), p.11.
9. W. Kohn and L.J. Sham, Phys. Rev. A **140**, 1133 (1965).
10. Y. Li, J.B. Krieger, M.R. Norman and G.J. Iafrate, Phys. Rev. B **44**, 10437 (1991); J.B. Krieger, Y. Li and G.J. Iafrate, Phys. Rev. A **45**, 101 (1992); and Phys. Rev. A **47**, 165 (1993).
11. E. Engel and S.H. Vosko, Phys. Rev. A **47**, 2800 (1993); Phys. Rev. B **47**, 13164 (1993); Phys. Rev. B **50**, 10498 (1994).
12. A. Görling and M. Levy, Phys. Rev. B **47**, 13105 (1993); Phys. Rev. A **50**, 196 (1994).
13. T. Kotani, Phys. Rev. B **50**, 14816 (1994); Phys. Rev. Lett. **74**, 2989 (1995); T. Kotani and H. Akai, Phys. Rev. B **54**, 16502 (1996).
14. D.M. Bylander and L. Kleinman, Phys. Rev. Lett. **74**, 3660 (1995); Phys. Rev. B **52**, 14566 (1995); Phys. Rev. B **54**, 7891 (1996); Phys. Rev. B **55**, 9432 (1997).
15. T. Grabo and E.K.U. Gross, Chem. Phys. Lett. **240**, 141 (1995).
16. E. Engel, S. Keller, A. Facco Bonetti, H. Müller, and R.M. Dreizler, Phys. Rev. A **52**, 2750 (1995).
17. T. Kreibich, E.K.U. Gross, and E. Engel, Phys. Rev. A **57**, 138 (1998).
18. M. Städele, J.A. Majewski, P. Vogl, and A. Görling Phys. Rev. Lett. **79**, 2089 (1997).
19. E. Engel, A. Facco Bonetti, S. Keller, R.M. Dreizler, and I. Andreikovics, Phys. Rev. A **58**, 964 (1998).
20. V. Sahni, J. Gruenebaum, and J.P. Perdew, Phys. Rev. B **26**, 4371 (1982).
21. R.T. Sharp and G.K. Horton, Phys. Rev. **90**, 317 (1953); J.D. Talman and W.F. Shadwick, Phys. Rev. A **14**, 36 (1976).
22. R. Colle and O. Salvetti, Theoret. Chim. Acta (Berl.) **37**, 329 (1975).
23. S.J. Chakravorty, S.R. Gwaltney, E.R. Davidson, F.A. Parpia, and C. Froese Fischer, Phys. Rev. A **47**, 3649 (1993).
24. C. Filippi, C.J. Umrigar, and X. Gonze, Phys. Rev. A **54**, 4810 (1997).
25. E. Engel and R.M. Dreizler, to be publ. in J. Comput. Chem. (1998).
26. J. Harris and R.O. Jones, J. Phys. F **4**, 1170 (1974).
27. L.J. Sham, Phys. Rev. B **32**, 3876 (1985).
28. J. Sucher, Phys. Rev. A **22**. 348 (1980).
29. O. Gunnarsson and B.I. Lundqvist, Phys. Rev. B **13**, 4274 (1976); D.C. Langreth and J.P. Perdew, Phys. Rev. B **15**, 2884 (1977).
30. Y. Ishikawa and K. Koc, Phys. Rev. A **50**, 4733 (1994); Phys. Rev. A **53**, 3966 (1996); Phys. Rev. A **56**, 1295 (1997).
31. A.K. Rajagopal and J. Callaway, Phys. Rev. B **7**, 1912 (1973).
32. A.K. Rajagopal, J. Phys. C **11**, L943 (1978).
33. A.H. MacDonald and S.H. Vosko, J. Phys. C **12**, 2977 (1979).
34. E. Engel, H. Müller, C. Speicher, and R.M. Dreizler, in [4], p. 65.
35. in *Density Functional Theory II*, edited by R.F. Nalewajski, Vol. 181 of *Topics in Current Chemistry* (Springer, Berlin, 1996), p.1.
36. P. Hohenberg and W. Kohn, Phys. Rev. B **136**, B864 (1964).
37. A.O.G. Källén, in *Handbuch der Physik, Band V, Teil 1* (Springer, Berlin, 1958).

38. J. Sucher, J. Phys. B **21**, L585 (1988).
39. M. Gell-Mann and F. Low, Phys. Rev. **84**, 350 (1951).
40. J. Sucher, Phys. Rev. **107**, 1448 (1957).
41. P.J. Mohr, Phys. Rev. A **32**, 1949 (1985).
42. P.J. Mohr, in *Relativistic, quantum electrodynamic and weak interaction effects in atoms*, edited by W.R. Johnson, P.J. Mohr, and J. Sucher (AIP, New York, 1989), p.47.
43. S.A. Blundell, P.J. Mohr, W.R. Johnson, and J. Sapirstein, Phys. Rev. A **48**, 2615 (1993).
44. I. Lindgren, H. Persson, S. Salomonson, and L. Labzowsky, Phys. Rev. A **51**, 1167 (1995).
45. M.V. Ramana and A.K. Rajagopal, Phys. Rev. A **24**, 1689 (1981).
46. M. Levy and J.P. Perdew, Phys. Rev. A **32**, 2010 (1985).
47. D.C. Langreth and J.P. Perdew, Phys. Rev. B **21**, 5469 (1980).
48. J.P. Perdew, M. Ernzerhof, and K. Burke, J. Chem. Phys. **105**, 9982 (1996).
49. O. Gunnarsson and B.I. Lundqvist, Phys. Rev. B **13**, 4274 (1976).
50. D.C. Langreth and J.P. Perdew, Phys. Rev. B **15**, 2884 (1977).
51. A.L. Fetter and J.D. Walecka, *Quantum Theory of Many-Particle Systems* (McGraw-Hill, New York, 1971).
52. M.V. Ramana and A.K. Rajagopal, Adv. Chem. Phys. **54**, 231 (1983).
53. A. Facco Bonetti, E. Engel, R.M. Dreizler, I. Andrejkovics, and H. Müller, Phys. Rev. A **58**, 993 (1998).
54. L.J. Sham and M. Schlüter, Phys. Rev. Lett. **51**, 1888 (1983).
55. B.A. Shadwick, J.D. Talman, and M.R. Norman, Comp. Phys. Commun. **54**, 95 (1989).
56. J.B. Krieger, Y. Li and G.J. Iafrate, Phys. Lett. **146** A, 256 (1990).
57. H.P. Kelly, Phys. Rev. **131**, 684 (1963).
58. E. Engel, S. Keller, and R.M. Dreizler, Phys. Rev. A **53**, 1367 (1996).
59. H.P. Kelly, Phys. Rev. **136**, B896 (1964).
60. K.G. Dyall, I.P. Grant, C.T. Johnson, F.A. Parpia, and E.P. Plummer, Comp. Phys. Commun. **55**, 425 (1989).

Ab-Initio ZORA Calculations

S. Faas[a], J.G. Snijders[b*] and J.H. van Lenthe[a]

[a] *Theoretical Chemistry Group, Debye Institute, Utrecht University, Padualaan 14, 3584 CH Utrecht, The Netherlands*

[b,*] *Laboratory of Chemical Physics, University of Groningen, Nijenborgh 4, 9747 AG Groningen, The Netherlands*

Abstract

In this paper we present the first application of the ZORA (Zeroth Order Regular Approximation of the Dirac Fock equation) formalism in Ab Initio electronic structure calculations. The ZORA method, which has been tested previously in the context of Density Functional Theory, has been implemented in the GAMESS-UK package. As was shown earlier we can split off a scalar part from the two component ZORA Hamiltonian. In the present work only the one component part is considered. We introduce a separate internal basis to represent the extra matrix elements, needed for the ZORA corrections. This leads to different options for the computation of the Coulomb matrix in this internal basis. The performance of this Hamiltonian and the effect of the different Coulomb matrix alternatives is tested in calculations on the radon en xenon atoms and the AuH molecule. In the atomic cases we compare with numerical Dirac Fock and numerical ZORA methods and with non relativistic and full Dirac basis set calculations. It is shown that ZORA recovers the bulk of the relativistic effect and that ZORA and Dirac Fock perform equally well in medium size basis set calculations. For AuH we have calculated the equilibrium bond length with the non relativistic Hartree Fock and ZORA methods and compare with the Dirac Fock result and the experimental value. Again the ZORA and Dirac Fock errors are of the same order of magnitude.

1. Introduction

The inclusion of relativistic effects is essential in quantum chemical studies of molecules containing heavy elements. A full relativistic calculation, i.e. based upon Quantum Electro Dynamics, is only feasible for the smallest systems. In the SCF approximation it involves the solution of the Dirac Fock equation. Due to the four component complex wave functions and the large number of basis functions needed to describe the small component Dirac spinors, these computations are much more demanding than the corresponding non-relativistic ones. This limits Dirac Fock calculations, which can be performed using e.g. the MOLFDIR package [1], to small molecular systems, UF_6 being a typical example, see e.g. [2].

There are many problems in e.g. catalysis in which relativity may play a deciding role in the chemical reactivity. These problems generally involve large organic molecules which cannot be handled within the Dirac Fock framework. It is therefore necessary to reduce the work by making additional approximations. Generally used approaches are based on the Pauli expansion or on the Douglas Kroll transformation [3].

In this paper we use a regular approximation of the Dirac Fock formalism known as

*To whom correspondence should be addressed.

A. Hernández-Laguna et al. (eds.), Quantum Systems in Chemistry and Physics, Vol. 1: Basic Problems and Model Systems, 251–261.

ZORA (Zeroth Order Regular Approximation). The ZORA or CPD method, which was first formulated by Chang, Pellisier and Durand [4] and Heully et al. [5], and then generalized by E. van Lenthe, Snijders and Baerends [6], within the framework of molecular Density Functional Theory, was previously investigated in an Ab Initio context [7] using the numerical Dirac Fock program GRASP2 [8]. It can be an attractive alternative, to a full Dirac Fock calculation, since it transforms the four component Dirac equation to a two component form. It was shown to recover a large part of the relativistic effect.

In this paper we present the first implementation of the ZORA formalism in a molecular Ab Initio program package (GAMESS-UK [9]). We show that the two component ZORA Hamiltonian can be split in a scalar (one component) part, which in practice only leads to a correction on the non-relativistic one electron operator, and a two component part which includes the spin orbit coupling. Here we consider only the one component part of the ZORA Hamiltionian. The implementation of the full two component Hamiltonian has recently been finished and results thereof will be presented in a forthcoming paper. The one component (scalar) ZORA method is tested on the xenon and radon atoms. We compare these results with numerical Dirac, ZORA and non-relativistic calculations and with non relativistic and full Dirac basis set calculations, taken from the MOLFDIR package [1]. Prelimanary results of the scalar ZORA method within a molecular context are given in a calculation on the AuH molecule.

2. Theory

It was previously shown [7] that one can, starting from the Dirac-Fock equation, derive the two component scaled ZORA equation

$$\frac{1}{\mathcal{S}}(\vec{\sigma}\cdot\vec{p}\mathrm{B}\vec{\sigma}\cdot\vec{p}+V_c-K_{\psi\psi})\psi_i=\varepsilon_i^{scaled}\psi_i \tag{1}$$

with the scaling factor

$$\mathcal{S}=1+\langle\psi_i|\vec{\sigma}\cdot\vec{p}\frac{1}{c^2}\mathrm{B}^2\vec{\sigma}\cdot\vec{p}|\psi_i\rangle \tag{2}$$

and the notations

$$V_c=V_{nuc}+J_{\psi\psi}\qquad\text{and}\qquad\mathrm{B}=\frac{c^2}{2c^2-V_c} \tag{3}$$

$J_{\psi\psi}$ and $K_{\psi\psi}$ are the usual Coulomb and exchange operators.

Equation (1) is obtained by using an expansion in $E/(2c^2-V_c)$ on the Dirac Fock equation. This expansion is valid even for a singular Coulombic potential near the nucleus, hence the name regular approximation. This is in contrast with the Pauli method, which uses an expansion in $(E-V)/2c^2$. Everything is written in terms of the two component ZORA orbitals, instead of using the large and small component Dirac spinors. This is an extra approximation with respect to the original formalism.

As in [9] the T^{zora} operator

$$\mathrm{T}^{zora}\equiv(\vec{\sigma}\cdot\vec{p})\mathrm{B}(\vec{\sigma}\cdot\vec{p}) \tag{4}$$

can be split in a scalar and a two-component (spin orbit) part. First we interchange the $(\vec{\sigma}\cdot\vec{p})$ and B operators, correcting with $[(\vec{\sigma}\cdot\vec{p}), \mathrm{B}] = -(i/c^2)\mathrm{B}^2\vec{\sigma}\cdot\vec{\nabla}V$. We then use the relation $(\vec{\sigma}\cdot\vec{p})^2 = p^2$ and bring one of the momentum operators back, again correcting with the corresponding commutator: $[\vec{p}, \mathrm{B}] = -(i/c^2)\mathrm{B}^2\vec{\nabla}V$. We now write

$$\mathrm{T}^{zora} \equiv (\vec{\sigma}\cdot\vec{p})\mathrm{B}(\vec{\sigma}\cdot\vec{p}) = \vec{p}\cdot\mathrm{B}\vec{p} + \frac{1}{c^2}\mathrm{B}^2\vec{\sigma}\cdot(\vec{\nabla}V_c \times \vec{p}) \tag{5}$$

where the following equation is used

$$(\vec{\sigma}\cdot\vec{\nabla}V)(\vec{\sigma}\cdot\vec{p}) = \vec{\nabla}V\cdot\vec{p} + i\vec{\sigma}\cdot(\vec{\nabla}V \times \vec{p}) \tag{6}$$

Using only the one component part of the T^{zora} operator we arrive at the scalar scaled ZORA equation

$$\frac{1}{\mathcal{S}}(\vec{p}\cdot\mathrm{B}\vec{p} + V_c - K_{\psi\psi})\psi_i = \varepsilon_i^{scaled}\psi_i \tag{7}$$

with the scalar scaling factor, where we have used similar approximations

$$\mathcal{S} = 1 + \langle\psi_i|\vec{p}\cdot\frac{1}{c^2}\mathrm{B}^2\vec{p}|\psi_i\rangle \tag{8}$$

3. Implementation in GAMESS-UK

The scalar ZORA method has been implemented in the standard non relativistic Ab Initio electronic structure program GAMESS-UK [8]. The technical details of this implementation will be given in the following section. Comparing the Schrödinger equation with the ZORA equation (7) one sees that application of the ZORA method has resulted in a potential dependent correction on the kinetic energy term.

$$\frac{1}{2}p^2 \rightarrow \vec{p}\cdot\frac{c^2}{2c^2 - V_c}\vec{p}$$

Schrődinger ZORA (9)

Thus, the kinetic energy matrix is replaced by the ZORA corrected kinetic energy matrix. Since this correction depends on the Coulomb matrix, and thus on the orbitals, it has to be evaluated every SCF cycle. The energy is now, again due to the occurence of the potential in the inverse operator, linear instead of quadratic in the orbitals, which may require tighter convergence criteria. One can now use all standard Ab Initio techniques to solve the ZORA equation by replacing the kinetic energy integrals with their ZORA corrected version. Note that the scaling factor does not affect the orbitals and can be applied after convergence of the SCF procedure.

3.1. Calculation of the matrix elements of the ZORA kinetic energy operator

Following the approach of [9] the matrix elements

$$(\mathrm{T}^{zora})_{\mu\nu} = \langle\phi_\mu|\vec{p}\cdot\frac{c^2}{2c^2 - V_c}\vec{p}|\phi_\nu\rangle \tag{10}$$

can be calculated by inserting resolutions of the identity

$$\sum_\kappa |\phi_\kappa\rangle \mathrm{S}_\phi^{-1}\langle\phi_\kappa| \tag{11}$$

where S_ϕ^{-1} is the inverse metric of the Gaussian basis $\{\phi\}$ which enters due to the non orthogonality of the basisfunctions. To recover the correct non-relativistic limit the kinetic energy is split off.

$$\begin{aligned}(T^{zora})_{\mu\nu} &= \frac{1}{2}\langle\phi_\mu|p^2|\phi_\nu\rangle - \frac{1}{2}\sum_\lambda\langle\phi_\mu|\vec{p}|\phi_\lambda\rangle \mathrm{S}_\phi^{-1}\langle\phi_\lambda|\vec{p}|\phi_\nu\rangle \\ &+ \frac{1}{2}\sum_{\lambda,\kappa}\langle\phi_\mu|\vec{p}|\phi_\lambda\rangle \mathrm{S}_\phi^{-1}\langle\phi_\lambda|1 - V_c/2c^2|\phi_\kappa\rangle^{-1}\mathrm{S}_\phi^{-1}\langle\phi_\kappa|\vec{p}|\phi_\nu\rangle\end{aligned} \tag{12}$$

where, in the last term, an internal projection, i.e. replacing the matrix of an inverse operator by the inverse matrix of the operator, is used. This is only exact in a complete basis. Also in a complete basis the first two terms cancel exactly. In the region where $V_c/2c^2$ is negligable, far away from the nuclei, the last two terms cancel.

The potential, $V_c = V_{nuc} + J_{\psi\psi}$, can be obtained from a non-relativistic fock matrix builder and the total matrix $\langle\phi_\lambda|1 - V_c/2c^2|\phi_\kappa\rangle^{-1}$ is subsequently calculated by a standard matrix inversion.

Since our Gaussian basis functions depend on the electronic (x_e) and nuclear (X_N) coordinates only in the special combination: $\phi = \phi(x_e - X_N)$, one can write

$$\langle\phi_\mu|\vec{p}|\phi_\lambda\rangle = -i\langle\phi_\mu|\frac{d}{dx_e}|\phi_\lambda\rangle = i\langle\phi_\mu|\frac{d}{dX_N}|\phi_\lambda\rangle \tag{13}$$

The matrix elements of the momentum operator in (12) are now written in terms of derivative overlap integrals which can be obtained from the gradient package. These integrals are standard available in Ab Initio programs capable of performing a geometry optimisation.

Now, we are able to calculate all ingredients of the ZORA corrections.

3.2. The Internal Basis

In the last term of (12), the basis used in the resolution of the identity and thus in the matrix inversion (the internal basis), is not necessarily the basis used in the electronic structure calculation (the external basis). The momentum operator in $\langle \mathrm{s}|\vec{p}|\mathrm{p}\rangle$ working on a high s exponent function will give a p-type function with the same (high) exponent. To be able to represent this matrix element correctly, one needs to have this high exponent p-function in the internal basis. In general each s type function will require a p function, a p function will require a s and a d function and so on. Thus we introduce a separate internal basis set in which these functions are included. The best solution would be the introduction of an internal basis set defined by $\vec{p} \cdot \{\chi\}$, where $\{\chi\}$ is the external basis. Since this would give a very large internal basis set, we have chosen to construct it in a different way. First all functions present in the external basis are copied to the internal basis. The internal basis is augmented with $(l+1)$-functions with all the exponents of l-functions larger than the highest $(l+1)$-exponent in the original external basis. All 1-electron parts of (12) are now computed in the internal basis, which has to be done only once, or only one time each geometry optimisation step.

3.3. *The Coulomb Matrix*

The internal basis can still be sizeable. Therefore the calculation of Coulomb matrix $(J_{\psi\psi})$, present in V_c, can be computationally very demanding. At the present time there are several possibilities implemented which will give different (in computational effort and accuracy) estimates of the Coulomb matrix.

- Complete neglect of the Coulomb operator in the inverse matrix (the bare nuclear option).
- Projection of the density matrix onto the internal basis, which is, due to the way the internal basis is constructed, no extra approximation, and build the full Coulomb matrix. A further approximation can be made by calculating only the intra-atomic components of this matrix, which is not expected to produce a serious loss in accuracy, since $V_c/2c^2$ is only large near the nuclei. We will refer to these alternatives as the full and atomic Coulomb ZORA option. Of course these two options are equivalent in the atomic calculations presented in the next section.
- A significantly cheaper, but theoretical not well justified, alternative is to calculate the Coulomb matrix in the smaller external basis and project it subsequently to the internal basis (the projected ZORA option)

$$\sum_{\kappa\lambda} \langle \phi_\mu | \chi_\kappa \rangle S_\chi^{-1} \langle \chi_\kappa | J | \chi_\lambda \rangle S_\chi^{-1} \langle \chi_\lambda | \phi_\nu \rangle \tag{14}$$

  where $\{\phi\}$ and $\{\chi\}$ represent the internal and external basis set respectively. The approximation here is the fact that the resolution of the identity in the external basis used in (14) is not an identity operator in the internal basis. This may produce an underestimation of the Coulomb matrix. Consider e.g. a situation where the external basis does not contain f functions but the internal basis does. This situation occurs if the external basis goes only up to d-functions and the internal basis is constructed as described in the previous section. In this case the overlaps between an internal f-function and all external basis functions will be equal to zero, and there will be no contribution to the internal Coulomb matrix.

Experience has shown that it is not necessary to update the Coulomb matrix (in the inverse operator) every SCF cycle. Therefore we have chosen to compute the internal Coulomb matrix with a direct scf fock matrix builder, thereby avoiding the use of large two electron integral files.

If we would have chosen to construct the internal basis by $\{\phi\} = \vec{p} \cdot \{\chi\}$ the $(1 - V_c/2c^2)$ matrix could have been written as

$$\langle \phi_\mu | (1 - V_c/2c^2) | \phi_\nu \rangle = \langle \vec{p} \cdot \chi_\mu | \vec{p} \cdot \chi_\nu \rangle - \frac{1}{2c^2} \langle \vec{p} \cdot \chi_\mu | V_{nuc} | \vec{p} \cdot \chi_\nu \rangle - \frac{1}{2c^2} \sum_{\lambda\lambda'} P_{\lambda\lambda'} \langle \vec{p} \cdot \chi_\mu | \langle \chi_\lambda | \frac{1}{r_{12}} | \chi_{\lambda'} \rangle | \vec{p} \cdot \chi_\nu \rangle \quad (15)$$

Again we rewrite the momentum operator as derivatives to nuclear coordinates and now all parts have been written in terms of second derivatives of overlap, nuclear potential and Coulomb matrices. These are available in the GAMESS-UK package. However, this option is, not yet, implemented.

It is interesting to note that the Coulomb matrix and the matrix of the nuclear potential present in V_c are opposite in sign. This means that an underestimation, or complete neglect, of the Coulomb matrix will lead to a larger V_c and thus to an overestimation of the relativistic effect. If V_c is negligable compared to $2c^2$ the ZORA equation reduces to the non relativistic Schrödinger equation.

4. Results

In order to test the various approximations of the Coulomb matrix, all electron basis set and numerical scalar scaled ZORA calculations have been performed on the xenon and radon atom. The numerical results have been taken from a previous publication [7], where it should be noted that the scalar orbital energies presented here are calculated by averaging, over occupation numbers, of the two component (i.e. spin orbit split) results. Tables (1) and (2) give the orbital energies for the numerical (s.o. averaged) and basis set calculations for the various Coulomb matrix approximations. The results from table

Table 1 *Xenon, comparison of orbital energies for numerical Dirac and ZORA and non relativistic calculations with basis set ZORA calculations in different Coulomb matrix approximations in the UGBS basis set*

Method	Dirac (num)	ZORA (num)	bare nucl.	proj. Coul.	full Coul.	Non Rel. (num)
1*s*	1277.4	1275.6	1285.9	1275.6	1275.8	1224.4
2*s*	202.48	202.27	203.90	202.18	202.32	188.44
2*p*	181.70	181.60	182.43	181.13	181.27	177.78
3*s*	43.013	42.974	43.316	42.927	42.990	40.176
3*p*	36.103	36.089	36.246	35.963	36.026	35.222
3*d*	25.731	25.728	25.814	25.853	25.737	26.119
4*s*	8.4305	8.4232	8.4868	8.4088	8.4270	7.8563
4*p*	6.1393	6.1376	6.1578	6.1092	6.1254	6.0083
4*d*	2.6647	2.6653	2.6647	2.6879	2.6670	2.7779
5*s*	1.0102	1.0094	1.0157	1.0078	1.0094	0.9444
5*p*	0.4574	0.4574	0.4564	0.4548	0.4555	0.4529

Table 2 *Xenon, comparison of orbital energies for numerical Dirac and ZORA and non relativistic calculations with basis set ZORA calculations in different Coulomb matrix approximations in the SV 3-21G basis set*

Method	Dirac (num)	ZORA (num)	bare nucl.	proj. Coul.	full Coul.	Non Rel. (num)
1*s*	1277.4	1275.6	1267.4	1265.7	1258.2	1224.4
2*s*	202.48	202.27	201.90	201.81	200.18	189.34
2*p*	181.70	181.60	182.93	182.13	181.34	177.78
3*s*	43.013	42.974	42.997	42.967	42.591	40.176
3*p*	36.103	36.089	36.248	36.107	35.899	35.222
3*d*	25.731	25.728	26.342	26.202	26.067	26.119
4*s*	8.4305	8.4232	8.4294	8.4198	8.3504	7.8563
4*p*	6.1393	6.1376	6.1305	6.1030	6.0715	6.0083
4*d*	2.6647	2.6653	2.7900	2.7737	2.7595	2.7779
5*s*	1.0102	1.0094	0.9974	0.9972	0.9911	0.9444
5*p*	0.4574	0.4574	0.4433	0.4424	0.4421	0.4580

(1) are produced with an Universal Gaussian Basis Set (UGBS) taken from Malli et al [10]. This is a large (26s22p17d) uncontracted basis set. Table (2) presents the same calculations within the standard non relativistic SV 3-21G basis set. Comparing the numerical non relativistic and spin orbit averaged scaled ZORA and Dirac results one sees that, except for the 1 s orbital, the scaled ZORA method is capable of recovering more than 98% of the relativistic effect. For the 1s the ZORA result is somewhat less accurate but even then the ZORA results corrects 96% of the error made by the non relativistic calculation. For a more extensive comparison between these methods we refer to our earlier paper [7].

Now, turning to the Univeral Gaussian Basis Set (UGBS) results, we can compare the different approximations of 'internal' Coulomb matrix. If we look at the projected and full Coulomb results we see that the projected option, being much cheaper than the full but lacking theoretical foundation, performs suprisingly well. We see that the full and projected Coulomb results are of the same quality in the core region, both recovering the bulk of the relativistic effect (compared with the Dirac values) and coming very close to the numerical ZORA results. However, if one looks at the valence orbitals, it is clear that the full Coulomb option is superior to the projected one. This is due to the 'missing' f orbitals in the internal-external overlap, as described in the basis set section. The bare-nuclear option is in this case clearly the most inaccurate. It constitutes an overestimation of the relativistic effect, in case of the 1s for as much as 20%. If one now looks at the SV 3-21G calculations it becomes apparent that this overcorrection compensates, at least in the core region, for the basis set defficiencies. In this case it may seem that the full Coulomb method is the most inaccurate one. This is, of course, an artefact of the use of a poor basis set.

It is interesting to compare the ZORA results with full Dirac basisset results. Since we want to use the ZORA formalism in molecular calculations we have chosen a basis set from Huzinaga [11], which is not quite as large as the UGBS but better than the SV 3-21G. This basis had to be adapted, since it was originally developed for non relativistic calculations, meaning that it is used in an uncontracted way and some functions had to be averaged due to the occurence of linear dependencies. We construct two basis sets by augmenting this basis with an high s exponent (basis I) and with 3 high s and 2 high p exponents (basis II). Table (3) lists the performance of the full Dirac (MOLFDIR) method in both basis sets. The performance of the projected and full Coulomb ZORA methods is given in Table (4). Note that the full Dirac basis set results are, again, averaged over the occupation numbers to give a spin free result. Looking at the different basis set results one sees that the error of the methods is of the same order of magnitude (comparing with the corresponding numerical results) except for the 2p orbital. In this case both ZORA methods give a considerably worse result compared with MOLFDIR, although even here ZORA gives a large improvement on the non relativistic result. Examining the effect of the extra functions we can see that the inclusion of the extra s functions produced a considerable improvement of the 1s and 2s orbitals (with about 30 and 4 a.u.). However, the addition of basis functions did not solve the 2p problem. The two component ZORA Hamiltonian, the implementation of which we have only recently finished, gives, in the same basis, $2p_{1/2}$: 639.28 a.u. and $2p_{3/2}$:540.96 a.u., which averages to 573.73 a.u.. This deviation is of the same order in magnitude as for the full Dirac results. In this case we see that the spin averaged two component calculation and the scalar ZORA method, who should produce the same result in first order perturbation theory, do not give the same result as the scalar ZORA method. The spin orbit splitting of the radon 2p orbital is clearly to large. We will adress

Table 3 *Radon, comparison of orbital energies for numerical Dirac and non relativistic calculations with Molfdir calculations in basis set I & II*

Method	Dirac (num)	Molfdir basis I	Moldir basis II	Non-Rel. (num)
$1s$	3644.8	3602.1	3635.7	3230.3
$2s$	669.39	664.11	668.07	556.91
$2p$	574.84	574.41	575.25	536.68
$3s$	166.97	166.54	165.22	138.42
$3p$	139.45	140.61	139.64	128.68
$3d$	109.68	111.72	109.90	110.70
$4s$	41.349	40.802	40.884	33.921
$4p$	32.086	32.879	32.217	29.491
$4d$	20.881	20.385	20.994	21.331
$4f$	9.0409	10.056	9.1373	10.108
$5s$	8.4168	7.8887	8.0911	6.9058
$5p$	5.5864	5.7515	5.2303	5.2252
$5d$	2.0854	1.9299	2.1061	2.3263
$6s$	1.0727	0.9350	0.9697	0.8740
$6p$	0.4360	0.4484	0.4412	0.4280

Table 4 *Radon, comparison of numerical ZORA orbital energies with results from project and full Colomb ZORA calculations in the basis sets I & II*

Method	ZORA (num)	ZORA proj. Coul basis I	ZORA proj. Coul. basis II	ZORA full Coul. basis I	ZORA full Coul. basis II
1*s*	3635.7	3604.0	3633.9	3597.1	3634.0
2*s*	668.07	664.56	668.32	663.55	668.49
2*p*	574.31	568.66	567.46	568.51	567.24
3*s*	168.13	166.78	167.01	166.56	167.08
3*p*	139.34	139.52	138.11	139.49	138.05
3*d*	109.64	111.95	110.17	111.87	110.10
4*s*	41.278	40.902	41.259	40.852	41.281
4*p*	32.278	32.644	31.860	32.636	31.851
4*d*	20.877	20.472	21.098	20.453	21.073
4*f*	9.0449	10.032	9.1276	10.047	9.1441
5*s*	8.4025	7.9132	8.1490	7.9072	8.1550
5*p*	5.5838	5.7057	5.5656	5.7053	5.5654
5*d*	2.0865	1.9438	2.1300	1.9417	2.1268
6*s*	1.0707	0.9389	0.9776	0.9384	0.9783
6*p*	0.43605	0.4418	0.4327	0.4418	0.4327

this problem in a future paper where we will investigate the performance of the full, two component, ZORA Hamiltonian. The difference in accuracy between the projected and full Coulomb ZORA method is not noticeable anymore. This is a result of the fact that the difference between internal and external basis is very small (g-functions are, at this moment, not available in GAMESS-UK).

Finally we have tested the scalar ZORA Hamiltonian in a calculation on the AuH molecule. On Au we use a basis set from Huzinaga [11] uncontracted and augmented with, even tempered generated, high s and a high p exponents. On H we use the standard SV 3-21G basis set. The bondlenghts for AuH are calculated with the non relativistic HF, the atomic and full ZORA methods. The DHF results in table (5) is taken from [13]. Comparing the non relativistic and ZORA methods we see a bond length contraction of 0.3 Å. The ZORA results are in good agreement with experiment. The ZORA methods

Table 5 *Comparison of equilibrium bondlengths for non relativistic ZORA and Dirac Fock calculations with the experimental value*

Method	R_e (angstrom)
Non Rel. HF	1.78
ZORA (atom)	1.59
ZORA (full)	1.56
DHF	1.57
Exp.	1.52

predict a bond length which is somewhat too large. The error in the DHF result is of the same order of magnitude. The DHF bond is too long. In the case of spin orbit coupled calculations on AuH, it has been shown that correlation will shorten the bond [13] and [14]. We see, at least in the calculation of R_e, no significant difference between the atomic and full Coulomb ZORA options.

5. Conclusion

We have presented an implementation of the scaled ZORA method in GAMESS-UK, which results in an one component spin free formalism. To overcome problems with the representation of the extra matrix elements, needed to compute the ZORA corrected one electron integrals, we introduce a separate internal basis set, which allows different approximations for the Coulomb matrix in this basis. The scalar scaled ZORA Hamiltonian has been tested in an Ab Initio SCF context on the xenon and radon atoms, for the different Coulomb matrix options and the results are compared with numerical non-relativistic, ZORA and Dirac calculations and it recovers a large part of the relativistic effect. Furthermore the ZORA method is compared with the full Dirac basis set results from the MOLFDIR package and one can conclude that, in a basis commonly used in molecular calculations the ZORA methods perform just as well as the four component Dirac Fock calculation. We performed calculations on the bond length of AuH. It was calculated with the full and atomic Coulomb ZORA option, and both are in equally good agreement with experiment. This justifies, at least in this case, the neglect of the inter-atomic Coulomb components in the ZORA kinetic energy operator.

Acknowledgements

We would like to thank Professor I.P. Grant and Professor Marketos for providing us with the GRASP2 [12] package, Professor W.C. Nieuwpoort and Dr. W.A. de Jong for making the MOLFDIR program available to us and Arno Blok for performing some of the calculations presented here.

References

1. L. Visscher, O. Visser, P.J.C. Aerts, H. Merenga and W.C. Nieuwpoort, *Comp. Phys. Comm.*, **81** (1994) 120.
2. G.L. Malli and J. Styszynski, *J. Chem. Phys.*, **104** (1996) 3.
3. M. Douglas and N.M. Kroll, *Ann. Phys.*, **82** (1974) 89.
4. Ch. Chang, M. Pelissier and Ph. Durand, *Phys. Scr.*, **34** (1986) 394.
5. J-L. Heully, I. Lindgren, E. Lindroth, S. Lundquist, and A-M. Mårtensson-Pendrill, *J. Phys. B,* **19** (1986) 2799.
6. E. van Lenthe, E.J. Baerends and J.G. Snijders, *J. Chem. Phys.* **99**, (1993) 4597.
7. S. Faas, J.G. Snijders, J.H. van Lenthe, E. van Lenthe and E.J. Baerends, *Chem. Phys. Let.*, **246** (1995) 632.
8. GRASP^2 F.A. Parpia, I.P. Grant and C.F. Fischer, *Comp. Phys. Comm.*, **94** (1996) 249.
9. GAMESS-UK is a package of ab initio programs written by M.F. Guest, J.H. van Lenthe, J. Kendrick, K. Schoffel, P. Sherwood, and R.J. Harrison, with contributions from R.D. Amos, R.J. Buenker, M. Dupuis, N.C. Handy, I.H. Hillier, P.J. Knowles, V. Bonacic-Koutecky, W. von Niessen, V.R. Saunders, and A.J.

Stone. The package is derived from the original GAMESS code due to M. Dupuis, D. Spangler and J. Wendoloski, *NRCC Software Catalog,* **Vol. 1**, Program No. QG01 (GAMESS) (1980).
10. E. van Lenthe, R. van Leeuwen, E.J. Baerends and J.G. Snijders, *Int. J. Quant. Chem.,* **57** (1996) 281–293.
11. G.L. Malli, J. Styszynski and A.B.F. Da Silva, *Int. J. Quant. Chem,* **55** (1995) 213–225.
12. S. Huzinaga, J. Andzelm, M. Klobukowski, E. Radzio-Andzelm, Y. Sakai and H. Tatewaki, *Gaussian basis sets for molecular calculations* Elsevier, Amsterdam (1994).
13. T. Saue, L. Visscher, K. Faegri Jr, O. Gropen and W. C Nieuwpoort, personal communication.
14. C.L. Collins, K.G. Dyall, H.F. Schaefer III, *J. Chem. Phys.,* **102** (1995) 5.

Relativistic Oscillator Strengths for Excited-State Transitions in Halogen Atoms. Regularities

C. Lavín[a], A.M. Velasco[b] and I. Martín[a]

[a]*Departamento de Química Física, Facultad de Ciencias, Universidad de Valladolid, E-47005 Valladolid, Spain*

[b]*Istituto di Chimica Quantistica ed Energetica Molecolare del Consiglio Nazionale delle Ricerche, Via Risorgimento 35, I-56126 Pisa, Italy*

Abstract

Relativistic Quantum Defect Orbital (RQDO) calculations, with and without explicit account for core-valence correlation, have been performed on several electronic transitions in halogen atoms, for which transition probability data are particularly scarce. For the atomic species iodine, we supply the only available oscillator strengths at the moment. In our calculations of *f*-values we have followed either the *LS* or $J_c\,l$ coupling schemes.

1. Introduction

In recent years, the description, understanding, and prediction of the physical processes that take place in the universe have been the object of increasing attention. In this context, neutral atomic chlorine is an important constituent of the interstellar medium and reliable estimates of oscillator strengths and transition probabilities can be used to constrain models of interstellar clouds [1,2]. However, the reproduction in the laboratory of the very extreme conditions to which the constituents of the universe are subjected is, if not impossible, at least extremely difficult, and reliable computational procedures are, thus, required. Iodine and bromine are very frequently used in electrodeless metal-halide lamps serving as light sources for the study of rare-earth and other metallic spectra. Since the spectrum of both the metal and the halogen will appear when such a lamp is excited, it is essential that the user has available a complete and accurate description of the spectrum of the halogen in order to separate the halogen lines from those of the metal under investigation [3,4].

Spectral lines are often characterized by their wavelength and intensity. The line intensity is a source-dependent quantity, but it is related to an atomic constant, the transition probability or oscillator strength. Transition probabilities are known much less accurately than wavelengths. This imbalance is mainly due to the complexity of both theoretical and experimental approaches to determine transition probability data. Detailed descriptions of the spectra of the halogens have been made by Radziemski and Kaufman [5] for *Cl I*, by Tech [3] for *Br I* and by Minnhagen [6] for *I I*. However, the existing data on *f*-values for those atomic systems are extremely sparse.

The analysis [5,7] of the *F I* and *Cl I* spectra shows that *LS* coupling is valid for the deepest $2p^4\,ns$, np and $3p^4\,ns$, np configurations, respectively. The observed [3] distribution of the levels in each of the *Br I nl* configurations indicates that the coupling energies of the $4p^4$ core dominate the structure of the *Br I* levels. In all configurations involved in this work one might say that the coupling is 'midway between' the *LS*, $J_c\,j$

A. Hernández-Laguna et al. (eds.), Quantum Systems in Chemistry and Physics, Vol. 1: Basic Problems and Model Systems, 263–272.

and $J_c l$ coupling schemes. The result of the coupling studies [6] for $5p^4 ns$, np configurations in *I I* reveals that the best characterization of *ns* is obtained by means of the $J_c l$ or $J_c j$ schemes, which are identical for *ns*. For *np* the coupling is intermediate between $J_c l$ and $J_c j$ but closer to $J_c l$.

In this work, we present oscillator strengths for the lines corresponding to allowed transitions between excited states in neutral fluorine, chlorine, bromine and iodine. The calculations have been carried out with the Relativistic Quantum Defect Orbital (RQDO) procedure [8]. In our calculations of oscillator strengths we have followed the *LS* coupling scheme for *F I* and *Cl I* and both the *LS* and $J_c l$ coupling schemes for *Br I* and *I I*. No comparative data from other sources have been found in the literature for many of the fine structure transitions of these atomic systems. However, the regular behaviour of oscillator strengths has proven to be a useful tool for analysing *f*-value data [9].

One of the principal regularities concerns homologous atoms, i.e., atoms with the same outer electron structure [10]. It has been found that, for certain analogous groups of spectral lines, the *f*-value remains approximately constant throughout a family of homologous atoms. The RQDO results show that multiplets and individual lines for transitions in halogen atoms remain the same as long as the coupling scheme remains constant.

2. Method of Calculation

The absorption oscillator strength for an electric dipole transition between an initial state $|i\rangle$ and a final state $|j\rangle$ is given by

$$f_{ij} = \frac{2}{3}\frac{(E_j - E_i)}{2J + 1} S \tag{1}$$

where $(E_j - E_i)$ is the transition energy (in atomic units), $2J + 1$ is the degeneracy of initial level and $\boldsymbol{S}$ is the so-called line strength (in atomic units). In a one-configuration treatment, such as ours, the angular and radial parts of the line strength can be separated. Within the *LS* coupling, the line strength is written, as follows:

$$\boldsymbol{S}(\gamma \boldsymbol{S} LJ, \gamma' \boldsymbol{S}' L' J')$$

$$= \left| \delta_{SS} \delta_{cc'} \boldsymbol{R}_{lin}(\boldsymbol{S}LJ, \boldsymbol{S}'L'J') \boldsymbol{R}_{multi}(\gamma L, \gamma' L')(-1)^{1-l_>} \sqrt{l_>} \int \boldsymbol{R}_{nl}(r) \boldsymbol{R}_{n'l'}(r) r\, dr \right|^2 \tag{2}$$

where J and J', L and L', S and S', and γ and γ' represent the quantum numbers required to complete the specification of the states $|i\rangle$ and $|j\rangle$, respectively. The subscripts c and c' refer to the core. The integral in (2) contains the radial parts of the initial and final wavefunctions, only. The line factor, R_{lin}, is given by the expression:

$$\boldsymbol{R}_{lin} = \sqrt{2J+1}\sqrt{2J'+1}(-1)^{S+J+L'+1} \begin{Bmatrix} \boldsymbol{S} & J & L \\ 1 & L' & J' \end{Bmatrix} \tag{3}$$

where the symbol in braces is a *6-j* symbol. $(R_{lin})^2$ gives the relative strength of the lines within a multiplet. In (2), nl and $n'l'$ are, respectively, the principal and orbital angular momentum quantum numbers of the jumping electron and $l_>$ indicates the larger of the two orbital quantum numbers, l and l', involved in the transition.

The multiplet factor, R_{mult}, depends upon the two particular configurations involved in the transition. For a $3s^2 3p^4 nl$-$3s^2 3p^4 n'l'$ the multiplet factor adopts the form,

$$R_{mult} = \sqrt{2L+1}\sqrt{2L'+1}(-1)^{L_c+L+l'+1}\begin{Bmatrix} L_c & l & L \\ 1 & L' & l' \end{Bmatrix} \tag{4}$$

The expression for the line strength in the $J_c l$ coupling scheme is the following,

$$\boldsymbol{S} = (2J+1)(2J'+1)$$

$$\times \begin{Bmatrix} K & \frac{1}{2} & J \\ J' & 1 & K' \end{Bmatrix}^2 (2K+1)(2K'+1)\begin{Bmatrix} J_c & 1 & K \\ 1 & K' & l' \end{Bmatrix}^2 l_> \left| \int R_{nl'} R_{n'l'} r \, dr \right|^2 \tag{5}$$

where J_c is the total angular momentum quantum number of the electronic core in the $3p^4\ ^3P$ state; K results from the coupling of the orbital angular momentum l of the outer electron with J_c, and J is obtained through the addition of the spin of the valence electron to K.

The relativistic version (RQDO) of the quantum defect orbital formalism has been employed to obtain the wavefunctions required to calculate the radial transition integral. The relativistic quantum defect orbitals corresponding to a state characterized by its experimental energy E^x are the analytical solutions of the quasirelativistic second-order Dirac-like equation [8]

$$\left[-\frac{d^2}{dr^2} + \frac{\Lambda(\Lambda+1)}{r^2} - \frac{2Z'_{net}}{r}\right]\Psi_k^{RD} = 2e^{RD}\Psi_k^{RD}, \tag{6}$$

where the parameter Λ, which accounts for polarization and penetration effects of the valence electron, is related to the relativistic principal quantum number, $\tilde{n}$, the principal and orbital angular momentum quantum numbers, n and l, respectively, the relativistic quantum defect, δ', and an integer, c, chosen to ensure the normalizability of the wavefunction and its correct nodal structure

$$\Lambda = \tilde{n} - n + l - \delta' + c, \tag{7}$$

Z'_{net} is the scaled Z_{net}, the screened nuclear charge at large radial distances,

$$Z'_{net} = Z_{net}(1 + \alpha^2 E^x), \tag{8}$$

with α being the fine-structure constant. To determine the relativistic quantum defect, the following expression is employed,

$$e^{RD} = -\frac{(Z'_{net})^2}{2(\tilde{n} - \delta')^2} = E^x\frac{(1 + \alpha^2 E^x/2)}{(1 + \alpha^2 E^x)^2}. \tag{9}$$

The relativistic quantum defect orbitals lead to closed-form analytical expressions for the transition integrals. This allows us to calculate transition probabilities and oscillator strengths by simple algebra and with little computational effort.

As previously done [11], we have employed two differents forms of the transition operator, $Q(r) = r$, and a core-polarization corrected expression [12], given by:

$$Q(r) = r\left\{1 - \frac{\alpha_c}{r^3}[1 - \exp(-r/r_c)]^3\right\} \tag{10}$$

where α_c is the core-polarizability and r_c is a cutoff radius.

We have carried out calculations for transitions for which empirical energy data were available [3,5,6,7]. The core polarizabilities have been take from Fraga et al. [13]. Since there is no analytical way of obtaining the cutoff radius r_c, we have chosen a value equal to the core mean radius, calculated in accord with an expression given by Chichkov & Shevelko [14].

Table 1 *Oscillator strengths for the $3p^4(^3P)3s$-$3p^4(^3P)3p$ and $3p^4(^3P)3p$-$3p^4(^3P)4s$ fine structure transitions in F I*

Transition	RQDO[a]	RQDO[b]	CC[a]	Expt[b]
$2p^4(^3P)3s$-$2p^4(^3P)3p$ transitions				
$^4P_{5/2}$-$^4P_{5/2}$	0.1915	0.1861	0.21	0.249
$^4P_{5/2}$-$^4P_{3/2}$	0.0824	0.0801	0.089	0.220
$^4P_{3/2}$-$^4P_{5/2}$	0.1229	0.1194	0.13	0.10
$^4P_{3/2}$-$^4P_{3/2}$	0.0366	0.0356	0.039	
$^4P_{3/2}$-$^4P_{1/2}$	0.1146	0.1115	0.12	0.145
$^4P_{5/2}$-$^4D_{7/2}$	0.3741	0.3649	0.42	0.376
$^4P_{5/2}$-$^4D_{5/2}$	0.0844	0.0823	0.095	0.061
$^4P_{5/2}$-$^4D_{3/2}$	0.0094	0.0092	0.011	0.0072
$^4P_{3/2}$-$^4D_{5/2}$	0.2958	0.2887	0.33	0.313
$^4P_{3/2}$-$^4D_{3/2}$	0.1505	0.1470	0.17	0.120
$2p^4(^3P)3p$-$2p^4(^3P)4s$ transitions				
$^4P_{5/2}$-$^4P_{5/2}$	0.1832	0.1836		
$^4P_{5/2}$-$^4P_{3/2}$	0.0756	0.0759		
$^4P_{3/2}$-$^4P_{5/2}$	0.1189	0.1191		
$^4P_{3/2}$-$^4P_{3/2}$	0.0340	0.0341		
$^4P_{1/2}$-$^4P_{3/2}$	0.2141	0.2147		
$^4D_{7/2}$-$^4P_{5/2}$	0.2816	0.2817		
$^4D_{5/2}$-$^4P_{5/2}$	0.0854	0.0854		
$^4D_{5/2}$-$^4P_{3/2}$	0.1941	0.1944		
$^4D_{3/2}$-$^4P_{5/2}$	0.0143	0.0143		
$^4D_{3/2}$-$^4P_{3/2}$	0.1493	0.1495		

[a]RQDO this work.
[b]RQDO this work with polarization.
[c]Critical Compilation, Wiese et al. [15].
[d]Bengtson et al. [16].

3. Results

In Tables 1–4, we report oscillator strengths for some fine structure transitions in neutral fluorine, chlorine, bromine and iodine, respectively. Two sets of RQDO *f*-values are shown, those computed with the standard dipole length operator $Q(r) = r$, and those where core-valence correlation has been explicitly introduced, Eq. (10). As comparative data, we have included in the tables *f*-values taken from critical compilations [15,18], results of length and velocity *f*-values by Ojha and Hibbert [17], who used large configuration expansions in the atomic structure code CIV3, and absolute transition probabilities measured through a gas-driven shock tube by Bengtson et al. converted

Table 2 *Oscillator strengths for the* $3p^4(^3P)4s$-$3p^4(^3P)4p$ *and* $3p^4(^3P)4p$-$3p^4(^3P)5s$ *fine structure transitions in Cl I*

Transition	RQDO[a]	RQDO[b]	CIV3(f_l)[c]	CIV3(f_v)[c]	CC[d]	Exp[e]
$3p^4(^3P)4s$-$3p^4(^3P)4p$ transitions						
$^4P_{5/2}$-$^4P_{5/2}$	0.2053	0.1895	0.3355	0.2701	0.220	
$^4P_{5/2}$-$^4P_{3/2}$	0.0887	0.0821	0.1636	0.1307	0.095	
$^4P_{5/2}$-$^4P_{5/2}$	0.1311	0.1210	0.0870	0.0727	0.140	
$^4P_{3/2}$-$^4P_{3/2}$	0.0392	0.0363	0.0629	0.0516	0.042	
$^4P_{3/2}$-$^4P_{1/2}$	0.1239	0.1148	0.1773	0.1442	0.130	
$^4P_{5/2}$-$^4D_{7/2}$	0.4032	0.3754	0.5392	0.4092	0.390	
$^4P_{5/2}$-$^4D_{5/2}$	0.0911	0.0850	0.0468	0.0351	0.080	0.056
$^4P_{5/2}$-$^4D_{3/2}$	0.0102	0.0095	0.0006	0.0040	0.010	
$^4P_{3/2}$-$^4D_{5/2}$	0.3197	0.2982	0.3818	0.2987	0.310	1.600
$^4P_{3/2}$-$^4D_{3/2}$	0.1633	0.1528	0.1799	0.1374	0.160	0.093
$3p^4(^3P)4p$-$3p^4(^3P)5s$ transitions						
$^4P_{5/2}$-$^4P_{5/2}$	0.1804	0.1804	0.1931	0.0500		
$^4P_{5/2}$-$^4P_{3/2}$	0.0718	0.0712	0.0157	0.0038		
$^4P_{3/2}$-$^4P_{5/2}$	0.1184	0.1183	0.1372	0.0374		
$^4P_{3/2}$-$^4P_{3/2}$	0.0327	0.0328	0.0130	0.0030		
$^4P_{3/2}$-$^4P_{1/2}$	0.0902	0.0911	0.0706	0.0195		
$^4P_{1/2}$-$^4P_{3/2}$	0.2102	0.2106	0.0940	0.0251		
$^4D_{7/2}$-$^4P_{5/2}$	0.2819	0.2807	0.2641	0.0981		
$^4D_{5/2}$-$^4P_{5/2}$	0.0861	0.0857	0.0292	0.0115		
$^4D_{5/2}$-$^4P_{3/2}$	0.1912	0.1909	0.0051	0.0008		
$^4D_{3/2}$-$^4P_{5/2}$	0.0147	0.0146	0.0005	0.0002		
$^4D_{3/2}$-$^4P_{3/2}$	0.1504	0.1500	0.0097	0.0022		

[a] RQDO without explicit polarization correction, this work.
[b] RQDO with explicit polarization correction, this work.
[c] Ojha & Hibbert, length and velocity forms [17].
[d] Critical compilation, Wiese et al. [18].
[e] Bengtson et al. [16].

Table 3 *Oscillator strengths for the $4p^4(^3P)5s$-$4p^4(^3P)5p$ and $4p^4(^3P)5p$-$4p^4(^3P)6s$ fine structure transitions in Br I*

Transition		*LS*-coupling		$J_c\,l$-coupling		
		RQDO[a]	RQDO[b]	RQDO[a]	RQDO[b]	Expt[c]
$4p^4(^3P)5s$-$4p^4(^3P)5p$ transitions						
LS notation	$J_c\,l$ notation					
$^4P_{5/2}$-$^4P_{5/2}$	$[2]_{5/2}$-$[2]_{5/2}$	0.2124	0.1919	0.2832	0.2433	
$^4P_{5/2}$-$^4P_{3/2}$	$[2]_{5/2}$-$[2]_{3/2}$	0.0918	0.0833	0.0204	0.0173	0.0723
$^4P_{3/2}$-$^4P_{5/2}$	$[2]_{3/2}$-$[2]_{5/2}$	0.1348	0.1219	0.0299	0.0258	
$^4P_{3/2}$-$^4P_{3/2}$	$[2]_{3/2}$-$[2]_{3/2}$	0.0406	0.0368	0.1219	0.1053	
$^4P_{3/2}$-$^4P_{1/2}$	$[2]_{3/2}$-$[1]_{1/2}$	0.1304	0.1193	0.1565	0.1369	
$^4P_{5/2}$-$^4D_{7/2}$	$[2]_{5/2}$-$[3]_{7/2}$	0.4086	0.3717	0.4086	0.3717	0.16
$^4P_{5/2}$-$^4D_{5/2}$	$[2]_{5/2}$-$[3]_{5/2}$	0.0920	0.0839	0.0205	0.0187	0.0096
$^4P_{5/2}$-$^4D_{3/2}$	$[2]_{5/2}$-$[1]_{3/2}$	0.0101	0.0093	0.1819	0.1682	0.0654
$^4P_{3/2}$-$^4D_{5/2}$	$[2]_{3/2}$-$[3]_{5/2}$	0.3253	0.2964	0.4337	0.3952	
$^4P_{3/2}$-$^4D_{3/2}$	$[2]_{3/2}$-$[1]_{3/2}$	0.1674	0.1544	0.0314	0.0289	0.1251
$4p^4(^3P)5p$-$4p^4(^3P)6s$ transitions						
$^4P_{5/2}$-$^4P_{5/2}$	$[2]_{5/2}$-$[2]_{5/2}$	0.1961	0.1947	0.2572	0.2560	
$^4P_{5/2}$-$^4P_{3/2}$	$[2]_{5/2}$-$[2]_{3/2}$	0.0782	0.0782	0.0169	0.0170	
$^4P_{3/2}$-$^4P_{5/2}$	$[2]_{3/2}$-$[2]_{5/2}$	0.1294	0.1283	0.0283	0.0282	
$^4P_{3/2}$-$^4P_{3/2}$	$[2]_{3/2}$-$[2]_{3/2}$	0.0360	0.0359	0.1055	0.1055	
$^4P_{1/2}$-$^4P_{3/2}$	$[1]_{1/2}$-$[2]_{3/2}$	0.2413	0.2397	0.2855	0.2842	
$^4D_{7/2}$-$^4P_{5/2}$	$[3]_{7/2}$-$[2]_{5/2}$	0.2947	0.2920	0.2947	0.2920	
$^4D_{5/2}$-$^4P_{5/2}$	$[3]_{5/2}$-$[2]_{5/2}$	0.0894	0.0885	0.0199	0.0197	
$^4D_{5/2}$-$^4P_{3/2}$	$[3]_{5/2}$-$[2]_{3/2}$	0.1978	0.1970	0.2637	0.2626	
$^4D_{3/2}$-$^4P_{5/2}$	$[1]_{3/2}$-$[2]_{5/2}$	0.0156	0.0154	0.2811	0.2771	
$^4D_{3/2}$-$^4P_{3/2}$	$[1]_{3/2}$-$[2]_{3/2}$	0.1628	0.1613	0.0305	0.0302	

[a]RQDO this work.
[b]RQDO this work with polarization.
[c]Bengtson et al. [16].

into oscillator strengths. In Table 5, we compare our results for the oscillator strengths of the multiplet transitions in halogen atoms with *f*-values calculated within the Opacity Project (OP) [19] and with results of critical compilations [15, 18]. The systematic trends of the multiplet oscillator strengths for analogous transitions in homologous atoms are shown graphically in Figures 1 and 2. In our calculations of oscillator strengths we have followed the *LS* coupling scheme for *F I* and *Cl I* and both the *LS* and $J_c\,l$ coupling schemes for *Br I* and *I I*. The level designation in the $J_c\,l$ scheme is given according to the notation $np^4(^{2S+1}L_{Jc})n'l[K]_J$.

An analysis of the tables leads us to point out the following main features:

- An overall good agreement is found between the RQDO *f*-values, recommended *f*-values by Wiese et al. [15] and the experimental [16] results existing for

Table 4 *Oscillator strengths for the* $5p^4(^3P)6s$-$5p^4(^3P)6p$ *and* $5p^4(^3P)6p$-$5p^4(^3P)7s$ *fine structure transitions in I I*

Transition		*LS*-coupling		$J_c\,l$-coupling	
		RQDO[a]	RQDO[b]	RQDO[a]	RQDO[b]
$5p^4(^3P)6s - 5p^4(^3P)6p$ transitions					
LS notation	$J_c\,l$ notation				
$^4P_{5/2}$-$^4P_{5/2}$	$[2]_{5/2}$-$[2]_{5/2}$	0.2197	0.1925	0.2930	0.2566
$^4P_{5/2}$-$^4P_{3/2}$	$[2]_{5/2}$-$[2]_{3/2}$	0.0944	0.0828	0.0210	0.0184
$^4P_{3/2}$-$^4P_{5/2}$	$[2]_{3/2}$-$[2]_{5/2}$	0.1390	0.1219	0.0309	0.0271
$^4P_{3/2}$-$^4P_{3/2}$	$[2]_{3/2}$-$[2]_{3/2}$	0.0414	0.0364	0.1242	0.1091
$^4P_{3/2}$-$^4P_{1/2}$	$[2]_{3/2}$-$[1]_{1/2}$	0.1345	0.1195	0.1614	0.1434
$^4P_{5/2}$-$^4D_{7/2}$	$[2]_{5/2}$-$[3]_{7/2}$	0.4249	0.3765	0.4249	0.3765
$^4P_{5/2}$-$^4D_{5/2}$	$[2]_{5/2}$-$[3]_{5/2}$	0.0956	0.0847	0.0212	0.0188
$^4P_{5/2}$-$^4D_{3/2}$	$[2]_{5/2}$-$[1]_{3/2}$	0.0104	0.0094	0.1864	0.1690
$^4P_{3/2}$-$^4D_{5/2}$	$[2]_{3/2}$-$[3]_{5/2}$	0.3365	0.2981	0.4489	0.3976
$^4P_{3/2}$-$^4D_{3/2}$	$[2]_{3/2}$-$[1]_{3/2}$	0.1737	0.1571	0.0326	0.0295
$5p^4(^3P)6p$-$5p^4(^3P)7s$ transitions					
$^4P_{5/2}$-$^4P_{5/2}$	$[2]_{5/2}$-$[2]_{5/2}$	0.1939	0.1920	0.2586	0.2560
$^4P_{5/2}$-$^4P_{3/2}$	$[2]_{5/2}$-$[2]_{3/2}$	0.0766	0.0765	0.0170	0.0170
$^4P_{3/2}$-$^4P_{5/2}$	$[2]_{3/2}$-$[2]_{5/2}$	0.1257	0.1244	0.0279	0.0276
$^4P_{3/2}$-$^4P_{3/2}$	$[2]_{3/2}$-$[2]_{3/2}$	0.0344	0.0343	0.1032	0.1030
$^4P_{1/2}$-$^4P_{3/2}$	$[1]_{1/2}$-$[2]_{3/2}$	0.2372	0.2350	0.2846	0.2820
$^4D_{7/2}$-$^4P_{5/2}$	$[3]_{7/2}$-$[2]_{5/2}$	0.2964	0.2919	0.2964	0.2919
$^4D_{5/2}$-$^4P_{5/2}$	$[3]_{5/2}$-$[2]_{5/2}$	0.0887	0.0874	0.0197	0.0194
$^4D_{5/2}$-$^4P_{3/2}$	$[3]_{5/2}$-$[2]_{3/2}$	0.1949	0.1935	0.2599	0.2579
$^4D_{3/2}$-$^4P_{5/2}$	$[1]_{3/2}$-$[2]_{5/2}$	0.0158	0.0155	0.2852	0.2788
$^4D_{3/2}$-$^4P_{3/2}$	$[1]_{3/2}$-$[2]_{3/2}$	0.1664	0.1636	0.0312	0.0307

[a]RQDO this work.
[b]RQDO this work with polarization.

individual lines of the $3p^4$ (^3P) $3s$-$3p^4$ (^3P) $3p$ transitions in *F I*, particularly if we take into account that Bengtson et al. [16] report uncertainties in their values of 20–50% (Table 1).

- The RQDO *f*-values conform with the recommended values by Wiese et al. [18] for the $3p^4$ (^3P) $4s$-$3p^4$ (^3P) $4p$ transitions in *Cl I* better than those of the much more complex theoretical procedure of Ojha and Hibbert [17], who used large multiconfiguration expansions in the atomic structure code CIV3. On the other hand, the discrepancies between the length and velocity CIV3 oscillator strengths are not negligible. Wiese et al. [18] remark that, for the case of the $4s\ ^4P$-$4p\ ^4D$ multiplet, the stronger lines measured by Bengtson et al. [16] seem to be affected by self-absorption (Table 2).

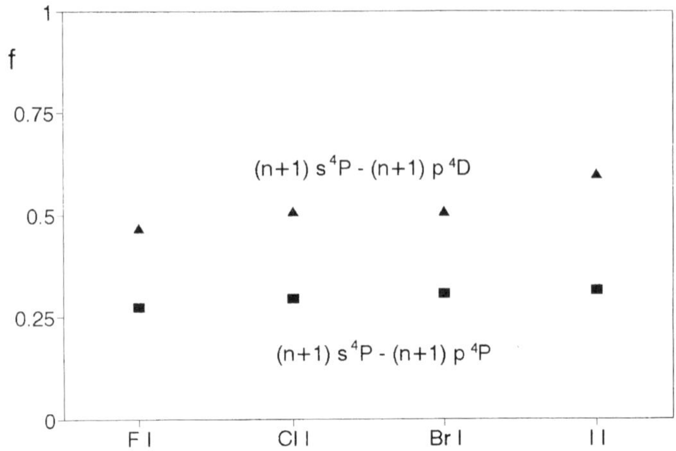

Fig. 1. Oscillator strengths for the ^4P-^4P and ^4P-^4D multiplets of the np^4 $(n+1)s$-np^4 $(n+1)p$ transition arrays of halogen atoms.

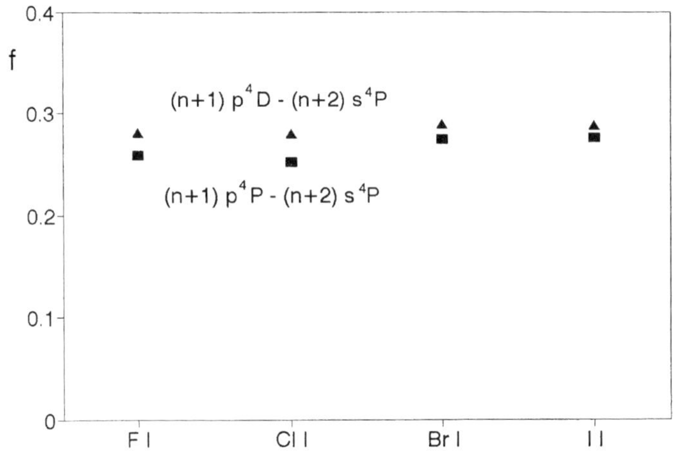

Fig. 2. Oscillator strengths for the ^4P-^4P and ^4D-^4P multiplets of the np^4 $(n+1)p$-np^4 $(n+2)s$ transition arrays of halogen atoms.

– Only the measurements of Bengtson are available for the $3p^4$ (^3P) $5s$-$3p^4$ (^3P) $5p$ fine structure transitions in *Br I* (Table 3). No comparative data from other sources have been found in the literature for fine structure transitions in *I I* (Table 4), and thus we have made use of the regular behaviour of the oscillator strengths for analysing *f*-value data.

We have studied how closely the *f*-values for analogous transitions in homologous atoms agree with each other, as predicted [9], (Table 5). The RQDO results show that multiplets and individual lines for transitions in halogen atoms remain the same as long

Table 5 *Oscillator strengths for the multiplet transitions in halogen atoms*

	Transition	RQDO	OP[c]	CC[d]
F I	3s ^{4}P-3p ^{4}P	0.2740[a]	0.319	0.29
Cl I	4s ^{4}P-4p ^{4}P	0.2942[a]		0.30
Br I	5s ^{4}P-5p ^{4}P	0.3050[a]		
		0.3061[b]		
I I	6s ^{4}P-6p ^{4}P	0.3145[a]		
		0.3152[b]		
F I	3s ^{4}P-3p ^{4}D	0.4679[a]	0.551	0.53
Cl I	4s ^{4}P-4p ^{4}D	0.5073[a]		0.48
Br I	5s ^{4}P-5p ^{4}D	0.5107[a]		
		0.6110[b]		
I I	6s ^{4}P-6p ^{4}D	0.5309[a]		
		0.6325[b]		
F I	3p ^{4}P-4s ^{4}P	0.2588[a]		
Cl I	4p ^{4}P-5s ^{4}P	0.2521[a]		
Br I	5p ^{4}P-6s ^{4}P	0.2743[a]		
		0.2741[b]		
I I	6p ^{4}P-7s ^{4}P	0.2705[a]		
		0.2756[b]		
F I	3p ^{4}D-4s ^{4}P	0.2806[a]		
Cl I	4p ^{4}D-5s ^{4}P	0.2796[a]		
Br I	5p ^{4}D-6s ^{4}P	0.2914[a]		
		0.2892[b]		
I I	6p ^{4}D-7s ^{4}P	0.2900[a]		
		0.2880[b]		

[a]RQDO this work, *LS* coupling.
[b]RQDO this work, $J_c\, l$ coupling.
[c]Opacity Project [19].
[d]Critical Compilation, Wiese et al. [15, 18].

as the coupling scheme remains constant. The disagreements between some of the *F I*, *Cl I*, *Br I* and *I I* *f*-values appear to be mostly due to changes in the coupling scheme, which lead to a redistribution of the individual line strengths, but leave the multiplet *f*-values (see Table 5 and Figures 1 and 2) nearly unaffected. Thus, for groups of lines starting from a given level, a strong disagreement in one direction is usually balanced by a similarly strong deviation in the other direction.

Overall, we seem to find reasons to be hopeful about the possibilities of the RQDO formalism for predicting spectral properties of interest in astrophysics and plasma physics. These reasons rest on the correctness of the results so far achieved for *F I* and *Cl I* [20, 21], and the low computational expense and avoidance of the numerous convergence problems which are common in the multiconfigurational approaches.

Acknowledgements

This work has been supported by the D.G.I.C.Y.T. of the Spanish Ministry of Education and Science, under Project No.PB94-1314-C03-03, and by the J.C.L. within Project VA21/97.

References

1. M. Jura: *ApJ* 190, L33 (1974); M. Jura and D.G. York: *ApJ* **219**, 861 (1978).
2. R.M. Schectman, S.R. Federman, D.J. Beideck and D.G. Ellis: *ApJ* **406**, 735 (1993).
3. J.L. Tech: *J. Res. Natl. Bur. Std.* **67 A**, 505 (1969).
4. C.J. Humphreys and E. Paul: *J. Opt. Soc. Am.* **62**, 432 (1972).
5. L.J. Radziemski and V. Kaufman: *J. Opt. Soc. Am.* **59**, 424 (1969).
6. L. Minnhagen: *J. Opt. Soc. Am.* **51**, 298 (1962).
7. S. Bashkin and J.O. Stoner: *Atomic Energy Levels and Gotrian Diagrams*, **Vol. 1**, North Holland (1975).
8. I. Martin and J. Karwowski: *J. Phys. B: At. Mol. Opt. Phys.* **24**, 1539 (1991); J. Karwowski and I. Martin: *Phys. Rev. A* **43**, 4832 (1991).
9. W.L. Wiese: *Phys. Scr.* **T65**, 188 (1996).
10. W.L. Wiese and A.W. Weiss: *Phys. Rev.* **175**, 50 (1968); W.L. Wiese: *Phys. Scr.* **35**, 896 (1987).
11. C. Lavín, P. Martin, I. Martin and J. Karwowski: *Int. J. Quant. Chem.* **S27**, 385 (1993).
12. D. Bielinska-Waz: *M. Sc. Thesis*, Institute of Physics, University Nicholas Copernicus, Torun, Poland (1992).
13. S. Fraga, J. Karwowski and K.M.A. Saxena: *Handbook of Atomic Data*, Amsterdam: Elsevier (1976).
14. B.N. Chichkov and V.B. Shevelko: *Phys. Scr.* **23**, 1055 (1981).
15. W.L. Wiese, M.W. Smith and B.M. Glennon: *Atomic Transition Probabilities*. **Vol. II**, Washington, NSRDS-NBS 22 (1966).
16. R.D. Bengtson, M.H. Miller, D.W. Koopman and T.D. Wilkerson: *Phys. Rev. A* **3**, 16 (1971).
17. P.C. Ojha and A. Hibbert: *Phys. Scr.* **42**, 424 (1990).
18. W.L. Wiese, M.W. Smith and B.M. Miles: *Atomic Transition Probabilities*. **Vol. II**, Washington, NSRDS-NBS 22 (1969).
19. Opacity Project, **Vol. I**, Institute of Physics, Bristol (1995).
20. A.M. Velasco, C. Lavín and I. Martin: *J. Quant. Spectr. Rad. Transfer* **57**, 509 (1997).
21. C. Lavín, A.M. Velasco and I. Martín: *Astron. Astrophys.* **328**, 426 (1997).

Extension of the Relativistic Quantum Defect Orbital Method to the Treatment of Many-Valence Electron Atoms. Atomic Transitions in Ar II

I. Martín[a], A.M. Velasco[b] and C. Lavín[a]

[a]*Departamento de Química Física, Facultad de Ciencias, Universidad de Valladolid, E-47005 Valladolid, Spain*

[b]*Istituto di Chimica Quantistica ed Energetica Molecolare del Consiglio Nazionale delle Ricerche, Via Risorgimento 35, I-56126 Pisa, Italy*

Abstract

Formulae for calculating transition probabilities in both the *LS* and $J_c l$ coupling schemes, within the context of the Relativistic Quantum Defect Orbital (RQDO) formalism, which yields one-electron functions, are given and applied to the complex atomic system Ar II. The application of a given coupling scheme to the different energy levels dealt with is justified.

1. Introduction

The Relativistic Quantum Defect Orbital (RQDO) method, as formulated by Martin and Karwowski [1] and Karwowski and Martin [2] has been applied in the last several years to the prediction of transition probabilities and oscillator strengths of atoms and atomic ions that are relevant in the fields of astrophysics and plasma physics. A recent summary of the method and some of its applications can be found in Martin [3]. Here, the numerical examples refer to atomic systems with up to three valence electrons, no matter how heavy some of the ionised atoms are. Encouraged by the results obtained for few valence- electron atoms, for which all the energy levels complied with the Russell-Saunders or *LS* spin-orbit coupling scheme, we have, very recently, extended our formalism in such a way that it can deal now with systems that possess an arbitrary number of valence electrons. This has carried along the modification of our computer codes in order to consider coupling schemes other than the *LS* one. For example, in our study of chlorine [4] we have found that the levels involved in some of the transitions that are relevant in astrophysics follow the energy-pair $J_c l$ coupling scheme.

For many years, the analysis and interpretation of atomic spectra was based almost entirely on the concepts of *LS* and *jj* coupling, no matter that as early as in the 1960's Racah [5], among other spectroscopists, showed an increased recognition of the importance of other types of spin-orbit coupling. For instance, when an outer electron is more highly excited than the others and, in particular, when its l quantum number is high, the 'symmetric' couplings *LS* and *jj* are, usually, no longer good approximations [6–8]. Pronounced changes in the level structure of the atom also often happen, the levels frequently showing a tendency to occur in pairs, with the J values of each pair differing by one unit. The alternative coupling schemes (often known as 'pair coupling') are generally referred to as '*jK* or $J_c l$' and '*LK*'. However, even at present, few spectroscopists employ these schemes. When the energy levels of a given atomic

A. Hernández-Laguna et al. (eds.), Quantum Systems in Chemistry and Physics, Vol. 1: Basic Problems and Model Systems, 273–288.

configuration do not follow a 'pure' coupling scheme, Racah [5], especially, emphasised the desirability of denoting energy levels in terms of the pure coupling scheme which most nearly approximates the actual coupling.

The different coupling schemes reflect the relative weighting of the electrostatic and magnetic forces among the electrons in an atom. When the electrostatic interactions are much stronger than those between the orbital motion and the spin in a given electron, we have the *LS* coupling scheme. The opposite limit corresponds to the *jj* coupling, where the concepts of orbital and spin momenta separately loose their significance, and only one (total) electron angular momentum, **j**, is conserved. Nevertheless, the balance between the two types of interactions, electrostatic and magnetic (spin-orbit) is generally different for different levels in a given atomic system. For instance, even in light atoms, where the *LS* scheme generally applies, in excited states in which the outer electron(s) are found at long mean distances from the core the electrostatic interaction between the core and the outer electron(s) is weak compared with the spin-orbit interaction in the latter. Nonetheless, it is strong compared to the spin coupling of the external electron(s) [5]. In these situations the energy levels of the atom tend to form pairs, as mentioned above, and generally occur in configurations where the outer electron(s) have large values of l. The pair of levels corresponds to two possible values of **J** obtained from the coupling of the spin of the excited electron(s), **s**, with **K**. **K** is the resultant of the total angular momentum of the parent ion, $\mathbf{J_c}$, and the orbital angular momentum of the external electron(s), $\boldsymbol{l}$ [5–7].

In the present work it is our aim to collect the main formulae involved in the calculation of atomic oscillator strengths in the *LS* and $J_c\,l$ coupling schemes, as well as to supply oscillator strengths corresponding to different transitions in the chlorine-like ion Ar II (Ar^+). Argon plasmas are relevant in fundamental studies as well as in technical applications. There have been attempts to improve the quality of the data related to spectral intensities in this ion. The ground-state valence configuration of Ar II is [Ne]$3s^2\ 3p^5$. The presently studied transitions involve energy levels where either of the aforementioned coupling schemes apply.

2. Computational Procedure

In this section we shall give the mathematical expressions for the properties object of the present calculations. One of the measurements of the intensity of an electronic transition is given by the oscillator strength. For an absorption transition between states i and f, it can be defined as follows

$$f_{if} = \frac{2(E_f - E_i)}{3} \langle \mu_{if} \rangle^2 \tag{1}$$

where E_i and E_f are the energies of the initial and final states, respectively. The symbol $\langle \mu_{if} \rangle$, known as the transition moment, contains the initial and final wavefunctions of the system undergoing the transition, as well as a transition operator. The former are represented below by the set of quantum numbers γ and by their corresponding J and M quantum numbers. In what follows, the primed values will refer to the upper state.

The transition operator, or electric multipole operator, is a tensor of rank k and it is given the symbol $\boldsymbol{P}_q^{(k)}$. We, thus, have

$$\langle\mu_{if}\rangle^2 = \sum_q |\langle\gamma' J' M'|\boldsymbol{P}_q^{(k)}|\gamma JM\rangle|^2. \tag{2}$$

In the case of $E1$, transitions, that is, those ocurring through the electric dipole mechanism, $k = 1$, and for an N electron system with $k = 1$, in atomic units (which are used throughout, except in the figures, where the energy values are given in cm^{-1}),

$$\boldsymbol{P}_q^{(1)} \equiv \sum_{i=1}^{N} r_q^{(1)}(i), \tag{3}$$

By Wigner-Eckart's theorem [6] Eq. (2) can be expressed in terms of a reduced matrix element that is independent of M and M',

$$|\langle\gamma JM|\boldsymbol{P}_q^{(1)}|\gamma' J' M'\rangle|^2 = \left[(-1)^{J-M}\begin{pmatrix} J & 1 & J' \\ -M & q & M' \end{pmatrix}\langle\gamma J\|\boldsymbol{P}^{(1)}\|\gamma' J'\rangle\right]^2 \tag{4}$$

where the symbol in parentheses is a Wigner's 3-j symbol [6–8]. From the properties of this, the general selection rules for E1 transitions are deduced. These are the following,

- $\Delta M = M' - M = 0, \pm 1$
- $\Delta J = J' - J = 0, \pm 1$
- Restriction: $J = J' = 0$ is not allowed. (5)

Another important quantity in the context of $E1$ transitions is the electric dipole line strength, defined as [7]

$$\mathbf{S} \equiv [(-1)^{J-M}, \langle\gamma J\|\boldsymbol{P}^{(1)}\|\gamma' J'\rangle]^2, \tag{6}$$

and related to the total absorption oscillator strength in the form:

$$f_{if} = \frac{2(E' - E)}{3(2J+1)}\mathbf{S} \tag{7}$$

Let us see now how the above expressions can be further developed in the LS and $J_c l$ coupling schemes.

2.1. *LS coupling scheme*

Here, the relevant angular momentum vectors and quantum numbers are $\mathbf{L}$ (L), the total orbital angular momentum of the atom, obtained as the vector coupling of those corresponding to the core and to the outer electron(s), $\mathbf{S}$ (S), the total spin, and $\mathbf{J}$ (J), the total angular momentum for a given atomic level [8]:

$$\mathbf{L} = \mathbf{L_c} + \boldsymbol{l}$$

$$\mathbf{J} = \mathbf{S} + \mathbf{L}. \tag{8}$$

The symbol for the energy level is denoted as ${}^{2S+1}L_J$. The line strength takes the form:

$$S = |\langle \gamma LSJ \| \boldsymbol{P}^{(1)} \| \gamma' L' S' J' \rangle|^2, \tag{9}$$

where the reduced matrix element in (9) can be written as:

$$\langle \gamma LSJ \| (\boldsymbol{P}^{(1)}) \| \gamma' L' SJ' \rangle^2$$
$$= \left[\delta_{SS'} \sqrt{2J+1}\sqrt{2J'+1}(-1)^{S+J+L'+1} \begin{Bmatrix} S & J & L \\ 1 & L' & J' \end{Bmatrix} \langle \gamma L \| \boldsymbol{P}^{(1)} \| \gamma' L' \rangle \right]^2, \tag{10}$$

being the symbol in braces a Wigner's 6-j symbol [6–8].

Some of the factors appearing in (10) can be collected in the so-called line factor, $\mathbf{D}_{\text{line}}$ [7]:

$$\mathbf{D}_{\text{line}} \equiv \sqrt{2J+1}\sqrt{2J'+1}(-1)^{S+J+L'+1} \begin{Bmatrix} S & J & L \\ 1 & L' & J' \end{Bmatrix}. \tag{11}$$

From the properties of the 6-j symbol in (11), the selection rules for electron transitions in this scheme are obtained [8],

- $\Delta S \equiv S' - S = 0$
- $\Delta L = L' - L = 0, \pm 1$
- $L = L' = 0$ is not allowed (12)

If a multiplet factor, $\mathbf{D}_{\text{multiplet}}$, is defined,

$$\mathbf{D}_{\text{multiplet}} = \sqrt{2L+1}\sqrt{2L'+1}(-1)^{Lc+L+l'+1} \begin{Bmatrix} Lc & 1 & L \\ 1 & L' & l' \end{Bmatrix}, \tag{13}$$

the reduced matrix element becomes, for one-electron transitions,

$$\langle \gamma L \| \boldsymbol{P}^{(1)} \| \gamma' L' \rangle^2 = [\mathbf{D}_{\text{multiplet}}(\gamma L, \gamma' L') \langle nl \| \mathbf{r}^{(1)} \| n' l' \rangle]^2 \tag{14}$$

where

$$\langle nl \| \mathbf{r}^{(1)} \| n' l' \rangle = (-1)^{l - l_{\text{max}}} \sqrt{l_{\text{max}}} I(nl, n'l'). \tag{15}$$

In Eq. (14), l_{max} is the maximum of the orbital angular momentum quantum numbers of the active electron in either the initial or final states, $I(nl, n'l')$ is the radial transition integral, that contains only the radial part of both initial and final wavefunctions of the jumping electron and a transition operator. Two different forms for this have been employed, the standard dipole-length operator, $\mathbf{P}(r) = \mathbf{r}$, and another derived from the former in such a way that it accounts explicitly for the polarization induced in the atomic core by the active electron [9],

$$\boldsymbol{P}(r) = \mathbf{r}\left\{1 - \frac{\alpha}{r^3}\left[1 - \exp\left(-\frac{r}{r_c}\right)\right]^3\right\}, \tag{16}$$

where α is the core dipole polarizability and r_c a cut-off radius that we regularly take of the order of magnitude of the mean radius of the core's outermost electron.
The line strength becomes:

$$\mathbf{S}(\gamma LSJ,\ \gamma' L' SJ')$$
$$= \delta_{SS'} |\mathbf{D}_{\text{line}}(LSJ,\ L'SJ')\mathbf{D}_{\text{multiplet}}(\gamma L,\ \gamma' L')(-1)^{l-l_{max}}\sqrt{l_{max}}I(nl,\ n'l')|^2. \quad (17)$$

When in the valence configuration of the atomic system there are N equivalent electrons expressions (9) and (17) should be multiplied by N and by the appropriate Coefficient of Fractional Parentage (CFP) [10,12]. In all the present transitions N has been taken equal to 5, as it is one of the five 3p electrons the one that experiences the transition. The CFP varies with the initial and final states. All the CFP values have been taken from Ref. 12.

2.2 $J_c l$ Coupling Scheme

The angular momentum vectors and quantum numbers that are well-defined in this coupling scheme are, respectively, **K** (K) [resulting from the coupling of the total angular momentum of the core electrons, $\mathbf{J_c}$ (J_c), and the orbital angular momentum of the outer electron(s), $\boldsymbol{l}$ (l)] and the total angular momentum of the atom, **J** (J), which results from the coupling of **s** (s), the spin of the outer electron(s), and **K** (K) [10–12],

$$\mathbf{K} = \mathbf{J_c} + \boldsymbol{l}$$
$$\mathbf{J} = \mathbf{s} + \mathbf{K}. \quad (18)$$

In this coupling scheme the symbol for a level is denoted as $^{2s+1}[K]_J$. The line strength here takes the form:

$$\mathbf{S} = |\langle J_c lKsJ \| \boldsymbol{P}^{(1)} \| J_c l' K' sJ' \rangle|^2. \quad (19)$$

For transitions involving a single electron,

$$\mathbf{S} =$$
$$(2J+1)(2J'+1)(2K+1)(2K'+1)l_{max}\begin{Bmatrix} K & \frac{1}{2} & J \\ J' & 1 & K' \end{Bmatrix}^2 \begin{Bmatrix} J_c & l & K \\ 1 & K' & l' \end{Bmatrix}^2 I^2(nl,\ n'l'), \quad (20)$$

where all the symbols have already been defined.

As in the case of LS coupling, when there are N equivalent electrons in the outer shell, both the line strength and the oscillator strength should be multiplied by N as well as by the corresponding CFP [10,12]. As in the LS scheme the two forms of the electric dipole length transition operator have been employed here in the calculation of the radial transition integral, $I(nl,\ n'l')$.

From the properties of the two 6-j symbols in Eq. (20) the specific selection rules for the $J_c l$ coupling scheme are derived as:

$$\bullet\ \Delta K \equiv K' - K = 0,\ \pm 1$$

$$\bullet\ \Delta J_c \equiv J_c' - J_c = 0. \tag{21}$$

We have seen that in these two different schemes (and this extends to all possible coupling schemes) we have the same resultant (total) momentum, **J**, and two intermediate ones, used to denote which scheme we are referring to. The latter allow an unambiguous classification of the energy levels, the total number of which with a given value of J being the same for all the coupling schemes [12].

2.3. *The Relativistic Quantum Defect Orbital (RQDO) method*

In order to calculate the radial matrix elements to whose square both the line strength and oscillator strength are proportional, the radial wavefunctions corresponding to the initial and final states in the transition must be known. In previous and the present calculations we have employed relativistic quantum defect orbitals (RQDO's). The RQDO formalism [1–3] is based on the decoupling of the Dirac second-order equation as well as on the interpretation of the resulting solutions previously proposed by Karwowski and Kobus [13]. These authors also showed that both the large and small components of the Dirac radial wavefunction may be recovered without any computational effort. By a reinterpretation of the parameters appearing in the much earlier non-relativistic Quantum Defect Orbital (QDO) method [14], the RQDO radial equation is written as a second-order Dirac-like equation in the following fashion [1–3]

$$\left[-\frac{d^2}{dr^2} + \frac{\Lambda(\Lambda+1)}{r^2} - \frac{2Z'_{net}}{r}\right]\Psi_k^{RQDO} = 2e^{RQDO}\Psi_k^{RQDO}, \tag{22}$$

where the parameter Λ that appears in the model Hamiltonian is connected with the relativistic quantites s, as defined by Karwowski and Kobus [13], and δ', the relativisitc quantum defect.

$$\Lambda = \begin{cases} s - 1 - \delta' + c \text{ if } j = l + 1/2 \\ -s - \delta' + c \text{ if } j = l - 1/2 \end{cases} \tag{23}$$

c is an integer [14] that both ensures the normalizability of the radial relativisitc quantum defect orbitals and determines their nodal structure.

$$Z'_{\text{net}} = Z_{\text{net}}(1 + \alpha^2 E^x), \tag{24}$$

where Z_{net} is the net nuclear charge acting on the valence electrons, α is the fine structure constant and E^x is the experimental (or independently calculated by other procedure) energy.

As in the non-relativistic case [14], the relativistic quantum defect is determined empirically. We have

$$e^{\text{RQDO}} = -\frac{(Z'_{\text{net}})^2}{2(\tilde{n} - \delta')^2}, \tag{25}$$

where ñ is the 'relativistic principal quantum number' [13], and δ' is obtained from [1–3]

$$\frac{E^x(1+\frac{1}{2}\alpha^2 E)}{(1+\alpha^2 E^x)^2} = -\frac{Z_{\rm net}^2}{2(\tilde{n}-\delta')^2}. \tag{26}$$

In this formulation all matrix elements, including the transition moments, may be expressed as closed-form analytical equations, thus avoiding numerical errors and convergence problems.

Kwato Njock *et al.* [15] have presented more recently a relativistic generalisation of the quantum defect orbital method. This formulation has some resemblances with the previous one [1,2] but is, in our view, unnecessarily complicated.

3. Results and Analysis

Prior to calculating oscillator strengths in a given transition array, we have determined which of the spin-orbit coupling schemes referred to in the preceding section is most appropriate for the levels involved. For the energy levels, we have employed the data reported by Bashkin and Stoner [17]. As we did in the case of the isoelectronic atom chlorine [4], we have employed a graphical technique reported by Quinet and co-workers [16] to determine whether the group of energy levels corresponding to a given (ground or excited) electronic state of the atomic system complies with the $J_c l$ scheme or, otherwise, with the *LS* scheme. These authors plotted the energies that correspond to a given value of the total angular momentum quantum number of the core, J_c, against a parameter h that is defined as follows

$$h = \tfrac{1}{2}[K(K+1) - J_c(J_c+1) - l(l+1)], \tag{27}$$

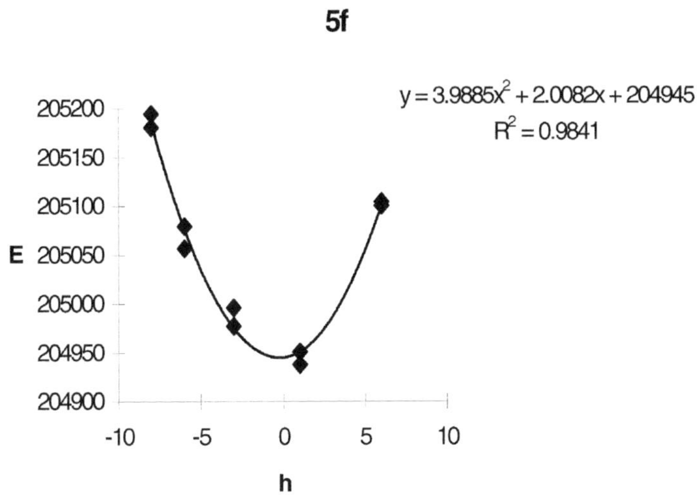

Fig. 1. Energy levels in cm^{-1} versus the h parameter, Eq. (27), for the $3s^2$ $3p^5$ 5f valence configuration of Ar II.

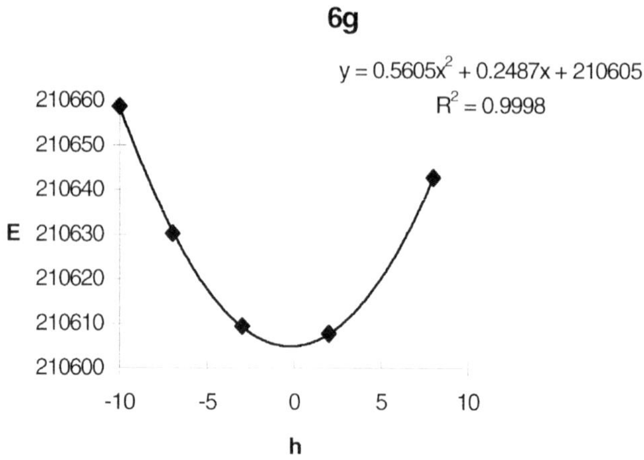

Fig. 2. Energy levels in cm^{-1} versus the h parameter, Eq. (27), for the $3s^2\ 3p^5\ 6g$ valence configuration of Ar II.

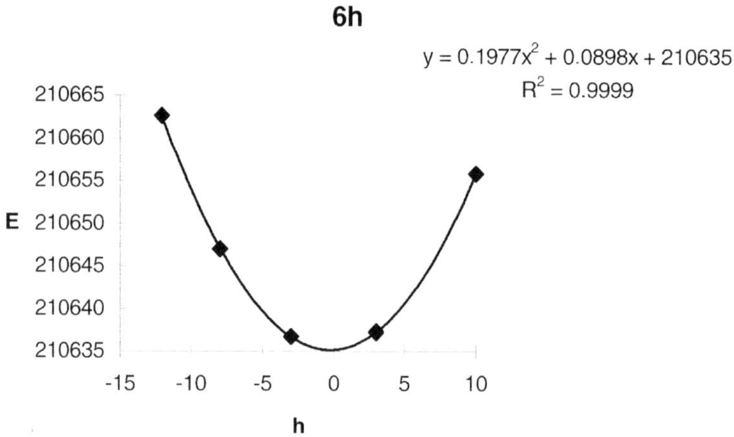

Fig. 3. Energy levels in cm^{-1} versus the h parameter, Eq. (27), for the $3s^2\ 3p^5\ 6h$ valence configuration of Ar II.

where all the symbols have been specified in Section 2.2. If the levels fit a parabola, the atomic state is best described by the $J_c l$ scheme [16], and the oscillator strengths should be calculated according to the expressions given in Section 2.2.

In our previous experience with chlorine [4], for which the ground state valence configuration is the same as that in Ar II, $3s^2\ 3p^5$, we found that all the excited configurations in which one of the 3p electrons is promoted to an upper $n'p$ or $n'd$ ($n' > 3$) state comply with the *LS* scheme. This was valid for two different states of the core, the ground, ^3P, and the excited ^1D states. The same has been found to occur in Ar II. However, in some of the states where one 3p electron is promoted to an orbital with higher n and l quantum numbers the levels showed a clear $J_c l$ structure. We observed that for a given value of l this was more obvious the higher the principal quantum

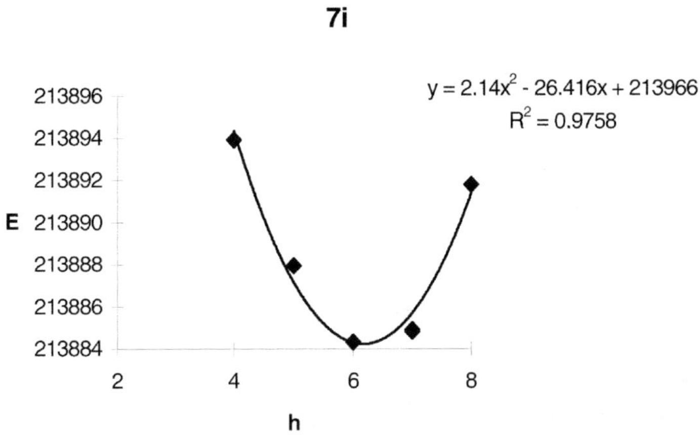

Fig. 4. Energy levels in cm^{-1} versus the h parameter, Eq. (27), for the $3s^2$ $3p^5$ 7i valence configuration of Ar II.

Table 1 *Oscillator strengths for 3p-4s fine-structure transitions in Ar II. LS coupling*

	RQDO[a]	RQDO[b]	CIV3[c]	CI[d]	PM[e]	MP[f]
$3p^2P_{3/2}$-$4s^2P_{3/2}$	0.1723	0.1659	0.1863	0.1514	0.1814	0.1612
$3p^2P_{3/2}$-$4s^2P_{1/2}$	0.0325	0.0320	0.0384	0.0319	0.0369	0.0322
$3p^2P_{1/2}$-$4s^2P_{3/2}$	0.0704	0.0675	0.0610	0.0504	0.0718	
$3p^2P_{1/2}$-$4s^2P_{1/2}$	0.1328	0.1301	0.1398	0.1152	0.1461	

[a]RQDO this work
[b]RQDO this work with polarization
[c]Hibbert and Hansen [20]
[d]Luyken [21]
[e]Statz et al. [22]
[f]Ganas [23]

number, as Minhagen had also observed over three decades ago [18]. In the case of Ar II we have centred our attention in levels involved in transitions for which comparative oscillator strengths were found in the literature. Of these, the ones that complied with a 'pair coupling' scheme were the excited states 5f, 6g, 6h and 7i. In Figures 1 to 4 we show that the corresponding E vs. h plots are well-defined parabolas. In each of the figures the fitting quadratic equation and the value of the quadratic regression coefficient, R^2, are inserted.

The oscillator strengths obtained for the different transitions studied in the present work with the RQDO methodology, and the use of the two forms of the transition operator, the standard one, and that corrected for core-valence polarization, are collected in Tables 1 to 8, where other data, from several theoretical and experimental sources, have been included for comparative purposes. The former comprise the large-scale configuration interaction performed with the use of the CIV3 computer package [19] by Hibbert and Hansen [20]; The configuration interaction (CI) procedure of

Table 2 *Oscillator strengths for 4s-4p fine-structure transitions in Ar II. LS coupling*

	RQDO[a]	RQDO[b]	CIV3[c]	CI[d]	PM[e]	PM[f]	EXPT[g]	EXPT[h]	EXPT[i]	EXPT[j]	EXPT[k]
$4s^4P_{5/2}$-$4p^4P_{5/2}$	0.2073	0.1813	0.2707	0.3144	0.3028	0.2501	0.2701	0.3019	0.2868	0.2389	0.2736
$4s^4P_{5/2}$-$4p^4P_{3/2}$	0.0899	0.0787	0.1315	0.1529	0.1431	0.1088	0.1300	0.1457	0.1370		
$4s^4P_{3/2}$-$4p^4P_{5/2}$	0.1306	0.1143	0.0866	0.0948	0.0773	0.1541	0.1136	0.0920	0.0847	0.0762	0.0829
$4s^4P_{3/2}$-$4p^4P_{3/2}$	0.0392	0.0343	0.0516	0.0595	0.0588	0.0464	0.0525	0.0576	0.0540		
$4s^4P_{3/2}$-$4p^4P_{1/2}$	0.1243	0.1090	0.1505	0.1717	0.1581	0.1475	0.1496	0.1656	0.1524	0.1374	0.1497
$4s^4P_{1/2}$-$4p^4P_{3/2}$	0.2419	0.2120	0.1758	0.1913	0.1599	0.2829	0.1713	0.1882	0.1660		
$4s^4P_{1/2}$-$4p^4P_{1/2}$	0.0491	0.0431	0.0388	0.0426	0.0359	0.0575	0.0359	0.0358	0.0359	0.0333	0.0356
$4s^4P_{5/2}$-$4p^4S_{3/2}$	0.1172	0.1046	0.0842	0.1104	0.1210	0.1530	0.0667	0.0917	0.1214	0.0655	0.0834
$4s^4P_{3/2}$-$4p^4S_{3/2}$	0.1166	0.1040	0.1127	0.1480	0.1623	0.1484	0.0860	0.1178	0.1543	0.0840	0.1045
$4s^4P_{1/2}$-$4p^4S_{3/2}$	0.1161	0.1036	0.1428	0.1892	0.2087	0.1458	0.1129	0.1527	0.1991	0.1152	0.1388
$4s^4P_{5/2}$-$4p^4D_{7/2}$	0.4268	0.3759	0.4490	0.5403	0.5255	0.5255	0.4425	0.5177	0.5104		
$4s^4P_{5/2}$-$4p^4D_{5/2}$	0.0973	0.0858	0.0501	0.0590	0.1207	0.0524	0.0447	0.0472	0.0573		
$4s^4P_{5/2}$-$4p^4D_{3/2}$	0.0110	0.0097	0.0025	0.0027	0.0137	0.0021	0.0021	0.0024	0.0027		
$4s^4P_{3/2}$-$4p^4D_{5/2}$	0.3363	0.2967	0.3765	0.4449	0.4072	0.4583	0.3599	0.4052	0.4169		
$4s^4P_{3/2}$-$4p^4D_{3/2}$	0.1735	0.1533	0.1537	0.1853	0.2113	0.1806	0.1614	0.1772	0.1764		
$4s^4P_{3/2}$-$4p^4D_{1/2}$	0.0273	0.0242	0.0180	0.0217	0.0334	0.0205	0.0181	0.0183	0.0266	0.0175	0.0165
$4s^4P_{1/2}$-$4p^4D_{3/2}$	0.2689	0.2377	0.3303	0.3954	0.3232	0.3992	0.3348	0.3472	0.3662		
$4s^4P_{1/2}$-$4p^4D_{1/2}$	0.2711	0.2398	0.2925	0.3508	0.3280	0.3510	0.2887	0.3307	0.3366	0.2527	0.2990

[a] RQDO this work
[b] RQDO this work with polarization
[c] Hibert and Hansen [20]
[d] Luyken [21]
[e] Statz et al., intermediate coupling [22]
[f] Statz et al., LS coupling [22]
[g] Vujovic and Wiese [28]
[h] Shumaker and Popenoe [24]
[i] Rudko and Tang [25]
[j] García and Campos [26]
[k] Wiese et al. [29]

Table 3 *Oscillator strengths for 4p-5s fine-structure transitions in Ar II. LS coupling*

	RQDO[a]	RQDO[b]	Expt[c]	Expt[d]
$4p^4S_{3/2}$-$5s^4P_{5/2}$	0.1443	0.1432	0.1011	0.0772
$4p^4S_{3/2}$-$5s^4P_{3/2}$	0.0926	0.0922	0.0595	0.0508
$4p^4S_{3/2}$-$5s^4P_{1/2}$	0.0439	0.0439	0.0501	0.0450
$4p^4P_{5/2}$-$5s^4P_{5/2}$	0.1561	0.1573	0.1443	0.2077
$4p^4P_{5/2}$-$5s^4P_{3/2}$	0.0625	0.0633		0.0344
$4p^4P_{3/2}$-$5s^4P_{5/2}$	0.1020	0.1026	0.1106	0.1450
$4p^4P_{3/2}$-$5s^4P_{3/2}$	0.0283	0.0286	0.0580	0.0345
$4p^4P_{3/2}$-$5s^4P_{1/2}$	0.0810	0.0824	0.0332	0.0631
$4p^4P_{1/2}$-$5s^4P_{3/2}$	0.1802	0.1811	0.1372	0.1766
$4p^4P_{1/2}$-$5s^4P_{1/2}$	0.0331	0.0337		0.0254
$4p^4D_{7/2}$-$5s^4P_{5/2}$	0.2481	0.2483		0.2292
$4p^4D_{5/2}$-$5s^4P_{5/2}$	0.0759	0.0759	0.0275	0.0333
$4p^4D_{5/2}$-$5s^4P_{3/2}$	0.1676	0.1684	0.1231	
$4p^4D_{3/2}$-$5s^4P_{5/2}$	0.0129	0.0129		
$4p^4D_{3/2}$-$5s^4P_{3/2}$	0.1308	0.1312	0.0738	0.1005
$4p^4D_{3/2}$-$5^4P_{1/2}$	0.0952	0.0960	0.0945	0.1200
$4p^4D_{1/2}$-$5^4P_{3/2}$	0.0414	0.0415		
$4p^4D_{1/2}$-$5s^4P_{1/2}$	0.1932	0.1946		0.1997
$4p^2D_{5/2}$-$5s^2P_{3/2}$	0.2322	0.2339		0.2206
$4p^2D_{3/2}$-$5s^2P_{3/2}$	0.0401	0.0403	0.0928	0.0953
$4p^2D_{3/2}$-$5s^2P_{1/2}$	0.1837	0.1858		0.1230
$4p^2P_{3/2}$-$5s^2P_{3/2}$	0.2089	0.2095		
$4p^2P_{3/2}$-$5s^2P_{1/2}$	0.0386	0.0389	0.1149	0.0921
$4p^2P_{1/2}$-$5s^2P_{3/2}$	0.0814	0.0818	0.1271	0.1420
$4p^2P_{1/2}$-$5s^2P_{1/2}$	0.1496	0.1512	0.1145	

[a]RQDO, this work
[b]RQDO, this work with polarization
[c]Aparicio [27]
[d]Rudko and Tang [25]

Luyken in intermediate coupling [21]; The perturbation (PM) calculations by Statz and co-workers [22], who for the 4s-4p transition complex followed both the *LS* and an intermediate coupling schemes; The analytical model potential (MP) employed by Ganas [23], and the theoretical analysis of their observations, performed in the framework of the Relativistic Hartree-Fock (HFR) approximation, by Quinet et al. [16]. The different experimental data sources are quoted at the foot of the tables. For the few transitions for which they were available, we have also included the data of the critical compilations (CC) by Vujnovic and Wiese [28] and by Wiese et al. [29].

In the levels for which the oscillator strengths are shown in Tables 4 to 6 the atomic core state is the excited, ^1D, whilst in the remaining tables the levels correspond to the ground state of the core, ^3P. After a careful inspection of the tables a few remarks can

Table 4 *Oscillator strengths for 4s'-4p' fine-structure transitions in Ar II. LS coupling*

	RQDO[a]	RQDO[b]	CIV3[c]	CI[d]	EXPT[e]	EXPT[f]	EXPT[g]	EXPT[h]
$4s'^2D_{5/2}$-$4p'^2F_{7/2}$	0.2478	0.2431	0.2703	0.3173	0.3351	0.3865	0.4800	0.3223
$4s'^2D_{5/2}$-$4p'^2F_{7/2}$	0.0124	0.0124	0.0173	0.0207	0.0229	0.0290	0.0303	0.0255
$4s'^2D_{3/2}$-$4p'^2F_{5/2}$	0.2600	0.2551	0.3325	0.2236	0.3145	0.3884	0.4974	0.2980
$4s'^2D_{5/2}$-$4p'^2D_{5/2}$	0.1674	0.1673	0.0851	0.1009	0.1442	0.1417		
$4s'^2D_{5/2}$-$4p'^2D_{3/2}$	0.0120	0.0120	0.0122	0.0146	0.0198	0.0433	0.0424	
$4s'^2D_{3/2}$-$4p'^2D_{5/2}$	0.0179	0.0179	0.0197	0.0081	0.0161	0.0165	0.0524	
$4s'^2D_{3/2}$-$4p'^2D_{3/2}$	0.1611	0.1610	01056	0.0321	0.0995	0.3430	0.3454	
$4s'^2D_{5/2}$-$4p'^2P_{3/2}$	0.1099	0.1090	0.0944	0.1428	0.1463	0.1829	0.1883	0.1658
$4s'^2D_{3/2}$-$4p'^2P_{3/2}$	0.0183	0.0181	0.0293	0.0194	0.0301	0.0565	0.0560	0.0495
$4s'^2D_{3/2}$-$4p'^2P_{1/2}$	0.0905	0.0901	0.1083	0.0903	0.1087	0.1792	0.1779	

[a]RQDO this work
[b]RQDO this work with polarization
[c]Hibert and Hansen [20]
[d]Luyken [21]
[e]Vujnovic andWiese [28]
[f]Wiese et al. [29]
[g]Rudko and Tang [25]
[h]Garcia and Campos [26]

Table 5. *Oscillator strengths for 4p'-5s' fine-structure transitions in Ar II. LS coupling*

	RQDO[a]	RQDO[b]	EXPT[c]
$4p'^2D_{5/2}$-$5s'^2D_{5/2}$	0.2492	0.2468	0.1938
$4p'^2D_{5/2}$-$5s'^2D_{3/2}$	0.0178	0.0176	
$4p'^2D_{3/2}$-$5s'^2D_{5/2}$	0.0266	0.0263	
$4p'^2D_{3/2}$-$5s'^2D_{3/2}$	0.2392	0.2368	
$4p'^2P_{3/2}$-$4d'^2D_{5/2}$	0.2128	0.2110	0.1628
$4p'^2P_{3/2}$-$4d'^2D_{3/2}$	0.0236	0.0234	
$4p'^2P_{1/2}$-$4d'^2D_{3/2}$	0.2518	0.2492	0.1913
$4p'^2F_{7/2}$-$4d'^2D_{5/2}$	0.1960	0.1952	0.2400
$4p'^2F_{5/2}$-$4d'^2D_{5/2}$			
$4p'^2F_{5/2}$-$4d'^2D_{3/2}$	0.1800	0.1794	0.2204

[a]RQDO this work
[b]RQDO this work with polarization
[c]Rudko and Tang [25]

be made. In the core ground configuration, ^3P, for those transitions that involve the less excited valence states (Tables 1 and 2) the explicit introduction of core polarization has stronger effects in the RQDO f-values, lowering their magnitude, than for those which take place between more highly excited valence sates (Tables 3, 7 and 8). Core

Table 6 *Oscillator strengths for 4p′-4d′ fine-structure transitions in Ar II. LS coupling*

	RQDO[a]	RQDO[b]	EXPT[c]
$4p'^2P_{3/2}$-$4d'^2D_{5/2}$	0.3032	0.2929	0.4113
$4p'^2P_{3/2}$-$4d'^2D_{3/2}$	0.0331	0.0320	
$4p'^2P_{1/2}$-$4d'^2D_{3/2}$	0.3376	0.3265	0.4672
$4p'^2P_{3/2}$-$4d'^2P_{3/2}$	0.3642	0.3517	0.4461
$4p'^2P_{3/2}$-$4d'^2P_{1/2}$	0.0732	0.0707	0.0717
$4p'^2P_{1/2}$-$4d'^2P_{3/2}$	0.1486	0.1436	
$4p'^2P_{1/2}$-$4d'^2P_{1/2}$	0.2985	0.2884	
$4p'^2D_{5/2}$-$4d'^2D_{5/2}$	0.3270	0.3163	0.3319
$4p'^2D_{5/2}$-$4d'^2D_{3/2}$	0.0230	0.0222	
$4p'^2D_{3/2}$-$4d'^2D_{5/2}$	0.0350	0.0338	0.0811
$4p'^2D_{3/2}$-$4d'^2D_{3/2}$	0.3095	0.2995	
$4p'^2D_{5/2}$-$4d'^2P_{3/2}$	0.0908	0.0878	0.1106
$4p'^2D_{3/2}$-$4d'^2P_{3/2}$	0.1512	0.1462	0.0008
$4p'^2D_{3/2}$-$4d'^2P_{1/2}$	0.0759	0.0734	
$4p'^2D_{5/2}$-$4d'^2F_{7/2}$	0.5199	0.5032	0.6454
$4p'^2D_{5/2}$-$4d'^2F_{5/2}$	0.0258	0.0250	0.0714
$4p'^2D_{3/2}$-$4d'^2F_{5/2}$	0.5417	0.5244	0.6127
$4p'^2F_{7/2}$-$4d'^2D_{5/2}$	0.0259	0.0250	0.0295
$4p'^2F_{5/2}$-$4d'^2D_{5/2}$	0.0172	0.0166	0.0135
$4p'^2F_{5/2}$-$4d'^2D_{3/2}$	0.0236	0.0228	0.0364
$4p'^2F_{7/2}$-$4d'^2F_{7/2}$	0.1695	0.1636	0.2548
$4p'^2F_{7/2}$-$4d'^2F_{5/2}$	0.0062	0.0060	0.0167
$4p'^2F_{5/2}$-$4d'^2F_{7/2}$	0.0083	0.0080	0.0089
$4p'^2F_{5/2}$-$4d'^2F_{5/2}$	0.1654	0.1597	0.2493
$4p'^2F_{7/2}$-$4d'^2G_{9/2}$	0.7258	0.6986	0.9510
$4p'^2F_{7/2}$-$4d'^2G_{7/2}$	0.0207	0.0200	0.0287
$4p'^2F_{5/2}$-$4d'^2G_{7/2}$	0.7439	0.7159	0.9806

[a]RQDO this work
[b]RQDO this work with polarization
[c]Rudko and Tang [25]

polarization has also little influence in the f-values between states that correspond to the core excited configuration, ^1D (Tables 4 to 6), where the symbols for both the initial and final orbitals of the active electron are primed.

In the cases where the *LS* coupling has been found by us to be appropriate (Tables 1 to 6) the theoretical f-values with which ours seem to comform better are those obtained by Hibbert and Hansen [20] with CIV3, a far more expensive method than the RQDO, the degree of agreement being higher for the stronger fine-structure transitions within

Table 7 *Oscillator strengths for 5f-6g fine-structure transitions in Ar II. J_cl coupling*

	RQDO[a]	RQDO[b]	HFR[c]
$[5]_{11/2}$-$[6]_{13/2}$	1.0396	1.0366	1.0491
$[5]_{11/2}$-$[6]_{11/2}$	0.0136	0.0136	
$[5]_{9/2}$-$[6]_{11/2}$	1.0522	1.0491	1.0471
$[5]_{11/2}$-$[5]_{11/2}$	0.0887	0.0884	
$[5]_{11/2}$-$[5]_{9/2}$	0.0013	0.0013	
$[5]_{9/2}$-$[5]_{11/2}$	0.0016	0.0016	
$[5]_{9/2}$-$[5]_{9/2}$	0.0885	0.0883	0.0794
$[4]_{9/2}$-$[5]_{11/2}$	0.9377	0.9350	0.9772
$[4]_{9/2}$-$[5]_{9/2}$	0.0222	0.0174	
$[4]_{7/2}$-$[5]_{9/2}$	0.9585	0.9372	0.3953
$[4]_{9/2}$-$[4]_{9/2}$	0.1500	0.1495	0.1349
$[4]_{9/2}$-$[4]_{7/2}$	0.0034	0.0034	
$[4]_{7/2}$-$[4]_{9/2}$	0.0043	0.0043	
$[4]_{7/2}$-$[4]_{7/2}$	0.1497	0.1492	0.0361
$[3]_{7/2}$-$[4]_{9/2}$	0.8985	0.8985	0.3213
$[3]_{7/2}$-$[4]_{7/2}$	0.0258	0.0258	
$[3]_{5/2}$-$[4]_{7/2}$	0.9214	0.9193	0.8747
$[3]_{7/2}$-$[3]_{7/2}$	0.1879	0.1874	0.1040
$[3]_{7/2}$-$[3]_{5/2}$	0.0070	0.0070	
$[3]_{5/2}$-$[3]_{7/2}$	0.0093	0.0092	
$[3]_{5/2}$-$[3]_{5/2}$	0.1849	0.1844	0.2301
$[2]_{5/2}$-$[3]_{7/2}$	0.9076	0.9050	0.8951
$[2]_{5/2}$-$[3]_{5/2}$	0.0452	0.0451	
$[2]_{3/2}$-$[3]_{5/2}$	0.9489	0.9462	0.9505
$[2]_{5/2}$-$[2]_{5/2}$	0.1728	0.1723	0.1667
$[2]_{5/2}$-$[2]_{3/2}$	0.0126	0.0125	
$[2]_{3/2}$-$[2]_{5/2}$	0.0188	0.0187	
$[2]_{3/2}$-$[2]_{3/2}$	0.0753	0.0750	0.1986
$[1]_{3/2}$-$[2]_{5/2}$	1.0452	1.0422	1.0421
$[1]_{3/2}$-$[2]_{5/2}$	0.2322	0.2315	
$[1]_{1/2}$-$[2]_{3/2}$	1.1585	1.1552	1.1454

[a]RQDO this work
[b]RQDO this work with polarization
[c]Quinet et al. [16]

an nl-$n'l'$ complex. The similarities between the RQDO f-values and those of other theoretical and experimental procedures also vary considerably from a fine-structure transition to another within the same complex, as it occurs among the f-values of the different comparative sources, both theoretical and experimental. In the case of the

Table 8 *Oscillator strengths for 6h-7i fine-structure transitions in Ar II. $J_c l$ coupling*

	RQDO[a]	RQDO[b]	HFR[c]
$[7]_{15/2}$-$[8]_{17/2}$	1.9090	1.9070	1.9314
$[7]_{15/2}$-$[8]_{15/2}$	0.0139	0.0139	
$[7]_{13/2}$-$[8]_{15/2}$	1.9223	1.9202	1.9225
$[7]_{15/2}$-$[7]_{15/2}$	0.0800	0.0799	0.0805
$[7]_{15/2}$-$[7]_{13/2}$	0.0007	0.0007	
$[7]_{13/2}$-$[7]_{15/2}$	0.0008	0.0008	
$[7]_{13/2}$-$[7]_{13/2}$	0.0799	0.0798	0.0801
$[6]_{13/2}$-$[6]_{13/2}$	0.1362	0.1360	0.1361
$[6]_{13/2}$-$[6]_{11/2}$	0.0021	0.0020	
$[6]_{11/2}$-$[6]_{13/2}$	0.0024	0.0024	
$[6]_{11/2}$-$[6]_{11/2}$	0.1359	0.1357	0.1352
$[5]_{11/2}$-$[6]_{13/2}$	1.8207	1.8187	1.8231
$[5]_{11/2}$-$[6]_{11/2}$	0.0024	0.0024	
$[5]_{9/2}$-$[6]_{11/2}$	1.8409	1.8389	1.8621
$[5]_{11/2}$-$[5]_{11/2}$	0.1595	0.1594	0.1588
$[5]_{11/2}$-$[5]_{9/2}$	0.0024	0.0024	
$[5]_{9/2}$-$[5]_{11/2}$	0.0029	0.0029	
$[5]_{9/2}$-$[5]_{9/2}$	0.1589	0.1588	0.1585

[a]RQDO this work
[b]RQDO this work with polarization
[c]Quinet et al. [16]

RQDO oscillator strengths, these generally satisfy a feature that is characteristic of the *LS* coupling scheme: The sum of the f-values of the lines that start from the same level is approximately constant. Not all the remaining f-values collected in Tables 1 to 6 comply with this feature.

In the calculations performed in $J_c l$ coupling, Tables 7 and 8, the accord between the two sets of RQDO oscillator strengths and the HFR values [16] is rather good. The discrepancies observed for the transitions that involve the $5f[4]_{7/2}$ and $5f[3]_{7/2}$ levels (Table 7) may be due to the fact that these two levels are strongly mixed. Also, an inspection of Tables 7 and 8 reveals that the RQDO f-values satisfy a characteristic of the $J_c l$ scheme, that is, the strongest lines are those for which $\Delta K = \Delta J$.

Overall, we seem to find reasons to be hopeful about the possibilities of the RQDO formalism for predicting spectral properties of complex atoms. Very recently, some lifetime calculations in Yb II have also been successfully performed [30]. These reasons rest on the correctness of the results so far obtained, as well as the low computational expense and avoidance of the frequent convergence problems that are common in configuration interaction approaches.

Acknowledgements

This work has been supported by the D.G.I.C.Y.T. of the Spanish Ministry of Education within Project PB94-1314-C03-03, and by the J.C.L. within Project VA21/97. I. M. wishes to express her gratitude to M. A. Almaraz for her generous help during the edition of this manuscript.

References

1. I. Martin and J. Karwowski, *J. Phys. B: At. Mol. Opt. Phys*, 1539 (1991).
2. J. Karwowski and I. Martin, *Phys. Rev. A 43*, 4832 (1991).
3. I. Martin, From: The Relativistic Quantum Defect Orbital Method and Some of its Applications, in R. Mc Weeny and others (Eds.), *Quantum Systems in Chemistry and Physics. Trends in Methods and Applications*, Kluwer Academic Publishers, Dordrecht, 51 (1997).
4. C. Lavín, A.M. Velasco and I. Martin, *Astron. Astrophys. 328*, 426 (1997).
5. G. Racah, *J. Opt. Soc. Am. 50*, 408 (1960).
6. I.I. Sobelman, *Introduction to the Theory of Atomic Spectra*, Pergamon Press, Oxford (1972).
7. R.D. Cowan, *The Theory of Atomic Structure and Spectra*, University of California Press, Berkeley (1981).
8. I.I. Sobelman, *Atomic Spectra and Radiative Transitions*, Springer-Verlag, Berlin (1992).
9. D. Bielinska-Waz, I. Martin and J. Karwowski, *Acta Phys. Polon.* A 85, 805 (1994).
10. R.D. Cowan and K.L. Andrew, *J. Opt. Soc. Am. 55*, 502 (1965).
11. W.C. Martin and W.L. Wiese: From Atomic Spectroscopy, in G.W.F. Drake (Ed.), *Atomic, Molecular and Optical Physics Handbook*, American Institute of Physics, Press, Woodbury, N.Y., 135 (1996).
12. Z. Rudzikas, *Theoretical Atomic Spectroscopy*, Cambridge University Press, Cambridge (1997).
13. J. Karwowski and J. Kobus, *Int. J. Quant. Chem. 28*, 741 (1985); *ibid. 30*, 809 (1986).
14. G. Simons, *J. Chem. Phys. 60, 645* (1974); I. Martin and G. Simons, *J. Chem. Phys. 62*, 4799 (1975).
15. M.G. Kwato Njock, B. Oumarou, L.C. Owono, J.D. Kenmogne, S.G. Gana and M. Onana Boyono, *Phys. Letters A 184*, 352 (1994).
16. P. Quinet, P. Palmeri, E. Biémont and J.W. Brault, *Phys. Rev. A 49*, 2446 (1994).
17. S. Bashkin and J.O. Stoner, *Atomic Energy Levels and Grotrian Diagrams, Vol. 2*, North Holland-American Elsevier, Amsterdam (1978).
18. L. Minhagen, *Arkiv für Physik 21*, 415 (1962).
19. A. Hibbert, *Comput. Phys. Comm. 16*, 19 (1975).
20. A. Hibbert and J.E. Hansen, *J. Phys. B: At. Mol. Opt. Phys. 27*, 3325 (1994).
21. B. Luyken, *Physica 60*, 432 (1972).
22. H. Statz, F.A. Harrigan, S.H. Koozenakasi, C.L. Tang and G.F. Koster, *J. Appl. Phys. 36*, 2278 (1965).
23. P.S. Ganas, *J. Appl. Phys. 52*, 3769 (1981).
24. J.B. Shumaker, Jr. And C.H. Popenoe, *J. Opt. Soc. Am. 59*, 980 (1969).
25. R.I. Ruzko and C.L. Tang, *J. Appl. Phys. 38*, 4731 (1967).
26. G. García and J. Campos, *J. Quant. Spectrosc. Radiat. Transfer 34*, 85 (1985).
27. J. A. Aparicio, *Doctoral Thesis*, Department of Applied Physics, University of Valladolid, Spain (1996).
28. V. Vujnovic and W.L. Wiese, *J. Phys. Chem. Ref. Data 21*, 919 (1992).
29. W.L. Wiese, M.W. Smith and B.M. Miles, *Atomic Transition Probabilities II*, NSRDS-NBS 22, U.S. Govt. Printing Office, Washington D.C. (1969).
30. E. Biémont, J.-F. Dutrieux, I. Martin and P. Quinet, *J. Phys. B: At. Mol. Opt. Phys.*, **31**, 3321 (1998).

Part IV

Valence Theory

Hyperspherical Harmonics as Atomic and Molecular Orbitals in Momentum Space

V. Aquilanti, S. Cavalli, C. Coletti, D. Di Domenico and G. Grossi
Dipartimento di Chimica dell'Universitá, Via Elce di Sotto 8, 06123 Perugia, Italy

Abstract

The relationship between alternative separable solutions of the Coulomb problem in momentum space is exploited in order to obtain hydrogenic orbitals which are of interest for Sturmian expansions of use in atomic and molecular structure calculations and for the description of atoms in fields. In view of their usefulness in problems where a direction in space is privileged, as when atoms are in an electric or magnetic field, we refer to these sets as to the *Stark* and *Zeeman* bases, as an alternative to the usual *spherical* basis set. Fock's projection onto the surface of a sphere in the four dimensional hyperspace allows us to establish the connections of the momentum space wave functions with hyperspherical harmonics. Its generalization to higher spaces permits to build up multielectronic and multicenter orbitals.

1. Introduction

Traditional hydrogenic orbitals used in atomic and molecular physics as expansion bases belong to the $|nlm\rangle$ representation, which in configuration space corresponds to separation in polar coordinates, and in momentum space to a separation in spherical coordinates on the (Fock's) hypersphere S^3 [1]. The $|nlm\rangle$ basis will be called *spherical* in the following. *Stark* states $|n\mu m\rangle$ have also been used for atoms in fields and correspond to separation in parabolic coordinates an ordinary space and in cylindrical coordinates on S^3 (for their use for expanding molecular orbitals see ref. [2]). A third basis, to be termed *Zeeman* states and denoted $|n\lambda m\rangle$ has been introduced more recently by Labarthe [3] and has found increasing applications [4].

In 1935 Fock [5] discovered that the wavefunction of hydrogen atom $|nlm\rangle$ in momentum space, as can be obtained by Fourier transform from configuration space, can be related to four dimensional spherical harmonics, i.e. eigenfunctions of the Laplace operator on S^3. Thus the *principal quantum number n* (which labels the energy spectrum) can appropriately be interpreted as a *hyperangular momentum* quantum number, manifesting that the *hidden symmetry* giving rise to the *accidental degeneracy* emerging in the three-dimensional configuration space treatment is actually a four dimensional symmetry, which has been analyzed and discussed in various papers [6], reviews [7] and books [8]. The properly normalized hyperspherical harmonics, here appearing as Fourier transforms of hydrogenic Sturmians [2,9], find their role as atomic orbitals and expansion basis sets in an increasing number of applications [10]. Accordingly, the passage between the alternative sets can be explicitly viewed as a change in coupling schemes of hyperangular momenta. Relationships among coordinate sets and the explicit formulae for transformations between the spherical and the Stark and Zeeman bases are given in Section 2.

A. Hernandez-Laguna et al. (eds.), Quantum Systems in Chemistry and Physics, Vol I: Basic Problems and Model Systems, 291–301.

This paper considers the hyperspherical harmonics of the four dimensional rotation group $O(4)$ in the same spirit of previous investigations [2,11]), where the possibility is considered of exploiting different parametrizations of the S^n hypersphere to build up alternative Sturmian [12] basis sets. Their symmetry and completeness properties make them in fact adapt to solve quantum mechanical problems where the hyperspherical symmetry of the kinetic energy operator is broken by the interaction potential, but the corresponding perturbation matrix elements can be worked out explicitly, as in the case of Coulomb interactions (see Section 3). A final discussion is given in Section 4.

2. Momentum space: Alternative hydrogenic orbitals

As already noted in some preceding papers [10,13] from the theory of the $O(4)$ group, [8], the 3-sphere admits different systems of hyperspherical coordinates to which correspond alternative harmonics. In Ref. [14] we have classified the hyperspherical harmonics of the group $O(4)$, and shown that there are 15 distinct bases. In the following, we consider in more detail the spherical, the Stark and the Zeeman basis sets.

The Sturmian eigenfunctions in momentum space in spherical coordinates are, apart from a weight factor, a standard hyperspherical harmonic, as can be seen in the famous Fock treatment of the hydrogen atom in which the tridimensional space is projected onto the 3-sphere, i.e. a hypersphere embedded in a four dimensional space. The essentials of Fock analysis of relevance here are briefly sketched now.

The Fock transformation of variables consists in projecting the momentum vector **p** with coordinates p_x, p_y, p_z and modulus p in momentum space on a tetradimensional hypersphere of unit radius. The momentum $p_0 = \sqrt{-2\mathrm{E}}$ is directly related to the energy spectrum. A point on the hypersphere surface has coordinates:

$$\begin{aligned} u_1 &= \frac{2p_0 p_y}{p_0^2 + p^2} = \sin\chi \sin\vartheta \sin\varphi \\ u_2 &= \frac{2p_0 p_x}{p_0^2 + p^2} = \sin\chi \sin\vartheta \cos\varphi \\ u_3 &= \frac{2p_0 p_z}{p_0^2 + p^2} = \sin\chi \cos\vartheta \\ u_4 &= \frac{p_0^2 - p^2}{p_0^2 + p^2} = \cos\chi \end{aligned} \tag{1}$$

where ϑ and φ are the polar angles of the vector **p**:

$$\begin{aligned} p_x &= p \sin\vartheta \cos\varphi \\ p_y &= p \sin\vartheta \sin\varphi \\ p_z &= p \cos\vartheta \end{aligned} \tag{2}$$

This transformation is analogous to a stereographic projection from a hyperplane onto a hypersphere in a four dimensional space:

$$u_1^2 + u_2^2 + u_3^2 + u_4^2 = 1 \tag{3}$$

The spherical parametrization (1) can be graphically exhibited by the tree method [13] (see Fig. 1(a)). To compare the volume element in momentum space:

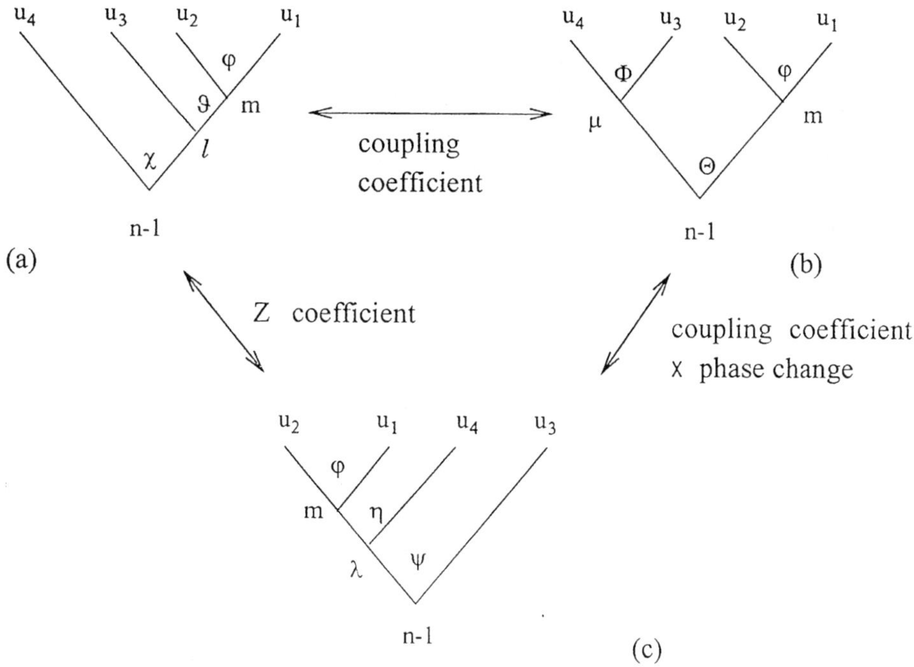

Fig. 1. Momentum space coordinates. Three parametrizations of the 3-sphere, corresponding to three different harmonics, are exhibited by the tree method [36,37,38], useful to represent graphically the relationship between Cartesian coordinates u_i and angular parametrization of the hypersphere. A 4-dimensional sphere of R^4 is parametrized either by 4 Cartesian coordinates or by a hyperradius ρ and 3 hyperangles. There are 4 'leaves' corresponding to the cartesian coordinates, connected to the branches which join at 3 nodes representing the angles. As a convention, we consider that the branch converging to the node from the left (right) represents the cosine (sine) of the hyperangles. Starting from a leaf and going down to the root of the tree, through the various nodes, we establish a relation between the coordinates and the hyperangles. (a) Fock parametrization according to the tree method. The tree shows Fock projection when the parametrization is according to the coordinate system (1) and the corresponding harmonics is $Y_{n-1,l,m}(\chi, \vartheta, \varphi)$. The tree (b) is an alternative representation with respect to the tree (a), it corresponds to a parametrization according to the angles in eq. 10. The corresponding harmonic is $(-)^{n-1}(\iota)^m (n/(2\pi^2))^{1/2} D^{(n-1)/2}_{(\mu+m)/2,(\mu-m)/2}(-\Phi - \varphi, 2\Theta, \varphi - \Phi)$. We can pass from tree a to b using the suitable Clebsch-Gordan coefficient (eq. 12). The tree (c) illustrates the hyperspherical parametrization that leads to the hyperspherical harmonics $Y_{n-1,\lambda,m}(\psi, \eta, \varphi)$. They are related to the harmonics of tree a through the Z coefficient defined in eg. (15). The connection between (b) and (c) requires a Clebsch Gordan coefficient and a phase change related to α (see eq. (14)).

$$\mathbf{dp} = p^2\, \mathrm{d}p \sin\vartheta\, \mathrm{d}\vartheta\, \mathrm{d}\varphi \tag{4}$$

with the solid angle element in the 4-dimensional space:

$$\mathrm{d}\Omega = \sin^2\chi \sin\vartheta\, \mathrm{d}\chi\, \mathrm{d}\vartheta\, \mathrm{d}\varphi \tag{5}$$

we use (1)

$$\frac{\mathrm{d}\chi}{\mathrm{d}p} = \frac{2p_0}{p_0^2 + p^2} \tag{6}$$

to get:

$$\mathbf{dp} = \left(\frac{p_0^2 + p^2}{2p_0}\right)^3 \mathrm{d}\Omega. \tag{7}$$

In the specific case of the hydrogen atom the eigenfunctions $\psi_{nlm}(\mathbf{p})$ in momentum space, which can be considered as Fourier transforms of their configuration space counterparts:

$$\psi_{nlm}(\mathbf{p}) = (2\pi)^{-3/2} \int \exp(-\iota\mathbf{r}\cdot\mathbf{p}) u_{nlm}(\mathbf{r})\, \mathbf{dr} \tag{8}$$

are just hyperspherical harmonics except for a weight function. That is:

$$\psi_{nlm}(\mathbf{p}) = (-)^{n-1-|m|}(\iota)^{l+m-|m|} \frac{4p_0^{5/2}}{(p_0^2 + p^2)^2} \cdot Y_{n-1,l,m}(\chi, \vartheta, \varphi) \tag{9}$$

The explicit expressions for $\psi_{nlm}(\mathbf{p})$ for the hydrogen atom were obtained for the first time by Podolsky and Pauling [15]. Making use of the tree method [13] we can see (Fig. 1a) that the part in m and l represents the spherical harmonic and the part in l and n a Gegenbauer polynomial. According to this view the quantum number n plays the role of a hyperangular momentum quantum number. The connection between alternative systems of hyperspherical coordinates is visualized by the tree method [13] in Fig. 1; from a group theoretical viewpoint, the alternative coupling schemes correspond to the subgroup reductions $O(4) \supset O(3) \supset O(2)$ and $O(4) \supset O(2) \times O(2)$.

Fig. 1 also helps in writing down the four-dimensional sphere parametrizations corresponding to the alternative representations for Fock projection.

The set of angles in Fig. 1(b) parametrizing the 3-sphere defines as follows a point on the 3-dimensional sphere [8]:

$$\begin{aligned} u_1 &= \sin\Theta \sin\varphi \\ u_2 &= \sin\Theta \cos\varphi \\ u_3 &= \cos\Theta \sin\Phi \\ u_4 &= \cos\Theta \cos\Phi \end{aligned} \tag{10}$$

where Θ, φ, Φ are the cylindrical angles describing the coordinates on a tetradimensional space (see also [8]; a similar system of angles is used by Kalnins et al. [6] who give a general treatment of the properties of the group $O(4)$).

Through the set of angles in Fig. 1(c), a point in the four-dimensional momentum space is defined in the following way:

$$\begin{aligned} u_1 &= \cos\psi\cos\eta\sin\varphi \\ u_2 &= \cos\psi\cos\eta\cos\varphi \\ u_3 &= \sin\psi \\ u_4 &= \cos\psi\sin\eta \end{aligned} \tag{11}$$

where ψ and η are the hyperangles which define the *Zeeman* basis.

2.1. *The connection between alternative basis sets*

In the following, we pay special attention to the connections among the spherical, Stark and Zeeman basis. Since in momentum space the orbitals are simply related to hyperspherical harmonics, these connections are given by orthogonal matrix elements similar (when not identical) to the elements of angular momentum algebra.

Let us now consider the overlap between the spherical and the Stark basis. For the latter, the momentum space eigenfunctions, which in configuration space correspond to variable separation in parabolic coordinates, are similarly related to alternative hyperspherical harmonics [2]. The connecting coefficient between *spherical* and *Stark* basis is formally identical to a usual vector coupling coefficient (from now on n is omitted from the notation):

$$\langle lm|\mu m\rangle = (-)^{(n-1+m-\mu)/2}\left\langle \frac{n-1+m}{2}, -\frac{\mu}{2}; \frac{n-1-m}{2}, \frac{\mu}{2}\middle| l0\right\rangle \tag{12}$$

for any allowed n, $n \geqslant l+1$ and $n \geqslant |m|+|\mu|+1$ [16]. This transformation was given for the first time in configuration space by Park [17], who did not specify the explicit form of the eigenfunctions and so did not define the phase convention. Progress was made by Hughes [18], by Tarter [19] (who calculated the elements by direct integration) and the correct phase difference is give by Englefield [8].

The relationship between the two bases is most clearly understood in the four dimensional momentum space. It is interesting to note that this connection is strictly linked to alternative parametrizations of the 3-sphere.

The Clebsch-Gordan coefficient in eg. (12), having at least one zero element, suggests that μ can be interpreted as a *helicity* quantum number. Physically this means that the chosen axis is the one corresponding to a zero component of the orbital angular momentum vector l and therefore lies in the plane of the orbit, as the Runge-Lenz vector [8,6]; an operation of this kind finds its analogues in several contexts: we mention the *space fixed — body fixed* transformations in molecular collisions [20], the Hund's case $(e) \to (c)$ transformation in molecular spectroscopy and atomic scattering [21,22] and the passage between symmetric and asymmetric coordinates in the hyperspherical treatment of the three-body problem [23,24].

The coefficient connecting the polar and Zeeman basis will be called Z:

$$\langle lm|\lambda m\rangle = (-)^{l+\lambda} Z^{nm}_{l\lambda} \tag{13}$$

The overlap between spherical and Zeeman states, was originally derived as a sum of the product of two vector coupling coefficients [3]:

$$Z^{n,m}_{l,\lambda} = \sum_{\mu}(-)^{\alpha}\left\langle \frac{n-1+m}{2}, -\frac{\mu}{2}; \frac{n-1-m}{2}, \frac{\mu}{2}|l, 0\right\rangle \times \left\langle \frac{n-1+m}{2}, -\frac{\mu}{2}; \frac{n-1-m}{2}, \frac{\mu}{2}|\lambda, 0\right\rangle \tag{14}$$

(the phase $\alpha = (n-1-m-\mu)/2 + l + \lambda$ is an integer: if it were omitted, the sum would be $\delta_{l\lambda}$).

By direct integration or by specializing overlap coefficients between alternative harmonics [25] we are able to write it directly as a single sum of the Racah type. This sum [26] is a hypergeometric function ${}_4F_3$ of unit argument:

$$Z^{n,m}_{l,\lambda} = \sqrt{\left(l+\frac{1}{2}\right)\left(\lambda+\frac{1}{2}\right)}[C(l)\cdot C(\lambda)]^{1/2}\cdot\frac{\Gamma\left(\frac{n-m-q(l)-q(\lambda)}{2}\right)}{\Gamma(m+1)\Gamma\left(\frac{m+n+1+q(l)+q(\lambda)}{2}\right)} \times {}_4F_3\left(\begin{matrix}\frac{m-1+q(l)}{2}, \frac{m+l+q(l)+1}{2}, \frac{m-\lambda+q(\lambda)}{2}, \frac{m+\lambda+1+q(\lambda)}{2} \\ m+1, \frac{m-n+1+q(l)+q(\lambda)}{2}, \frac{m+n+1+q(l)+q(\lambda)}{2}\end{matrix}; 1\right) \tag{15}$$

where

$$C(k) = \frac{\left(\frac{n+k-1}{2}+p(k)\right)!\left(\frac{k+m}{2}-q(k)\right)!\left(\frac{k-m}{2}-q(k)\right)!\Gamma\left(\frac{n-k}{2}+p(k)\right)}{\Gamma\left(\frac{n+k+2}{2}+p(k)\right)\Gamma\left(\frac{m+k+1}{2}+q(k)\right)\Gamma\left(\frac{k-m+1}{2}+q(k)\right)\left(\frac{n-k-1}{2}-p(k)\right)!},$$

and

$$p(k) = -\frac{1+(-1)^{k-n}}{4}, \quad q(k) = \frac{(-1)^{k-m}-1}{4}$$

(see Ref. [14] for the explicit expression and Table 1 for some numerical values).

It can also be connected with Racah polynomials [25], although it cannot be reduced to the ordinary Racah's or 6-j coefficient which performs angular momentum recoupling. Indeed, like a Racah's recoupling coefficient it is orthogonal with respect to summation on two angular momentum quantum numbers (l and λ), but contains a

Table 1 $Z_{l,\lambda}^{n,m}$ *matrix elements for some quantum numbers*

$nl\lambda m$	$Z_{l,\lambda}^{n,m}$	$nl\lambda m$	$Z_{l,\lambda}^{n,m}$
1000	1	4111	$\frac{1}{5}$
2100	1	4210	$\frac{2}{\sqrt{5}}$
211-1	-1	422-1	-1
2111	1	4221	1
3000	$\frac{1}{3}$	4300	$\frac{2}{\sqrt{5}}$
3110	1	431-1	$-\frac{2\sqrt{6}}{5}$
3200	$\frac{2\sqrt{2}}{3}$	4311	$\frac{2\sqrt{6}}{5}$
321-1	-1	432-2	1
3211	1	4320	$-\frac{1}{\sqrt{5}}$
322-2	1	4322	1
322-1	-1	433-3	-1
3220	$-\frac{1}{3}$	433-1	$\frac{1}{5}$
3222	1	4331	$-\frac{1}{5}$
4100	$\frac{1}{\sqrt{5}}$	4333	1
411-1	$-\frac{1}{5}$		

projection quantum number. The Z coefficient can be compactly written as a *generalized* 6-j symbol [27]:

$$Z_{l,\lambda}^{n,m} = (-)^{(l+\lambda)/2+1+p(l)+p(\lambda)}\sqrt{\left(l+\frac{1}{2}\right)\left(\lambda+\frac{1}{2}\right)}\begin{Bmatrix} p(\lambda)-\frac{1}{4} & \frac{n-1}{2} & \frac{\lambda}{2}-\frac{1}{4} \\ p(l)-\frac{1}{4} & \frac{m-1}{2} & \frac{l}{2}-\frac{1}{4} \end{Bmatrix} \tag{16}$$

Note that $Z_{l,\lambda}^{n,m}$ *is zero when* $n+1+\lambda+m$ *is even* and shows the symmetries

$Z^{n,m}_{l,\lambda} = Z^{n,-m}_{l,\lambda}$ and $Z^{n,m}_{l,\lambda} = Z^{n,m}_{\lambda,l}$. Moreover it enjoys most properties of ordinary 6-j symbols, such as several recurrence relationships [28], including a 3-term one which allows accurate and efficient calculations [29] even for high values of the arguments.

3. Multielectronic Atomic Orbitals

Hyperspherical harmonics are now explicitly considered as expansion basis sets for atomic and molecular orbitals. In this treatment the key role is played by a generalization of the famous Fock projection [5] for hydrogen atom in momentum space, leading to the connection between hydrogenic orbitals and four-dimensional harmonics, as we have seen in the previous section. It is well known that the hyperspherical harmonics are a basis for the irreducible representations of the rotational group on the four-dimensional hypersphere S^3: from this viewpoint hydrogenoid orbitals can be looked at as representations of the four-dimensional hyperspherical symmetry [14].

The generalization of Fock treatment to higher dimensions [11] allows us to study atomic and molecular structure from the point of view of the broken symmetry of hyperspheres. As a matter of fact atoms and molecules can be considered as arising from the breaking of the hyperspherical symmetry of d-dimensional hydrogenoid atom [$d = 3(\mathrm{N}-1)$ for N body Coulomb problems] due to the introduction of charged particles (electrons or nuclei respectively). Thus in configuration space, Sturmian basis functions [11] (multidimensional hydrogenic orbitals of fixed energy) can be used as expansion bases sets to build up atomic and molecular orbitals. Since the counterparts of d-dimensional Sturmian functions in momentum space are $(d+1)$-dimensional hyperspherical harmonics, the possibility of exploiting different parametrizations of the $(d+1)$-dimensional sphere can be considered. In fact one can choose among alternative hyperspherical harmonics pertaining to different subgroup chain reductions of the original $(d+1)$-dimensional rotation group and thus possessing different symmetry properties. The corresponding overlap coefficients can be written in terms of (generalized) vector coupling or recoupling coefficients and, due to the duality between configuration and momentum space, can also be used to connect alternative Sturmian bases in direct space.

The hyperspherical method, from a formal viewpoint, is general and thus can be applied to any N-body Coulomb problem. Our analysis of the three body Coulomb problem exploits considerations on the symmetry of the seven-dimensional rotational group. The matrix elements which have to be calculated to set up the secular equation can be very compactly formulated. All intervals can be written in closed form as matrix elements corresponding to coupling, recoupling or transformation coefficients of hyperangular momenta algebra.

The results that we will present elsewhere (see however [30]) represent an exploration about the power of this formulation, with particular reference to convergence of alternative sets. This point of view is crucial to work out approximation and truncation techniques for the basis sets and to make this formulation not only formally complete, but also computationally efficient.

4. Conclusion and Further Remarks

As is well known, conventional hydrogenoid *spherical* orbitals are strictly linked to tetradimensional harmonics when the atomic orbitals for the tridimensional hydrogen atom are considered in momentum space. We have therefore studied an alternative representation, providing the *Stark* and *Zeeman* basis sets, related to the *spherical* one by orthogonal transformation, see eqs. (12) and (15). The latter can also be interpreted as suitable *timber* coefficients relating different *tree* structures of hyperspherical harmonics for R^4 (Fig. 1).

The analysis of alternative representations for hydrogen atom wavefunctions can be extended to any mathematical dimension and therefore can be of interest beyond the one-electron case. The outline provided in a previous work [10] suggests that what we have done here can also be exploited in higher dimensional problems [11]. In general, one has orthonormal bases of two kinds – the (hyper)harmonics and the (hyper)hydrogenoid position wavefunctions: since the multidimensional extension of Fock projection relates the harmonics to hydrogenoid momentum space wavefunctions, both can be used to expand the *d*-dimensional plane wave and to Fourier transform between position and momentum space. Transformation between the generalizations to hyperspaces of the spherical and Stark bases similar to the one discussed here (Sec. 2.1) will require the introduction of generalized angular momentum coupling coefficients (Hahn polynomials), whose properties are described in ref. [31].

In configuration space, the many body Coulomb problem, describing atomic and molecular structure, is isomorphic to that for a multidimensional hydrogen atom with an anisotropic charge [32]. This encourages (see [10]) multidimensional hydrogenoid Sturmian expansions [33,34] consisting of a (hyper)-radial part and of a hyperspherical harmonic. As outlined above, the corresponding eigenfunctions in momentum space are hyperspherical harmonics belonging to a space of one dimension higher. The calculation of matrix elements for the secular equation for the spectrum involves radial and angular integrals. Guided by the previous observation of the duality between multidimensional hydrogenoid Sturmians and hyperspherical harmonics, it can be shown [10] that not only angular integrals, but also radial ones can be obtained as (generalized) vector coupling coefficients.

Therefore, the numerical solution of the secular equation for different representations for three-body problems should enable us to make interesting discoveries concerning on the various quantum numbers arising in the labeling of the various kind of harmonics. Alternative subgroup reductions, generalising those that we encountered in Sec. 2, and corresponding to alternative expansion in hyperspherical harmonics, can also be dealt with by a generalization of angular momentum algebra [10]. Explicitly, as already noted, the three-body problem requires a mapping of the 6-sphere embedded in a seven dimensional Euclidean space. To test the effectiveness of the different parametrizations in some specific cases we have already illustrated the above considerations by some numerical examples again for the test case for H_2^+ [10,35], but relaxing the usual restriction to the fixed nuclei scheme as adopted e.g. in [2] (see however [35]).

Acknowledgements

This research is supported by the Italian National Research Council (CNR), by the Ministero dell'Universitá e della Ricerca Scientifica e Tecnologica (MURST), and by the European Union within the Training and Mobility of Researchers Network 'Potential Energy Surfaces for Molecular Spectroscopy and Dynamics' [Contract no. FMRX-CT96-088].

References

1. I. Shibuya and C.E. Wulfman, Proc. Roy. Soc. **A286**, 376 (1965).
2. V. Aquilanti, S. Cavalli, C. Coletti and G. Grossi, Chem. Phys. **209**, 405 (1996). Notations and phase conventions are as in this reference.
3. J.J. Labarthe, J. Phys. B **14**, L-467 (1981); D.R. Herrick, Phys. Rev. A **26**, 323 (1982); D. Delande and J.C. Gay, Europhys. Lett. **5**, 303 (1988).
4. D. Delande and J.C. Gay, Phys. Rev. Lett. **59**, 1809 (1987); J.C. Gay in *Spectrum of Atomic Hydrogen* edited by G.W. Series (World Scientific, Singapore, 1988); F. Penent, D. Delande and J.C. Gay, Phys. Rev. A. **37**, 4707 (1988). These authors also emphasize that relationships among bases are given by four dimensional rotations.
5. V. Fock, Z. Phys. **98**, 145 (1935).
6. C.E. Wulfman, in *Group Theory and its Applications* edited by E.M. Loebl, Vol. 2, p. 145, (Academic Press, New York, 1971); E.G. Kalnins, W. Miller and P. Winternitz, S.I.A.M. J. Appl. Math. **30**, 630 (1976).
7. M. Bander and C. Itzykson, Rev. Mod. Phys **38**, 330 (1966).
8. M.J. Englefield, *Group Theory and the Coulomb Problem*; Wiley-Interscience; New York, (1972); B.G. Wybourne, *Classical Groups for Physicists*; Wiley and Sons; New York, 1972; L.C. Biedenharn and J.D. Louck, *Angular Momentum in Quantum Physics. Theory and Applications*, p. 335–364; Addison-Wesley Publishing Co.; Reading Mass., 1981; B.R. Judd, *Angular Momentum Theory for Diatomic Molecules*; Academic Press: New York, 1975.
9. J. Avery and D.R. Herschbach, Int. J. Quantum Chem., **41**, 673, (1992).
10. V. Aquilanti, S. Cavalli, C. Coletti, D. De Fazio, G. Grossi, in *New Methods in Quantum Theory*; C.A. Tsipis, V.S. Popov, D. R. Herschbach, J.S. Avery Eds., Kluwer (1996), pp. 233–250.
11. V. Aquilanti, S. Cavalli and C. Coletti, Chem. Phys. **214**, 1 (1997).
12. V. Aquilanti and J. Avery, Chem. Phys. Lett. **267**, 1 (1997), and references therein.
13. V. Aquilanti, S. Cavalli and G. Grossi, J. Chem. Phys. **85**, 1362 (1986).
14. V. Aquilanti, S. Cavalli and C. Coletti, Phys. Rev. Lett. **80**, 3209 (1998).
15. B. Podolsky and L. Pauling, Phys. Rev., **34**, 109 (1929).
16. In Ref. [2] relationships among alternative notation for 'parabolic quantum numbers are presented. Expression (12) is preferred for the form of the coefficient rather than $\langle lm|\mu m\rangle = (-)^{(n-1+m-\mu/)2} \cdot \langle (n-1)/2,\ (m-\mu)/2;\ (n-1)/2,\ (m+\mu)/2|l,\ m\rangle$, which is normally found in literature and from which it follows by Regge's symmetry [2]. This leads to the physical interpretation of the transformation as one to a body fixed reference frame, where the quantization axis is such that the orbital angular momentum has zero projection and μ appears as a *helicity* quantum number.
17. D. Park, Z. Phys. **36**, 155 (1960).
18. J.W.B. Hughes, Proc. Phys. Soc., **91**, 810 (1967).
19. C.B. Tarter, J. Math. Phys., **11**, 3192 (1970).
20. V. Aquilanti, L. Beneventi, G. Grossi, F. Vecchiocattivi, J. Chem. Phys., **89**, 751 (1988).
21. V. Aquilanti, G. Grossi, J. Chem. Phys., **73**, 1165 (1980).
22. V. Aquilanti, S. Cavalli, G. Grossi, Z. Phys. D, F. Hund Festschrift, **36**, 215 (1996).
23. V. Aquilanti, S. Cavalli, G. Grossi, Theor. Chim. Acta, **79**, 283 (1991).
24. V. Aquilanti, S. Cavalli, Few-Body Systems, Suppl. **6**, 573 (1992).

25. A.F. Nikiforov, S.K. Suslov and V.B. Uvarov, *Classical Orthogonal Polynomials of a Discrete Variable*, Springer-Verlag: Berlin, (1991); S.K. Suslov, Sov. J. Nucl. Phys. **38**, 829 (1983).
26. C. Coletti, *Tesi di Dottorato*, Università di Perugia, (1997). Further details are given here.
27. J. Raynal, J. Nucl. Phys. **19**, 467 (1978).
28. D.A. Varshalovich, A.N. Moskalev and V.K. Khersonskii, *Quantum Theory of Angular Momentum*, World Scientific: Singapore, (1988).
29. K. Schulten and R.G. Gordon, J. Math. Phys. **16**, 1961 (1975).
30. D. Di Domenico, *Tesi di Laurea*, Università di Perugia, (1997).
31. V. Aquilanti, S. Cavalli and D. De Fazio, J. Phys. Chem **99**, 15694 (1996).
32. H. Klar, J. Phys. B: Atom. Molec. Phys., **7**, L436 (1974).
33. J. Avery, Hyperspherical Harmonics, Applications in Quantum Theory; Kluwer Academic Publishers: Dordrecht, The Netherlands, (1989).
34. R.C. Whitten, J. S. Sims, Phys. Rev. A, **9**, 1586 (1974).
35. V. Aquilanti, G. Grossi, A. Laganá, E. Pelikan, H. Klar, Lett. Nuovo Cim., **41**, 541 (1984).
36. Yu. F. Smirnov and K.V. Shitikova, Sov. J. Part. Nucl., **8**, 344 (1976).
37. N. Ya. Vilenkin, G. I. Kuznetsov, Ya. A. Smorodinskii, Sov. J. Nucl. Phys., **2**, 645 (1966).
38. N. Ya. Vilenkin, Fonctions Spéciales et Théorie de la Représentation des Groupes; Dunod: Paris, (1969).

An Overview of the CASVB Approach to Modern Valence Bond Calculations

Thorstein Thorsteinsson[a] and David L. Cooper[b]
[a]*Chemistry Laboratory IV, Copenhagen University, Universitetsparken 5, DK-2100 Copenhagen Ø, Denmark*
[b]*Department of Chemistry, University of Liverpool, P.O. Box 147, Liverpool L69 7ZD, United Kingdom*

Abstract

We discuss all of the key features of our current CASVB methodology for modern valence bond calculations on ground and excited states. The CASVB strategy may be used to generate compact representations of CASSCF wavefunctions or, alternatively, to perform the fully-variational optimization of various general types of VB wavefunction. We report also a new application, namely to the fourteen π electrons of a planar dimethylenecyclobutadiene chain with three rings.

1. Introduction

The CASVB strategy [1–9] uses a very efficient algorithm for the transformation of CASSCF [10] structure spaces, for the interpretation of CASSCF wavefunctions, and for the fully-variational optimization of VB wavefunctions. Important features for the quality of the final description include the unbiased optimization of both the VB orbitals and the mode of spin coupling, and also flexibility in the choice of the form of wavefunction.

Many other methods, of course, take advantage of invariances with respect to orbital transformations to obtain alternative representations of wavefunctions, such as for example the commonly employed localization procedures for the doubly-occupied MOs from Hartree-Fock calculations [11–15]. We give here a brief account of procedures that particularly seek a *valence bond* representation of MO wavefunctions.

In an early application to butadiene [16], and later to the ground and excited states of benzene [17], Berry analyzed MO-based wavefunctions using valence bond concepts, simply by considering the overlaps with nonorthogonal VB structures. Somewhat closer than this to a CASVB type of approach, are the procedures employed by Linnett and coworkers, in which small CI wavefunctions were transformed (exactly) to nonorthogonal representations [18–20]. The main limitation in their case was on the size of systems that may be treated (the authors considered no more than four-electron systems), both because this non-linear transformation must exist, and because it must be possible to obtain it with reasonable effort.

The first paper to consider a transformation of specifically a CASSCF wavefunction to modern VB structures is probably that of McDouall and Robb [21]. An initial localization of the CASSCF MOs [22] is essential in their method. From this localized

A. Hernández-Laguna et al. (eds.), Quantum Systems in Chemistry and Physics, Vol. 1: Basic Problems and Model Systems, 303–326.

representation of the CASSCF, the nonorthogonal structures are found by a one-step transformation, taking into account the orbital parameters only to first order. Their method therefore strongly depends on the localized-orbital representation being of high quality. A more satisfactory solution to the problem of structure transformations has been found by Hiberty [23], although he considers the problem from a totally different perspective. His work emphasizes the projection of Hartree-Fock wavefunctions, or small CI expansions, onto a *classical*-VB many-electron basis consisting typically of a full CI expansion in the valence space (or selected parts thereof). His algorithm transforms separately the α and β parts of the structure space, which leads to acceptable efficiency (*cf.* Section 2.5).

Hirao has also recently considered the transformation of CASSCF wavefunctions to valence bond form [24, 25]. An orthogonal VB orbital basis was first considered, in which case the CASSCF CI vector may be found by re-solving the CI problem. Later he considered also the transformation to a classical VB representation. The transformation of the CASSCF space was achieved by calculating all overlap terms, $\langle \Phi_I^{\mathrm{CASSCF}} | \Phi_J^{\mathrm{CASVB}} \rangle$, and solving the subsequent linear problem, using a Davidson-like iterative scheme.

It must be pointed out that the quality of (truncated) VB representations of MO wavefunctions, as in the variational case, is strongly related to the freedom of the orbitals to deform and to overlap in the bond-forming directions. Such considerations date back to the improvements made by Coulson and Fischer [26] to the Heitler-London [27] description of H_2. It is the opinion of the present authors, and indeed of many others, that neither the orthogonal nor classical VB procedures give sufficiently compact representations for practical applications. The philosophy of both the variational spin-coupled approach [28], as well as of the CASVB strategy, is to allow unprejudiced optimization of the orbitals. Although the orbitals resulting from, say, a variational optimization often turn out to be fairly localized, we by no means consider the generation of localized (*e.g.* atomic-orbital-like) orbitals a goal in itself (*cf.* Ref. [7]).

We mention here also two schemes that have utilized the properties of CASSCF spaces for realizing *variational* optimization of VB wavefunctions (*cf.* the discussion in Section 2.2). Murphy and Messmer [29] used a non-linear parametrization similar to Linnett's of a hierarchy of increasingly correlated valence bond wavefunctions, suggesting the 'orbital-relaxed' GVB (GVB/R) wavefunction as the best compromise between compactness and quality. Malcolm and McDouall [30,31] have utilized instead the variational character of the biorthogonal method, when expanded in a full CI. The resulting wavefunction is formally equivalent to a CASSCF, with the valence bond orbitals being found by optimizing a fully nonorthogonal perfect-pairing structure.

The main purpose of the present article is to provide, in one place, an overview of all of the key features of our current CASVB methodology. The CASVB code has already been applied to a wide range of problems [1–9,32,33] and its use in further fully-variational modern-VB calculations are reported elsewhere in this Volume. Although we have chosen to concentrate here on methodology, we do present also a new application (to a fourteen-electron system).

2. Key Aspects of the CASVB Strategy

2.1. Decomposing the transformation of a many-electron space

We consider in this Section a many-electron space that is closed under linear transformations of the defining orbitals, so that

$$\{\phi'\} = \{\phi\}\mathbf{O} \Rightarrow \{\Phi'\} = \{\Phi\}\mathbf{T}(\mathbf{O}). \tag{1}$$

We have used { } for the row vectors of the respective entities, while we denote by ϕ and Φ the orbitals and many-electron functions, and by $\mathbf{O}$ and $\mathbf{T}(\mathbf{O})$ the two corresponding linear transformations, respectively. Various types of many-electron space for which such transformations may be carried out have been described by Malmqvist [34]. In general, $\mathbf{O}$ may be non-unitary, possibly with subsidiary conditions imposed for ensuring that the corresponding transformation of the N-electron space exists (*e.g.* a block-diagonal form according to orbital subsets or irreducible representations).

A quite general method for obtaining $\mathbf{T}(\mathbf{O})$ involves the decomposition of $\mathbf{O}$ into simpler 'orbital updates', for example as

$$\mathbf{O} = \mathbf{O}_{11}(\lambda_1)\mathbf{O}_{12}(\lambda_2)\mathbf{O}_{13}(\lambda_3)\ldots\mathbf{O}_{mm}(\lambda_{m^2}), \tag{2}$$

where we choose $\mathbf{O}_{\mu\nu}(\lambda)$ to describe a simple change of just one orbital:

$$\mathbf{O}_{\mu\nu}(\lambda)\colon\ \phi_\nu \to \phi_\nu + \lambda\phi_\mu. \tag{3}$$

A factorization of $\mathbf{O}$ of this form can be carried out using standard matrix algorithms such as LU decomposition or Gaussian elimination with back-substitution [2,34,35]. The main advantage is the straightforward evaluation of the structure transformation corresponding to Eq. (3), which is just

$$\mathbf{T}(\mathbf{O}_{\mu\nu}(\lambda)) = (\mathbf{1} + \lambda\mathbf{E}^{(\alpha)}_{\mu\nu})(\mathbf{1} + \lambda\mathbf{E}^{(\beta)}_{\mu\nu}) = \mathbf{1} + \lambda\mathbf{E}^{(1)}_{\mu\nu} + \lambda^2\mathbf{E}^{(2)}_{\mu\nu}, \tag{4}$$

in terms of the spin orbital single-replacement operators $\mathbf{E}^{(\alpha)}_{\mu\nu}$ and $\mathbf{E}^{(\beta)}_{\mu\nu}$, or, equivalently, in terms of the spin-averaged single- and double-replacement operators $\mathbf{E}^{(1)}_{\mu\nu}$ and $\mathbf{E}^{(2)}_{\mu\nu}$. Considering Eq. (2), the complete transformation of the CI space may be realized by m^2 applications of the replacement operators in Eq. (4). The treatment of these operators is well established and there is no difference from their application in standard procedures based on orthogonal orbitals. The first form in Eq. (4) is appropriate for a many-electron space defined in terms of Slater determinants, and in such a case the transformation may be separated into independent α and β parts [2]. The spin-averaged replacement operators are, of course, more natural in connection with a many-electron basis consisting of configuration state functions.

We shall in the following restrict out attention to transformations of CI spaces of CASSCF type. We may then divide $\{\phi\}$ into subsets consisting of core, active, and virtual orbitals, $\{\phi\} = \{\{\phi^{\text{core}}\}, \{\phi^{\text{act}}\}, \{\phi^{\text{virtual}}\}\}$, so that Eq. (1) may be fulfilled for any transformation affecting only the active space. (Also, certain transformations involving the core orbitals leave the wavefunction invariant, but those will not be considered here). In order to specify the CASSCF completely, the three numbers

(N, m, S) may be provided, signifying the numbers of active electrons and active orbitals, and total spin quantum number, respectively.

The (infinite) latitude for choosing different representations of the CASSCF wavefunction may be exploited in order to bring it into a VB-like form. We adopt here the partitioning

$$\Psi_{\rm CAS} = c_{\rm VB}\Psi_{\rm VB} + c^{\perp}_{\rm RES}\Psi^{\perp}_{\rm RES}, \tag{5}$$

in which we take the component $\Psi_{\rm VB}$ to be of VB type (see later), while the 'residual' is the orthogonal complement, with $\langle\Psi_{\rm VB}|\Psi^{\perp}_{\rm RES}\rangle = 0$. We could define the optimal representation of $\Psi_{\rm CAS}$ as the one that maximizes the weight of the valence bond component. Equivalently, we may adopt the criterion

$$\text{maximize} \qquad S_{\rm VB} = \frac{\langle\Psi_{\rm CAS}|\Psi_{\rm VB}\rangle}{\langle\Psi_{\rm VB}|\Psi_{\rm VB}\rangle^{1/2}}, \tag{6}$$

since such a formulation proves more convenient when considering the VB optimization problem below. Our initial applications of this strategy have shown that, with a suitably chosen $\Psi_{\rm VB}$, more than 99% of a CASSCF wavefunction may be brought to VB-like form. When this is the case, one might view $\Psi^{\perp}_{\rm RES}$ as merely a 'perturbation' to the (modern) valence bond description, and many of the most important features of the electron correlation taken into account by the CASSCF may be rationalized from an interpretation of just $\Psi_{\rm VB}$.

2.2. *An alternative viewpoint: addressing the nonorthogonality problem*

We consider here a slightly different perspective to the one adopted in Section 2.1, more strongly associated with the *fully-variational* optimization of VB wavefunctions.

A natural starting point for modern valence bond applications lies in the optimization of the spin-coupled wavefunction [28], which consists of a single 'covalent' configuration of N singly occupied orbitals:

$$\Psi_{\rm VB} = A(\Phi^{\rm core}\phi_1^{\rm VB}\phi_2^{\rm VB}\ldots\phi_N^{\rm VB}\Theta^N_{SM}). \tag{7}$$

Such a spin-coupled wavefunction is optimized with respect to the core wavefunction (if applicable), as well as to the nonorthogonal valence bond orbitals,

$$\phi_\mu^{\rm VB} = \sum_i {\rm o}_{i\mu}\chi_i, \tag{8}$$

and to the spin-coupling coefficients, c_{Sk}, which define the optimal N-electron spin eigenfunction Θ^N_{SM} with spin S and projection M:

$$\Theta^N_{SM} = \sum_k^{f^N_S} c_{Sk}\Theta^N_{SM;k}. \tag{9}$$

For the construction of spin eigenfunctions see, for example, Ref. [36]. The spin-coupled wavefunction may be extended by adding further configurations, in which case we may speak of a multiconfigurational spin-coupled (MCSC) description. In the following, we collect the VB structure coefficients, irrespective of the form of the VB wavefunction, into the vector $\mathbf{c}^{\mathrm{VB}}$.

The evaluation of matrix elements of a given N-electron operator, $\hat{Q}$:

$$\langle \Psi_{\mathrm{VB}} | \hat{Q} | \Psi_{\mathrm{VB}} \rangle, \tag{10}$$

can be termed the basic problem in valence bond calculations. An extensively used approach for its solution involves the expansion of Ψ_{VB} in terms of Slater determinants, leading to a summation over $(N-p)$-dimensional cofactors [37], for a p-particle operator, $\hat{Q}$. A further saving, relative to a simple determinant-based algorithm, can be achieved by the use of projected spin functions [38].

An alternative way of attacking this problem lies in the reexpression of Ψ_{VB} using a different orbital basis [1–9]:

$$\Psi_{\mathrm{VB}} = \sum_I c_I \Phi_I^{\mathrm{VB}} = \sum_I d_I \Phi_I^{\mathrm{MO}}. \tag{11}$$

In the notation of Eq. (1), $\{\Phi^{\mathrm{VB}}\}$ signifies the many-electron functions defined in terms of the VB-orbital basis, while $\{\Phi^{\mathrm{MO}}\}$ is a set of many-electron functions defined in terms of orthogonal orbitals. In this way, the nonorthogonality problem has been essentially 'bypassed', so that the computational effort is instead determined by

(1) the transformation from the $\{\Phi^{\mathrm{VB}}\}$ to the $\{\Phi^{\mathrm{MO}}\}$,
(2) the increased length of the expansion necessary for the $\{\Phi^{\mathrm{MO}}\}$, and
(3) the evaluation of matrix elements in the MO basis, $\langle \Phi_I^{\mathrm{MO}} | \hat{Q} | \Phi_J^{\mathrm{MO}} \rangle$.

We have outlined in the previous Section an efficient solution to the problem posed in (1). For the second point, one quickly realizes that the space $\{\Phi^{\mathrm{MO}}\}$ must take a full CI (or, CASSCF) form, if no particular restrictions are to be placed on the valence bond orbitals. For the single-configuration spin-coupled wavefunction, there is for this reason an important link to (N, N, S) CASSCF wavefunctions.

For the variational description of a ground state, Ψ_{VB} is of course optimized by minimizing its energy expectation value

$$E_{\mathrm{VB}} = \frac{\langle \Psi_{\mathrm{VB}} | \hat{H} | \Psi_{\mathrm{VB}} \rangle}{\langle \Psi_{\mathrm{VB}} | \Psi_{\mathrm{VB}} \rangle} \tag{12}$$

with respect to the free parameters. This optimization is carried out in the CASVB strategy by means of a two-step procedure:

(1) optimize active-space parameters, *i.e.* $\mathbf{O}$ and $\mathbf{c}^{\mathrm{VB}}$,
(2) optimize $\{\phi^{\mathrm{MO}}\}$ with respect to core-active, core-virtual, and active-virtual orbital rotations.

Step (1) defines a particular set of CASSCF CI coefficients, so that, by using this CI vector, step (2) may be carried out by most current CASSCF programs, such as the one available in the MOLPRO *ab initio* package [39–41].

It is, of course, also possible to optimize the parameters in $\mathbf{O}$ and $\mathbf{c}^{\mathrm{VB}}$ (step (1) above), so as to minimize the energy, *without* relaxing the CASSCF orbital spaces. This defines an 'energy-based' valence bond interpretation of the CASSCF solution, analogous to the overlap-based criterion embodied in Eq. (6). This is particularly useful for the direct comparison of the two types of optimization criterion. We have generally found very close agreement between the descriptions generated by the energy-based and (cheaper) overlap-based criteria.

2.4. Expressions for derivatives

In order to ensure reliable convergence, we generally employ VB optimization procedures that require first and second derivatives with respect to all of the variational parameters. Expressions for these derivatives are most easily derived by considering the first- and second-order changes in Ψ_{VB} with respect to the VB parameters defined by $\mathbf{O}$ and $\mathbf{c}^{\mathrm{VB}}$. These may be generated from combinations of $\mathbf{E}_{\mu\nu}^{(1)}$ (*cf.* Eqs. (3) and (4)) and the analogous operator for the structure space, $\mathbf{E}_{I1}$. For the first-order variations in orbitals and structure coefficients, we find:

$$\mathbf{E}_{\mu\nu}^{(1)}\mathbf{c}^{\mathrm{VB}} \qquad \text{and} \qquad \mathbf{E}_{I1}\mathbf{c}^{\mathrm{VB}}, \tag{13}$$

respectively. The second-order variations may be derived from (*cf.* Ref. [42]):

$$(\mathbf{E}_{\mu\nu}^{(1)}\mathbf{E}_{\sigma\tau}^{(1)} - \delta_{\nu\sigma}\mathbf{E}_{\mu\tau}^{(1)})\mathbf{c}^{\mathrm{VB}} \tag{14}$$

associated with simultaneous orbital changes $\mu \to \nu$ and $\sigma \to \tau$, and

$$\mathbf{E}_{\mu\nu}^{(1)}\mathbf{E}_{I1}\mathbf{c}^{\mathrm{VB}}, \tag{15}$$

associated with a combination of orbital excitation and a change of structure coefficients. No second-order changes of the structure coefficients arise, because the complete structure space is spanned by the first-order variations.

Inserting the first- and second-order wavefunction variations into either of the expressions Eq. (6) or (12) yields the complete expressions for the gradient and Hessian. In doing this, it becomes clear that the computationally most expensive parts are related to the terms containing first-order variations in both bra and ket, *e.g.*

$$\mathbf{c}^{\mathrm{VB}\dagger}\mathbf{E}_{\mu\nu}^{(1)\dagger}\mathbf{T}(\mathbf{s})\mathbf{E}_{I1}\mathrm{c}^{\mathrm{VB}}, \tag{16}$$

to the second-order change in wavefunction normalization, and to

$$\mathbf{c}^{\mathrm{VB}\dagger}\mathbf{E}_{\mu\nu}^{(1)\dagger}\mathbf{H}\mathbf{E}_{\sigma\tau}^{(1)}\mathrm{c}^{\mathrm{VB}}, \tag{17}$$

contributing to the second-order variation in energy. An exception to this general finding, however, occurs for the combinations of first-order orbital changes, since terms of the form $\mathbf{c}^{\mathrm{VB}\dagger}\mathbf{E}_{\mu\nu}^{(1)\dagger}\mathbf{T}(\mathbf{s})\mathbf{E}_{\sigma\tau}^{(1)}\mathrm{c}^{\mathrm{VB}}$ are related to the terms $\mathbf{c}^{\mathrm{VB}\dagger}\mathbf{T}(\mathbf{s})\mathbf{E}_{\mu\nu}^{(1)}\mathbf{E}_{\sigma\tau}^{(1)}\mathrm{c}^{\mathrm{VB}}$ by a straightforward two-index transformation.

2.5. Efficiency considerations

For a general application of $\mathbf{T(O)}$ to an (N, m, S) CASSCF CI vector using a determinant-based algorithm [2], the iteration count can be shown to be

$$N_{\mathrm{L}} = N_{\mathrm{det}} \times [N_\alpha(m - N_\alpha) + N_\beta(m - N_\beta) + N_\alpha + N_\beta], \tag{18}$$

with $N_\alpha = \frac{1}{2}N + S$ and $N_\beta = \frac{1}{2}N - S$, and

$$N_{\mathrm{det}} = \binom{m}{N_\alpha} \times \binom{m}{N_\beta} \qquad \text{for } N_\alpha \neq N_\beta, \text{ or}$$

$$N_{\mathrm{det}} = \frac{1}{2}\binom{m}{N_\alpha} \times \left[\binom{m}{N_\alpha} + 1\right] \qquad \text{for } N_\alpha = N_\beta. \tag{19}$$

For the off-diagonal (*i.e.* $\mu \neq \nu$) replacement operators (giving rise to the two first terms in Eq. (18)), each iteration consists of a multiplication and an addition, while the effect of the diagonal operators can be achieved by simple multiplies.

For comparison, the count for the more straightforward structure transformation algorithm employed by Hiberty and coworkers [23] can in the general case be shown to be

$$N_{\mathrm{L}}^{\mathrm{Hiberty}} = N_{\mathrm{det}} \times \left[\binom{m}{N_\alpha} + \binom{m}{N_\beta}\right], \tag{20}$$

although this number may be reduced for particularly simple types of CI expansion.

For the energy-based optimization, the effect of the electronic Hamiltonian on a CI vector must be realized. In the most efficient approach for achieving this [43–45], the associated computational effort is approximately

$$N_{\mathrm{H}} = N_{\mathrm{det}} \times \left[\tfrac{1}{4}N_\alpha^2(m - N_\alpha)^2 + \tfrac{1}{4}N_\beta^2(m - N_\beta)^2 + N_\alpha N_\beta(m - N_\alpha)(m - N_\beta)\right] \tag{21}$$

multiplications and additions, the three terms in this expression being related to the $\alpha\alpha$, $\beta\beta$ and $\alpha\beta$ parts of the two-electron integrals, respectively.

Values for N_{L} and N_{H} are shown for general (N, N, 0) CASSCF spaces in Figure 1. (The utilization of point group symmetry would, however, reduce these numbers). Both quantities show approximately exponential growth for this range of N, but this is not in itself significantly worse than the increase in the dimension of the CASSCF CI space. Compared to our more usual procedure [37,38] involving the calculation of the fourth-order density matrix, $\mathrm{D}^{(4)}$, the CASVB strategy's scaling properties are clearly very favourable.

Given that the applications of **T(O)** and/or **H** are the most demanding steps, our algorithm centres on obtaining the required first and second derivatives with as few such applications as possible. This may be achieved by constructing each term sequentially left-to-right (or right-to-left), so that at most two vectors need be stored in memory at any one time. The computational effort associated with one iteration, employing a full second-order optimization scheme, will then be approximately

$$N_{\mathrm{var}} \times N_{\mathrm{H}} + N_{\mathrm{var}}^{\mathrm{struc}} \times N_{\mathrm{L}} \tag{22}$$

for the energy-based optimization criterion, and

$$N_{\mathrm{var}}^{\mathrm{struc}} \times N_{\mathrm{L}} \tag{23}$$

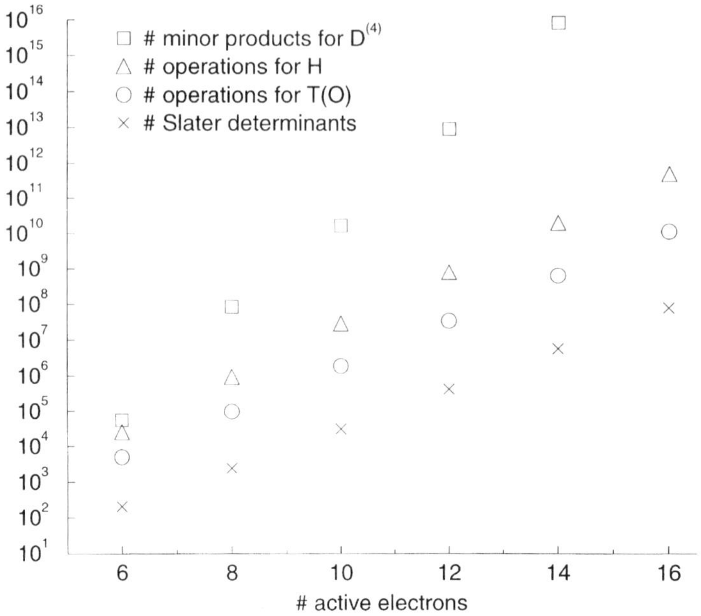

Fig. 1. Key values of operation counts for the nonorthogonal treatment of N-electron systems with $S = 0$. The operation count for $D^{(4)}$ is appropriate to a spin-coupled wavefunction, while the other quantities relate to an (N, N, 0) CASSCF space.

for the overlap-based case. $N_{\rm var}^{\rm struc}$ is the number of free variables associated with the VB structure space, while $N_{\rm var}$ is the total number of free variational parameters.

3. Wavefunction Optimization

We consider in this Section particular aspects relating to the optimization of a CASVB wavefunction. As for most procedures involving the optimization of orbitals, special attention should be given to the choice of optimization strategy. The optimization problem is in this case non-linear, so that an exact second-order scheme is preferable in order to ensure reliable convergence. A particularly useful account of various second-order optimization schemes has been presented by Helgaker [46].

Our MIN($E_{\rm VB}$) optimization scheme is based on a second-order model of the form

$$E_{\rm VB}(\mathbf{x} + \delta\mathbf{x}) \approx E_{\rm VB}(\mathbf{x}) + \mathbf{g}^{\rm T}\delta\mathbf{x} + \tfrac{1}{2}\delta\mathbf{x}^{\rm T}\mathbf{G}\delta\mathbf{x}, \tag{24}$$

where $\mathbf{g}$ is the (column) vector of first derivatives of E with respect to the variational parameters and $\mathbf{G}$ is the corresponding matrix of second derivatives (Hessian). We have opted for the method due to Fletcher [47,48], in which an equation of the form

$$\delta\mathbf{x} = -(\mathbf{G} - \alpha\mathbf{I})^{-1}\mathbf{g}, \qquad \|\delta\mathbf{x}\| \leqslant h \tag{25}$$

is solved for an update, $\delta\mathbf{x}$. A main advantage of a stabilized Newton-Raphson scheme, such as Eq. (25), is the natural step-size (h) control, but other strategies, such as for example augmented Hessian methods (see, for example, Refs. [49, 50]), may be just as suitable. Analogous considerations apply also to our MAX(S_{VB}) optimization scheme.

There are even more possibilities for optimization procedures if one considers also a partitioning of the optimization problem. Work in this area is currently ongoing, and we present for the first time, in Section 5, results obtained using a Davidson-like strategy for the (linear) optimization of the VB structure coefficients [51,52]. Further details of this new procedure will be presented in a future publication.

3.1. Elimination of redundant or constrained parameters

We have always aimed in CASVB for the simplest possible elimination procedure, bearing in mind that the number of variational prameters can be quite considerable in practical applications.

We shall see later that all the various types of constraints lead to conditions on the first-order variation of **x** (see Eq. (24)) of the form

$$\mathbf{C}^{(i)\mathrm{T}}\delta\mathbf{x} = 0, \qquad i = 1 \ldots. \tag{26}$$

Collecting the orthogonal complement to the $\mathbf{C}^{(i)}$ in the ($N_{\mathrm{parm}} \times N_{\mathrm{free}}$)-dimensional matrix **L** (note that the constraints vectors may be linearly dependent), we can express the requirements in Eq. (26) as a linear transformation to a new set of variables, $\mathbf{x}'$:

$$\delta\mathbf{x} = \mathbf{L}\delta\mathbf{x}'. \tag{27}$$

The second-order expression, Eq. (24), may be transformed correspondingly, to give

$$E_{\mathrm{VB}}(\mathbf{x} + \delta\mathbf{x}') \approx E_{\mathrm{VB}}(\mathbf{x}) + (\mathbf{g}')^{\mathrm{T}}\delta\mathbf{x}' + \tfrac{1}{2}(\delta\mathbf{x}')^{\mathrm{T}}\mathbf{G}'\delta\mathbf{x}', \tag{28}$$

with

$$\mathbf{g}' = \mathbf{L}^{\mathrm{T}}\mathbf{g} \tag{29}$$

and

$$\mathbf{G}' = \mathbf{L}^{\mathrm{T}}\mathbf{G}\mathbf{L}. \tag{30}$$

As such, solving the reduced problem of Eq. (28), and back-transforming, ensures that the update, $\delta\mathbf{x}$, satisfies all the necessary conditions.

Linear constraints:
It is useful distinguish between linear constraints that take the form

$$\mathbf{C}^{(i)\mathrm{T}}\mathbf{x} = 0, \qquad i = 1, \ldots, \tag{31}$$

from which Eq. (26) follows directly, and non-linear constraints consisting of normalization conditions and orthogonality relations.

Linear constraints include various simple cases such as forcing a single coefficient to zero (in which case **C** has only one non-zero coefficient) or setting pairs of coefficients equal or opposite. However, more complicated conditions, such as symmetry requirements for orbitals, as considered in the next Section, are also included in this category. For example, Eq. (38) in Section 3.3, leads to m–$n(\zeta_{\mu\mu})$ linear constraints, corresponding to the orthogonal complement of $\hat{R}$'s $\zeta_{\mu\mu}$ eigenspace. (These various quantities are defined later).

Non-linear constraints — normalization conditions and orthogonality relations: The normalization condition for an orbital ϕ_μ, can be expressed in the form

$$\langle \phi_\mu + \delta\phi_\mu | \phi_\mu + \delta\phi_\mu \rangle - 1 = 0 \tag{32}$$

(the condition for Ψ_{VB} can be treated in an analogous manner). This leads to the requirement (to first order) that $\langle \phi_\mu | \delta\phi_\mu \rangle = 0$, which can trivially be expressed in the form of Eq. (26). However, due to the second-order term in Eq. (32), the orbitals must be renormalized after each update.

The elimination of the constraint(s) associated with $\langle \phi_\mu | \phi_\nu \rangle = 0$ can similarly be derived from the condition

$$\langle \phi_\mu + \delta\phi_\mu | \phi_\nu + \delta\phi_\nu \rangle = 0, \tag{33}$$

which requires that

$$\langle \delta\phi_\mu | \phi_\nu \rangle + \langle \phi_\mu | \delta\phi_\nu \rangle = 0 \tag{34}$$

to first-order. As for the case of the normalization condition, an error can be introduced that is second-order in the step size, so that (exact) orthogonality must be reimposed after each update in an iteration. This must be achieved without affecting the simple orbital constraints outlined above and, to this end, a 'pair-wise symmetric' orthogonalization procedure has been implemented [6].

3.2. *Symmetry adaptation of VB wavefunctions*

Defining the (idempotent) projection operator

$$\hat{P}^{(i)} = (d_i/g) \sum_R \chi^{(i)}(R)^* \hat{R}, \tag{35}$$

(*cf.* for example Ref. [53]), the condition for the VB wavefunction to belong to the irreducible representation $\Gamma^{(i)}$ can be written simply as

$$\hat{P}^{(i)} \Psi_{\mathrm{VB}} = \Psi_{\mathrm{VB}}. \tag{36}$$

Here the $\hat{R}$ form the set of linear coordinate transformations that leave the nuclear framework invariant, $\chi^{(i)}$ are the characters associated with the d_i-dimensional irreducible representation $\Gamma^{(i)}$, and g is the order of the point group, $\mathcal{G}$.

We shall consider first the case of a spin-coupled wavefunction, as the requirements for an MCSC wavefunction may be easily derived from this. Our key assumption will be that the operators $\hat{R}$ induce *permutations* of the valence bond orbitals:

$$\hat{R}\phi_\mu = \zeta_{\mu\nu}(R)\phi_\nu \qquad \forall R \in \mathcal{G}, \tag{37}$$

where $\zeta_{\mu\nu}(R) = \pm 1$ for real orbitals (special cases in which more general transformation properties of the orbitals lead to the correct symmetry may also be envisaged). In practice, it is sufficient to satisfy Eq. (37) just for the *generators* of $\mathcal{G}$. Since a complete set of spin eigenfunctions for N electrons form a basis for S_N (the permutation group of degree N), correct symmetry properties may be ensured by combining Eq. (37) with appropriate restrictions on the form of the spin function [4,54,55]. If an MCSC wavefunction is employed, an additional requirement is that the set of spatial configurations must be closed under the orbital permutations induced by the $\hat{R}$.

It is convenient to consider separately the diagonal and off-diagonal orbital relations in Eq. (37). To satisfy a *diagonal* relation of the form

$$\hat{R}\phi_\mu = \zeta_{\mu\mu}(R)\phi_\mu, \tag{38}$$

we can simply diagonalize the matrix representation of $\hat{R}$, and restrict the expansion of ϕ_μ to be in terms of only the right-hand eigenvectors associated with the eigenvalue $\zeta_{\mu\mu}(R)$. (Alternatively, it is possible to define a projection operator for ϕ_μ, analogous to the one defined in Eq. (35)). For an *off-diagonal* ($\mu \neq \nu$) orbital relation,

$$\hat{R}\phi_\mu = \zeta_{\mu\nu}(R)\phi_\nu, \tag{39}$$

one may simply eliminate ϕ_ν from the optimization problem, substituting it by $\zeta^*_{\mu\nu}(R)\hat{R}\phi_\mu$.

An alternative to the above procedure is to utilize the fact that

$$\Psi^{(i)}_{\mathrm{VB}} = \hat{P}^{(i)}\Psi_{\mathrm{VB}} \tag{40}$$

must have the correct symmetry properties, irrespective of the form of Ψ_{VB}. Application of this operator in the VB orbital basis may prove somewhat cumbersome, because $\hat{P}^{(i)}\Psi_{\mathrm{VB}}$ has multiconfigurational form in this case. However, applying $\hat{P}^{(i)}$ to a CI vector in a symmetry-adapted MO basis is particularly simple, especially if only subgroups of D_{2h} are considered. The effect of $\hat{P}^{(i)}$ may be realized in CASVB according to

$$\mathbf{c}^{(i)}_{\mathrm{VB}} = \mathbf{T}(\mathbf{O}^{-1})\mathbf{P}^{(i)}_{\mathrm{MO}}\mathbf{T}(\mathbf{O})\mathbf{c}^{\mathrm{VB}}. \tag{41}$$

Here, $\mathbf{c}^{(i)}_{\mathrm{VB}}$ is the CI vector in the basis of VB structures, projected such that it transforms according to the irreducible representation $\Gamma^{(i)}$. Because even the standard CASVB approach involves an expansion of Ψ_{VB} in terms of structures formed from orthogonal molecular orbitals (the transformation given in Eq. (41)), this implementation is completely straightforward.

Symmetry projection may be applied after an optimization procedure is completed, so as to 'remedy' a symmetry-broken solution. Alternatively it may be applied *during* an optimization procedure by substituting $\Psi^{(i)}_{\mathrm{VB}}$ for Ψ_{VB} in Eq. (6) or Eq. (12). An attractive third option is to employ the projection operator in combination with orbital and structure-coefficient constraints. For examples of these different possibilities we refer to previously published accounts [4].

3.3. *Excited state optimization*

We consider in the following the optimization of modern valence bond wavefunctions for states that are second or higher within a particular symmetry. If CASSCF solutions are available for each electronic state, $\{\Psi_{\rm CAS}^{(1)}, \Psi_{\rm CAS}^{(2)}, \ldots\}$, application of the overlap-based criterion,

$$\text{maximize} \qquad S_{\rm VB} = \frac{\langle \Psi_{\rm CAS}^{(i)} | \Psi_{\rm VB}^{(i)} \rangle}{\langle \Psi_{\rm VB}^{(i)} | \Psi_{\rm VB}^{(i)} \rangle^{1/2}}, \tag{42}$$

will lead to the $i^{\rm th}$ VB solution, $\Psi_{\rm VB}^{(i)}$. Such a procedure has the obvious advantage of simplicity, since the CASSCF solution $\Psi_{\rm CAS}^{(i)}$ may in most cases may be found as just the $i^{\rm th}$ root of the appropriate full CI problem.

For energy-based criteria, it is generally accepted that non-linear optimization of the $i^{\rm th}$ root is best achieved by seeking a saddle point of order $i - 1$ [56,57]. The most natural generalization of a restricted-step Newton-Raphson optimization procedure to saddle-point optimization is, we believe, the 'trust region image minimization' (TRIM) algorithm proposed by Helgaker [58], and our version of this scheme has proved to be very well behaved for this type of problem. As for ground states, we consistently find good agreement between analogous overlap-based and energy-based results [8].

The modern VB descriptions of excited states investigated so far with CASVB can be classified into one of three categories:

(1) Recoupling of electron spins.
(2) 'Valence' → 'valence' orbital excitation, so that the excited state has doubly-occupied orbitals (so-called 'ionic' configurations).
(3) 'Valence' → 'virtual' orbital excitation, including those leading to Rydberg states.

The optimization of excited states of type (1) or (3) is usually straightforward, but excited states of type (2) may require care in the choice of reference function [9].

Optimization using the energy-based criterion is somewhat more involved than its overlap-based counterpart, due mainly to the necessity for saddle-point optimization. An important advantage of the energy-based optimization, however, is the possibility of partitioning the optimization problem. By employing a two-step (or, in general n-step) optimization procedure, convergence may be achieved onto stationary points corresponding to particular types of orbital excitations. Using such a strategy, we have obtained a similar functionality to the procedure described by Doggett *et al.* [59–61] and, as such, we have been able to reproduce their results. Olsen *et al.* [56] have stressed the importance for two-step procedures of taking into account the coupling between the optimization problems, and we foresee a need to address this question for treatments of more complicated systems.

3.4. *Practical aspects*

As is the case for standard orthogonal-orbital MCSCF calculations, the optimization of VB wavefunctions can be a complicated task, and a program such as CASVB should therefore not be treated as a 'black box'. This is true, to a greater or lesser extent, for most procedures that involve orbital optimization (and, hence, non-linear optimization problems), but these difficulties are compounded in valence bond theory by the possibility of linear dependence. In a modern-VB optimization involving no restrictions on the orbital overlaps, the possibility of two (or more) orbitals becoming identical, or nearly identical, cannot be excluded. Such a situation will lead to singularities in the optimization problem, and possibly also to numerical instabilities. While problems of the latter type might be overcome by a clever choice of algorithm (such as the use of a singular value decomposition [35] of the orbital transformation matrix, **O**), this is not, at present, a problem that has been solved in a completely satisfactory manner. Applications for which one should be particularly aware of such difficulties include the description of 'inner' electrons, and, in some cases, of lone-pair orbitals.

In practice, the success of a CASVB calculation depends strongly on the way in which the optimization is carried out, and we emphasize in this Section some particular aspects that have not been considered in Sections 3.1 and 3.2:

(1) choice of core orbital space;
(2) irreducible representations spanned by active orbitals;
(3) the guess for the VB orbitals;
(4) constraints on the spin coupling.

The electron density described by a core orbital space will of course strongly affect the nature of the active orbitals. The form of the inactive orbitals may be influenced by placing symmetry restrictions on them, or by invoking an initial orbital localization [11–15]. The localized orbitals that are not of interest for the VB description may then be placed in the core in the subsequent CASSCF or fully variational VB calculation and, if necessary, some or all of them may be frozen.

Symmetry restrictions may also be placed on the active orbitals in order to determine the nature of the resulting modern valence bond solution. This is exemplified by the common use of $\sigma - \pi$ separation for planar molecules (*cf.* Section 5). In earlier applications to ozone and diborane [2,4] it was also seen that the distribution of active orbitals among the irreducible representations was the deciding factor for the types of VB solution possible. It should also be borne in mind here that the nature of the lowest-lying CASSCF solution may not always coincide with that of the optimal fully-variational modern VB wavefunction.

For the valence bond orbitals themselves, it is generally natural to specify a starting guess in the AO basis. Such a guess might, of course, not lie entirely inside the space spanned by the active space, and it must therefore be projected onto the space of the active MOs. This is achieved trivially in CASVB, by multiplication by the inverse of the matrix of MO coefficients.

One of the most useful types of constraint is the restriction of the spin coupling to just a single mode. Many molecular systems are described rather well by the perfect pairing mode of spin coupling, for example. A useful alternative, especially when this is not the case, is to base the structure coefficients on the CASSCF wavefunction in the VB orbital basis:

$$\mathbf{c}^{\mathrm{VB}} = \mathbf{P}^{\mathrm{VB}}\mathbf{T}(\mathbf{O}^{-1})\mathbf{c}^{\mathrm{CAS}}, \tag{43}$$

in which $\mathbf{P}^{\mathrm{VB}}$ sets to zero all coefficients for non-valence-bond structures [2]. This approach has the advantage of being less biased than, say, a perfect-pairing constraint, and has led to very satisfactory convergence characteristics in many applications.

4. Wavefunction Analysis

We focus in this Section on particular aspects relating to the direct interpretation of valence bond wavefunctions. Important features of a description in terms of modern valence bond concepts include the orbital shapes (including their overlap integrals) and estimates of the relative importance of the different structures (and modes of spin coupling) in the VB wavefunction. We address here the particular question of defining nonorthogonal weights, as well as certain aspects of spin correlation analysis.

4.1. Nonorthogonal weights

Consider a wavefunction given as a linear combination of structures Ψ_I:

$$\Psi = \sum_I c_I \Psi_I. \tag{44}$$

The weights of structures in an orthogonal case are defined simply as $w(I) = |c_I|^2$. However, some ambiguity is inevitable in the nonorthogonal case, because of the non-vanishing overlap matrix, $(\mathbf{S})_{IJ} = \langle \Psi_I | \Psi_J \rangle$.

We list here the main features of the three most commonly used definitions for nonorthogonal weights – for further details we refer to our previously published account [6]:

Scheme	Definition	Remarks
Chirgwin-Coulson weights [62]	$w^{\mathrm{CC}}(I) = \sum_J c_J^* c_I \mathbf{S}_{JI} = \langle \Psi \vert \Psi_I \rangle c_I$	May lie outside the range [0;1].
Löwdin weights [63]	$w^{\mathrm{L}}(I) = \left\vert \sum_J c_J (\mathbf{S}^{1/2})_{IJ} \right\vert^2$	Not linear, *i.e.* $w^{\mathrm{L}}(I) + w^{\mathrm{L}}(J) \neq w^{\mathrm{L}}(I+J)$, in general.
Inverse-overlap weights [64,65]	$w^{\mathrm{I}} = \vert \mathbf{c}_I \vert^2 / (\mathbf{S}^{-1})_{II}$	Not linear. Not normalized to unity unless scaled.

Some additional computational overhead is associated with the calculation of $w^{\mathrm{L}}(I)$ and $w^{\mathrm{I}}(I)$, because of the requirement for $\mathbf{S}^{1/2}$ and $\mathbf{S}^{-1}$, respectively.

Nonorthogonal weights are used within CASVB:

(a) for estimating the importance of individual structure in the total VB wavefunction, and
(b) for estimating the importance of components in the CASSCF wavefunction expressed in the VB basis, *i.e.* in $\mathbf{T(O)c}_{\mathrm{CAS}}$.

For (a), of course, the choice of spin basis may be very important for highlighting different features of the spin coupling, with our most common choices being the Rumer, Kotani, Serber, or projected spin function bases. Transformation between these (complete) bases is, in any case, very straightforward [36,66,67].

The various schemes for evaluating nonorthogonal weights all require various components of the structure overlap matrix, which in the CASVB strategy may be obtained from

$$\mathbf{S} = \mathbf{T}^{\dagger}(\mathbf{O})\mathbf{T}(\mathbf{O}) = \mathbf{T}(\mathbf{O}^{\dagger}\mathbf{O}) = \mathbf{T}(\mathbf{s}). \tag{45}$$

Here, $\mathbf{S}$ is the overlap matrix between structures in the full CI expansion, while $\mathbf{s}$ denotes the VB orbital overlap matrix. In cases where Ψ_{VB} is used in combination with a projection operator (*e.g.* $\hat{P}^{(i)}$ in Eq. (35)), the expression for $\mathbf{S}$ may be slightly more involved, but it can always be expressed in terms of the two transformations $\mathbf{T}(\mathbf{O}^{\dagger})$ and $\mathbf{T}(\mathbf{O})$.

Computation of the overlap matrix between valence bond structures, as required for $w^{\mathrm{CC}}(I)$ or $w^{\mathrm{L}}(I)$, then requires N_{VB} applications of $\mathbf{T}(\mathbf{s})$, after which diagonalization of $\mathbf{S}_{\mathrm{VB}}$ easily yields both its inverse and square root.

The implementation of weights for functions in the full CI space differs from the above, since it is impractical to form the full overlap matrix, at least for very large cases. Instead we must rely on one or more full-CI transformations, making use of Eq. (45), and the relations

$$\mathbf{S}^{1/2} = \mathbf{T}(\mathbf{s}^{1/2}) \tag{46}$$

and

$$\mathbf{S}^{-1} = \mathbf{T}(\mathbf{s}^{-1}). \tag{47}$$

The Chirgwin-Coulson and Löwdin weights may in this way be calculated from a single transformation of $\mathbf{c}_{\mathrm{CAS}}$. Evaluation of inverse-overlap weights is somewhat more involved, because the diagonal elements of $\mathbf{S}^{-1}$ must be known. In a straightforward strategy, one would therefore require N_{CAS} applications of $\mathbf{T}(\mathbf{s}^{-1})$. The problem may be greatly simplified, however, by decreasing the *resolution* of the weights. The number of full-CI transformations required coincides with the length of the expansion, Eq. (44), used for the CASSCF wavefunction. It can be shown [7] that this simplification does not involve any loss of accuracy.

For the weights of the CASSCF wavefunction in the VB basis, one may also consider analyzing $\Psi_{\mathrm{RES}}^{\perp}$ (*cf.* Eq. (5)). This not only focuses directly on the changes that occur on augmenting Ψ_{VB} with the remainder of the full CI space, but the resulting weights also prove to be significantly more stable than when they are evaluated on the basis of just Ψ_{CAS} [7].

4.2. Spin correlation analysis

Spin correlation analysis [68–71] is an attractive means of representing the coupling between electron spins, having the important advantage of being independent of the spin basis used. We shall consider in the following the particular case of a spin-coupled wavefunction, defined as in Eq. (7). For such a case, spin expectation values may be defined for any *subgroup* of electrons. Of particular utility are the values for pairs of electrons, for which we can make use of the identity.

$$\langle\Psi_{\mathrm{VB}}|(\hat{s}_\mu+\hat{s}_\nu)^2|\Psi_{\mathrm{VB}}\rangle=\langle\Psi_{\mathrm{VB}}|(1-\hat{P}_{\mu\nu})|\Psi_{\mathrm{VB}}\rangle, \tag{48}$$

in which $\hat{P}_{\mu\nu}$ interchanges the orbitals ϕ_μ^{VB} and ϕ_ν^{VB} [70]. Alternatively, we may consider expectation values over the total spin function, Θ_{SM}^N, as is done for the application presented in Section 5.

In the implementation of Eq. (48) within CASVB, we use the fact that the effect of an orbital permutation is very straightforward to realize in the determinant basis. Just as for more general transformations, the permutation may be decomposed into separate α and β parts, and the transformation $\mathbf{P}^\alpha\times\mathbf{P}^\beta$ carried out either in two steps, or as a single pass through all the determinants. This procedure is quite inexpensive, even for a CI vector in the complete CASSCF space. In our implementation of the full-CI structure transformation (described in Section 2.1), we have employed a decomposition of $\mathbf{O}$ with full pivoting, in order to improve numerical accuracy.

4.3. *Weights in irreducible representations*

An extraordinarily simple, but nevertheless quite useful procedure involves an analysis of the distribution of Ψ_{VB} among the irreducible representations in the point group. In CASVB this is easily achieved by transforming the VB wavefunction to the CASSCF MO basis giving weights according to

$$w(\Gamma^{(i)})=\sum_{I\in\Gamma^{(i)}}(\mathbf{T}(\mathbf{O})\mathbf{c}^{\mathrm{VB}})_I^2. \tag{49}$$

In the case of a symmetry-broken solution, these weights can not only diagnose the problem, but also quantify its extent. As for the projection operator considered in Section 3.2, this analysis is used primarily with Abelian point groups.

5. New Application

For our first ever fourteen-electron example, we chose a dimethylenecyclobutadiene chain with three rings, as shown in Figure 2. Spin-coupled (SC) calculations have been reported for different spin multiplicities of this system by Raos *et al.* [71], who treated only ten π electrons as active, optimizing two doubly-occupied orthogonal orbitals for the four π electrons of the terminal C-C linkages, and freezing all of the σ orbitals at

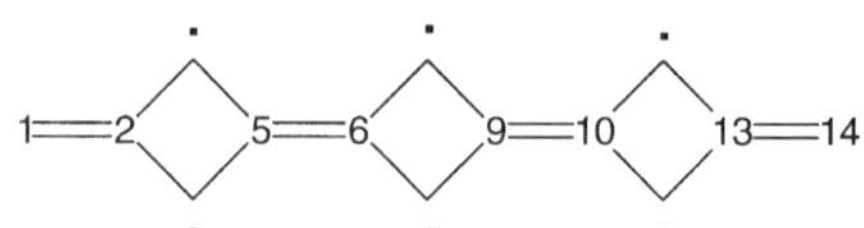

Fig. 2. Schematic representation of a planar dimethylenecyclobutadiene chain with three rings.

the RHF level. An interesting feature of such systems is the occurrence of so-called 'anti-pairs' [72]: these may be described, to a first approximation, as in- and out-of-phase combinations of deformed p_π orbitals from opposite corners of a ring, with predominantly triplet coupling of the associated electron spins.

We have adopted the same basis set and model geometry as in the previous SC study, but present instead modern VB representations of (14, 14, S) CASSCF wavefunctions using our MAX(S_{VB}) optimization scheme. Coordinates were chosen so that the molecular plane is $z = 0$ with the x-axis pointing along the chain. The π orbitals transform as $6B_{1u} + 5B_{2g} + 2B_{3g} + A_u$ and the doubly-occupied σ orbitals as $14A_g + 12B_{3u} + 8B_{2u} + 6B_{1g}$. Initially, we insisted that the $1b_{1u}$ and $1b_{2g}$ orbitals remain doubly-occupied in all configurations, thereby generating (10, 12, S) CASSCF expansions, in which all the 'core' orbitals were optimized simultaneously with the active ones. The resulting energies for all possible term symbols for $S = 0, 1, 2, 3$ are listed in Table 1.

Subsequent (14, 14, S) CASSCF calculations, in which we removed the previous restriction on the $1b_{1u}$ and $1b_{2g}$ orbitals, were carried out for the two lowest-energy term symbols for each value of S (see Table 1). The relative CASSCF energies (ΔE_{CAS}) are reported in Table 2 for the lowest root of each spin multiplicity. These values, which are fairly similar to those from the ten-electron SC calculations of Raos *et al.* [71], add further weight to their assertion that the ground state of this system should be a triplet.

Table 1 *CASSCF total energies (in hartree) for the system shown in Figure 2.*

	Singlet	Triplet	Quintet	Septet
(10, 12, S)				
A_g	−535.433 852	−535.425 910	−535.398 316	−535.234 946
B_{3u}	−535.432 967	−535.426 548	−535.377 600	−535.235 160
B_{2u}	−535.456 744	−535.496 909	−535.485 612	−535.398 473
B_{1g}	−535.488 455	−535.477 045	−535.428 142	−535.352 963
(14, 14, S)				
B_{2u}	−535.534 642	−535.579 050	−535.565 607	−535.474 411
B_{1g}	−535.572 841	−535.558 387	−535.508 502	−535.435 366

Table 2 *Summary of (14, 14, S) CASSCF and 14-electron CASVB results for the system shown in Figure 2.*

State	N_{CSF}	N_{det}	f_S^{14}	S_{VB}	ΔE_{CAS} (kJ mol^{-1})	ΔE_{SC} (kJ mol^{-1})[a]
$^1B_{1g}$	687 616	2 944 256	429	0.994173	16	15
$^3B_{2u}$	1 253 103	2 257 265	1 001	0.993613	0	0
$^5B_{2u}$	753 501	1 004 162	1 001	0.992189	35	42
$^7B_{2u}$	217 245	250 661	637	0.992295	275	289

a Ten-electron spin-coupled calculations from Ref. [71]

The dimensions of the various (14, 14, S) CASSCF spaces are shown in Table 2, in which we list the numbers of configuration state functions (N_{CSF}) and the corresponding numbers of determinants (N_{det}). Direct interpretation of such CASSCF wavefunctions is by no means a trivial task. The CASVB procedure, on the other hand, has been used here to generate for each state a *single* spatial configuration (*cf.* Eq. (7)) consisting of fourteen singly-occupied, nonorthogonal π orbitals (plus doubly-occupied σ orbitals taken directly from the CASSCF wavefunction). We used the full spin space for each multiplicity, with no constraints on the form of the orbitals or on the mode of spin coupling. The resulting values of S_{VB} exceed of 0.992 in each case (see Table 2), indicating that an extraordinary proportion of the full (14, 14, S) CASSCF wavefunctions may be expressed in this simple form.

Although no such constraints were imposed in the optimizations, we find that eight of the π orbitals resemble deformed p_π functions, each associated with one of the carbon atoms along the x-axis, and it proves convenient to number the orbitals according to the corresponding centres in Figure 2. The six remaining π orbitals take the form of three anti-pairs, with one associated with each ring. From left-to-right, orbitals ϕ_3, ϕ_7 and ϕ_{11} are the 'in-phase' combinations, and ϕ_4, ϕ_8 and ϕ_{12} the 'out-of-phase' ones. Symmetry-unique orbitals for the triplet state are plotted in Figure 3 as contours in the plane 1 bohr above the molecular plane, with projected positions of the nuclei denoted by their chemical symbols. On the whole, the plots are fairly similar to those presented in the previous study [71], except that edge effects were apparent in the ten-electron SC study, due to orthogonalization of active orbitals to the two doubly-occupied π orbitals.

The overlaps between the π orbitals are shown in the upper triangle of Table 3; the successive rows in each box relate to $S = 0$, 1, 2, and 3. Not surprisingly, the largest values (highlighted) occur for orbitals associated with neighbouring atoms. The septet state is clearly different from the others with, for example, much reduced overlaps in the 'terminal' C—C units. Even so, the values of $\langle\phi_2|\phi_1\rangle$ for the other states are somewhat smaller than is typical for polyenes (>0.65), and the values of $\langle\phi_6|\phi_5\rangle$ are significantly smaller again, suggesting that the 'internal' C—C units are not associated with conventional, semi-localized π bonds. Further insight is obtained by examining the mode of spin coupling.

The dimensions of the spin spaces for the active electrons (f_S^{14} in Table 2, *cf.* Eq. (9)) are certainly not small. It proved difficult to find a spin basis in which very few of the coefficients were large and so we adopted instead a spin correlation scheme (*cf.* Section 4.2). In the present work, we exploited the way in which expectation values of the two-electron spin operator $\hat{s}_i \cdot \hat{s}_j$, evaluated over the total spin eigenfunction Θ_{SM}^N, depend on the coupling of the individual spins associated with orbitals ϕ_i and ϕ_j. Negative values indicate singlet character and positive values triplet character. Special cases of the expectation value are:

Fig. 3. Symmetry-unique π orbitals for the triplet state, shown as contours in the plane 1 bohr above the molecular plane, with projected positions of the nuclei denoted by their chemical symbols.

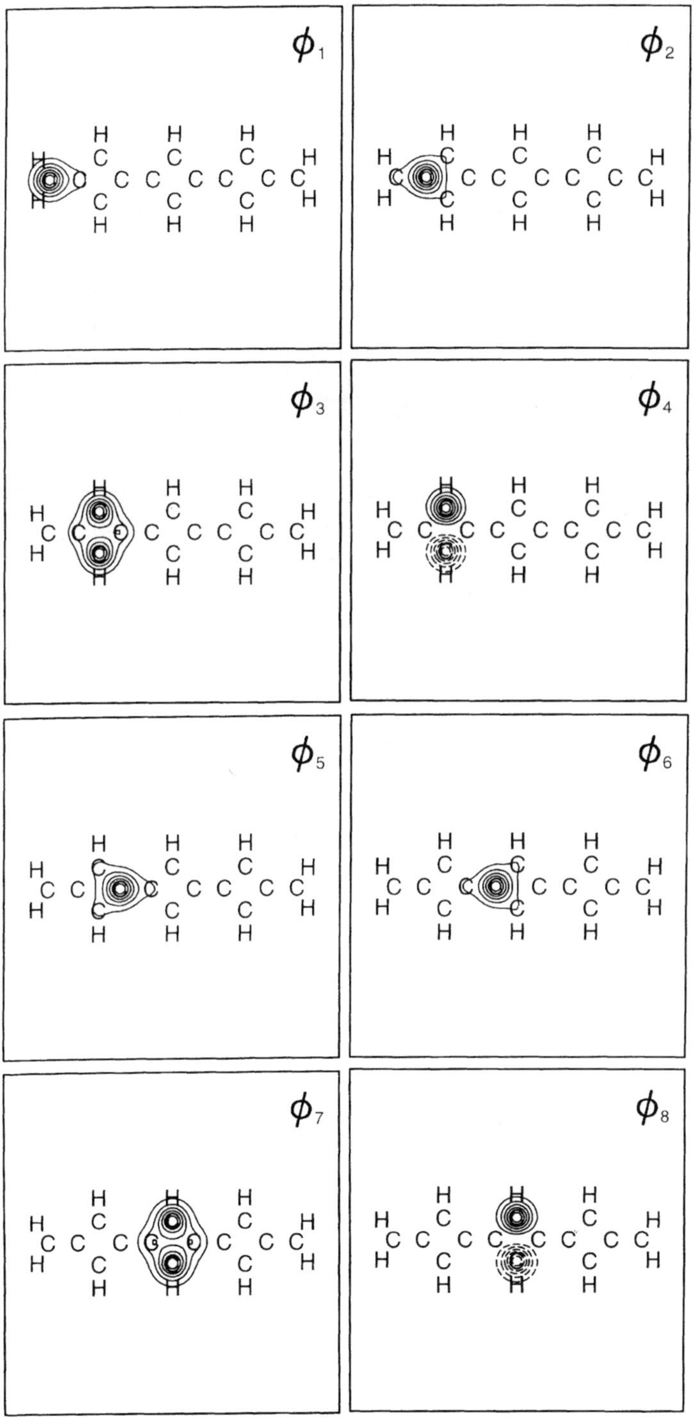

ϕ_1
ϕ_2
ϕ_3
ϕ_4
ϕ_5
ϕ_6
ϕ_7
ϕ_8

Table 3 *Overlap (upper triangle) and spin correlation (lower triangle) matrices; successive values in the boxes relate to S = 0, 1, 2, and 3*

	1	2	3	4	5	6	7	8	9	10	11	12	13	14
1		***0.57*** ***0.58*** ***0.58*** ***0.34***	0.14 0.13 0.14 0.00	0.00 0.00 0.00 0.00	−0.02 0.00 0.00 −0.24	−0.05 −0.06 −0.04 −0.12	0.01 0.00 0.01 −0.07	0.00 0.00 0.00 0.00	0.01 0.00 0.00 −0.01	0.01 −0.01 −0.04 0.04	0.00 0.00 −0.02 0.01	0.00 0.00 0.00 0.00	0.01 0.00 −0.01 0.01	0.00 0.00 −0.01 −0.02
2	***−0.63*** ***−0.63*** ***−0.63*** ***−0.30***		***0.57*** ***0.53*** ***0.57*** ***0.73***	0.00 0.00 0.00 0.00	0.19 0.18 0.20 0.25	−0.11 −0.15 −0.10 −0.07	−0.04 −0.02 −0.07 −0.01	0.00 0.00 0.00 0.00	0.02 0.01 −0.01 −0.08	0.02 0.02 0.02 −0.02	−0.01 0.00 0.01 0.02	0.00 0.00 0.00 0.00	−0.01 0.00 0.00 0.03	0.01 0.00 −0.01 0.01
3	0.17 0.18 0.18 0.21	−0.25 −0.26 −0.26 −0.46		0.00 0.00 0.00 0.00	***0.69*** ***0.70*** ***0.67*** ***0.62***	0.08 0.07 0.07 0.19	−0.07 0.01 −0.13 0.20	0.00 0.00 0.00 0.00	−0.04 −0.04 −0.08 0.00	0.02 0.01 0.04 −0.05	0.02 0.01 0.03 0.01	0.00 0.00 0.00 0.00	−0.01 0.00 0.01 0.02	0.00 0.00 −0.02 0.01
4	0.14 0.15 0.15 0.22	−0.18 −0.19 −0.19 −0.31	***0.24*** ***0.24*** ***0.24*** ***0.24***		0.00 0.00 0.00 0.00	0.00 0.00 0.00 0.00	0.00 0.00 0.00 0.00	0.03 0.03 0.01 0.01	0.00 0.00 0.00 0.00	0.00 0.00 0.00 0.00	0.00 0.00 0.00 0.00	0.00 0.00 0.00 0.00	0.00 0.00 0.00 0.00	0.00 0.00 0.00 0.00
5	−0.21 −0.21 −0.20 0.01	0.17 0.17 0.16 0.14	−0.45 −0.46 −0.43 −0.14	−0.33 −0.34 −0.33 −0.07		***0.46*** ***0.41*** ***0.49*** ***0.50***	0.05 0.07 0.07 0.16	0.00 0.00 0.00 0.00	−0.09 −0.13 −0.09 −0.12	−0.03 −0.05 −0.02 −0.14	0.02 0.01 0.04 −0.05	0.00 0.00 0.00 0.00	0.02 0.02 0.02 −0.02	0.01 −0.01 −0.04 0.04
6	0.12 0.13 0.09 0.05	−0.10 −0.12 −0.09 −0.11	0.11 0.18 0.13 0.07	0.09 0.19 0.12 0.05	***−0.29*** ***−0.32*** ***−0.32*** ***−0.47***		***0.59*** ***0.61*** ***0.67*** ***0.70***	0.00 0.00 0.00 0.00	0.15 0.14 0.23 0.28	−0.09 −0.13 −0.09 −0.12	−0.04 −0.04 −0.08 0.00	0.00 0.00 0.00 0.00	0.02 0.01 −0.01 −0.08	0.01 0.00 0.00 −0.01
7	−0.12 −0.13 0.00 0.12	0.08 0.09 0.00 −0.01	−0.07 −0.17 0.04 0.09	−0.06 −0.20 0.06 0.13	0.13 0.16 0.03 0.14	−0.45 −0.43 −0.30 −0.27		0.00 0.00 0.00 0.00	***0.59*** ***0.61*** ***0.67*** ***0.70***	0.05 0.07 0.07 0.16	−0.07 0.01 −0.13 0.19	0.00 0.00 0.00 0.00	−0.04 −0.02 −0.07 −0.01	0.01 0.00 0.01 −0.07
8	−0.16 −0.16 0.02 0.15	0.10 0.12 −0.01 −0.03	−0.10 −0.22 0.07 0.11	−0.07 −0.26 0.10 0.15	0.15 0.18 0.01 0.13	−0.41 −0.36 −0.23 −0.21	***0.24*** ***0.24*** ***0.25*** ***0.25***		0.00 0.00 0.00 0.00	0.00 0.00 0.00 0.00	0.00 0.00 0.00 0.00	0.03 0.03 0.01 0.01	0.00 0.00 0.00 0.00	0.00 0.00 0.00 0.00
9	0.08 0.10 0.06 0.04	−0.05 −0.08 −0.05 −0.03	0.03 0.14 0.08 0.01	0.01 0.16 0.09 0.01	−0.09 −0.15 −0.13 −0.09	0.21 0.22 0.21 0.14	−0.45 −0.43 −0.30 −0.27	−0.41 −0.36 −0.23 −0.21		***0.46*** ***0.41*** ***0.49*** ***0.50***	0.08 0.07 0.07 0.19	0.00 0.00 0.00 0.00	−0.11 −0.15 −0.10 −0.07	−0.05 −0.06 −0.04 −0.12
10	0.00 −0.07 −0.07 −0.02	−0.01 0.05 0.05 0.03	0.08 −0.09 −0.10 0.02	0.12 −0.12 −0.11 0.04	−0.01 0.08 0.09 0.09	−0.09 −0.15 −0.13 −0.09	0.13 0.16 0.03 0.14	0.15 0.18 0.01 0.13	***−0.29*** ***−0.32*** ***−0.32*** ***−0.47***		***0.69*** ***0.70*** ***0.67*** ***0.62***	0.00 0.00 0.00 0.00	0.19 0.18 0.20 0.25	−0.02 0.00 0.00 −0.24
11	−0.05 0.08 0.08 0.15	0.05 −0.06 −0.06 −0.05	−0.21 0.12 0.13 0.10	−0.30 0.15 0.15 0.13	0.08 −0.09 −0.10 0.02	0.03 0.14 0.08 0.01	−0.07 −0.17 0.04 0.09	−0.10 −0.22 0.07 0.11	0.11 0.18 0.13 0.07	−0.45 −0.46 −0.43 −0.14		0.00 0.00 0.00 0.00	***0.57*** ***0.53*** ***0.57*** ***0.73***	0.14 0.13 0.14 0.00
12	−0.08 0.09 0.09 0.18	0.08 −0.07 −0.07 −0.06	−0.30 0.15 0.15 0.13	−0.42 0.19 0.18 0.17	0.12 −0.12 −0.11 0.04	0.01 0.16 0.09 0.01	−0.06 −0.20 0.06 0.13	−0.07 −0.26 0.10 0.15	0.09 0.19 0.12 0.05	−0.33 −0.34 −0.33 −0.07	***0.24*** ***0.24*** ***0.24*** ***0.24***		0.00 0.00 0.00 0.00	0.00 0.00 0.00 0.00
13	0.00 −0.04 −0.04 −0.09	−0.01 0.03 0.03 0.05	0.05 −0.06 −0.06 −0.05	0.08 −0.07 −0.07 −0.06	−0.01 0.05 0.05 0.03	−0.05 −0.08 −0.05 −0.03	0.08 0.09 0.00 −0.01	0.10 0.12 −0.01 −0.03	−0.10 −0.12 −0.09 −0.11	0.17 0.17 0.16 0.14	−0.25 −0.26 −0.26 −0.46	−0.18 −0.19 −0.19 −0.31		***0.57*** ***0.58*** ***0.58*** ***0.34***
14	0.01 0.05 0.05 0.20	0.00 −0.04 −0.04 −0.09	−0.05 0.08 0.08 0.15	−0.08 0.09 0.09 0.18	0.00 −0.07 −0.07 −0.02	0.08 0.10 0.06 0.04	−0.12 −0.13 0.00 0.12	−0.16 −0.16 0.02 0.15	0.12 0.13 0.09 0.05	−0.21 −0.21 −0.20 0.01	0.17 0.18 0.18 0.21	0.14 0.15 0.15 0.22	***−0.63*** ***−0.63*** ***−0.63*** ***−0.30***	

$$\langle \hat{s}_i \cdot \hat{s}_j \rangle = \begin{cases} -\frac{3}{4} & \text{if the spins associated with orbitals } \phi_i \text{ and } \phi_j \text{ are exactly singlet coupled} \\ 0 & \text{if the spins associated with orbitals } \phi_i \text{ and } \phi_j \text{ are completely uncoupled} \\ \frac{1}{4} & \text{if the spins associated with orbitals } \phi_i \text{ and } \phi_j \text{ are exactly triplet coupled} \end{cases}$$

The differences of the actual values of $\langle \hat{s}_i \cdot \hat{s}_j \rangle$ (lower triangle of Table 3) from the pure-singlet value of $-\frac{3}{4}$ also give the proportions of triplet character. The values highlighted in the Table correspond to those we might expect from Figure 2 to be heavily singlet coupled and to those associated with the anti-pairs. The septet state again appears to be somewhat different from the others, with much reduced singlet character in the terminal C—C units. The most striking feature, for all the states, is the proportion of triplet character in the internal C—C units – 46% in the singlet state, for example. It is thus clear that each anti-pair is indeed very strongly triplet coupled, with $\langle \hat{s}_i \cdot \hat{s}_j \rangle$ values close to $\frac{1}{4}$, and that the internal C—C units play a key role in coupling these triplets via a 'superexchange' mechanism.

Further insight into the mode of spin coupling and, in particular, into the role played by the formal C═C bonds linking the rings, is provided by the weights of the different modes in the Serber basis [36]. We list in Table 4 the accumulated weights of modes with particular values for $S_{1,2}$, $S_{5,6}$, $S_{9,10}$ and $S_{13,14}$, where $S_{i,j}$ is the total spin of electrons i and j. The Table shows that no one set of these values is overwhelmingly dominant and, in particular, that the mode represented by Figure 2 is not in fact the largest contributor for any of the spin multiplicities, consistently coming second. For the triplet ground state, as well as for the singlet and quintet excited states, the largest contribution corresponds to triplet coupling in one of the internal C—C units, and third place is taken by triplet coupling of both of these. The septet state is clearly rather different from the others, with large contributions from modes with triplet coupling in the terminal C—C units. Such effects in the terminal units were, of course, not apparent in the ten-electron SC study [71].

Table 4 *Accumulated weights in the Serber basis, expressed as percentages;* $S_{i,j}$ *is the total spin of electrons i and j.*

$S_{1,2}S_{5,6}S_{9,10}S_{13,14}$	$^1B_{1g}$	$^3B_{2u}$	$^5B_{2u}$	$^7B_{2u}$
0010 & 0100	42.3	42.4	43.5	6.9
0000	18.2	21.4	20.9	18.2
0110	17.1	13.9	13.5	2.6
1010 & 0101	9.5	9.1	8.5	4.4
1000 & 0001	7.5	8.0	7.6	26.8
1100 & 0011	2.5	2.5	3.0	16.7
1110 & 0111	1.6	1.4	1.7	7.0
1001	0.7	0.7	0.7	8.8
1101 & 1011	0.6	0.6	0.6	7.9
1111	0.0	0.0	0.0	0.9

6. Conclusions

Modern valence bond methods take as their staring point one or more spatial configurations constructed from fully-opitimized, nonorthogonal orbitals, plus expansions in the full spin space. In general, they combine useful accuracy with highly visual interpretation. We have described here the key features of our CASVB approach, which represents a relatively new development in modern VB theory, in which compact representations of CASSCF wavefunctions may be generated or, alternatively, fully-variational optimization of various general types of VB wavefunction may be performed, for ground and excited states.

We have reported also a new application, namely to the fourteen π electrons of a planar dimethylenecyclobutadiene chain with three rings. The various spin multiplicities are characterized by antiferromagnetic spin couplings: the 'internal' C—C linkages play a key role in coupling the three high-spin units via a 'superexchange' mechanism. We predict a triplet ground state.

Our CASVB code is incorporated as a standard feature in MOLPRO [41], and we are actively seeking ways to make all of the methodology more widely available via other packages.

Acknowledgements

This project has received financial support from the EEC under the TMR programme (contract ERBFMRXCT960088). TT gratefully acknowledges the Danish Natural Science Research Council for financial support.

References

1. T. Thorsteinsson: *Development of Methods in Spin-Coupled Theory* (Ph.D. Thesis), University of Liverpool, Liverpool (1995).
2. T. Thorsteinsson, D.L. Cooper, J. Gerratt, P.B. Karadakov and M. Raimondi: *Theor. Chim. Acta* **93**, 343 (1996).
3. T. Thorsteinsson and D.L. Cooper: *Theor. Chim. Acta* **94**, 233 (1996).
4. T. Thorsteinsson, D.L. Cooper, J. Gerratt and M. Raimondi: *Theor. Chim. Acta* **95**, 131 (1997).
5. T. Thorsteinsson, D.L. Cooper, J. Gerratt and M. Raimondi: A New Approach to Valence Bond Calculations: CASVB, in R. McWeeny, J. Maruani, Y.G. Smeyers and S. Wilson (Eds.), *Quantum Systems in Chemistry and Physics: Trends in Methods and Applications*, Kluwer Academic Publishers, Dordrecht (1997).
6. D.L. Cooper, T. Thorsteinsson and J. Gerratt: *Int. J. Quant. Chem.* **65**, 439 (1997).
7. T. Thorsteinsson and D.L. Cooper: *J. Math. Chem.* **23**, 105 (1998).
8. D.L. Cooper, T. Thorsteinsson and J. Gerratt: *Adv. Quant. Chem.* **32**, 51 (1998).
9. T. Thorsteinsson and D.L. Cooper: *Int. J. Quant. Chem.* **70**, 637 (1998).
10. B.O. Roos: *Adv. Chem. Phys.* **69**, 399 (1987).
11. S.F. Boys: *Rev. Mod. Phys.* **32**, 296 (1960).
12. J.M. Foster and S.F. Boys: *Rev. Mod. Phys.* **32**, 300 (1960).
13. C. Edminston and K. Ruedenberg: *Rev. Mod. Phys.* **35**, 457 (1963).
14. C. Edminston and K. Ruedenberg: *J. Chem. Phys.* **43**, S97 (1963).
15. J. Pipek and P.G. Mezey: *J. Chem. Phys.* **90**, 4916 (1989).
16. R.S. Berry: *J. Chem. Phys.* **30**, 936 (1959).

17. M. Craig and R.S. Berry: *J. Am. Chem. Soc.* **89**, 2801 (1967).
18. D.M. Hirst and J.W. Linnett: *J. Am. Chem. Soc.* **83**, 2643 (1961).
19. W.H. Kirchnoff, J. Farren and J.W. Linnett: *J. Chem. Phys.* **42**, 1410 (1965).
20. D.M. Hirst and J.W. Linnett: *J. Chem. Phys.* **43**, S74 (1965).
21. J.J.W. McDouall and M.A. Robb: *Chem. Phys. Lett.* **142**, 131 (1987).
22. J.J.W. McDouall and M.A. Robb: *Chem. Phys. Lett.* **132**, 319 (1986).
23. P.C. Hiberty and C. Leforestier: *J. Am. Chem. Soc.* **100**, 2012 (1978).
24. K. Hirao, H. Nakano, K. Nakayama and M. Dupuis: *J. Chem. Phys.* **105**, 9227 (1996).
25. K. Hirao, H. Nakano and K. Nakayama: *J. Chem. Phys.* **107**, 9966 (1997).
26. C.A. Coulson and I. Fischer: *Phil. Mag.* **40**, 386 (1949).
27. W. Heitler and F. London: *Z. Phys.* **44**, 455 (1927).
28. D.L. Cooper, J. Gerratt and M. Raimondi: *Chem. Rev.* **91**, 929 (1991).
29. R.B. Murphy and R.P. Messmer: *J. Chem. Phys.* **98**, 7958 (1993).
30. N.O.J. Malcolm and J.J.W. McDouall: *J. Comp. Chem.* **15**, 1357 (1994).
31. N.O.J. Malcolm and J.J.W. McDouall: *J. Comp. Chem.* **15**, 1365 (1994).
32. P.B. Karadakov, D.L. Cooper and J. Gerratt: *J. Am. Chem. Soc.* **120**, 3975 (1998).
33. P.B. Karadakov, D.L. Cooper and J. Gerratt: *Theor. Chem. Acc.* **100**, 222 (1998).
34. P. Å. Malmqvist: *Int. J. Quant. Chem.* **30**, 479 (1986).
35. W.H. Press, S.A. Teukolsky, W.T. Vetterling and B.T. Flannery: *Numerical Recipes* (1st ed.), Cambridge University Press, New York (1986).
36. R. Pauncz: *Spin Eigenfunctions*, Plenum Press, New York (1979).
37. D.L. Cooper, J. Gerratt, M. Raimondi, M. Sironi and T. Thorsteinsson: *Theor. Chim. Acta* **85**, 261 (1993).
38. B. Friis-Jensen, D.L. Cooper and S. Rettrup: *Theor. Chem. Acc.* **99**, 64 (1998).
39. H.-J. Werner and P.J. Knowles: *J. Chem. Phys.* **82**, 5053 (1985).
40. P.J. Knowles and H.-J. Werner: *Chem. Phys. Lett.* **115**, 259 (1985).
41. MOLPRO is a package of *ab initio* programs written by H.-J. Werner and P.J. Knowles, with contributions from J. Almlöf, R.D. Amos, A. Berning, D.L. Cooper, M.J.O. Deegan, F. Eckert, S.T. Elbert, C. Hampel, R. Lindh, W. Meyer, A. Nicklaß, K. Peterson, R. Pitzer, A.J. Stone, P.R. Taylor, M.E. Mura, P. Pulay, M. Schütz, H. Stoll and T. Thorsteinsson.
42. J. Paldus: *J. Chem. Phys.* **61**, 5321 (1974).
43. J. Olsen, B.O. Roos, P. Jørgensen and H.J. Aa. Jensen: *J. Chem. Phys.* **89**, 2185 (1988).
44. S. Zarrabian, C. R. Sarma and J. Paldus: *Chem. Phys. Lett.* **155**, 183 (1989).
45. R.J. Harrison and S. Zarrabian: *Chem. Phys. Lett* **158**, 393 (1989).
46. T. Helgaker: Optimization of Minima and Saddle Points, in: B.O. Roos (Ed.), *Lecture Notes in Quantum Chemistry 58: European Summer School in Quantum Chemistry*, Springer-Verlag, Berlin, 295 (1992).
47. S.M. Goldfeld, R.E. Quant and H.F. Trotter: *Econometrica* **34**, 541 (1966).
48. R. Fletcher: *Practical Methods of Optimization* (2nd ed.), Wiley, New York (1987).
49. B.H. Lengsfield III: *J. Chem. Phys.* **73**, 382 (1980).
50. H.J. Aa. Jensen and P. Jørgensen: *J. Chem. Phys.* **80**, 1204 (1984).
51. E.R. Davidson: *J. Comput. Phys.* **17**, 87 (1975).
52. S. Rettrup: private communication.
53. G. Hamermesh: *Group theory and its application to physical problems* (Section 3–18), Addison-Wesley, Massachusetts (1962).
54. J. Gerratt: *Adv. Atom. Mol. Phys.* **7**, 141 (1971).
55. S.C. Wright, D.L. Cooper, J. Gerratt and M. Raimondi: *J. Phys. Chem.* **96**, 7943 (1992).
56. J. Olsen, D.L. Yeager and P. Jørgensen: *Adv. Chem. Phys.* **54**, 1 (1983).
57. J.T. Golab, D.L. Yeager and P. Jørgensen: *Chem. Phys.* **78**, 175 (1983).
58. T. Helgaker: *Chem. Phys. Lett.* **182**, 503 (1991).
59. G.D. Fletcher, G. Doggett and A.S. Howard: *Phys. Rev. A* **46**, 5459 (1992).
60. G. Doggett, G.D. Fletcher and F.R. Manby: *J. Mol. Struc.* **300**, 191 (1993).
61. F.R. Manby, G. Doggett and G.D. Fletcher: *J. Mol. Struc.* **343**, 63 (1995).
62. B.H. Chirgwin and C.A. Coulson: *Proc. Roy. Soc. Lond.* **A201**, 196 (1950).

63. P.-O. Löwdin: *Arkiv. Mat. Astr. Fysik* **35A**, 9 (1947).
64. G.A. Gallup and J.M. Norbeck: *Chem. Phys. Lett.* **21**, 495 (1973).
65. G.A. Gallup, R.L. Vance, J.R. Collins and J.M. Norbeck: *Adv. Quant. Chem.* **16**, 229 (1982).
66. G. Raos, J. Gerratt, D.L. Cooper and M. Raimondi: *Mol. Phys.* **79**, 197 (1993).
67. P.B. Karadakov, J. Gerratt, D.L. Cooper and M. Raimondi: *Theor. Chim. Acta* **90**, 51 (1995).
68. G. Raos, J. Gerratt, D.L. Cooper and M. Raimondi: *Chem. Phys.* **186**, 233 (1994).
69. G. Raos, J. Gerratt, D.L. Cooper and M. Raimondi: *Chem. Phys.* **186**, 251 (1994).
70. D.L. Cooper, R. Ponec, T. Thorsteinsson and G. Raos: *Int. J. Quant. Chem.* **57**, 501 (1996).
71. G. Raos, S.J. McNicholas, J. Gerratt, D.L. Cooper and P.B. Karadakov: *J. Phys. Chem. B* **101**, 6688 (1997).
72. S.C. Wright, D.L. Cooper, J. Gerratt and M. Raimondi: *J. Phys. Chem.* **96**, 7943 (1992).

Modern Valence-Bond Description of the Mechanisms of Six-Electron Pericyclic Reactions

Peter B. Karadakov[a], David L. Cooper[b], Thorstein Thorsteinsson[c] and Joseph Gerratt[d]*

[a]*Department of Chemistry, University of Surrey, Guildford, Surrey GU2 5XH, U.K.*
[b]*Department of Chemistry, University of Liverpool, P.O. Box 147, Liverpool L69 7ZD, U.K.*
[c]*Chemistry Laboratory IV, Copenhagen University, Universitetsparken 5, DK-2100 Copenhagen Ø, Denmark*
[d]*School of Chemistry, University of Bristol, Cantock's Close, Bristol BS8 1TS, U.K.*

Abstract

The combination of modern valence bond theory, in its spin-coupled (SC) form, and intrinsic reaction coordinate calculations utilizing a complete-active-space self-consistent field (CASSCF) wavefunction, is demonstrated to provide quantitative and yet very easy-to-visualize models for the electronic mechanisms of three gas-phase six-electron pericyclic reactions, namely the Diels-Alder reaction between butadiene and ethene, the 1,3-dipolar cycloaddition of fulminic acid to ethyne, and the disrotatory electrocyclic ring-opening of cyclohexadiene.

The SC descriptions of the electronic mechanisms of these three reactions are shown to substantiate the use of the long-established reaction schemes with full- and half-arrows, well-known from organic chemistry textbooks, in a context which is very meaningful, albeit slightly different from the classical interpretation. The half-arrows now indicate changes in the shapes of individual orbitals, accompanying the breaking of the bonds in which they participate in the reactant(s), and their re-engagement in new bonds within the product(s), rather than the movements of individual electrons. The full-arrows correspond to relocations of orbital, rather than electron pairs.

The SC results strongly suggest that the Diels-Alder reaction between butadiene and ethene and the ring-opening of cyclohexadiene pass through aromatic conformations, while in the case of the 1,3-dipolar cycloaddition of fulminic acid to ethyne, the reacting system remains distinctly nonaromatic throughout the course of the reaction.

1. Introduction

Pericyclic reactions, according to the definition of Woodward and Hoffmann [1,2], are 'reactions in which all first-order changes in bonding relationships take place in concert on a closed curve'. An organic chemistry textbook might represent a six-electron reaction of this type, such as the [$_\pi 6$] disrotatory electrocyclic ring-opening of cyclohexadiene, using one of the following two schemes (A and B):

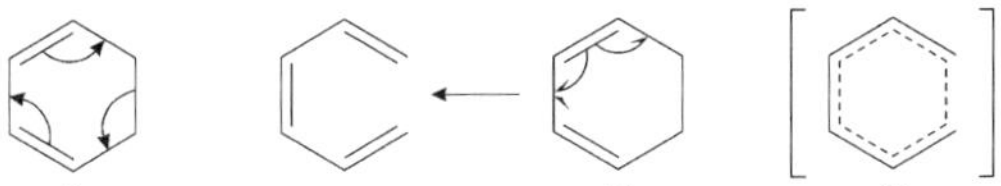

A B C

---

*Joseph Gerratt passed away on 16 October 1997.

A. Hernández-Laguna et al. (eds.), Quantum Systems in Chemistry and Physics, Vol. 1: Basic Problems and Model Systems, 327–344.

(Other familiar examples include the $[_{\pi}4_s + _{\pi}2_s]$ pericyclic reaction between butadiene and ethene, known as the Diels-Alder reaction, and the $[3_s, 3_s]$ sigmatropic shift in 1,5-hexadiene, known as the Cope rearrangement.) Scheme A, which is much more popular with organic chemists, depicts a 'heterolytic' mechanism, in which the electron pairs making up the two π bonds and the σ bond closing the cyclohexadiene ring shift simultaneously, following the three full arrows, to form the three π bonds in hexatriene. In the alternative 'homolytic' scheme, B, the three cyclohexadiene bonds not present in the open chain break up at the same time, and the participating electrons recouple within new bonds, as indicated by the six half-arrows. The fact that almost all current theoretical explanations for the mechanisms of pericyclic reactions are based on molecular orbital (MO) theory, and typically utilize delocalized MOs, has made many chemists regard the types of schemes illustrated by A and B as nothing more than simple electron-counting aids that have little, if anything, in common with the true reaction mechanism, which is deemed to pass through a transition structure (TS) involving considerable electron delocalization (see scheme C).

However, despite their proven explanatory and predictive capabilities, all well-known MO models for the mechanisms of pericyclic reactions, including the Woodward-Hoffmann rules [1,2], Fukui's frontier orbital theory [3] and the Dewar-Zimmerman treatment [4–6] share an inherent limitation: They are based on nothing more than the simplest MO wavefunction, in the form of a single Slater determinant, often under the additional oversimplifying assumptions characteristic of the Hückel molecular orbital (HMO) approach. It is now well established that the accurate description of the potential surface for a pericyclic reaction requires a much more complicated *ab initio* wavefunction, of a quality comparable to, or even better than, that of an appropriate complete-active-space self-consistent field (CASSCF) expansion. A wavefunction of this type typically involves a large number of configurations built from orthogonal orbitals, the most important of which (*i.e.* those in the active space) have fractional occupation numbers. Its complexity renders the re-introduction of qualitative ideas similar to the Woodward-Hoffmann rules virtually impossible.

At this point, it is appropriate to draw a parallel with the straightforward MO explanations for the aromaticity of benzene using approaches based on a single closed-shell Slater determinant, such as HMO and restricted Hartree-Fock (RHF), which also have no equivalent within more advanced multi-configuration MO constructions. The relevance of this comparison follows from the fact that aromaticity is a primary factor in at least one of the popular treatments of pericyclic reactions: Within the Dewar-Zimmerman approach [4–6], allowed reactions are shown to pass through aromatic transition structures, and forbidden reactions have to overcome high-energy antiaromatic transition structures.

As is well-known, modern valence-bond (VB) theory in its spin-coupled (SC) form (for a recent review, see Ref. 7) provides an alternative description of benzene [8–10] which, in qualitative terms, is no less convincing and is arguably even more intuitive than the MO picture with delocalized orbitals. The six π electrons are accommodated within a single product of six nonorthogonal orbitals, the spins of which are coupled in all five possible ways that lead to an overall six-electron singlet. The simultaneous optimization of the orbitals and of the weights of the five six-electron singlet spin couplings results in well-localized orbitals, similar in shape to $C(2p_\pi)$ atomic orbitals (AOs), but with small symmetrical bulges towards neighbouring carbon atoms, while the optimal spin-coupling pattern is dominated by two equivalent Kekulé structures, with much smaller contributions from three equivalent Dewar (or para-bonded) structures. In this way, SC theory reintroduces the familiar qualitative elements of the classical VB description of benzene in terms of resonance structures. However, this description is not postulated as in classical VB theory, but stems from a variational *ab initio* wavefunction constructed without any preliminary assumptions about the shapes of the orbitals (apart from the σ–π separation) or the coupling of their spins. The quality of the wavefunction comes very close to that of a '6 in 6' π-space CASSCF ansatz.

SC theory is equally well-suited to describing molecular systems in their equilibrium geometries, as well as away from equilibrium. The fact that the nonorthogonal SC orbitals are usually well-localized and resemble distorted AOs or hybrid AOs, together with the completely flexible fashion of coupling together the associated electron spins so as to yield the required overall spin multiplicity, allows the SC wavefunction to accommodate a very wide variety of the bond rearrangements and fragmentation patterns that can occur in chemical reactions. These features result in descriptions which, very much as in the case of benzene, combine ease of interpretation with near-CASSCF quality.

We show in this article that SC theory provides models for the electronic mechanisms of three well-known six-electron pericyclic reactions that justify the use of the long-established schemes with full- and half-arrows (such as A and B) in a context which is very meaningful, albeit slightly different from the classical interpretation. The half-arrows indicate changes in the shapes of individual orbitals, accompanying the breaking of the bonds in which they participate in the reactant(s), and their re-engagement in new bonds within the product(s), rather than movements of individual electrons. The full-arrows correspond to relocations of orbital, rather than electron pairs.

The article is organized as follows. In the next Section we present a brief outline of the theoretical background for the present work. Section 3 contains summaries of the SC models for the electronic mechanisms of the gas-phase Diels-Alder reaction between butadiene and ethene [11] and the 1,3-dipolar cycloaddition of fulminic acid to ethyne [12]. In Section 4 we provide, for the first time, a description of the SC model for the electronic mechanism of the gas-phase disrotatory electrocyclic ring-opening of cyclohexadiene. Conclusions and final comments are presented in Section 5.

2. Theoretical Background

The SC wavefunction appropriate for the description of a six-electron pericyclic reaction can be written as

$$\Psi_{00}^{6} = \hat{\mathcal{A}}\left[\left(\prod_{i=1}^{n}\varphi_i\alpha\,\varphi_i\beta\right)\left(\prod_{\mu=1}^{6}\psi_\mu\right)\Theta_{00}^{6}\right]. \tag{1}$$

where φ_i and ψ_μ stand for the 'core' (or inactive) and SC (or active) orbitals, respectively. Both types of orbitals are represented by MO-style expansions in terms of a suitable AO basis set involving functions on all atoms in the system. The number of core orbitals is $n = 20$ for the Diels-Alder reaction between butadiene and ethene, $n = 15$ for the 1,3-dipolar cycloaddition of fulminic acid to ethyne, and $n = 19$ for the electrocyclic ring-opening of cyclohexadiene. The double occupancy of the core orbitals allows one to assume, without any loss of generality, that they are orthogonal between themselves and to the SC orbitals. The SC orbitals are all singly occupied and may freely overlap with each other.

The singlet spin function Θ_{00}^{6} for the valence electrons (where the two subscripts indicate the eigenvalues of $\hat{S}^2$ and $\hat{S}_z$ for the active space, $S = M = 0$) is expressed as a linear combination of all five linearly-independent spin-coupling modes for a singlet system of six electrons:

$$\Theta_{00}^{6} = \sum_{k=1}^{5} C_{0k}\Theta_{00;k}^{6}. \tag{2}$$

The spin-coupling coefficients C_{0k}, together with the coefficients determining the orbitals φ_i and ψ_μ within the selected basis, are treated as variational parameters.

A complete set of spin eigenfunctions, *e.g.* $\{\Theta_{00;k}^{6}|k = 1, 2, \ldots, 5\}$ in the case of a six-electron singlet, can be constructed by means of one of several available algorithms. The most commonly used ones are those due to Kotani, Rumer and Serber [13]. Once the set of optimized values of the coefficients defining a spin-coupling pattern is available [see C_{0k} in Eq. (2)], it can be transformed easily [14] to a different spin basis, or to a modified set reflecting a change to the order in which the active orbitals appear in the SC wavefunction [see Eq. (1)].

When discussing the SC models for the electronic mechanisms of the Diels-Alder reaction between butadiene and ethene [11], and of the disrotatory electrocyclic ring-opening of cyclohexadiene, we shall make use of the Rumer spin basis, which allows a straightforward translation of our results into a form that resembles traditional VB ideas.

Each of the five Rumer spin eigenfunctions for a six-electron singlet represents a product of three singlet two-electron spin functions:

$$\begin{aligned}
{}^{\mathrm{R}}\Theta_{00;1}^{6} &= (\alpha\beta)(\alpha\beta)(\alpha\beta) \equiv (1-2,\ 3-4,\ 5-6),\\
{}^{\mathrm{R}}\Theta_{00;2}^{6} &= (\alpha(\alpha\beta)\beta)(\alpha\beta) \equiv (1-4,\ 2-3,\ 5-6),\\
{}^{\mathrm{R}}\Theta_{00;3}^{6} &= (\alpha\beta)(\alpha(\alpha\beta)\beta) \equiv (1-2,\ 3-6,\ 4-5),\\
{}^{\mathrm{R}}\Theta_{00;4}^{6} &= (\alpha(\alpha\beta)(\alpha\beta)\beta) \equiv (1-6,\ 2-3,\ 4-5),\\
{}^{\mathrm{R}}\Theta_{00;5}^{6} &= (\alpha(\alpha(\alpha\beta)\beta)\beta) \equiv (1-6,\ 2-5,\ 3-4).
\end{aligned} \tag{3}$$

In the first of the two notations used in this equation, a singlet pair is associated with each combination of an α and a β spin function enclosed within a matching pair of left and right parentheses. The second notation indicates all singlet pairs explicitly. It is easiest to visualize the spin-coupling patterns (3) by means of the familiar five resonance structures for benzene (two Kekulé and three Dewar structures), pictures of which can be found in many standard organic chemistry textbooks.

The Rumer spin basis is nonorthogonal, which creates ambiguities in the definition of numerical estimates for the relative importance of the individual ${}^{\mathrm{R}}\Theta^{6}_{00;k}$ [see Θ^{6}_{00} in Eq. (2)]. In this paper, we shall use the Chirgwin-Coulson expressions [15] (it is assumed that all $\Theta^{6}_{00;k}$ are normalized):

$$ {}^{\mathrm{R}}P_{0k} = {}^{\mathrm{R}}C_{0k} \sum_{l=1}^{5} \langle {}^{\mathrm{R}}\Theta^{6}_{00;k} | {}^{\mathrm{R}}\Theta^{6}_{00;l} \rangle {}^{\mathrm{R}}C_{0l}. \tag{4} $$

For a detailed discussion of nonorthogonal weights in modern VB wavefunctions, see Ref. 16.

In the case of the 1,3-dipolar cycloaddition of fulminic acid to ethyne, none of the three common spin bases leads to any particular interpretational advantages. For this reaction, we report the composition of the optimal spin-coupling pattern (2) in the Kotani spin basis, which is orthonormal; as a result, the weights of the individual spin functions making up Θ^{6}_{00} are given simply by the squares of the corresponding spin-coupling coefficients, ${}^{\mathrm{K}}C^{2}_{0k}$.

The Kotani spin functions for a six-electron singlet ${}^{\mathrm{K}}\Theta^{6}_{00;k}$ may be constructed by successive coupling of six one-electron spin functions (α or β) to an overall singlet according to the rules for addition of angular momenta. Each spin function is uniquely defined by the series of partial resultant spins of the consecutive groups of 1, 2, ..., 5 electrons, which can be used as an extended label for the spin function

$$ \begin{aligned} {}^{\mathrm{K}}\Theta^{6}_{00;1} &\equiv (\tfrac{1}{2}1\tfrac{3}{2}1\tfrac{1}{2}), \\ {}^{\mathrm{K}}\Theta^{6}_{00;2} &\equiv (\tfrac{1}{2}1\tfrac{1}{2}1\tfrac{1}{2}), \\ {}^{\mathrm{K}}\Theta^{6}_{00;3} &\equiv (\tfrac{1}{2}0\tfrac{1}{2}1\tfrac{1}{2}), \\ {}^{\mathrm{K}}\Theta^{6}_{00;4} &\equiv (\tfrac{1}{2}1\tfrac{1}{2}0\tfrac{1}{2}), \\ {}^{\mathrm{K}}\Theta^{6}_{00;5} &\equiv (\tfrac{1}{2}0\tfrac{1}{2}0\tfrac{1}{2}). \end{aligned} \tag{5} $$

Spin functions ${}^{\mathrm{R}}\Theta^{6}_{00;1}$ and ${}^{\mathrm{K}}\Theta^{6}_{00;5}$ are identical and define the 'perfect-pairing' mode for a six-electron-singlet.

In order to be able to construct a model for the electronic mechanism of a chemical reaction, one needs a sequence of geometries to describe the structural changes in the reacting system along a realistic reaction path. The most consistent approach would, of course, be to use one and the same level of theory for the calculation of the required geometries and for the interpretation of the electronic reaction mechanism. However, although the properties of the fully-variational SC wavefunction make it an obvious choice for the second of these two tasks, we do not at present possess codes that are capable of calculating reaction paths at this level of theory. Most of the SC calculations described in the present work were performed using a very efficient new code, CASVB [17,18], which is described elsewhere in this Volume and which is currently available within MOLPRO [19]. It would be relatively straightforward to incorporate the CASVB methodology into a package capable of calculating reaction paths at the CASSCF level, and thereby perform all computational work at precisely the same level of theory. This is an avenue that is currently being explored but, even before this combination becomes available, we can make use of the fact that the SC wavefunction with N active orbitals recovers, in many cases, an impressive proportion of the correlation energy incorporated into the corresponding 'N in N' CASSCF construction. This suggests that the geometries along the reaction path to be used in the SC model for the electronic reaction mechanism could reasonably be taken from preliminary 'N in N' CASSCF calculations.

The transition structure (TS) provides the most appropriate starting point for the calculation of the segment of the reaction path within which the electronic structure of the reacting system switches from a form more suited to the reactant(s) to a form that can be better associated with the product(s). The changes in the geometry of the reacting system before and after the TS can be traced by following the minimum-energy path (MEP) [20,21], which is also known as the intrinsic reaction coordinate (IRC) [22,23]. The MEP or IRC is defined as the union of steepest-descent paths, expressed in mass-weighted Cartesian coordinates, emerging from the saddle point corresponding to the TS and leading towards reactant(s) and product(s). For all three reactions described in the present paper, we employed the IRC algorithm of Gonzales and Schlegel [24,25] at the '6 in 6' CASSCF level of theory. These calculations, as well as all other non-SC computational work, were performed using GAUSSIAN94 [26].

As we have indicated, the fully-variational SC calculations along the '6 in 6' CASSCF IRC curves were performed with CASVB, as implemented in MOLPRO [19]. An older SC code [27] was employed only at the transition structures and at the ends of the calculated IRC segments, mainly in order to assist orbital plotting.

3. The Diels-Alder Reaction and a 1,3-Dipolar Cycloaddition: Summaries

The gas-phase Diels-Alder reaction between butadiene and ethene follows a synchronous pathway, while the 1,3-dipolar cycloaddition of fulminic acid to ethyne proceeds in a concerted, almost synchronous fashion, in spite of the different nature of the two bonds closing the isoxazole cycle.

Our results for the TS geometries for both reactions, obtained at the CASSCF(6,6)/4-31G level in the case of the Diels-Alder reaction [11] and at the CASSCF(6,6)/6-31G* level in the case of the 1,3-dipolar cycloaddition [12], are in close agreement with the previous work of Houk and coworkers [28,29] and of McDouall *et al.* [30], respectively. All of our subsequent CASSCF and SC calculations were performed using the same basis sets as for the TS optimizations.

Along the whole '6 in 6' CASSCF IRC segments calculated for these two reactions, SC calculations (carried out at geometries corresponding to points on the respective IRCs) recover not less than *ca.* 93% of the '6 in 6' CASSCF correlation energy. This justifies our use of this combination of approaches to study the evolution of the electronic structure of each of the two reacting systems.

Throughout the Diels-Alder reaction, all six nonorthogonal SC orbitals resemble well-localized sp^x hybrids, each of which remains permanently attached to a single carbon atom. Fig. 1 shows the shapes of the symmetry-unique SC orbitals at the TS (central column) and at two IRC points well away from it in the directions of product (leftmost column) and reactants (rightmost column). The remaining orbitals can be obtained from those shown in the Figure by means of reflections in the TS symmetry plane: the reflections of ψ_1, ψ_2 and ψ_6 result in ψ_4, ψ_3 and ψ_5, respectively.

In the rightmost column (reactant-like orbitals), we can see that the two butadiene π-bonds are formed by the pairs of orbitals (ψ_1, ψ_2) and (ψ_3, ψ_4), respectively. Orbital ψ_2 has a small additional bulge towards ψ_3, which is an indication of the much weaker π-orbital interaction across the central carbon-carbon bond. Orbital ψ_6 and its symmetry-related counterpart, ψ_5, are responsible for the π bond on the ethene fragment.

The tail of ψ_1 in the direction of ψ_2 becomes much less pronounced at the TS (central column of orbitals). This orbital resembles a distorted carbon $2p_\pi$ AO in the rightmost column, but it begins to attain the form of a distorted sp^3 hybrid at the TS. Orbital ψ_6 is reshaped in a similar fashion and becomes less distorted towards ψ_5 and more towards ψ_1. The changes in ψ_1 and ψ_6 are mirrored by identical changes in ψ_4 and ψ_5. All of this indicates that the bonds ψ_1–ψ_2, ψ_5–ψ_6 and ψ_3–ψ_4 are breaking up at the TS, in parallel with the formation of new bonds ψ_1–ψ_6 and ψ_4–ψ_5. The shapes of orbital ψ_2 and its symmetry-related counterpart, ψ_3, are particularly interesting: These orbitals appear to be almost equally distorted towards one another (as the reaction proceeds further, these distortions lead to the formation of the cyclohexene π bond) and towards their former partners in the butadiene π bonds (ψ_1 and ψ_4, respectively). Consequently, ψ_2 and ψ_3 closely resemble in form the SC orbitals for benzene [8–10].

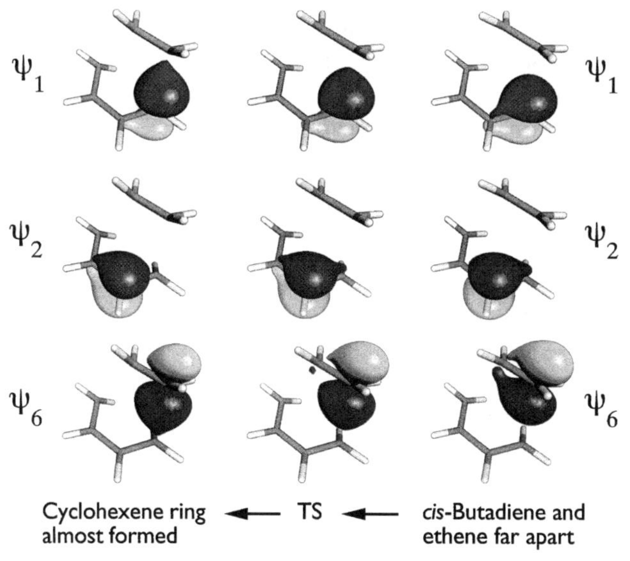

Figure 1. Symmetry-unique SC orbitals for the gas-phase Diels-Alder reaction along the CASSCF(6,6) IRC at IRC ≈ -0.6 amu$^{1/2}$ bohr *(leftmost column), TS (IRC = 0) and IRC* $\approx +0.6$ amu$^{1/2}$ bohr *(rightmost column). Three-dimensional isovalue surfaces, corresponding to* $\psi_\mu = \pm 0.08$, *were drawn from virtual reality modelling language (VRML) files produced by MOLDEN [31].*

The new bonding interactions, suggested by the changes in the shapes of the orbitals observed at the TS, are much more noticeable in the leftmost column, which corresponds to a much later stage of the reaction. The shapes of ψ_1 and ψ_6 are now quite similar and much more sp^3-like, while ψ_2 is clearly engaged together with ψ_3 in a new π bond.

At each end of the IRC segment, the spin function for the active electrons [Θ_{00}^6 in Eq. (2)] is dominated by just one of its five components. The first Kekulé-type Rumer spin eigenfunction (1 – 2, 3 – 4, 5 – 6), which couples the spins of the orbitals engaged in the butadiene and ethene π bonds to singlet pairs, is the most important one well before the TS. The second Kekulé-type Rumer spin eigenfunction (1 – 6, 2 – 3, 4 – 5), in which the singlet pairs correspond to the orbitals engaged in the newly-formed two σ bonds and one π bond in cyclohexene, prevails well after the TS. The most important changes in the composition of Θ_{00}^6 occur in the close vicinity of the TS where, within a very narrow IRC segment, the spin-coupling pattern switches from a form appropriate for the reactants to a form more suited to the product. Very close to the TS, the two Kekulé-type Rumer spin functions attain equal weights and, at the same moment, the values of all overlaps between neighbouring SC orbitals become almost equal and nearly identical to the corresponding overlap for benzene. In combination with the observation that, at the TS, orbitals ψ_2 and ψ_3 are very similar in shape to the SC orbitals for benzene (as well as other numerical results [11]) this provides a clear indication that the gas-phase Diels-Alder reaction passes through a geometry, very close to the TS, at which it may be considered to be aromatic.

Our analysis of the changes in the SC wavefunction along the reaction path strongly suggests that the two π bonds on the butadiene fragment and the ethene π bond break simultaneously, and that the formation of the two new σ bonds that close the cyclohexene ring and of the cyclohexene π bond also takes place almost in parallel. If we wish to express all of this using full- or half-arrows, it would be most appropriate to use half-arrows, as in scheme B from the Introduction:

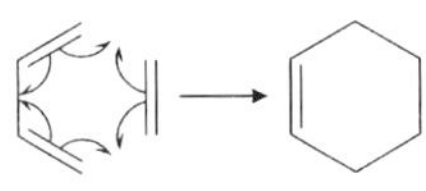

D

Having in mind the SC results, it is justified to say that these half-arrows indicate (*i*) the simultaneous processes of bond-breaking and formation, during which the orbitals change in shape, but remain on one and the same atomic centre, and (*ii*) the recoupling of spins, so that the initial singlet pairs associated with the butadiene and ethene π bonds are replaced by singlet pairs coupling the orbitals participating in the new σ bonds and in the cyclohexene π bond.

We now turn to the gas-phase 1,3-dipolar cycloaddition of fulminic acid to ethyne. The concerted, almost synchronous nature of this reaction might create the impression that the electronic mechanism of this process should be very similar to that of the Diels-Alder reaction. Such an expectation is reinforced by frontier orbital theory, which treats both reactions in very much the same way (see Ref. 32). The only significant differences are related to the fact that the lowest unoccupied MO (LUMO) for a linear 1,3-dipole has a node at one of the terminal heavy atoms. This LUMO is ignored in frontier orbital theory, which considers instead the next lowest unoccupied MO.

In spite of these anticipated great similarities, the results of our SC calculations [12] reveal a completely different picture. When the reacting molecules are far apart (see the rightmost column in Fig. 2), the orbitals on the HCNO fragment (ψ_1, ψ_3, ψ_5 and ψ_6) reproduce the well-known SC model for the electronic structure of 1,3-dipoles [33,34], according to which the central heavy atom is 'hypervalent'. The nitrogen atom in

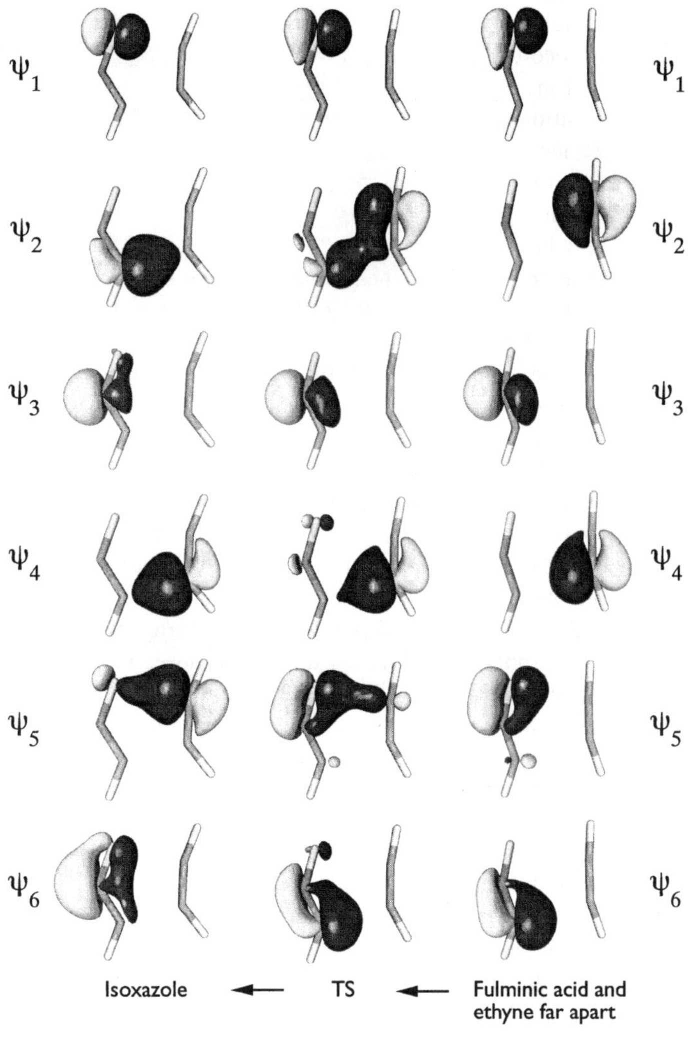

Figure 2. SC orbitals for the gas-phase 1,3-dipolar cycloaddition of fulminic acid to ethyne along the CASSCF(6,6) IRC at IRC ≈ -1.2 amu$^{1/2}$ bohr *(leftmost column), the TS (IRC = 0) and IRC* $\approx +1.2$ amu$^{1/2}$ bohr *(rightmost column). The plot details are as for Fig. 1, except that the isovalue surfaces correspond to* $\psi_\mu = \pm 0.1$.

HCNO appears to take part in more than four covalent bonds: an almost triple bond between C and N (ψ_3 and ψ_6 account for one of the components of this bond, which would be of π symmetry in linear HCNO), a σ bond between N and O, and a highly polar bond between N and O (of π symmetry in linear HCNO, described here by ψ_1 and ψ_5). The remaining two orbitals, ψ_2 and ψ_4, form the ethyne bond (of π symmetry in linear C_2H_2) that is broken during the reaction.

The clearest representation of the spin-coupling pattern [Θ^6_{00} in Eq. (2)] at this stage of the cycloaddition process is achieved by reordering the SC orbitals [14] so as to arrange in pairs the orbitals involved in the three bonds:

$$\Psi^6_{00} = \hat{\mathcal{A}}[(core)\psi_1\psi_2\psi_3\psi_4\psi_5\Theta^6_{00}] = \hat{\mathcal{A}}[(core)\psi_2\psi_4\psi_6\psi_3\psi_5\psi_1\Theta^6_{00}{'}] \tag{6}$$

where

$$\Theta^6_{00}{'} = \sum_{k=1}^{5} C_{0k}{'}\Theta^6_{00;k} \tag{7}$$

The spin-coupling pattern $\Theta^6_{00}{'}$ for the reordered orbital set is largely dominated by its 'perfect-pairing' component [${}^{K}\Theta^6_{00;5} \equiv (\frac{1}{2}0\frac{1}{2}0\frac{1}{2})$ in the Kotani basis, see Eq. (5)], in which the spins associated with the pairs of orbitals (ψ_2, ψ_4), (ψ_6, ψ_3) and (ψ_5, ψ_1) are coupled to singlets. The composition of $\Theta^6_{00}{'}$ changes very little throughout the cycloaddition process; this indicates that the orbital pairs (ψ_2, ψ_4), (ψ_6, ψ_3) and (ψ_5, ψ_1) are preserved all the way from reactants to product. As a consequence, all essential features of the SC model for the electronic mechanism of the reaction between HCNO and C_2H_2 are contained entirely within the variations of the shapes of the SC orbitals.

For each of the pairs (ψ_2, ψ_4) and (ψ_5, ψ_1), the shape of one orbital changes in a very dramatic fashion along the reaction coordinate. Orbital ψ_2, which enters the reaction as one of two orbitals involved in the in-plane ethyne 'π' bond, becomes similar at the TS to a linear combination of two sp^x hybrids, one at its 'original' ethyne carbon and another at the HCNO carbon. After the TS, this orbital loses its connection with the ethyne carbon and shifts entirely onto the HCNO carbon. The other orbital within this pair, ψ_4, undergoes a 're-hybridization' reminiscent of that for orbital ψ_6 in the case of the Diels-Alder reaction (see Fig. 1). As a result, ψ_2 and ψ_4 become responsible after the TS (see the leftmost column in Fig. 2) for one of the two bonds closing the isoxazole ring. SC orbital ψ_5 starts as one of the orbitals from the in-plane polar 'π' N–O bond in HCNO, but then, in a similar fashion to ψ_4, attains at the TS a shape that resembles a mixture of two sp^x hybrids, one of which is at its 'original' location (the oxygen atom) and the other at the incoming ethyne carbon. After the TS, this orbital becomes associated primarily with the ethyne carbon. The shape of the second orbital in this pair, ψ_2, changes very little throughout the reaction. After the TS, ψ_5 and ψ_1 account for the second bond closing the isoxazole ring. The most 'sedentary' orbital pair consists of ψ_3 and ψ_6, which initially form the in-plane 'π' C–N bond in HCNO. In comparison to some of the others, they shift very little, so as to form a nonbonding pair on the isoxazole nitrogen, partially polarized towards the oxygen atom.

Thus, the SC model for the electronic mechanism of the gas-phase 1,3-dipolar cycloaddition of fulminic acid to ethyne involves the simultaneous relocation of three orbital pairs, which can be represented by a scheme with full-arrows:

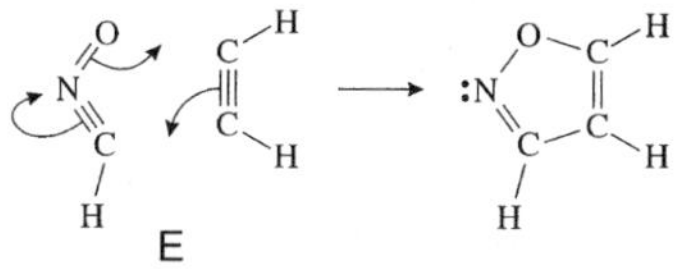

E

The SC description of the 1,3-dipolar cycloaddition of fulminic acid to ethyne is very different from that of the Diels-Alder reaction between butadiene and ethene. The changes in the SC wavefunction during the Diels-Alder reaction are in good agreement with the principles of classical VB theory. From reactants to product, each of the six SC orbitals remains distinctly associated with a single atomic centre, and while the changes in the shapes of the orbitals provide very useful information about the evolution of the bonding interactions, an equally important role is played by the spin-coupling pattern. This last switches gradually from a form more suited to reactants to a form appropriate for the product, becoming very reminiscent, near the TS, of classical VB resonance for benzene. However, such a classical VB element is all but absent from the SC model for the 1,3-dipolar cycloaddition. Instead, the bond rearrangement is achieved by the movement of singlet-coupled orbital pairs through space, during which at least one of the orbitals within a pair becomes completely detached from the atomic centre with which it is associated initially and ends up localized about another centre. The ability of the SC wavefunction to produce a description of this type follows from a purely MO element in its construction: the SC orbitals are, in fact, singly-occupied nonorthogonal MOs. The composition of the spin-coupling pattern remains essentially unchanged during the most important part of the 1,3-dipolar cycloaddition process. This suggests that the reacting system remains distinctly nonaromatic before, at, and after the TS.

It would be difficult, at this stage, to present a properly argued preference between schemes A and B in the Introduction as the more faithful representation for the disrotatory electrocyclic ring-opening of cyclohexadiene. Chemical intuition (at least that of the authors) would suggest scheme B, but scheme A is on the textbook pages and, as has been shown by the SC description of the 1,3-dipolar addition of fulminic acid to ethyne, a 'heterolytic' scheme remains a definite possibility.

4. The Disrotatory Electrocyclic Ring-Opening of Cyclohexadiene

Starting from the TS for the disrotatory electrocyclic ring-opening of cyclohexadiene, we followed the IRC, with twelve points in the direction of cyclohexadiene and another twelve points in the direction of hexatriene, all in steps of about 0.1 amu$^{1/2}$ bohr. All calculations were performed at the CASSCF(6,6)/4-31G level of theory. The TS has C_s symmetry and the carbon-carbon bond length equalization is not so well expressed as in the case of the Diels-Alder TS. The lengths of the carbon-carbon bonds alternate as 1.410–1.391–1.409–1.391–1.410 (in Å, starting from one end of the opening ring) against 1.397–1.391–1.397 for butadiene and 1.404 for ethene at the Diels-Alder TS [in Å, CASSCF(6,6)/4-31G results from Ref. 11]. The distance between the carbon atoms at the two ends of the opening ring is 2.291Å. It should be mentioned that the RHF/6-31G* TS for the disrotatory ring-opening of cyclohexadiene [35] exhibits smaller alternation of the carbon-carbon bond-lengths: 1.390–1.387–1.396–1.387–1.390 (all values in Å). The bond equalization at the CASSCF(6, 6)/4-31G level occurs after the TS, between 0.1 and 0.2 amu$^{1/2}$ bohr in the direction of hexatriene (see Fig. 3, which shows the variation of the lengths of the symmetry-unique carbon-carbon bonds along the IRC, with the carbon atoms numbered clockwise from the appropriate end of the open chain).

The RHF, SC and CASSCF(6,6) energy profiles, all calculated along the CASSCF(6,6) IRC using a 4-31G basis, are shown in Fig. 4. The SC and CASSCF(6,6) curves are sufficiently close to allow the use of CASSCF geometries to derive a SC model for the reaction mechanism. The proportion of CASSCF(6,6) correlation energy recovered by the SC wavefunction ranges between 97.0% at the cyclohexadiene end of the IRC segment, 93.6% at the TS, and 97.1% at the hexatriene end of the IRC segment.

The evolution of the shapes of the SC orbitals with the progress of the cyclohexadiene ring-opening is illustrated by Fig. 5. The symmetry-unique SC orbitals ψ_1–ψ_3 are shown as three-dimensional isovalue surfaces at the cyclohexadiene end of the IRC segment (leftmost column of orbitals), at the TS (central column of orbitals) and at the hexatriene end of the IRC segment (rightmost column of orbitals). The reflections of ψ_1, ψ_2 and ψ_3 in the symmetry plane which is retained throughout this IRC interval, result in ψ_6, ψ_5 and ψ_4, respectively.

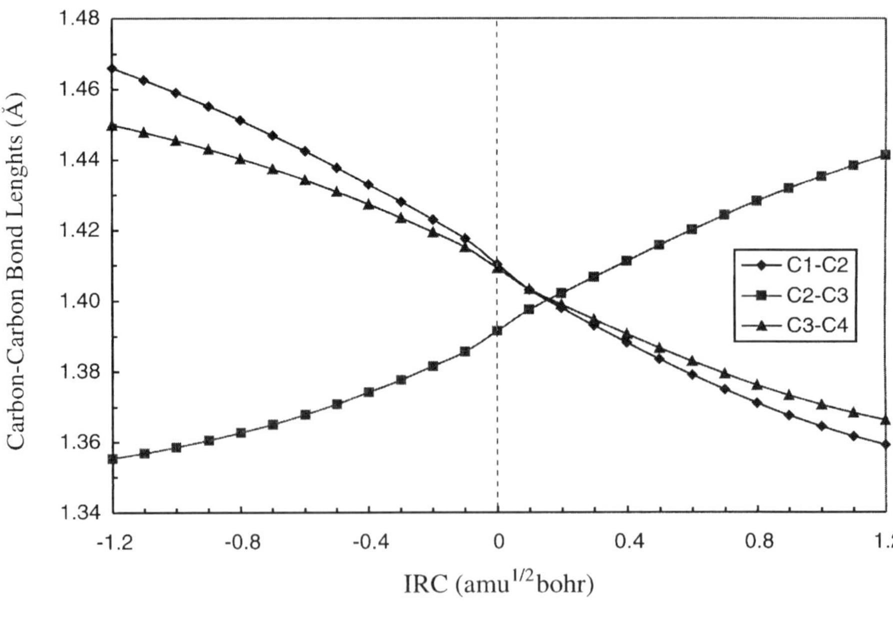

Figure 3. Variation of symmetry-unique carbon-carbon bond lengths (in Å) along the CASSCF(6,6) IRC for the gas-phase disrotatory ring-opening of cyclohexadiene.

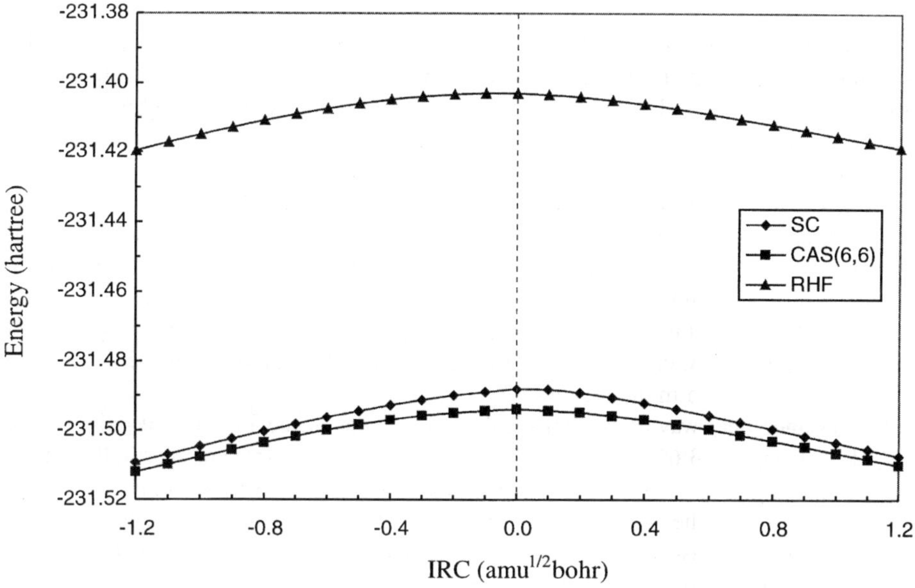

Figure 4. RHF, SC and CASSCF(6,6) energy profiles for the disrotatory ring-opening of cyclohexadiene along the CASSCF(6,6) IRC (all calculated with a 4-31G basis).

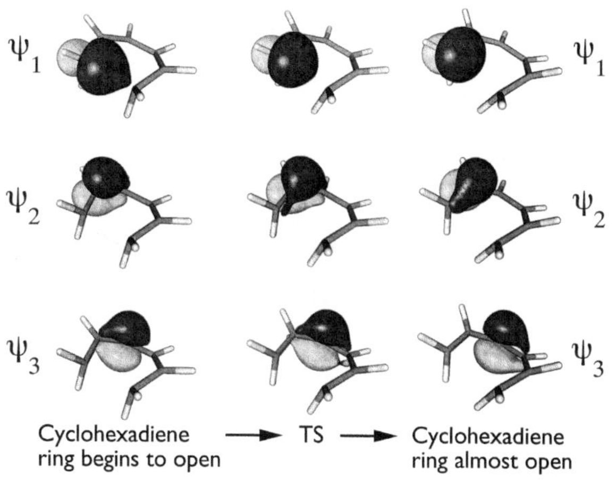

Figure 5. Symmetry-unique SC orbitals for the disrotatory ring-opening of cyclohexadiene along the CASSCF(6,6) IRC at IRC ≈ -1.2 amu$^{1/2}$ bohr *(leftmost column), TS (IRC = 0) and IRC* $\approx +1.2$ amu$^{1/2}$ bohr *(rightmost column). The plot details are as for Fig. 1, except that the isovalue surfaces correspond to* $\psi_\mu = \pm 0.075$.

The length of the breaking carbon-carbon bond corresponding to the leftmost column of orbitals is 2.092Å, yet orbital ψ_1 which, together with its symmetry-related counterpart, ψ_6, is responsible for this bond, still has the shape of a hybrid with significant s content. Orbitals ψ_2 and ψ_3 have the familiar silhouettes of distorted, almost π SC orbitals (compare to orbital ψ_2 from the leftmost column of Fig. 1 for the Diels-Alder reaction), engaged in a (nearly) π bond between the second and third carbon atoms. The other (nearly) π bond is formed by orbitals ψ_4 and ψ_5. This picture of the bonding finds further support in the corresponding spin-coupling pattern (see Fig. 6) and from the overlaps between neighbouring SC orbitals (see Fig. 7). At the leftmost (negative) end of the IRC interval, Θ^6_{00} [see Eq. (2)] is largely dominated by the second Kekulé-type Rumer spin function ${}^{R}\Theta^6_{00;4} = (\alpha(\alpha\beta)(\alpha\beta)\beta) \equiv (1-6, 2-3, 4-5)$ [see Eq. (3)] which couples the spins of orbitals ψ_1 and ψ_6, ψ_2 and ψ_3, and ψ_4 and ψ_5 in singlet pairs. At this early stage of the reaction, the largest nearest-neighbour orbital overlaps are $\langle\psi_1|\psi_6\rangle$ and $\langle\psi_2|\psi_3\rangle = \langle\psi_4|\psi_5\rangle$. The relatively high overlap between orbitals ψ_3 and ψ_4 already hints at the central hexadiene π bond, which will establish itself between these orbitals at a later stage of the process.

The distance between the two terminal carbon atoms at the TS (central point on the IRC segment) is already 2.291Å, which results in very perceptible changes in the orbital shapes, spin-coupling pattern and overlaps between neighbouring orbitals. Orbital ψ_1 (see the central column of orbitals in Fig. 5) becomes less distorted towards the orbital at the other terminal carbon, ψ_6, and this is reflected in a decrease in their overlap (see Fig. 7). Orbitals ψ_2 and ψ_3 (and their symmetry-related counterparts, ψ_5 and ψ_4) attain shapes which are very similar to those of ψ_2 from the TS of the Diels-Alder reaction (see the central column of orbitals in Fig. 1) and of a SC orbital for benzene [8–10]. These changes are accompanied by a tendency towards equalization of the nearest-

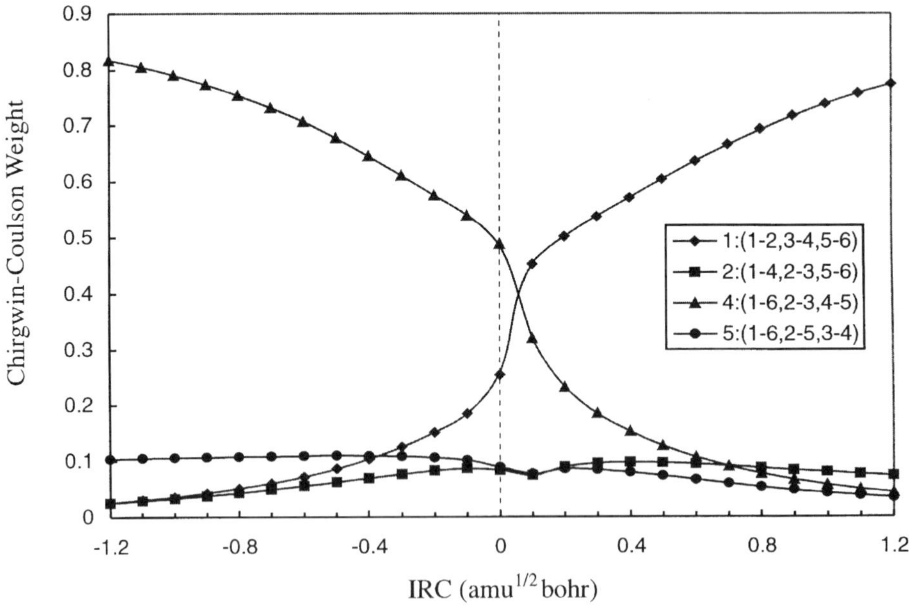

Figure 6. Composition of the active space spin-coupling pattern [Θ^6_{00} in Eq. (2)] from the SC wavefunction for the disrotatory ring-opening of cyclohexadiene along the CASSCF(6,6) IRC, expressed in terms of Chirgwin-Coulson weights P_{0k} [see Eq. (4)] in the Rumer basis [see Eq. (3)].

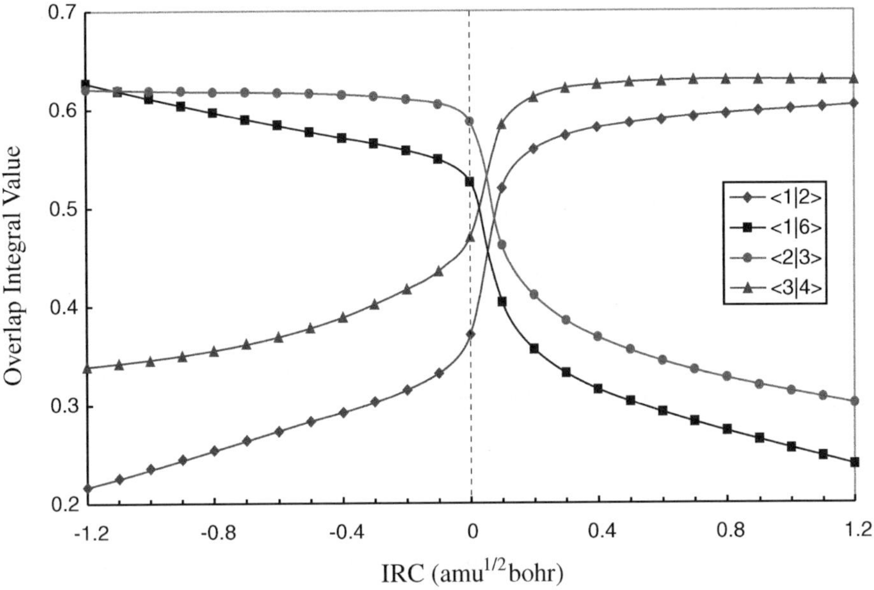

Figure 7. Variation of the symmetry-unique overlap integrals $\langle\psi_\mu|\psi_\nu\rangle$ between neighbouring SC orbitals along the CASSCF(6,6) IRC for the disrotatory ring-opening of cyclohexadiene.

neighbour orbital overlaps, which reaches its fullest extent after the TS, at about 0.1 amu$^{1/2}$ bohr along the positive side of the IRC segment. The spin-coupling pattern at the TS is still dominated by the second Kekulé-type Rumer spin function (1 – 6, 2 – 3, 4 – 5), but already has appreciable input from the first Kekulé-type Rumer spin function, ${}^{\mathrm{R}}\Theta^{6}_{00;1} = (\alpha\beta)(\alpha\beta)(\alpha\beta) \equiv (1-2, 3-4, 5-6)$. As shown in Fig. 6, the Chirgwin-Coulson weights of the two Kekulé-type Rumer spin functions become identical at *ca.* 0.1 amu$^{1/2}$ bohr along the IRC leaf leading towards hexatriene.

Thus, the SC description of the reacting system becomes most similar to that of benzene [8–10] not at the CASSCF/4-31G TS, but after it, at a further 0.1 to 0.2 amu$^{1/2}$ bohr along the IRC towards hexatriene, when the carbon-carbon bond lengths within the chain become almost equal (see Fig. 3), the two Kekulé-type Rumer spin functions are in near-perfect 'resonance' (see Fig. 6), and all overlaps between neighbouring SC orbitals attain values (see Fig. 7) close to that for benzene (0.52, see Ref. 10). The obvious conclusion, just as in the case of the Diels-Alder reaction [11], is that the gas-phase disrotatory electrocyclic ring-opening of cyclohexadiene passes through a geometry at which the characteristic features of the electronic structure of the reacting system leave little, if any, doubt about its aromaticity.

The rightmost column of orbitals in Fig. 5 corresponds to the final point along the calculated IRC interval in the direction of hexadiene. The distance between the two terminal carbon atoms is already 2.490Å and, as a consequence, the shape of orbital ψ_1 is much more π-like. Together with ψ_2, this orbital forms one of the hexatriene π bonds. The second of these bonds is formed by ψ_3 and ψ_4, and the third by ψ_5 and ψ_6.

Accordingly, the most important spin function becomes (1 – 2, 3 – 4, 5 – 6), which couples the spins associated with the bonded orbitals in singlet pairs (see Fig. 6). The fact that the largest overlaps in Fig. 7 are $\langle\psi_1|\psi_2\rangle = \langle\psi_5|\psi_6\rangle$ and $\langle\psi_3|\psi_4\rangle$ completes the picture of the now almost completely formed *cis*-1,3,5-hexatriene chain.

The SC description of the disrotatory electrocyclic ring-opening of cyclohexadiene turns out to be very similar to that of the Diels-Alder reaction between butadiene and ethene. The most significant difference is that the aromatic intermediate forms not at the TS, but at a later stage of the opening of the cyclohexadiene ring. Just as in the case of the Diels-Alder reaction, the most appropriate simple description of the electronic mechanism of the gas-phase disrotatory electrocyclic ring-opening of cyclohexadiene is provided by scheme B (see the Introduction), in which the half-arrows indicate the simultaneous breaking of the two π and one σ bond in the ring, accompanied by the formation of three new π bonds in the open chain.

5. Conclusions and Final Comments

The SC descriptions of the electronic mechanisms of the three six-electron pericyclic gas-phase reactions discussed in this paper (namely, the Diels-Alder reaction between butadiene and ethene [11], the 1,3-dipolar cycloaddition of fulminic acid to ethyne [12], and the disrotatory electrocyclic ring-opening of cyclohexadiene) take the theory much beyond the HMO and RHF levels employed in the formulation of the most popular MO-based treatments of pericyclic reactions, including the Woodward-Hoffmann rules [1,2], Fukui's frontier orbital theory [3] and the Dewar-Zimmerman model [4–6]. The SC wavefunction maintains near-CASSCF quality throughout the range of reaction coordinate studied for each reaction but, in contrast to its CASSCF counterpart, it is very much easier to interpret and to visualize directly.

It is tempting to call the resulting pictures of varying orbital shapes, orbital overlaps and spin-coupling patterns, which may have to undergo significant changes in order to be able to accommodate the differences between the electronic structures of reactant(s) and product(s), a modern rendition of classical VB ideas. Such a view certainly seems to be appropriate for the Diels-Alder reaction between butadiene and ethene, as well as for the disrotatory electrocyclic ring-opening of cyclohexadiene. However, as demonstrated by the 1,3-dipolar cycloaddition of fulminic acid to ethyne, the SC approach can, if required, break the limitations of traditional VB thinking and allow orbital relocations between different atoms in the reacting systems.

One very important aspect of SC theory, as applied to the description of pericyclic reactions, lies in the demonstration that reactions of this type can proceed through one of two rather different electronic mechanisms, which can be illustrated by the traditional organic chemistry textbook schemes with full- and half-arrows (see schemes A and B in the Introduction). Our analysis strongly suggests that only reactions which follow a scheme with half-arrows, such as the Diels-Alder reaction between butadiene and ethene or the disrotatory electrocyclic ring-opening of cyclohexadiene, may pass through an aromatic reaction intermediate. In the case of the Diels-Alder reaction, this intermediate is sufficiently close to the TS for us to claim that the TS is aromatic. However, the aromatic intermediate in the disrotatory electrocyclic ring-opening of cyclohexadiene occurs after the TS, upon further opening of the cyclohexadiene ring, which questions the importance of the TS in an electronic reaction mechanism. Certainly, for many reactions, the most profound changes in the electronic structure of the reacting system do indeed occur in a very narrow IRC interval around the TS, but this does not rule out possible exceptions. As was shown by the SC description of the electrocyclic isomerization of cyclobutene to *cis*-butadiene [36], the major changes in the electronic structure of the opening ring occur well after the TS, for both the conrotatory and disrotatory pathways.

We believe that reactions that follow a scheme with full-arrows, such as the gas-phase 1,3-dipolar cycloaddition of fulminic acid to ethyne, are most likely to remain nonaromatic along the whole reaction coordinate.

At the RHF level of theory, which uses a wavefunction that is relatively straightforward to interpret, the subtle differences between the half- and full-arrow reaction schemes would remain well-hidden within the doubly-occupied, usually delocalized orbitals. While it can be argued that the application of an orbital localization procedure could produce a semblance of the SC description for the 1,3-dipolar cycloaddition of fulminic acid to ethyne, the double-occupancy restriction makes it impossible to obtain the analogue of a half-arrow SC mechanism using an RHF wavefunction.

All three reactions discussed in this paper are of the thermally-allowed type, according to the Woodward-Hoffmann rules. It appears much easier, within a higher-level treatment such as the SC approach, to establish and explain the factors facilitating an allowed reaction than to account for the barriers in the paths of thermally-forbidden reactions. The aromatic intermediate structures that we observe along the reaction pathways for the Diels-Alder reaction between butadiene and ethene and for the disrotatory electrocyclic ring-opening of cyclohexadiene, lend some support to notions of aromatic stabilization of the TS in thermally-allowed reactions, as suggested by Dewar and Zimmerman [4–6]. On the other hand, the SC descriptions of the conrotatory and disrotatory pathways in the electrocyclic isomerization of cyclobutene to *cis*-butadiene [36] did not discover even hints of antiaromaticity in the conformations along the thermally-forbidden disrotatory pathway. This may be due to the fact that antiaromaticity is a concept that is much more elusive than aromaticity, especially if evidence is sought for it in the results of *ab initio* calculations. Another problem is related to the fact that the establishment of a series of conformations to define a forbidden reaction pathway is a task which is far from trivial. Such considerations reach beyond the scope of the current paper, but this is certainly an area that we plan to revisit in the near future.

References

1. R.B. Woodward and R. Hoffmann: *Angew. Chem. Int. Ed. Engl.* **8**, 781 (1969).
2. R.B. Woodward and R. Hoffmann: *The Conservation of Orbital Symmetry*, Verlag Chemie, Weinheim (1970).
3. K. Fukui: *Acc. Chem. Res.* **4**, 57 (1971).
4. M.J.S. Dewar: *Tetrahedron Suppl.* **8**, 75 (1966).
5. H.E. Zimmerman: *J. Am. Chem. Soc.* **88**, 1564 (1966).
6. H.E. Zimmerman: *Acc. Chem. Res.* **4**, 272 (1971).

7. J. Gerratt, D.L. Cooper, P.B. Karadakov and M. Raimondi: *Chem. Soc. Rev.* **26**, 87 (1997).
8. D.L. Cooper, J. Gerratt and M. Raimondi: *Nature* **323**, 699 (1986).
9. J. Gerratt: *Chem. Br.* **23**, 327 (1987).
10. D.L. Cooper, S.C. Wright, J. Gerratt, P.A. Hyams and M. Raimondi: *J. Chem. Soc. Perkin Trans. II*, 719 (1989).
11. P.B. Karadakov, D.L. Cooper and J. Gerratt: *J. Am. Chem. Soc.* **120**, 3975 (1998).
12. P.B. Karadakov, D.L. Cooper and J. Gerratt: **100**, 222 (1998).
13. R. Pauncz: *Spin Eigenfunctions*, Plenum Press, New York (1979).
14. P.B. Karadakov, J. Gerratt, D.L. Cooper and M. Raimondi: *Theor. Chim. Acta* **90**, 51 (1995).
15. B.H. Chirgwin and C.A. Coulson: *Proc. Roy. Soc. Lond. Ser. A* **201**, 196 (1950).
16. T. Thorsteinsson and D.L. Cooper: *J. Math. Chem.* **23**, 105 (1998).
17. T. Thorsteinsson, D.L. Cooper, J. Gerratt, P. B. Karadakov and M. Raimondi: *Theor. Chim. Acta* **93**, 343 (1996).
18. T. Thorsteinsson and D.L. Cooper: *Theor. Chim. Acta* **94**, 233 (1996).
19. H.-J. Werner, P.J. Knowles, with contributions from R.D. Amos, A. Berning, D.L. Cooper, M.J.O. Deegan, A.J. Dobbyn, F. Eckert, C. Hampel, T. Leininger, R. Lindh, A.W. Lloyd, W. Meyer, M.E. Mura, A. Nicklaß, P. Palmieri, K. Peterson, R. Pitzer, P. Pulay, G. Rauhut, M. Schütz, H. Stoll, A.J. Stone and T. Thorsteinsson: *MOLPRO (A Package of Ab Initio Programs)*.
20. D.G. Truhlar and A. Kuppermann: *J. Am. Chem. Soc.* **93**, 1840 (1971).
21. D.G. Truhlar and A. Kuppermann: *J. Chem. Phys.* **56**, 2232 (1972).
22. K. Fukui: *J. Chem. Phys.* **74**, 4161 (1970).
23. K. Fukui: in R. Daudel and B. Pullman (Ed.), *The World of Quantum Chemistry. Proceedings of the First International Congress of Quantum Chemistry, Menton, 1973*, D. Reidel, Dordrecht (1974), p. 113.
24. C. Gonzalez and H.B. Schlegel: *J. Chem. Phys.* **90**, 2154 (1989).
25. C. Gonzalez and H.B. Schlegel: *J. Phys. Chem.* **94**, 5523 (1990).
26. M.J. Frisch, G.W. Trucks, H.B. Schlegel, P.M.W. Gill, B.G. Johnson, M.A. Robb, J.R. Cheeseman, T. Keith, G.A. Petersson, J.A. Montgomery, K. Raghavachari, M.A. Al-Laham, V.G. Zakrzewski, J.V. Ortiz, J.B. Foresman, J. Cioslowski, B.B. Stefanov, A. Nanayakkara, M. Challacombe, C.Y. Peng, P.Y. Ayala, W. Chen, M.W. Wong, J.L. Andres, E.S. Replogle, R. Gomperts, R.L. Martin, D.J. Fox, J.S. Binkley, D.J. Defrees, J. Baker, J.P. Stewart, M. Head-Gordon, C. Gonzalez and J.A. Pople: *Gaussian 94, Revision D.4*, Gaussian, Inc., Pittsburgh, PA, 1995.
27. P.B. Karadakov, J. Gerratt, D.L. Cooper and M. Raimondi: *J. Chem. Phys.* **97**, 7637 (1992).
28. Y. Li and K.N. Houk: *J. Am. Chem. Soc.* **115**, 7478 (1993).
29. K.N. Houk, Y. Li, J. Storer, L. Raimondi and B. Beno: *J. Chem. Soc. Faraday Symp.* **90**, 1599 (1994).
30. J.J.W. McDouall, M.A. Robb, U. Niazi, F. Bernardi and H.B. Schlegel: *J. Am. Chem. Soc.* **109**, 4642 (1987).
31. G. Schaftenaar: *MOLDEN (A Pre- and Post Processing Program of Molecular and Electronic Structure)*, CAOS/CAMM Center, the Netherlands.
32. I. Fleming: *Frontier Orbitals and Organic Chemical Reactions*, Wiley, Chichester (1994).
33. D.L. Cooper, J. Gerratt, M. Raimondi and S.C. Wright: *Chem. Phys. Lett.* **138**, 296 (1987).
34. D.L. Cooper, J. Gerratt, M. Raimondi and S.C. Wright: *J. Chem. Soc. Perkin Trans. II*, 1187 (1989).
35. J.E. Baldwin, V.P. Reddy, L.J. Schaad and B.A. Hess: *J. Am. Chem. Soc.* **110**, 8554 (1988).
36. J.M. Oliva, J. Gerratt, P.B. Karadakov and D.L. Cooper: *J. Chem. Phys.* **107**, 8917 (1997).

A Topological Study of Electron Transfer and Three-Electron Bond

Xénophon Krokidis and Alain Sevin
Laboratoire de Chimie Théorique, Université Pierre et Marie Curie, 4 Place Jussieu, 75252 Paris Cedex 05, France

Abstract

A topological study of the electron transfer in $\{Li + Cl_2\}$ system and of the three-electron bond created through this transfer has been achieved, based on the topological concepts of Bonding Evolution Theory. Our results suggest that the dual cusp catastrophe characterizes the diabatic surface crossings which are subjacent in the classical adiabatic analysis of the overall reaction path.

1. Introduction

Electron transfer in metal plus dihalogen or related systems has been the subject of many experimental and theoretical works, and is generally considered as taking place through the so-called 'harpooning mechanism' [1]. The latter process has been characterized by the crossing of the neutral covalent Potential Energy Surface (PES), asymptotically linked to $\{Li(^2S) + Cl_2(^1\Sigma_g^+)\}$, with the ionic Charge Transfer (CT) PES, linked to $\{Li^+(^1S) + Cl_2^-(^2\Sigma_u^+)\}$ at infinite separation. We thus see that the Cl_2^- anionic species plays a central role. It is well established that upon electron capture, dihalogens X_2 afford 'vertical' anionic species of very low negative Electron Affinity (EA), and through an important bond relaxation, yield stable 3-electron bonds, stabilized by resonance, according to $(X^{\cdot}{\cdots}X^-) \leftrightarrow (X^-{\cdots}X^{\cdot})$ [2]. Consequently both vertical and relaxed Cl_2^- anions remain central for the study of electron transfer in our system.

The study of chemical reactions requires the definition of simple concepts associated with the properties of the system. Topological approaches of bonding, based on the analysis of the gradient field of well-defined local functions, evaluated from any quantum mechanical method are close to chemists' intuition and experience and provide method-independent techniques [4–7]. In this work, we have used the concepts developed in the Bonding Evolution Theory [8] (BET, see Appendix B), applied to the Electron Localization Function (ELF, see Appendix A) [9]. This method has been applied successfully to proton transfer mechanism [10,11] as well as isomerization reaction [12]. The latter approach focuses on the evolution of chemical properties by assuming an isomorphism between chemical structures and the *molecular graph* defined in Appendix C.

The aim of the present study is double: i) to show that BET can be used as a tool for analyzing the adiabatic PESs and localizing the diabatic crossings which govern the overall electron changes; ii) to provide a topological description of the three-electron bonds.

Our study has been performed in two steps: first, quantum mechanical calculations

A. Hernández-Laguna et al. (eds.), Quantum Systems in Chemistry and Physics, Vol. 1: Basic Problems and Model Systems, 345–359.

have been carried out in order to get density matrices. In the second step, these outputs have been used as starting material for the topological study. In this exploratory report, we will restrict ourselves to the study of the C_{2v} geometry, the full analysis being reported elsewhere [13].

2. Methodology

The quantum mechanical calculations have been performed with the Gaussian 94 [14] series of programs. In this paper, we will restrict ourselves to report MP2/6-311G(3df) [15,16] results, though DFT/B3LYP results obtained with the same basis set are given in Table 1 for the sake of comparison. In all calculations involving doublet species, we have checked that spin contamination remained very small and not likely to affect significantly the corresponding calculated energies. The calculated geometries and energies of the stationary points are given in Table 1. It is worth recalling that these calculations were not aimed at describing the energetics of the $Li + Cl_2$ system as accurately as possible, but at providing reliable data for further topological exploitation using BET method.

The topological analysis has been performed with the TopMoD series of programs written in our laboratory [17]. These programs use as input the *wfn* file generated by GAUSSIAN94, with natural orbital population. The calculations are then carried out in four steps: i) evaluation of the ELF function over a 3D grid; ii) location of the critical points of the ELF function; iv) integration of charge density over the basins; iii) identification of the various basins and assignment of the corresponding grid points.

Table 1 *Experimental and calculated structures and energies of the stationary points. In this table d_c is egal to the MP2 equilibrium value of the d distance.* [(a)] *Cl_2 non relaxed.*

		MP2		DFT		Exp.
Species	d(Å); θ(°)	E(a.u)	ΔE(eV)	E(a.u)	ΔE(eV)	ΔE(eV)[32]
Li		−7.432	0.00	−7.491	0.00	1.84
Li(^2P)			1.84			1.84
Li^+			5.02		5.62	5.4
Cl		−459.632	0.00	−460.167	0.00	
Cl^-			−3.55		−3.62	3.62
	MP2 DFT					
Cl_2 (eq.)	1.9854/2.0112	−919.355	0.00	−920.422	0.00	
Cl_2^- (vertical)		−919.375	−0.55	−920.458	−0.98	
Cl_2^-	2.596/2.2723	−919.443	−2.40	−920.527	2.86	2.4–2.5
	$d = \infty$	−926.787	0.00	−927.913	0.00	
$Li+Cl_2$ [(a)]						
	$d_c = 1.94$; $\theta = 90.0$	−926.842	−1.50	−927.978	−1.69	

3. Quantum Mechanical Study

For reproducing as closely as possible diabatic conditions, we have fixed the Cl—Cl bondlength at its neutral equilibrium value. This way, the system depends on two parameters as shown in Figure 1. Previous experimental and theoretical studies on similar systems, [1,18] have shown that electron jump from Li to the acceptor molecule Cl_2, which has, once relaxed, a positive vertical electron affinity (see Table 1), is likely to take place at a distance *d*, (see the definition of this parameter in Figure 1) which is superior to the LiCl equilibrium distance (MP2 value 2.0425 Å). The description of this phenomenon in terms of MO and states will be briefly recalled in the next section.

3.1. Calculated MP2 PESs

The lowest energy $^2A_1$ and $^2B_2$ PESs calculated in the C_{2v} geometry are reported in Figure 2. They have been obtained by scanning the *d* distance by increments of 0.14 Å. The $^2A_1$ PES has been obtained starting from *long d* distance and then using the resulting initial SCF-MOs as starting guesses for decreasing *d* values (forward process). Conversely, the $^2B_2$ PES has been obtained by starting from a calculation at *short d* distance, and subsequent propagation of the resulting MOs as initial guesses for higher *d* values (backward process). Both behaviors nicely confirm the analysis developed in

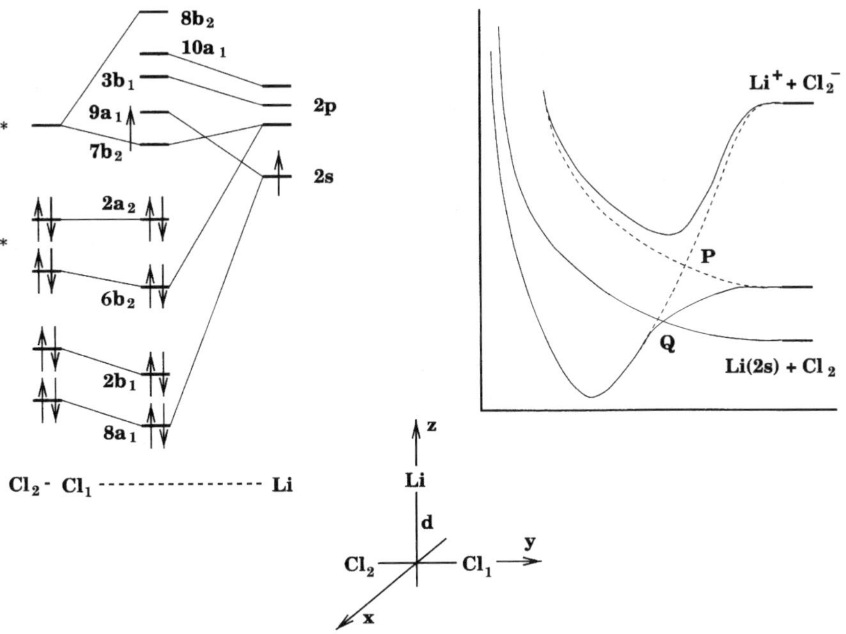

Fig. 1. Left part: Qualitative MO diagram in the C_{2v} geometry of the Li + Cl_2 *system. Right part: Associated state correlations diagram. Diabatic correlations are shown in broken lines.*

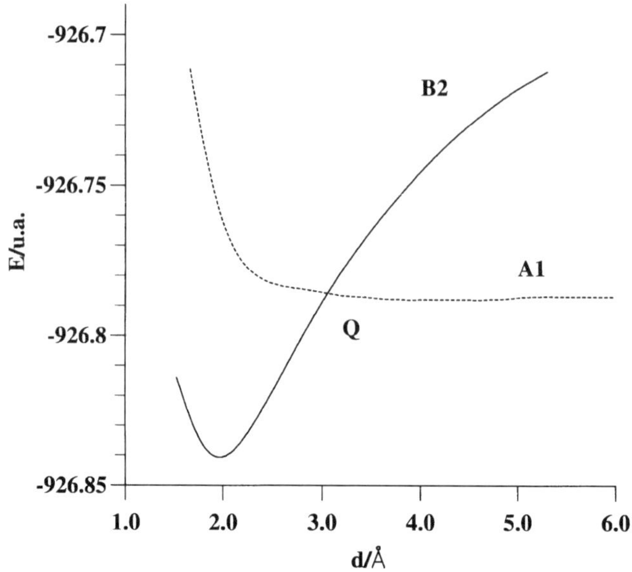

Fig. 2. *Calculated PESs. The $^2B_2$ PES has been obtained through a 'backward' process, i.e. starting from the left, (see text), while the $^2A_1$ PES has been obtained through a 'forward' process, i.e. from right to left. The scanning of both PESs has been carried out by constant increment of 0.14 Å, according to the case.*

Figure 1. The PESs cross at point Q ($\sim$ 3 Å). A stable complex ($^2B_2$ symmetry), is obtained at $d = 1.94$ Å, lying 1.50 eV below the asymptotic neutral limit (Table 1). In this complex, the spin density, calculated after the MP2 step is 0.0192 on Li, and 0.4904 on each Cl atom, thus showing that the electronic transfer is complete. The atomic charges, calculated at the MP2 level, are 0.6375 on Li and -0.3188 on Cl, showing that back-donation from the lone pairs of Cl_2^- towards Li^+ is present, via the $6b_2$ MO, displayed in the left part of Figure 1. Other studies have shown that upon complete structure optimization, one gets a true stable complex [19–21], so that the situation which is reached through electron transfer might adiabatically evolve towards a more stable one, but this further evolution lies beyond the scope of the present study. These results show that two types of PESs, one covalent and one ionic, might be obtained in the region of diabatic crossing. It is noteworthy that a similar behavior, not reported here, has been obtained using other techniques such as DFT or CASSCF at various levels [13].

4. The BET Analysis

The Bonding Evolution Theory, briefly presented in Appendix B, provides a description of the bonding features of a system, along with their evolution accompanying a reaction path. It relies on the variation of the ELF topological profile as a function of nuclear coordinates. The ELF makes a partition of the molecular space into open sets having a clear chemical meaning (see Appendix A), and corresponds to the classical chemists' vision of structures, in terms of bonds, lone pairs etc. Obviously, this partition depends on the electronic state under scrutiny. In coming sections, we examine the electron transfer in the case of C_{2v} geometry, using the MP2 results, keeping in mind that DFT and CASSCF calculations yield the same description.

4.1. *Electron transfer*

Figure 3 displays the molecular partition of the fragments for the three states previously discussed in the quantum mechanical section, at $d = 6$ Å. Figure 3 A and 3 B respectively display the $^2A_1$ and $^2B_2$ covalent states, and Figure 3 C shows the ionic $^2B_2$ Charge Transfer state. It is worth examining the striking features of the molecular partitions in each case. In the $^2A_1$ molecular partition, the disynaptic basin V(Cl_1, Cl_2), indicated by an arrow, corresponds to the Cl—Cl bond [22]. Two basins are found around Li, one corresponding to its core C(Li), and the second one, V(Li), to its valence odd electron (L shell). The $^2B_2$ covalent state is characterized by two monosynaptic basins, V_1(Li) and V_2(Li), located on both sides of the C(Li) basin in the molecular plane. They correspond to the half-filled $2p$ AO of Li. As when dealing with the previous state, the Cl atoms are bonded through a disynaptic basin, still noted V(Cl_1, Cl_2). In the ionic state, the Cl atoms are linked by a (3, −1) saddle point, or,

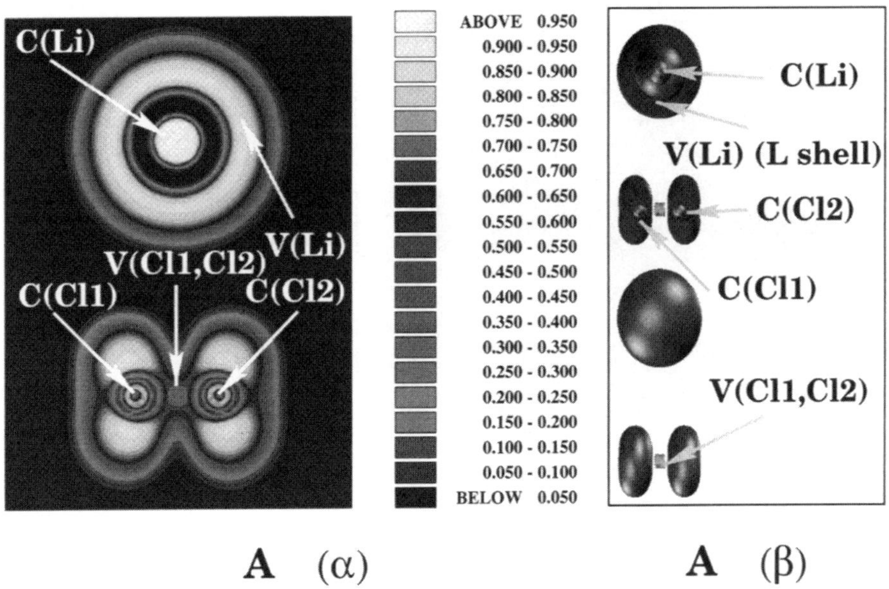

Fig. 3. Part A: display of the ELF molecular partition of the $^2A_1$ state in C_{2v} geometry, $d = 6.00$ Å. A(a): is the projection of the ELF in the molecular plane. A(b) (upper part): is a cut of the ELF isosurface ($\eta = 0.8$). A(b) (lower part): shows the entire envelope of the same ELF isosurface. Part B: corresponds to the $^2B_2$ state. part C: charge transfer state.

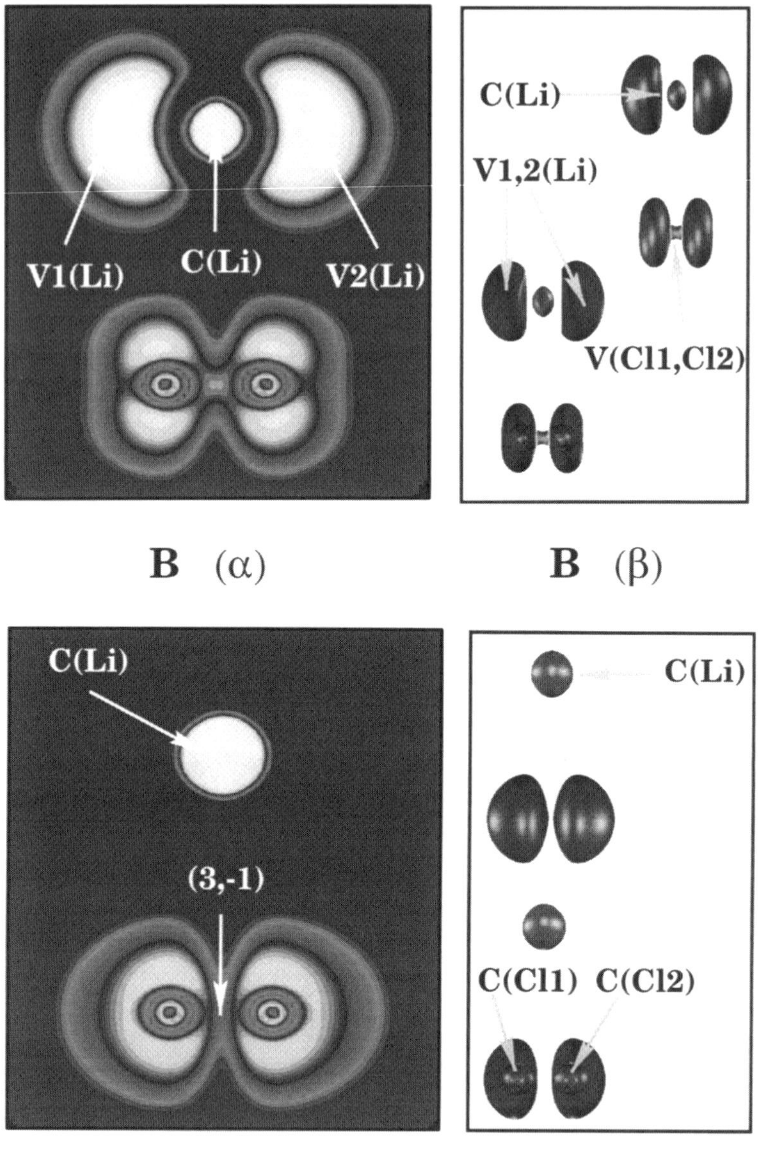

Fig. 3. (*continued*)

alternatively by a critical point of index I = 1 (see Appendix C), which is typical of a 3-electron bond. Around Li, one only finds the core basin C(Li). In all three states, the valence electrons of the Cl lone pairs are contained into a torus centered on the atom and perpendicular to the Cl—Cl axis. In Figure 4 A, is displayed the molecular partition

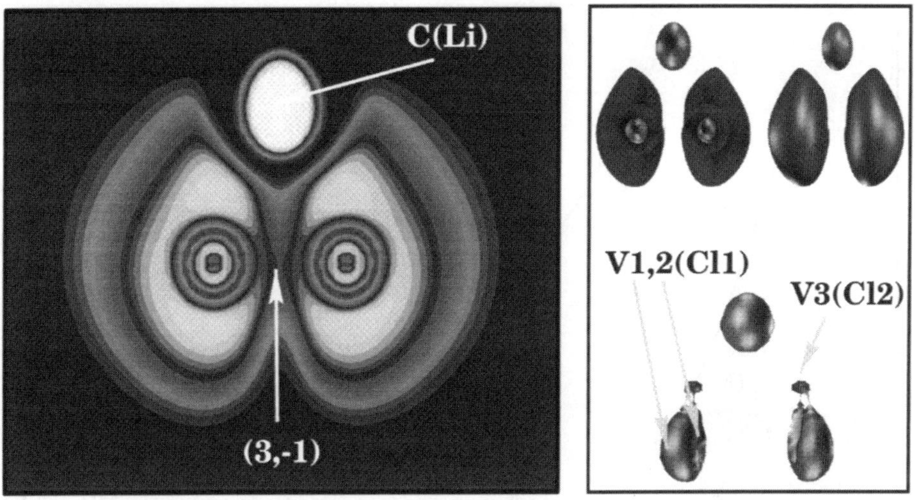

Fig. 4. Display of the ELF partition of the C_{2v} complex defined in Table 1 (MP2 calculation).

of the short distance complex. Between both Cl atoms, there exists a (3, −1) critical point, and each Cl valence torus now splits into three basins. Two of them, labelled $V_1(Cl_i)$ and $V_2(Cl_i)$ ($i = 1, 2$), are located on both sides of the molecular plane, while the third one, labelled $V_3(Cl_i)$, lies in the molecular plane and is directed towards Li. This structure might be considered as a mixture of both covalent and ionic $^2B_2$ states, in which $V_3(Cl_1)$ and $V_3(Cl_2)$ are remnant of the $V_1(Li)$ and $V_2(Li)$ basins found at large distance, Figure 3 B. It is noteworthy that the presence of a (3, −1) critical point between both Cl atoms, might be related to the existence of a 3e-electron bond which is well documented for the Cl_2^- species. In Figure 5, a schematical description of the reaction, yielding the previously examined complex, is displayed as a function of the control parameter d. The crossing at point Q, previously described in Figure 1, is first studied. It takes place at $d = 3.06$ Å. For both crossing states, the ELF η has been calculated. In Figure 5 D, the structure of the $^2A_1$ state is given. The slightly polarized V(Li) and the C(Li) basins are shown, along with the still existing $V(Cl_1, Cl_2)$ basin. Figure 5 E shows the $^2B_2$ ionic state at point Q. As expected, there is no common topological feature between Figure 5 D and Figure 5 E, in agreement with the fact that both states have different symmetries. Let us now focus our attention on the $^2B_2$ state (Figure 5 E). By comparison with Figure 5 C, one sees that the latter state is mostly ionic, thus showing that a crossing with the steeply descending CT state has already occurred. As in Figure 1, this crossing point is noted P. The BET method has been used for localizing the avoided crossing at point P. Adiabatic calculations, starting from $d = 6.00$ Å have been carried out along the covalent $^2B_2$ PES. For 4.88 Å$< d \leqslant 6$ Å, all molecular partitions are equivalent to the one of Figure 5 B (or Fig. 3 B as well), thus showing that the system lies in a region of structural stability (see Appendix D). At $d^* = 4.88$ Å, and in a narrow range (~ 0.12 Å), three catastrophes occur quite simultaneously. The first catastrophe corresponds to the breaking of the Cl—Cl bond.

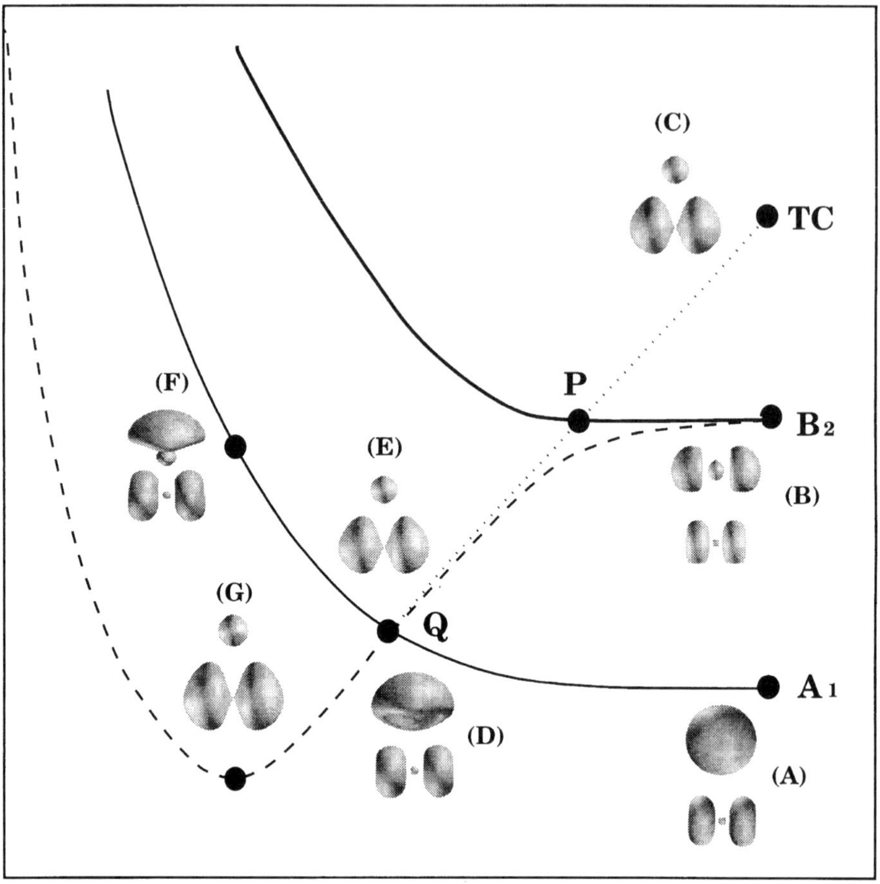

Fig. 5. Display of the ELF for various values of d in the C_{2v} geometry, calculated with the TopMoD series of programs.

The disynaptic basin $V(Cl_1, Cl_2)$ of the covalent Cl—Cl bond yields two monosynaptic ones and a $(3, -1)$ saddle point. After this catastrophe, a reorganization of the electron localization around the Cl cores takes place, which splits the valence basins of the Cl atoms into two basins, $V_1(Cl_i)$ and $V_2(Cl_i)$ ($i = 1, 2$). The latter process has no great importance in the scope of localizing the crossing itself. The second and third catastrophes involve the $V_1(Li)$ and $V_2(Li)$ basins, and correspond to the harpooning process: the electronic density they contain (Figure 5 B, or Fig. 3 B) is absorbed by the Cl valence basins. Once this process has occurred, the system remains topologically equivalent to the complex of Figure 4 A. Another interesting feature consists in considering the evolution of the CT state, in going from large distance to point P. In the vicinity of point P, the two torus basins of the Cl atoms split, yielding the two additional small basins of Figure 4 A, labelled $V_3(Cl_1)$, $V_3(Cl_2)$, directed towards Li^+ along both Cl-Li axes. The latter behavior corresponds to the back-donation mentioned in the discussion of Figure 1. The integration of the electronic density within the basins of a molecular partition yields their population. The calculated results are given in Table 2. The aforementioned changes in the topology exactly reveal the crossing of the CT and of the $^2B_2$ covalent PESs. The existence of the crossing at point Q explains the ionic nature of the complex at its equilibrium distance (Figures 5 G or 4 A). In conclusion, the BET allows for localizing the avoided crossings by comparing the topology of the complex to that of the asymptotic species. Moreover, when the crossing is symmetry-allowed, it preserves the symmetry of the crossing states.

4.2. *Modelization of the process*

In our study the avoided crossings are associated with the topological changes accompanying the chemical transformation, i.e. the breaking of the Cl—Cl bond which takes place at a particular d^* value of the control parameter, in the molecular partition of the starting configuration. More precisely, the *dual cusp* catastrophe which characterizes the breaking of a covalent bond [8] involves the $V(Cl_1, Cl_2)$ attractor located at $\mathbf{r}^{(c)}$. The Taylor development of $\eta(\mathbf{r}; d)$, around $\mathbf{r}^{(c)}$, after a change of variables can be written: [23]

$$\eta(x; a, b) \doteq -\left[\frac{1}{4}x^4 + \frac{1}{2}ax^2 + bx\right] \tag{1}$$

(see Appendix D for the meaning of $\doteq$). In this equation, x is a generalized space variable along the Cl—Cl axis. The three-fold degenerate critical point, $\mathbf{r}^{(c)}$ which is obtained for $a = b = 0$ is located at the origin $x = 0$, middle of Cl—Cl. The observed topological behavior, for example in C_{2v} symmetry, implies that b is constant during the process, and equal to the value found for the bifurcation state. Accordingly, in the latter equation, $b = b^*$ is determined by the condition for which x_0, which locates the attractor of the Cl—Cl bond for a given geometry, becomes three-fold degenerate. [23]

$$\left.\frac{d\eta}{dx}\right|_{x_0} = \left.\frac{d^2\eta}{dx^2}\right|_{x_0} = 0 \tag{2}$$

This condition yields $b = 2x_0^3$. For the corresponding value of a, namely $a(d^*)$, one gets: $a(d^*) = -3x_0^2 = (-\frac{27}{4}b^2)^{\frac{1}{3}}$. Hence the splitting of the $V(Cl_1, Cl_2)$ basin takes place

Table 2 *Basin populations of the complexes.* $V_t(Cl_i)$ *with* $i = 1, 2$ *means the total valence population of Cl atoms.*

	C_{2v}
$C(Cl_1)$	10.07
$C(Cl_2)$	10.07
$C(Li)$	2.03
$V(Li)$	
$V_t(Cl_1)$	7.40
$V_t(Cl_2)$	7.40

as $a(d)$ passes from $a(d) > a(d^*)$ to $a(d) < a(d^*)$, via $a(d) = a(d^*)$. The latter finding means that, for $a(d^*)$, either the attractor (in the C_{2v} symmetry), or in other cases, a wandering point of the $V(Cl_1,Cl_2)$ basin, i.e., a point which satisfies the condition $\nabla_r \eta \neq 0$, becomes degenerate, with one zero eigenvalue. Then, after a small perturbation, expressed by a, and corresponding to a variation of d, two monosynaptic basins and a $(3, -1)$ saddle point between them appear. The C_{2v} symmetry implies that $b^* = 0$, while for other geometries $b^* \neq 0$, which describes the disymmetry of the bond breaking. Figure 6 displays the partition of the control space, according to the number of maxima (full circles) and saddle points (full triangles) of the η function. Regions I.1 and II.2 correspond to the molecular partitions in which the $V(Cl_1, Cl_2)$ basin exists, (one maximum), while regions I.2 and II.1 correspond to molecular partitions in which the Cl—Cl bond is broken (two maxima and a $(3, -1)$ saddle point). The domains having $b < 0$ and $b \geqslant 0$ differ by the way the θ angle is defined (see Figure 6). In the present study, only the case $b = 0$ has been explored. Thus, the electron transfer is considered as a trajectory in the control space, in the sense described above. The cusp curve represents the transitions from region I.1 to I.2, and characterizes the avoided PES crossings.

5. Conclusion

In a context of *irreversible* conditions, electron transfer in the $\{Li + Cl_2\}$ system realizes and illustrates the intuitive knowledge we have of a 'catastrophic decay'. When Li and Cl_2 approach themselves, in the C_{2v} geometry which is the simplest reference for any discussion, we observe a crossing between the covalent PES of the neutral ground state, and the ionic $\{Li^+ + Cl_2^-\}$ CT PES. Through this crossing, electron jump takes place from the covalent to the ionic PES, with a probability which might be

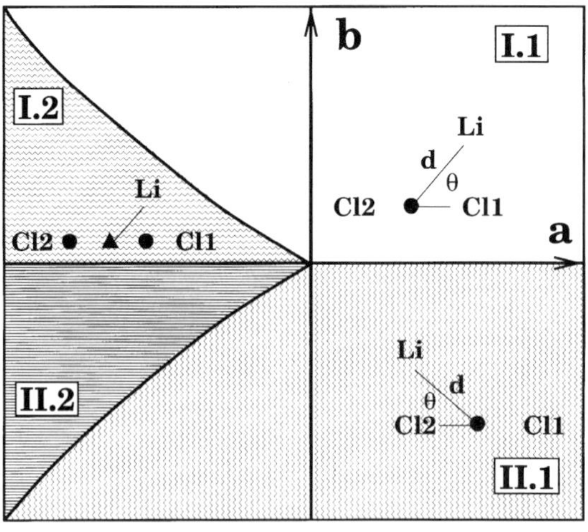

Fig. 6. Partition of the control space according to the number of maxima and saddle points in the η function [33].

estimated by the classical Landau-Zener model [3]. In the usual adiabatic description this crossing becomes more or less avoided in any other geometry, for symmetry reason, though offering the possibility of a progressive electron transfer, but the question might be raised of the pertinence of this result. In particular, once the electron has been trapped by Cl_2, yielding the formation of a 3e-bond, stable at all Cl—Cl distance, an irreversible further chemical evolution gives LiCl + Cl via a strongly exothermic process. It is therefore difficult to speak in terms of potentially reversible process in this case, and the diabatic language seems more appropriate. The topological analysis of ELF η, based on the calculated MP2 results, nicely comforts this analysis by providing a firm basis for topological analysis of the electronic events, in the general catastrophe theory framework. It thus becomes possible to identify the invariant corresponding to this type of process, namely a dual cusp, whose universal unfolding variables might easily be related to the actual geometrical constraints. After the resulting general modelization, it becomes possible to find the crossing zone in any geometry. The topological study presented here illustrates the fact that it is possible to extract the 'memory' of the diabatic crossings by the analysis of calculated adiabatic PESs through the localization of the dual cusp catastrophe in this actual case. This type of investigation which, as shown by our study, is largely method-independent, will be developed by further studies of various irreversible processes.

6. Appendix

6.1. A: Electron Localization Function (ELF)

The first topological approach of bonding was made by Richard Bader in his theory of Atoms in Molecules, in which he applied topological concepts to the electronic density $\rho(\mathbf{r})$ [4]. An alternative topological approach based on the density has been proposed by Mezey [24–26] who considers the change of the shape of bonding isosurfaces. This enables to recognize formal species along the path which characterize the steps of the reaction. However, this method rely on properties which are not able to provide a description of the bonding such as it is done by the density laplacian or by ELF.

Apart from structure considerations, Bader's theory has been applied to unimolecular reactions and dissociative processes. In the most general case of bond dissociation, the study of the electronic density does not allow for identifying any change in the structure of attractors, at the exception of systems possessing a non nuclear attractor, such as Li_2 [27]. The Bonding Evolution Theory (see Appendix B), provides an alternative method for classifying and studying chemical processes which overcomes the previous inconvenient [28]. It is based on the Silvi and Savin's [5] description of bonding, resulting from the topological analysis of Becke and Edgecombe's ELF [9]. The latter function is defined by:

$$\eta(\mathbf{r}) = \frac{1}{1 + \left(\dfrac{D(\mathbf{r})}{D_h(\mathbf{r})}\right)^2} \tag{3}$$

For a single determinantal wavefunction built from orbitals labelled ϕ_i, the quantity $D(\mathbf{r})$ defined according to:

$$D(\mathbf{r}) = \frac{1}{2}\sum_i |\nabla\phi_i(\mathbf{r})|^2 - \frac{1}{8}\frac{|\nabla\rho(\mathbf{r})|^2}{\rho(\mathbf{r})} \tag{4}$$

is the excess of local kinetic energy, due to Pauli's repulsion. [29] $D_h(\mathbf{r}) = C_F\rho(\mathbf{r})^{\frac{5}{3}}$ is the Thomas-Fermi kinetic energy density which acts here as a renormalization factor, and C_F is the Fermi constant ($C_F = 2.871$ a.u.). The range of values of η is $0 \leqslant \eta \leqslant 1$. In principle, the ELF can be calculated from the exact wavefunction, if available, or from experimental results. In practice $\eta(\mathbf{r})$ is calculated from the *natural orbital populations*, with no restriction on the quantum mechanical method used for obtaining them. In the Silvi-Savin theory of bonding, a partition of the molecular space into basins of attractors having a clear chemical signification is obtained. These basins are either *core* basins located around the nuclei with $Z \geqslant 2$, or *valence* basins, some of which being associated with bonds. In this context, a classification of bonds has been proposed.

6.2. *B: Main features of the Bonding Evolution Theory (BET)*

This theory consists in a set of topological concepts characterizing the chemical changes taking place along a reaction path. These concepts, by nature, are independent of the actual quantum mechanical method used for calculating the ELF. Accordingly, the BET is compatible with any quantum mechanical scheme. Moreover, the topological description of a process is structurally stable, in the sense that the number and type of critical points of ELF, as well as their connectivity and evolution remain constant for any quantum method yielding a wavefunction for suitably describing the system in any configuration. For distinguishing the valence basins among themselves, the concept of *synaptic order* σ has been introduced [22]. This is the number of core basins with which a given valence basin shares a common boundary, as shown in Table 3. From a chemical point of view, monosynaptic basins are related to lone pairs, while di- and polysynaptic basins are related to bonds. In terms of BET [8], a chemical process is considered as a succession of different regions of structural stability in the control space, (the space of nuclear coordinates) passing through a *bifurcation state*, (see Appendix D). The BET

Table 3 *Nomenclature of valence basins. The expression in parenthesis corresponds to the list of core basins sharing a common boundary with the valence basin under scrutiny*

Synaptic order σ	Nomenclature	Symbol
0	Asynaptic	V
1	Monosynaptic	$V(X_i)$
2	Disynaptic	$V(X_i, X_j)$
$\geqslant 3$	Polysynaptic	$V(X_i, X_j, \ldots)$

introduces the concept of *morphic number* μ, as being the number of basins of the molecular partition of the system [8] in each region of structural stability. It generalizes the concept of molecular graph, (Appendix C), first introduced by R. Bader [4]. By definition, a step in a chemical process, takes place when a change occurs in the topological structure of the gradient field of the ELF associated with the system. The latter changes are typical of physical or chemical events such as bond breaking and forming, or electron switch from one PES to another, around a conical intersection. This approach provides a classification of chemical processes according to topological arguments. It is worth recalling that this classification is likely to remain independent of the quantum chemical method which is used for obtaining the PESs.

7. C: Topological Concepts

The gradient field of a local, well-defined function $F(\mathbf{r};\, c_\alpha)$, may be considered as a velocity vector field, as shown by the following relation.

$$\frac{d\mathbf{r}}{d\tau} = \nabla_{\mathbf{r}} F(\mathbf{r};\, c_\alpha) \tag{5}$$

This corresponds to a system of equations, which is called *gradient dynamical system* [30]. In this equation τ is an effective time and c_α represents the set of control parameters, in our case the set of nuclear coordinates. The points for which $\nabla_{\mathbf{r}} F(\mathbf{r};\, c_\alpha) = 0$ are of particular importance. They called *critical points* and are characterized by the couple (r, s), formed by the rank r, which is the number of non-zero eigenvalues, and the signature s, which is the excess of positive eigenvalues of the *hessian matrix*, the elements of which are the second derivatives of $F(\mathbf{r};\, c_\alpha)$. In $\mathbb{R}^3$, a critical point of $r = 3$ is called *non-degenerate*, and according to the signature, four types of critical points are distinguished: i) $(3, -3)$, attractor, ii and iii) $(3, +1)$ and $(3, -1)$, saddle points; iv) $(3, +3)$, repellor. Alternatively, a critical point is characterized by its index I, which is the number of positive eigenvalues of the hessian matrix, calculated at the critical point, labelled $\mathbf{r}^{(c)}$. With this definition, an attractor corresponds to $I = 0$, a repellor to $I = 3$, while the saddle points $(3, +1)$ and $(3, -1)$ correspond to $I = 2$ and $I = 1$, respectively. These critical points satisfy the Poincaré-Hopf relationship [30]

$$\sum (-1)^I \mathbf{r}^{(c)} = 1 \text{ for finite systems, and 0 for periodic systems} \tag{6}$$

In the previous equation, the sum runs over all critical points of the gradient dynamical system. In the Bonding Evolution Theory, the critical points form the *molecular graph*. In this graph, they are represented according to the dimension of their unstable manifold. Thus, critical points of $I = 0$, are associated with a dot, these with $I = 1$ are associated with a line, these with $I = 2$ by faces, and finally these with $I = 3$ by 3D cages.

8. D: Elements of Catastrophe Theory

Catastrophe theory [31] studies the change of the critical points $(\mathbf{r}^{(c)};\, c_\alpha)$ (see Appendix C), of a gradient dynamical system as a function of the changes of the control parameters c_α, in the special case where the dimension $\mathbf{k}$ of the control space W, also called the *codimension* fulfills the relationship: $\mathbf{k} \leqslant 4$. In this context, the evolution of the critical points can be studied by considering the behavior of the hessian matrix H of $F(\mathbf{r};\, c_\alpha)$ (see Appendix C). The configuration of the control parameters c_α^* for which the determinant of H is zero, i.e., $detH_{ij}(c_\alpha^*)|_{\mathbf{r}=\mathbf{r}^{(c)}} = 0$ is called a *bifurcation state*. The set of c_α for which the hessian matrix of a given critical point is non-zero defines the *domain of stability* of the critical point. A small perturbation of $F(\mathbf{r};\, c_\alpha^*)$, brings the system from a domain of stability to another one. If none of the critical points of the system change during a variation of c_α, the system is located in a *domain of structural stability*. R. Thom's theorem [31] states that in the neighborhood of $(\mathbf{r}^{(c)};\, c_\alpha^*)$, after a smooth change of the variables the function F can be written in the form: $F(\mathbf{r};\, c_\alpha) \doteq Q + u$ where $\doteq$ means: 'equal after a smooth change of variables'. In this equation, u is the *universal unfolding* of the singularity, it is a polynomial function of degree higher than 2 of a canonical form. The latter polynomial depends on l variables associated with the l zero eigenvalues of the hessian matrix. l is called the *corank*, while Q is a quadratic expression of the r-l variables, r being the rank of the hessian matrix. The unfolding contains all the information about how $F(\mathbf{r};\, c_\alpha)$ may change around the critical point when the control parameters change.

References

1. D.R. Herschbach, in *Reactive Scattering in Molecular Beams*, Advances in Chemical Physics, 1966, **10**, 319.
2. (a) L. Pauling, *J. Am. Chem. Soc.*, 1931, **53**, 3225.; (b) P. C. Humbel, P. Archirel, *J. Phys. Chem.*, 1994, **98**, 11697. and references cited therein.
3. (a) E.E. Nikitin, in *Theory of Atomic and Molecular Processes in Gases*, Oxford University, Oxford, 1974.; (b) E.E. Nikitin, *J. Chem. Phys.*, 1965, **43**, 744.; (c) A. Bjerre and E.E. Nikitin, *Chem. Phys. Letters*, 1967, **1**, 179.; (d) J.B. Delos and W.R. Thorson, *Phys. Rev. A*, 1973, **6**, 728.; (e) C.A. Mead and D.G. Truhlar, *J. Chem. Phys.*, 1982, **77**, 6090.; (f) M. Desouter Lecomte, D. Dehareng and J.C. Lorquet, *J. Chem. Phys.*, 1987, **86**, 1429 and references cited therein.
4. R.F. Bader, in *Atoms in Molecules: A Quantum Theory*, Oxford Univ. Press, Oxford, 1990.
5. B. Silvi and A. Savin, *Nature*, 1994, **371**, 683.
6. T.S. Slee, *J. Am. Chem. Soc.* 1986, **108**, 7541.
7. K. Collard and G.G. Hall, Int. *J. Quant. Chem.*, 1977, **12**, 623.
8. X. Krokidis, S. Noury and B. Silvi, *J. Phys. Chem.*, 1997, **101** 7277.
9. A.D. Becke and K.E. Edgecombe, *J. Chem. Phys.* 1990, **92**, 5397.
10. X. Krokidis, V. Goncalves, A. Savin and B. Silvi, *J. Phys. Chem. A* 1998, **102**, 5065.
11. X. Krokidis, R. Vuilleumier, D. Borgis and B. Silvi, *M. Phys*, 1999, **96**, 265.
12. X. Krokidis, B. Silvi and E. Alikhani *Chem. Phys. Let.*, 1998, **292**, 35.
13. X. Krokidis, B. Silvi, C. Dezarnaud-Dandine and A. Sevin, *N. J. Chem.*, 1998, **22**, 1341.
14. Gaussian 94, Revision C3. M.J. Frisch, G.W. Trucks, H.B. Schlegel, P.M.W. Gill, B.G. Johnson, M.A. Robb, J.R. Cheeseman, T. Keith, G.A. Peterson, J.A. Montgomery, K. Raghavachari, M.A. Al-Laham, V.G. Zakrezewski, J.V. Ortiz, J.B. Foresman, J. Cioslowski, B.B. Stefanov, A. Nanayakkara, M. Challacombe, C.Y. Peng, P.Y. Ayala, W. Chen, M.W. Wong, J.L. Andres, E.S. Replogle, R. Gomperts,

R.L. Martin, D.J. Fox, J.S. Binkley, D.J. Defrees, J. Baker, J.P. Stewart, M. Head-Gordon, C. Gonzalez and J.A. Pople, Gaussian, Inc., Pittsburg PA, 1995.
15. P.M. Kozlowski and E.R. Davidson, *J. Chem. Phys.*, 1994, **100**, 3672.
16. M.M. Francl, W.J. Pietro, W.J. Hehre, J.S. Binkley, M.S. Gordon, D.J. De Frees and J.A. Pople, *J. Chem. Phys.* 1982, **77**, 3654–3665 and references cited therein.
17. Top MoD, series of programs S. Noury, X. Krokidis, F. Fuster and B. Silvi, Computers & Chemistry, 1999, **23**, 597. This software is available upon request at the following email address: Xenophon.Krokidis@lct.jussieu.fr
18. (a) T.V. Hertel, *Advan. Chem. Phys.*, 1981, **45**, 341.; (b) T.V. Hertel, *Advan. Chem. Phys.*, 1982, **50**, 475.; (c) W.H. Breckenridge and H. Umemoto, *J. Chem. Phys.*, 1981, **75**, 698.; (d) W.H. Breckenridge and H. Umemoto, *J. Chem. Phys.*, 1981, **75**, 4153.; (e) R.P. Blickensderfer, K.D. Jordan, N. Adams and W.H. Breckenridge, *J. Phys. Chem.* 1982, **86**, 1930.; (f) W.H. Breckenridge and H. Umemoto, *Advan. Chem. Phys.*, 1982, **50**, 325.
19. F. Ramondo, N. Sanna and L. Bencivenni, *Journal of Molecular Structure (Theochem)*, 1992, **258**, 361.
20. J. Guan-Zhi and E.R. Davidson, *J. Phys. Chem.* 1992, **96**, 3683.
21. F.B.C. Machado, J. Guan-Zhi and E.R. Davidson, *J. Phys. Chem.* 1997, **97**, 5882, and references cited therein.
22. A. Savin, B. Silvi and F. Colonna, *Can. J. Chem.* 1986, 74, 1088.
23. (a) T. Poston and P.N. Stewart, in *Taylor Expansion and Catastrophes*, Battelle, Geneva and University of Warwick, Pitman Pub. London, 1976; (b) A.E.R. Woodcock and T.A. Poston, in *Study of the Elementary Catastrophes*, Lectures Notes in Mathematics, Springer-Verlag, Berlin, 1974.
24. P.G. Mezey, *Int. J. Quant. Chem., Quant. Biol. Symp.*, 1986, **12**, 113.
25. P.G. Mezey, *Int. Comput. Chem.*, 1987, **8**, 462.
26. P.G. Mezey, in *Shape in Chemistry: An Introduction to Molecular Shape and Topology*, VCH Publishers, New York, 1993.
27. (a) J. Cioslowski, *J. Phys. Chem.* 1990, **94**, 5496.; (b) D.L. Cooper, *Nature*, 1990, **346**, 789.; (c) G.I. Bersuker, C. Peng and J. E. Boggs, J. Phys. Chem. 1993, 97, 9323.
28. B. B. Stefanov and J. Cioslowski, *Can. J. Chem.*, 1996, bf 74, 1263.
29. A. Savin, J. Jepsen, O.K. Andersen, H. Preuss, and H.G. von Schnering, *Angew. Chem. Int. Ed. Engl.* 1992, **31**, 187.
30. R. Abraham and J.E. Marsden, in *Foundations of Mechanics*, Addison-Wesley, N. Y. 2nd edition, 1987, and references cited therein.
31. R. Thom, in *Stabilité Structurelle et Morphogénèse*, Interéditions, Paris, 1972.
32. (a) S.G. Lias, J.E. Bartness, J.F. Liebman, J.L. Holmes, R.D. Levin and W.G. Mallard, *Journal of Physical and Chemical Reference Data*, 1988, **117**, Supplement 1.; (b) C.E. Moore, in *Atomic Energy Levels*, Natl. Bur. Std. Washington, 1949, Vol. 1. (c) C.E. Moore, in *Atomic Energy Levels*, Natl. Bur. Std. Washington, 1952, Vol 2.
33. R. Gilmore, in *Catastrophe Theory for Scientists and Engineers*, Dover Publications Inc., New York, 1993.

BSSE-Free MCSCF Method for Strong Hydrogen Bonds: Investigation of H_2O-HCl and NH_3-HCl Complexes

A. Famulari*, M. Sironi and M. Raimondi

Dipartimento di Chimica Fisica ed Elettrochimica and Centro CNR-CSRSRC, Università degli Studi di Milano, via Golgi 19, 20133 Milano, Italy

Abstract

A Multi-Configuration extension of the Self Consistent Field for Molecular Interaction method – MCSCF-MI – is presented. Applications to the study of hydrogen bonded systems are reported. Potential energy, optimal geometry and the characteristic H-Cl stretching in the H_2O-HCl and NH_3-HCl complexes were examined. BSSE, which represents a major problem in the study of the interaction potentials of loosely bonded fragments, is avoided by the *a priori* SCF-MI technique. Equilibrium geometries, binding energies, and vibrational constants computed by both SCF-MI and MCSCF-MI methods turned out more stable than the corresponding standard SCF calculations upon variation of the quality of the basis set employed. In particular, the proton transfer process was examined. All the calculations predicted that the minimum corresponds to hydrogen bonded structures. In the case of NH_3-HCl good agreement with recent extensive MO-CI and DFT studies was established.

1. General Introduction

Hydrogen bonding, the interaction between a hydrogen atom bonded to an electronegative group and a lone electron pair on another system, is the key of many biological and chemical phenomena. In spite of its simplicity, it has became one of the most difficult challenges for both theoreticians and experimentalists. Due to the difficulty in obtaining experimental data for the geometry and energy of hydrogen bonded complexes, an accurate prediction of these quantities is mandatory.

As it is well known, the Basis Set Superposition Error (BSSE) affects calculations involving hydrogen bonds [1] and, more generally, intermolecular interaction investigations [2,3]. This issue of consistency, as first pointed out in 1968 [4], arises from the use of an incomplete basis set but it does not correspond to the basis set truncation error and it is due to the use of diffuse functions on neighbouring interacting particles, which leads to a non physical contribution to the interaction energy within the complex.

BSSE also affects the shape of the potential energy surface and the energy derivatives. There have been numerous attempts to find a general scheme to eliminate this error, and both *a posteriori* [2] and *a priori* [3] schemes are available. The counterpoise approach (CP) by Boys and Bernardi [5] and related methods are the most common *a posteriori* procedures. Within this method, the monomer electrons are described by the same basis functions as those used in the complex by means of the so

*To whom correspondence should be addressed: e-mail antoniro.famulari@unimi.it

A. Hernández-Laguna et al. (eds.), Quantum Systems in Chemistry and Physics, Vol. 1: Basic Problems and Model Systems, 361–379.

called 'ghost orbitals'. Several efforts have been also dedicated to design basis sets of small size that could be used with some confidence for molecular interactions [6,7].

Many authors [8–10] have demonstrated that the CP method undercorrects the BSSE. Moreover, Karlström and Sadlej [11] pointed out that addition of the partner orbitals to the basis set of a molecule not only lowers its energy, in accordance with the variation principle, but also affects the monomer properties (multipole moments and polarizabilities). Latajka and Scheiner [12] found that in a model ion-neutral system such as Li^+-OH_2, this secondary BSSE can be comparable in magnitude to the primary effect at both SCF and MP2 levels. The same authors also underlined the strong anisotropy of secondary error [13].

Generally, if the CP procedure is adopted, a point-by-point procedure is employed to locate minima in the potential energy surface. Only very recently, a method to perform CP corrected energy optimisation by analytic derivatives has been proposed [14]; for most of the reported cases, CP correction is included in a previously optimised geometry so that the final results are BSSE contaminated.

It is important to note that for accurate intermolecular interaction investigations, geometry relaxation effects should be considered. This requires a more elaborate definition of the CP corrected binding energy which needs, at each point on the PES, the calculation of five single point energies corresponding to those of the monomers – with and without 'ghost orbitals' – at the geometry of the complex, in addition to the usual SCF energy of the supermolecule [15]. Several authors have confirmed the importance of these additional corrections, especially in the case of large system interactions (see for example Hobza et al. [16] where a significant amount of the geometry relaxation energy is calculated).

We have already presented [17,18] the SCF-MI (Self Consistent Field for Molecular Interactions) method, based on the idea that BSSE can be avoided *a priori* provided the MOs of each fragment are expanded only using basis functions located on each subsystem. In the present work we propose a multiconfiguration extension (MCSCF-MI) of the same technique, particularly suited to deal with systems for which proton transfer processes must be considered.

2. The HCl Complexes

Recent developments in rotational spectroscopy [19] have provided an accurate study of a number of hydrogen bonded dimers in the gas phase. Experimental information is thus available nowadays on the molecular characteristics of such systems. Due to major advances in the field of computational chemistry, a number of theoretical calculations were performed (for an overview on the subject see references [1,20,21]).

The hydrogen chloride molecule, being a strong acid, interacts with a great number of bases: CO, C_2H_2, C_2H_4, PH_3, H_2S, HCN, H_2O and NH_3; its hydrogen bonded complexes are therefore of great interest (see for example [1,22]). Due to the strength of the hydrogen bond involved, the possibility of an effective proton exchange has been considered, and it remains especially interesting to establish if the interaction can be explained in terms of the strong polarisation of the monomers. The systems H_2O and NH_3 have been thoroughly studied by means of *ab initio* calculations; the SCF method or perturbative techniques such as MP2 were typically employed. Special attention was reserved to the BSSE elimination [1,23,24].

In the present work, calculations were performed with a series of standard basis sets, comparing the relative variation of the predicted properties when employing different quantum mechanical methods and the newly developed MCSCF-MI strategy.

2.1. The H_2O-HCl Complex

This represents one of the most studied complexes of the hydrogen chloride molecule. A source of experimental information on this system is the work by Legon and Willoughby [25]. The authors recorded a rotational spectra of a mixture of gases containing 2% HCl and less than 0.5% H_2O in Ar, for a total pressure of about 2 atm. The equilibrium geometry reported corresponds to a planar geometry. This is of special importance, as the majority of the computational studies predict a non planar complex with the HCl molecule with the C_2 axis of the water molecule forming different angles with the O and Cl line (see Figure 1).

Bacskay *et al* [26] have determined a bent geometry for the H_2O-HCl complex at the SCF level, with $\varphi = 19.7°$ and $\theta = 1.3°$. A bent geometry was also found at the MP2 level, with $\varphi = 44.7°$ and $\theta = 2.6°$ [27]. Other papers seem to confirm the non-planar geometry. Latajka and Scheiner [28] obtain analogous results employing both the SCF method and the SCF+MP2 technique and the CP correction of BSSE. The study by Almeida and Hinchliffe comes closer to the experimental results [22], finding a value of about 0 for the φ and θ angles. Table 1 shows other literature data relative to the equilibrium geometry.

Dissociation energies range between 5 and 6 kcal/mol: values obtained in some of the existing papers are reported in Table 2.

The evaluation of the change in frequency of the H-Cl bond ($\Delta\omega$) on passing from the isolated molecule to the complex, is an index of the strength of the bond. A *red shift* is observed, pointing to an overall weakening of the bond. Some of the pertinent data reported in the literature are shown in Table 3.

2.2. The NH_3-HCl Complex

The NH_3-HCl complex represents a useful model system for proton transfer reaction studies, which are of great relevance in chemical and biological phenomena. For this

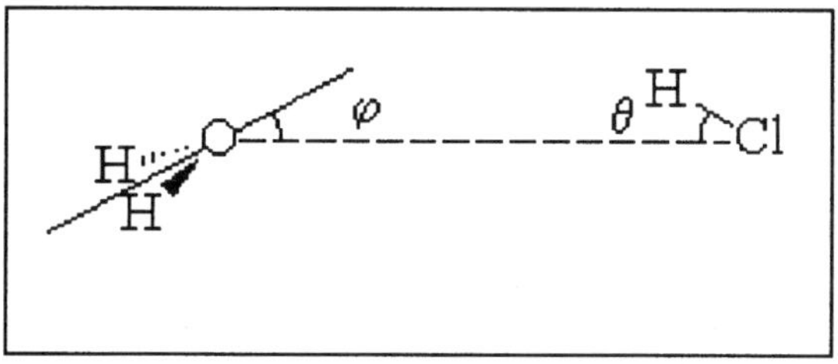

Fig. 1. Geometrical parameters for the H_2O-HCl complex.

Table 1 *Previously reported geometrical parameters for the H_2O-HCl complex. Distances are reported in Å and angles in degrees*

Method	r_{HCl}	r_{OCl}	θ	φ	Ref.
Exp.		3.2149	0	0	[25]
SCF		3.31	1.3	19.7	[26]
MP2	1.2829		2.6	44.7	[27]
SCF[(a)]	1.277	3.250	2.2	35.2	[28]
SCF + MP2[(b)]	1.289	3.281	3.4	46.8	[28]
EFG fog[(c)]	1.326	3.077	0.136	0.007	[22]
EFG fmg[(d)]		3.092	0.007	0.007	[22]

r_{HCl} = H-Cl distance; r_{OH} = O-H distance; α_{HOH} = HOH angle in water; θ and φ as defined in Figure 1.
(a)the 6-31G** basis set was employed;
(b)BSSE correction was performed with the CP technique.
(c)Electric Field Gradient, fully optimized geometry.
(d)Electric Field Gradient, frozen monomer geometry.

Table 2 *Previously reported binding energies for the H_2O-HCl complex*

Method	ΔE (kcal/mol)	Ref.
SCF	−5.04	[25]
SCF	−6.16	[28]
SCF + MP2	−7.36	[28]
SCF	−5.31	[28]
SCF + MP2[(a)]	−5.92	[28]
MP2	−6.68	[27]
MP3	−5.88	[27]

(a)BSSE corrected by the CP technique.

Table 3 *Previously estimated red shift values for the intramolecular stretching frequency of the HCl bond in the H_2O-HCl complex*

Method	$\Delta\omega$ (cm^{-1})	Ref.
Exp.	314	[29,30]
SCF	120.3	[26]
MP2	227.4	[27]

reason the system has been investigated by many authors [1,21–24,31–37]. The potential energy surface of this complex has recently been the object of a thorough study by ab initio and density functional theory calculations [23].

The investigations, which aimed to establish the character of the interaction as either ion-pair like (outer complex) or H-bonded-like (inner complex), started with the first theoretical ab initio calculation by Clementi [31] in the early 1967. Contrarily to Mulliken's suggestion [38], formulated in 1952 when there was no experimental evidence of a stable NH_3-HCl system in the gas phase, the RHF predicted an H-bonded complex without any barrier between the fragments and the complex.

All the data reported in the literature seem to confirm that the NH_3-HCl complex possesses C_{3V} symmetry (see Figure 2). Table 4 shows the geometrical parameters of the complex reported in the pertinent literature. Bacskay *et al* [26,27] confirm the C_{3V} symmetry of the complex both at the MP2 and SCF level. A moderate variation from the C_{3V} symmetry, however, results from EFG (Electric Field Gradient) calculations [22].

Interaction energies, reported in Table 5, range between 6 and 9 kcal/mol and result considerably higher than those regarding the H_2O-HCl complex.

Table 6 reports theoretical and experimental data relative to the *red shift* for the intramolecular stretching frequency of the HCl bond in the complex. This is greater than that observed for the H_2O-HCl system, a further evidence of the strength of this hydrogen bond. Experimental data, however, show large differences between the $\Delta\omega$ recorded in different matrixes. The first IR investigation was accomplished by Ault and Pimentel [30] in N_2 matrix but later, in a more complete study, Barnes et al. [35]

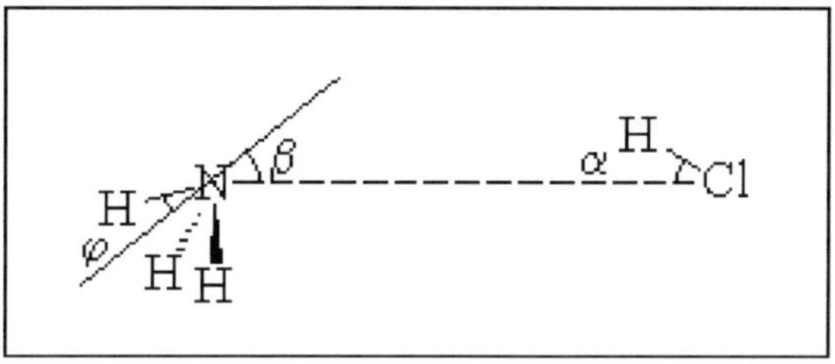

Fig. 2. Geometrical parameters for the NH_3-HCl complex.

Table 4 *Previously reported geometrical parameters for the NH_3-HCl complex. Distances are reported in Å and angles in degrees*

Method	r_{HCl}	r_{NH}	θ_{HNH}	r_{NCl}	φ	α	β	Ref.
Exp.				3.14		0		[19]
SCF				3.33		0	0	[26]
MP2	1.3146	1.0159	106.78	3.11		0	0	[27]
EFG fmg[(a)]				3.171	0.136	0	0.018	[22]
EFG fog[(b)]	1.388	0.99	112.1	3.045	47.98	0.090	0.544	[22]

r_{HCl} = H-Cl distance; r_{NH} = N-H distance in ammonia; θ_{HNH} = HNH angle in ammonia; r_{NCl} = N-Cl distance; φ, α and β as defined in Figure 2.
[(a)]Electric Field Gradient, frozen monomer geometry.
[(b)]Electric Field Gradient, fully optimized geometry.

Table 5 *Previously reported binding energies of the NH_3-HCl complex*

Method	ΔE (kcal/mol)	Ref.
Exp.	-8.0 ± 2.8	[33]
SCF	-6.52	[26]
HF SCF limit	-5.5	[23]
MP2	-9.53	[27]
MP3	-8.34	[27]
MP2	-11.0	[36]
ACCD	-9.2	[39]
CI	-9.0	[33]
MP2 limit	-7.6	[23]
DFT-LDA	-19.2	[23]
DFT-GGA	-13.0	[23]
EFG fmg(a)	-9.99	[22]
EFG fog(b)	-11.64	[22]
CASSCF	-5.2	[23]
MRCI	-8.0	[23]

(a)Electric Field Gradient, frozen monomer geometry.
(b)Electric Field Gradient, fully optimized geometry.

Table 6 *Previously estimated red shift values for the intramolecular stretching frequency of the HCl bond in the NH_3-HCl complex*

Method	$\Delta\omega$ (cm^{-1})	Ref.
Exp. Ar matrix	1517	[35]
Exp. N_2 matrix	2149	[30]
SCF	293.8	[26]
MP2	592.3	[27]
CI harmonic	~ 500	[40]
CI anharmonic	~ 700	[40]
DFT-LDA	1107	[23]
DFT-GGA	700	[23]

demonstrated that the stretching frequency of HCl shows great sensitivity to the environment. Poor prediction of this experimental data is therefore expected by theoretical approaches which do not include explicit consideration of the environment: see [23] for a complete discussion on this point.

3. Theory

3.1. The SCF-MI Wavefunction

The SCF-MI technique and its Multi Configuration extension – MCSCF-MI – were employed for the calculations performed in the present work. For a more detailed presentation of the SCF-MI theory see references [17,18]. As already emphasised in the introduction, the main characteristic of the scheme is that it permits the elimination of the BSSE in an *a priori* fashion, both at the SCF and at the correlated level; basis set and size consistency is also ensured.

The SCF-MI algorithm, recently extended to compute analytic gradients and second derivatives [18,41], furnishes the Hartree Fock wavefunction for the interacting molecules and also provides automatic geometry optimisation and vibrational analysis in the harmonic approximation for the supersystems. The full strategy has been implemented into GAMESS-US package [42].

The theory is here presented very briefly. The SCF-MI one determinant wavefunction of the supersystem AB is expressed as:

$$\Psi_{AB} = (N!)^{-1/2}\mathcal{A}[\Phi_1^A(1)\overline{\Phi}_1^A(2)\ldots\overline{\Phi}_{N_A}^A(2N_A)\Phi_1^B(2N_A+1)\overline{\Phi}_1^B(2N_A+2)$$
$$\ldots\ \overline{\Phi}_{N_B}^B(2N_A+2N_B)]$$

In order to take naturally into account interaction forces between the monomers the molecular orbitals of each fragment are left free to overlap with each other. The orbitals of fragment A ($\Phi_1^A \ldots \Phi_{N_A}^A$) and the orbitals of fragment B ($\Phi_1^B \ldots \Phi_{N_B}^B$) are expanded in two different subsets of the total basis set $\chi = \{\chi_k\}_{k=1}^{M}$ which is partitioned as:

$$\chi^A = \{\chi_p^A\}_{p=1}^{M_A} \text{ centred on A and } \chi^B = \{\chi_q^B\}_{q=1}^{M_B}\} \text{ centred on B.}$$

That is:

$$\Phi_i^A = \sum_{p=1}^{M_A} \chi_p^A \mathbf{T}_{pi}^A \qquad i = 1 \ldots N_A$$

$$\Phi_j^B = \sum_{q=1}^{M_B} \chi_q^B \mathbf{T}_{qj}^B \qquad j = 1 \ldots N_B$$

where the total number of electrons is $N = 2N_A + 2N_B$ and the dimension of the total basis set is $M = M_A + M_B$.

By assuming and maintaining the orbital coefficient variation matrix in block diagonal form,

$$\delta\mathbf{T} = \begin{bmatrix} \delta\mathbf{T}_A & 0 \\ 0 & \delta\mathbf{T}_B \end{bmatrix}$$

a complete elimination of BSSE in an *a priori* fashion is ensured.

The general stationary condition of the variational scheme leads to the coupled secular problems:

$$\begin{cases} \mathbf{F}'_A\mathbf{T}_A = \mathbf{S}'_A\mathbf{T}_A\mathbf{L}_A \\ \mathbf{T}^{\dagger}_A\mathbf{S}'_A\mathbf{T}_A = \mathbf{I}_A \end{cases} \qquad \begin{cases} \mathbf{F}'_B\mathbf{T}_B = \mathbf{S}'_B\mathbf{T}_B\mathbf{L}_B \\ \mathbf{T}^{\dagger}_B\mathbf{S}'_B\mathbf{T}_B = \mathbf{I}_B \end{cases}$$

where the effective Fock and overlap matrixes $\mathbf{F}'$and $\mathbf{S}'$ are Hermitian and possess the correct asymptotic behaviour.

3.2. *The MCSCF-MI wavefunction*

A two-configuration MCSCF-MI wavefunction corresponding to the covalent – A + BH – and ionic – $AH^+ + B^-$ – systems is constructed in the form:

$$\Psi = C_1\Psi_{ABH} + C_2\Psi_{AH^+B^-}$$

where the optimal orbitals for the covalent and ionic structures have been determined separately. The solutions corresponding to the covalent and ionic structures are obtained by means of different partitions of the basis set, which take account of the possible proton transfer from one fragment to the other. The resulting orbitals, largely *non-orthogonal* but resulting from solution of Roothaan like equations, allow to obtain the MCSCF-MI wave function with many characteristics of both Valence Bond and Molecular Orbital theories. The BSSE is avoided by definition and a straightforward physical interpretation is possible.

The final non-orthogonal CI problem, implicit in the MCSCF-MI approach, is solved variationally according to standard VB techniques [43,44].

4. Calculations

The SCF, SCF-MI and MCSCF-MI techniques were employed to determine the optimal geometry.

As regards SCF and SCF-MI calculations, the GAMESS-US program was employed, in which the SCF-MI algorithm including evaluation of analytic gradient, geometry optimisation and force constant matrices computation is available [18,41,42].

With regard to the basis function expansions, several standard basis sets were employed (namely 6-31G, 6-31G**, 6-311G* and TZVP).

Two minima were found corresponding to ionic and covalent structures.

The weights of the covalent and ionic solutions in the MCSCF-MI wave function are strongly dependent upon the H-Cl distance. When such distance is close to its value in isolated hydrogen chloride, the covalent solution weighs more than the ionic one. The opposite trend is observed when the H-Cl distance is such as to justify a picture where the proton is transferred to the base. In all cases, the covalent function turned out more important, with the ionic structure much less stable. This result concerning the NH_3 + HCl system has been recently confirmed by Clementi et al. [23] for the complex in the gas phase.

Equilibrium geometries at the SCF, SCF-MI and MCSCF-MI levels were determined for each complex, corresponding to the two minima of the potential energy curves, employing the four different standard basis sets considered. The optimisation of the geometry at the MCSCF-MI level was performed starting from the SCF-MI minimum by a numerical procedure, varying one parameter at a time.

At the asymptotic geometries ($r_{OCl} = r_{NCl} = 100$ Å), SCF, SCF-MI and MCSCF-MI solutions coincide.

A potential energy curve was also computed for different values of the r_{OCl} distance in the H_2O-HCl complex, with HCl approaching H_2O along the C_{2v} axis. At each r_{OCl} distance, optimal SCF geometry was determined and used in the subsequent SCF-MI and MCSCF-MI calculations.

An analogous approach was employed for the study of NH_3-HCl, computing a potential energy curve as a function of the r_{NCl} distance.

Both these potential energy curves have been computed employing 6-31G** standard basis set, see Figures 3 and 4.

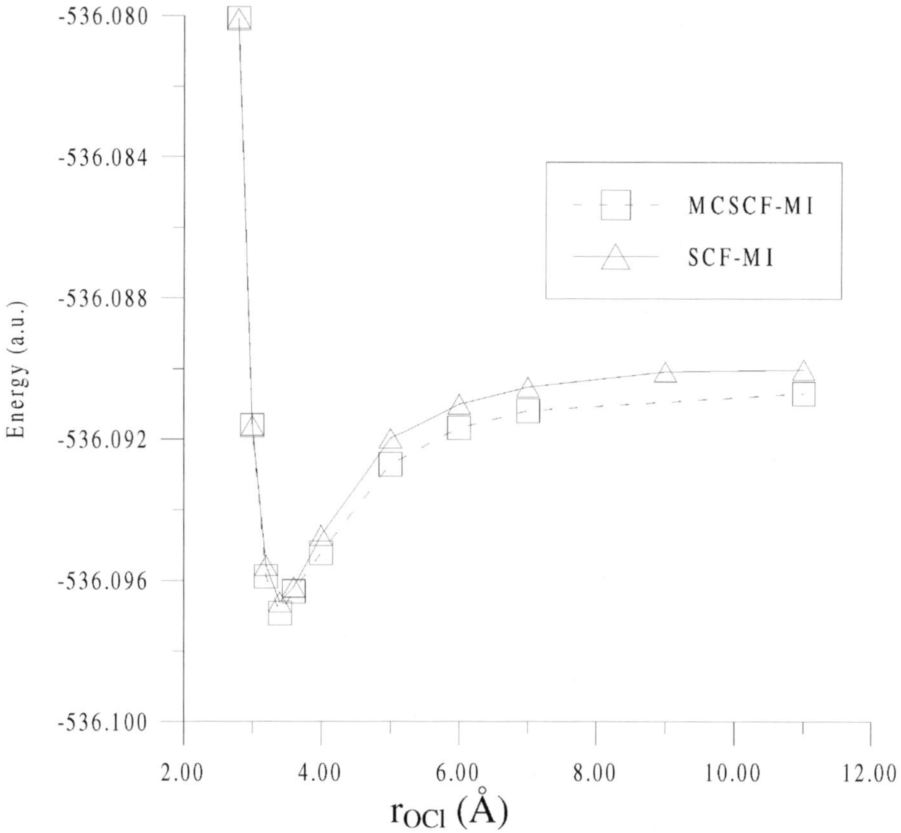

*Fig. 3. Potential energy curves for SCF-MI (covalent structure) and MCSCF-MI calculations for the H_2O-HCl complex. The 6-31G** standard basis was employed. Energies are in a.u., r_{OCl} distances are in Å.*

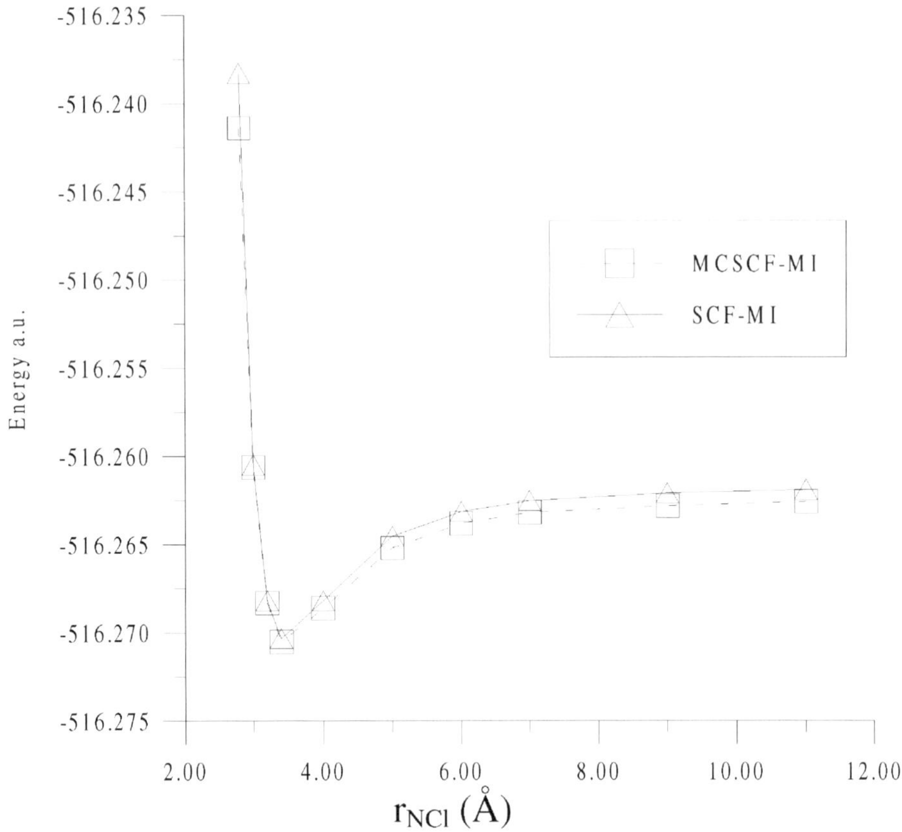

Fig. 4. Potential energy curves for SCF-MI (covalent structure) and MCSCF-MI calculations for the NH_3-HCl complex. The 6-31G** standard basis was employed. Energies are in a.u., r_{NCl} distances are in Å.

5. Results and Discussion

5.1. The H_2O-HCl complex

Optimal geometries computed at the SCF, SCF-MI and MCSCF-MI level are presented in Table 7.

All the calculations predict a complex characterised by a longer H-Cl bond distance when compared with isolated HCl. The r_{HCl} parameter shows small variations with the basis set employed only at the MCSCF-MI and SCF-MI level, while considerable variations are observed for the corresponding SCF calculations. Analogous considerations can be made for the r_{OCl} distance which shows good stability only when evaluated at the MCSCF-MI and SCF-MI level. In addition, the complex is always planar at the SCF-MI and MCSCF-MI level, while large deviations from planarity are predicted by the SCF calculations when the 6-31G** basis set is employed (for the definition of the geometry see Figure 1).

Table 7 *SCF, SCF-MI (covalent structure) and MCSCF-MI calculated equilibrium geometries for the H_2O-HCl complex. Distances are reported in Å and angles in degrees*

Basis	r_{OCl}	r_{HCl}	φ	θ
SCF				
TZVP	3.28	1.2823	0	0
6-311G*	3.19	1.2848	0	0
6-31G**	3.25	1.2774	33.07	1.07
6-31G	3.03	1.3278	0	0
SCF-MI				
TZVP	3.40	1.2758	0	0
6-311G*	3.40	1.2758	0	0
6-31G**	3.39	1.2658	0	0
6-31G	3.42	1.2958	0	0
VB				
TZVP	3.40	1.2758	0	0
6-311G*	3.40	1.2758	0	0
6-31G**	3.39	1.2658	0	0
6-31G	3.42	1.2958	0	0
asymptotic				
TZVP	100.0	1.2720	0	0
6-311G*	100.0	1.2713	0	0
6-31G**	100.0	1.2656	0	0
6-31G	100.0	1.2953	0	0

Table 8 reports equilibrium binding energies relative to the geometries described above. The differences between MCSCF-MI or SCF-MI energies and the SCF ones are obviously to be ascribed to the different basis set truncation effect and to BSSE.

Interaction energies show a less marked dependence on the basis set in the SCF-MI and MCSCF-MI calculations, where the results prove to be more stable while in acceptable accordance with the existing literature data.

MCSCF-MI and SCF-MI results, for both the equilibrium geometries and binding energies, are very close. In Figure 3 the potential energies are reported as a function of the r_{OCl} distance. The MCSCF-MI plot is almost parallel to the SCF-MI curve, showing that the ionic configuration in the MCSCF-MI calculation mostly introduces intramole-

Table 8 *SCF, SCF-MI and MCSCF-MI calculated binding energies relative to the equilibrium geometries (reported in Table 7) for the H_2O-HCl complex. All the values are in kcal/mol*

Basis	ΔE_{SCF}	$\Delta E_{SCF\text{-}MI}$	$\Delta E_{MCSCF\text{-}MI}$
TZVP	−5.43	−4.52	−4.39
6-311G*	−6.99	−5.18	−4.98
6-31G**	−6.11	−4.22	−3.99
6-31G	−9.86	−6.10	−5.70

cular electronic correlation localised in the HCl molecule. As the optimal r_{HCl} interatomic distance remains practically constant, the final result is a uniform shift of the potential energy curve.

As the optimal H-Cl distance remains very close to its value in the isolated molecule also when approaching H_2O, the proton exchange reaction $H_2O + HCl \rightarrow H_3O^+ + Cl^-$ results energetically unfavourable as further confirmed by the data reported in the following paragraph.

5.2. The NH_3-HCl complex

Comparable results are obtained for the NH_3-HCl complex. SCF geometrical parameters (r_{HCl} and r_{NCl}) vary considerably with the basis set employed, while the MCSCF-MI and SCF-MI results are more stable (see Table 9).

The C_{3V} symmetry for the complex is confirmed. The order of magnitude of the binding energy confirms the importance of the BSSE (see also reference 16). Again, SCF energies show a greater dependence on the basis set employed when compared with the MCSCF-MI and SCF-MI ones (see Table 10); it is to be noted that SCF-MI interaction energies are sufficiently close to the estimated HF limit of -5.5 kcal/mol [23].

The potential energy curve (Figure 4) relative to the approach of HCl to NH_3 has the same characteristics observed for the H_2O-HCl complex, with the MCSCF-MI curve essentially parallel to the SCF-MI curve.

Table 9 *SCF, SCF-MI (covalent structure) and MCSCF-MI calculated equilibrium geometries for the NH_3-HCl complex. Distances are reported in Å and angles in degrees*

Basis	r_{HCl}	r_{NCl}	φ	α	β
SCF					
TZVP	1.2956	3.28	69.41	0	0
6-311G*	1.3004	3.22	68.72	0	0
6-31G**	1.2936	3.23	68.88	0	0
6-31G	1.7899	2.90	71.85	0	0
SCF-MI					
TZVP	1.2875	3.40	68.93	0	0
6-311G*	1.2775	3.39	67.81	0	0
6-31G**	1.2775	3.39	67.72	0	0
6-31G	1.3075	3.42	73.91	0	0
VB					
TZVP	1.2875	3.40	68.91	0	0
6-311G*	1.2775	3.39	67.92	0	0
6-31G**	1.2775	3.39	67.93	0	0
6-31G	1.3075	3.42	73.94	0	0
asymptotic					
TZVP	1.2720	100.0	69.96	0	0
6-311G*	1.2713	100.0	68.61	0	0
6-31G**	1.2656	100.0	68.68	0	0
6-31G	1.2959	100.0	78.61	0	0

Table 10 *SCF, SCF-MI and MCSCF-MI calculated binding energies relative to the equilibrium geometries (reported in Table 7) for the NH_3-HCl complex. All the values are in kcal/mol*

Basis	ΔE_{SCF}	$\Delta E_{SCF\text{-}MI}$	$\Delta E_{MCSCF\text{-}MI}$
TZVP	−7.07	−5.20	−4.92
6-311G*	−8.85	−5.93	−5.61
6-31G**	−8.56	−5.44	−5.09
6-31G	−15.89	−6.35	−5.70

No proton transfer is predicted as the r_{HCl} value remains practically constant and very close to the free HCl bond length on varying the r_{NCl} distance. The general features of the hydrogen bond of NH_3-HCl are similar to those of the H_2O-HCl, the larger binding energy being explained by the higher polarizability of NH_3.

5.3. *Interconversion between ionic and covalent structures*

In this paragraph, potential energy curves for the proton transfer reactions – $H_2O + HCl \rightarrow H_3O^+ + Cl^-$ and $NH_3 + HCl \rightarrow NH_4^+ + Cl^-$ – are described.

All calculations were performed employing the TZVP basis set and the MCSCF-MI technique. The H-Cl distance was varied, moving the proton involved in the hydrogen bond either towards oxygen or nitrogen, keeping all other geometrical variables frozen at their values at the minimum energy geometry. The MCSCF-MI energy was computed for a number of values of the H-Cl distance.

5.4. *H_2O-HCl complex*

The calculated potential energy curve (see Figure 5) is characterised by the presence of one relative and one absolute minimum, corresponding to the ionic and covalent structures, respectively. At the absolute minimum the r_{HCl} is equal to 1.2758 Å, close to the bond length in isolated H-Cl, while at the relative minimum r_{HCl} is 2.4 Å, typical of the H_3O^+-Cl^- ionic system. The potential energy barrier which separates the two minima is located at a r_{HCl} distance of 2.0 Å.

The covalent structures turns out more stable by 62.23 kcal/mol and is separated by a high barrier of 70.01 kcal/mol from the $H_3O^+ + Cl^-$ system. The reverse conversion of $H_3O^+ + Cl^-$ into $H_2O + HCl$ presents a much lower barrier of 7.78 kcal/mol, (see Figure 5).

5.5. *NH_3-HCl complex*

The calculated MCSCF-MI potential energy surface is shown in Figure 6. Analysis of the data shows that the ionic structure again is less stable than the covalent one and than the free monomers.

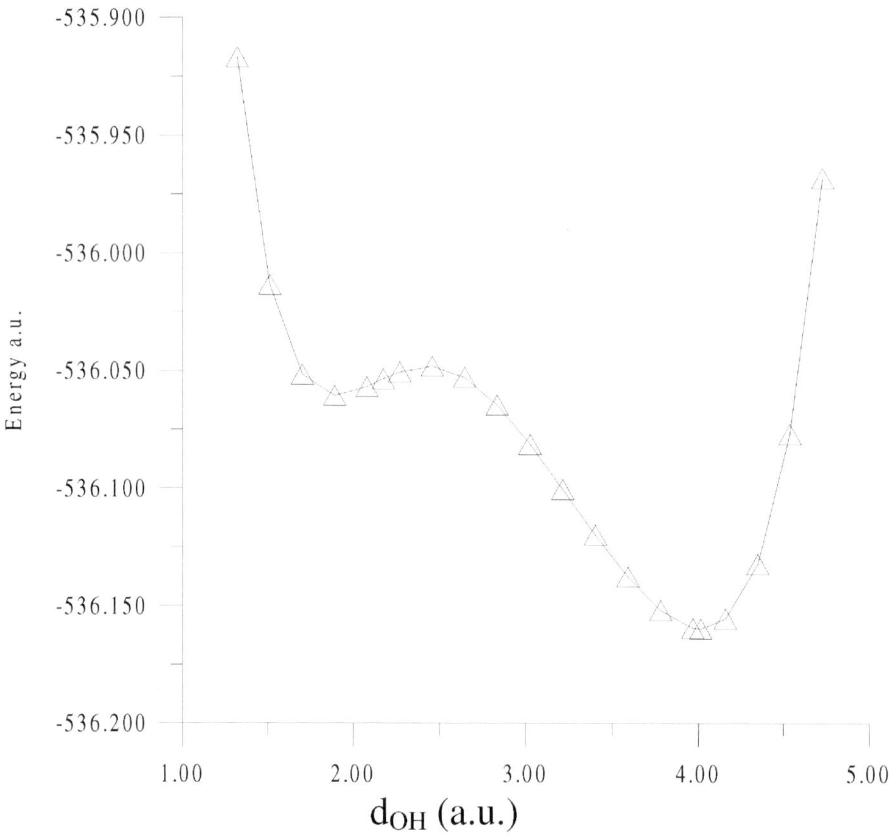

Fig. 5. Potential energy curve relative to the interconversion between ionic and covalent structure for the H_2O-HCl complex ($H_2O + HCl \rightarrow H_3O^+ + Cl^-$). d_{OH} represents the distance between the water molecule oxygen and the hydrogen of the HCl fragment involved in hydrogen bond along the C_{2v} axis ($d_{OH} = r_{OCl} - r_{HCl}$). A TZVP standard basis was employed. Energies and distances are in a.u.

The minimum at $r_{HCl} = 2.36$ Å corresponds to the ionic structure; the covalent structure is found at a $r_{HCl} = 1.2775$ Å; the maximum of the barrier is located at $r_{HCl} = 1.9$ Å. Comparison of Figures 5 and 6 shows that the potential well relative to the ionic structure is deeper for NH_3-HCl than for H_2O-HCl. The computed N-H distance of 1.0055 Å is typical for the N-H bond in the ammonium ion.

The two minima, see Figure 6, differ by 24.74 kcal/mol with a barrier of 45.73 kcal/mol separating the covalent system $NH_3 + HCl$ from the ionic structure $NH_4^+ + Cl^-$; a corresponding smaller barrier of 20.99 kcal/mol is found for the reverse $NH_4^+ + Cl^- \rightarrow NH_3 + HCl$ reaction. The covalent structure is still preferred in the gas phase.

6. Vibrational Analysis

Analytical or numerical force constants for the H-Cl stretching were calculated at the

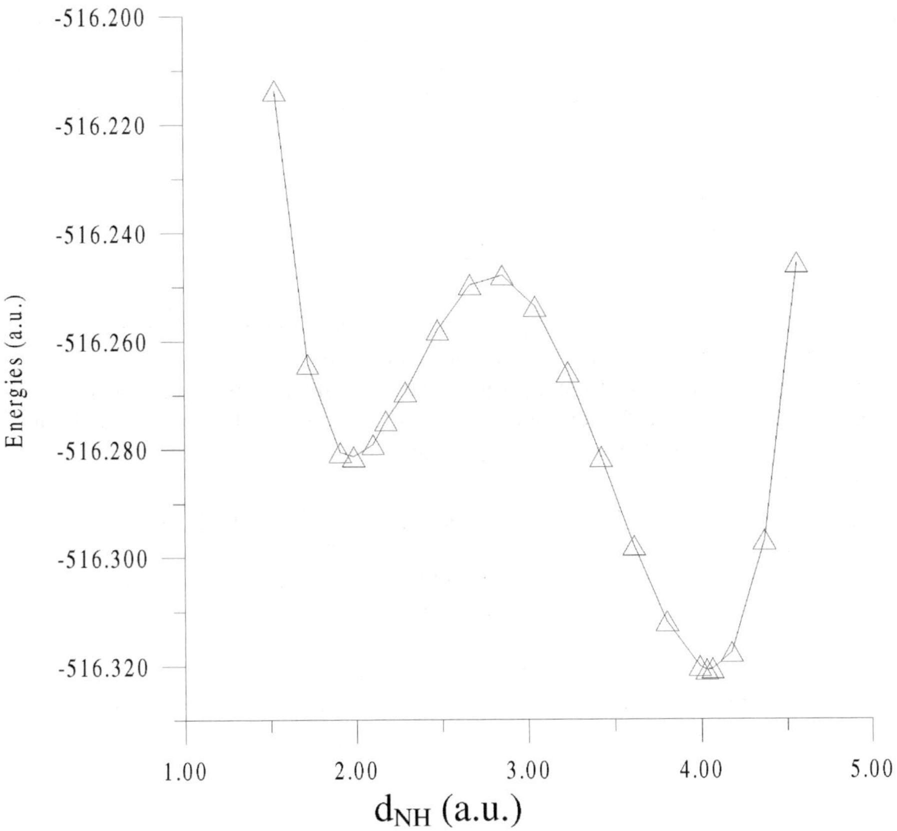

Fig. 6. Potential energy curve relative to the interconversion between ionic and covalent structure for the NH_3-HCl complex (NH_3 + HCl $\rightarrow$ NH_4^+ + Cl^-). *d_{NH} represents the distance between the ammonia molecule nitrogen and the hydrogen of the HCl fragment involved in hydrogen bond along the C_{3v} axis ($d_{NH} = r_{NCl} - r_{HCl}$). A TZVP standard basis was employed. Energies and distances are in a.u.*

SCF, SCF-MI and MCSCF-MI level. Frequencies ω were calculated directly from the force constants K by applying the harmonic approximation or by means of the Noumerov method [45,46]. In the latter case, frequencies were evaluated from the energy difference ΔE between the fundamental and the first excited states by applying Plank's law. This technique is undoubtedly more accurate than the harmonic approximation.

6.1. *H_2O-HCl complex*

Computed H-Cl stretching force constants of the complex, see Table 11, show a minor dependence on the basis set employed when the SCF-MI and MCSCF-MI procedures are applied. Moreover, the MCSCF-MI and SCF-MI results are very similar. The reliability of the SCF-MI and MCSCF-MI techniques is thus further demonstrated.

Table 11 *Red shift for the H-Cl stretching frequencies calculated by the harmonic approximation (H_2O-HCl complex). All the values are reported in cm^{-1}*

Basis	$\Delta\omega_{SCF}$	$\Delta\omega_{SCF\text{-}MI}$	$\Delta\omega_{MCSCF\text{-}MI}$	$\Delta\omega_{NOUMEROV}$	$\Delta\omega_{EXP}$
					314 [29,30]
TZVP	137	28	17	129	
6-311G*	197	44	33		
6-31G	458	23	9		

The difference between the asymptotic and the equilibrium H-Cl force constants, ΔK, varies with the basis set and the technique employed.

In all cases the calculated $\Delta\omega$ are positive and a *red shift* is predicted. The MCSCF-MI and SCF-MI results, differ considerably from the experimental value of 314 cm^{-1}, but it must be noted, on the other hand, that the SCF best result is obtained with the poorest (6-31G) basis set.

Noumerov analysis leads to more encouraging results. The energy values associated with the first 10 ($\nu = 0$–9) vibrational levels were computed. The first eight ($\nu = 0$–7) are relative to the absolute minimum of the potential energy curve employed (see Figure 5); the remaining two ($\nu = 8$–9) are relative to levels above the maximum.

The value of $\Delta\omega$ estimated from the first two eigenvalues is 129 cm^{-1}. This is still far from the experimental value of 314 cm^{-1}, but it represents an improvement with respect to the SCF, SCF-MI and MCSCF-MI results obtained by employing the harmonic approximation.

6.2. *NH_3-HCl complex*

Analogous results are obtained for this system, see Table 12. Also in this case the SCF-MI and MCSCF-MI results show a greater stability for what concerns the basis set if compared with the SCF method. The difference between the force constants relative to asymptotic and equilibrium geometries (ΔK) is positive. The ΔK for the NH_3-HCl complex is greater than that of the H_2O-HCl complex.

Parallel considerations can be made for the evaluation of vibration frequencies and $\Delta\omega$ by the harmonic approximation. A *red shift* is observed also in this case. Employing the Noumerov technique, the corresponding estimate of $\Delta\omega$ is 179 cm^{-1}, to be compared with a value of 129 cm^{-1} found for water; these results are not in accordance

Table 12 *Red shift for the H-Cl stretching frequencies calculated by the harmonic approximation (NH_3-HCl complex). All the values are reported in cm^{-1}*

Basis	$\Delta\omega_{SCF}$	$\Delta\omega_{SCF\text{-}MI}$	$\Delta\omega_{MCSCF\text{-}MI}$	$\Delta\omega_{NOUMEROV}$	$\Delta\omega_{EXP}$
					1517 [35]
					2149 [30]
TZVP	359	136	124	179	
6-31G**	421	121	114		

with the experimental data which depend strongly on the matrix (Ar or N_2) where the NH_3-HCl complex is trapped; it is gratifying that the computed $\Delta\omega$ predicts a larger *red shift* confirming a hydrogen bond stronger than in the H_2O-HCl system, as a consequence of the greater polarisability of the N-H bond.

7. Conclusions

Special attention has been dedicated to the study of the basis set superposition error (BSSE). The SCF-MI algorithm which excludes the BSSE from the SCF function was employed. A multi configuration version of it, particularly suited to study proton transfer effects, has been formulated. The use of these techniques has led to binding energy values which show a better stability against variation of the basis set, when compared with standard SCF results. For a more complete evaluation of the advantages of the *a priori* strategy to avoid BSSE see references [47–50], where applications to the study of the water properties are reported, and reference [51], where the Spin Coupled Valence Bond calculations for the He-LiH system are presented.

With regard to the vibrational analysis, the results confirm that the SCF-MI and MCSCF-MI BSSE free methods give reasonably stable results; the similarity of the SCF-MI and MCSCF-MI predictions (see Tables 9–12) confirm that the effect of the ionic structures – $NH_4^+ + Cl^-$ and $H_3O^+ + Cl^-$ – on the *red shift* of HCl is not particularly important in gas phase.

As the basis sets of the two fragments A and B which constitute the Van der Waals complex AB are kept separated, it can be questioned whether the SCF-MI results take the possibility of a charge transfer into consideration even if, as underlined by Chalasinski and Szczesniak [52], the charge transfer elude rigorous definition and should be invoked more as a conceptual visualisation rather than a well-defined additional physical effect. A plausible explanation why this might be possible is that the SCF-MI MOs of A and B are not forced to be orthogonal but are allowed to *overlap* so that the MOs of one system can extend over the space occupied by the other one and vice versa so long as sufficiently diffuse basis functions are included in the basis sets. For these reasons some degree of electronic charge transfer is permitted. Bader analysis on systems previously studied such as H_2O dimer, has demonstrated that the SCF and SCF-MI charge located on A and B is of the same order of magnitude; in fact the charge transfer calculated employing a basis set of TZVP++ quality, is of 0.009 and 0.004 electrons for SCF and SCF-MI wavefunction respectively [47,53]. In the present work we have repeated this analysis on the NH_3-HCl and H_2O-HCl systems where stronger hydrogen bonds are involved and larger charge transfers are expected. In fact, in the case of NH_3-HCl complex, the results show that a greater electron charge is transferred from the base (the H-acceptor molecule, NH_3) to the acid (the H-donor molecule, HCl). The amount of such charge transfer predicted by the SCF-MI wavefunction (0.01 electrons) is consistent with the values of 0.03 calculated employing the standard SCF wavefunction even with a small basis set of TZVP quality. The result demonstrate that our method correctly detects a charge transfer one order of magnitude higher with respect to the water dimer analysis and this is confirmed with a larger standard TZVP++ basis set.

The proton transfer process has been studied by the newly developed MCSCF-MI technique. An effective electron transfer is not observed: at the optimal geometrical conformation, the ionic SCF-MI function weighs much less than the covalent SCF-MI function as shown by the eigenvector of the MCSCF-MI wavefunction. Over most of the configuration space, the MCSCF-MI wavefunction can be viewed as a two-configuration SCF-MI function where the ionic solution has just the effect of introducing electronic correlation of intramolecular nature localised in the hydrogen chloride molecule.

As reported in [23], in order to rationalise the dramatic influence of the environment on the vibrational behaviour of HCl involved in the $NH_3 + HCl$ complex, Corongiu et al. [23] have accomplished DFT calculations by means of the self consistent reaction fields (SCFR) approach [54–56]; solvent effects were thus modelled by a continuum with a dielectric constant. These investigations demonstrated that even a weakly interacting environment can produce strong effects on the predicted properties, and particularly on geometrical and vibrational parameters of the H-Cl bond. A complete shift of the absolute minimum towards the ionic structure was observed when solvent effects are included. These results are confirmed in very recent literature where proton transfer reaction for nitric acid-ammonia complex [57] and gas-phase sulphuric acid hydrates [58] have been theoretically investigated. Our future studies will concentrate on the explicit treatment of solvent molecules interacting with the complex, by means of the generalised SCF-MI method [59] that still preserves the BSSE free scheme.

References

1. S. Scheiner: 'Calculating the Properties of Hydrogen Bonds by *ab initio* Methods', in K.B. Lipkowitz and D.B. Boyd (Eds.), *Reviews in Computational Chemistry II*, VCH Publisher, New York, 165 (1991); S. Scheiner: *Annu. Rev. Phys. Chem.* **45**, 23 (1994).
2. J.H. van Lenthe, J.C.G.M. van Duijneveldt-van de Rijdt and D.B. van Duijneveldt: *Adv. Chem. Phys.* **69**, 521 (1987).
3. F.B. van Duijneveldt, J.G.C.M. van Duijneveldt-van de Rijdt and J.H. van Lenthe: *Chem. Rev.* **94**, 1873 (1994).
4. N.R. Kestner: *J. Chem. Phys.* **48**, 252 (1968).
5. S.F. Boys and F. Bernardi: *Mol. Phys.* **19**, 553 (1970).
6. W. Kolos: *Theor. Chim. Acta* **51**, 219 (1979).
7. G. Alogna, C. Ghio, R. Cammi and J. Tomasi: *Int. J. Quantum Chem.* **32**, 207 (1987).
8. B.H. Wells and S. Wilson: *Mol. Phys.* **50**, 1295 (1983).
9. D.W. Schwenke and D.G. Truhlar: *J. Chem. Phys.* **82**, 2418 (1985).
10. E.R. Davidson and S. J. Chakravorty: *Chem. Phys. Lett.* **217**, 48 (1994).
11. G. Karlström and A.J. Sadlej: *Theor. Chim. Acta* **61**, 1 (1982).
12. Z. Latajka and S. Scheiner: *J. Chem. Phys.* **87**, 1194 (1987).
13. Z. Latajka and S. Scheiner: *Chem. Phys. Lett.* **140**, 338 (1987).
14. S. Simon, M. Duran and J. J. Dannenberg: *J. Chem. Phys.* **105**, 11024 (1996); P. Hobza and Z. Hables: *Theor. Chem. Acc.* **99**, 372 (1998).
15. S. S. Xantheas: *J. Chem. Phys.* **104**, 8821 (1996).
16. J. Sponer, J. Leszczynski and P. Hobza: *J. Phys. Chem.* **100**, 1965 (1996).
17. E. Gianinetti, M. Raimondi and E. Tornaghi: *Int. J. Quantum Chem.* **60**, 157 (1996).
18. A. Famulari, E. Gianinetti, M. Raimondi and M. Sironi: Int. *J. Quantum Chem.* **69**, 151 (1998).
19. A.C. Legon and D.J. Millen: *Proc. Roy. Soc.* **A417**, 21 (1988).
20. M.S. Gordon and J.H Jensen: *Acc. Chem. Res.* **29**, 536 (1996).
21. D. Hadzi, *Theoretical Treatments of Hydrogen Bonding*, Wiley Research Series in Theoretical Chemistry, John Wiley & Sons Ltd Publishers, Chichester-England (1997).
22. W. B. De Almeida and A. Hinchliffe: *Chem. Phys.* **137**, 143 (1989).

23. G. Corongiu, D. Estrin, G, Murgia, L. Paglieri, L. Pisani, G. Suzzi Valli, J. d. Watts and E. Clementi: *Int. J. Quantum Chem.* **59**, 119 (1996).
24. a) I. Mayer and Á. Vibók: *Int. J. Quantum Chem.* **40**, 139 (1991); b) Á. Vibók and I. Mayer: *Int. J. Quantum Chem.* **43**, 801 (1992); c) I. Mayer, Á. Vibók, G. Halász and P. Valiron: *Int. J. Quantum Chem.* **57**, 1049 (1996); d) G. Halász, Á. Vibók and S. Suhai: *Int. J. Quantum Chem.* **68**, 151 (1998); e) P. Valiron, Á. Vibók and I. Mayer: *J. Comp. Chem.* **14**, 401 (1993).
25. A.C. Legon and L.C. Willoughby: *Chem. Phys. Lett.* **95**, 449 (1983).
26. G.B. Bacskay, D.I. Kerdraon and N.S. Hush: *Chem. Phys.* **144**, 53 (1990).
27. G.B. Bacskay: *Mol. Phys.* **77**, 61 (1992).
28. Z. Latajka and S. Scheiner: *J. Chem. Phys.* **87**, 5928 (1987).
29. A.J. Barnes: *J. Mol. Struct.* **100**, 259 (1983).
30. B.S. Ault and G.C. Pimentel: *J. Phys. Chem.* **77**, 1649 (1973).
31. E. Clementi: *J. Chem. Phys.* **46**, 3851 (1967); Ibid. **47**, 2323 (1967).
32. E. Clementi and J.N. Gayles: *J. Chem. Phys.* **47**, 3837 (1967).
33. P. Goldfinger and G. Verhaegen: *J. Chem. Phys.* **50**, 1467 (1968).
34. R.C. Raffenetti and D.H. Phillips: *J. Chem. Phys.* **71**, 4534 (1979).
35. A.J. Barnes, T.R. Beech and Z. Mielke: *J. Chem. Soc. Far. Trans. II* **80**, 455 (1984).
36. Z. Latajka and S. Scheiner: *J. Chem. Phys.* **81**, 4014 (1984).
37. Z. Latajka and S. Scheiner: *J. Chem. Phys.* **82**, 4131 (1985).
38. R.S. Mulliken: *J. Phys. Chem.* **56**, 801 (1952).
39. P.G. Jasien and W.J. Stevens: *Chem. Phys. Lett.* **130**, 127 (1986).
40. Y. Bouteiller, C. Mijoule, A. Karpfen, H. Lischka and P. Schuster: *J. Phys. Chem.* **91**, 4464 (1987).
42. A. Famulari, Ph.D. Thesis (Dipartimento di Chimica Fisica ed Elettrochimica, Università degli Studi di Milano, 1997).
42. a) M.W. Schmidt, K.K. Baldridge, J.A. Boatz, S.T. Elbert, M.S. Gordon, J. Jensen, S. Koseki, N. Matsunaga, K.A. Nguyen, S.J. Su, T.L. Windus, M. Dupuis, J.A. Montgomery: *J. Comput. Chem.* **14**, 1347 (1993); b) GAMESS-US User's Guide.
43. M. Raimondi, W. Campion and M. Karplus: *Mol. Phys.* **34**, 1483 (1977).
44. a) D.L. Cooper, J. Gerratt and M. Raimondi: *Adv. Chem. Phys.* **69**, 319 (1987); b) J. Gerratt, D.L. Cooper, P.B. Karadakov and M. Raimondi: *Chem. Soc. Rev.* **26**, 87 (1997).
45. J.M. Blatt: *J. Comp. Phys.* **1**, 382 (1967).
46. J.W. Cooley: *Math. Comp.* **5**, 363 (1961).
47. A. Famulari, M. Raimondi, M. Sironi and E. Gianinetti: *Chem. Phys.* **232**, 275 (1998).
48. A. Famulari, M. Raimondi, M. Sironi and E. Gianinetti, *Chem. Phys.* **232**, 289 (1998).
49. M. Raimondi, A. Famulari, E. Gianinetti, M. Sironi, R. Specchio and I. Vandoni: *Adv. Quantum Chem.* **32**, 263 (1998).
50. A. Famulari, R. Specchio, M. Sironi and M. Raimondi: *J. Chem. Phys.* **108**, 3296 (1998).
51. a) F.A. Gianturco, S. Kumar, S.K. Pathak, M. Raimondi, M. Sironi, J. Gerratt and D.L. Cooper: *Chem. Phys.* **215**, 227 (1997); b) F.A. Gianturco, S. Kumar, S.K. Pathak, M. Raimondi and M. Sironi: *Chem Phys* **15**, 239 (1997); c) E. Bodo, S. Kumar, F.A. Gianturco, A. Famulari, M. Raimondi and M. Sironi: *J. Phys. Chem.* 4 **102**; 9330 (1998); d) E. Bodo, E. Buonomo, F.A. Gianturco, S. Kumar, A. Famulari, M. Raimondi and M. Sironi: *Chem. Phys.* **237**, 315 (1998).
52. G. Chalasinski and M.M. Szczesniak: *Chem. Rev.* **94**, 1723 (1994).
53. C. Gatti and A. Famulari: 'Interaction Energies and Densities. A Quantum Theory of Atom in Molecules insight on the Effect of Basis Set Superposition Error Removal', P.G. Mezey and B. Robertson (Eds.), Understanding Chemical Reactivity: Electron, Spin and Momentum Densities and Chemical Reactivity, Vol. 2, Kluwer book series (1999). In press.
54. L. Onsager: *J. Am. Chem. Soc.* **59**, 1486 (1936).
55. O. Tapia and O. Goscinski: *Mol. Phys.* **29**, 1653 (1975).
56. M.W. Wong, K.B. Wiberg and M.J. Frisch: *J. Am. Chem. Soc.* **112**, 4776 (1990); *Ibid.* **114**, 1645 (1992).
57. F.M. Tao: *J. Chem. Phys.* **108**, 193 (1998).
58. H. Arstila, K. Laasonen and A. Laaksonen: *J. Chem. Phys.* **108**, 1031 (1998).
59. E. Gianinetti, I. Vandoni, A. Famulari and M. Raimondi: *Adv. Quantum Chem.* **31**, 251(1998).

Part V

Nuclear Motion

Non-Adiabatic Molecular Hamiltonian. Canonical Transformation Coupling Electronic and Vibrational Motions

Ivan Hubač[a,b,c]*, Peter Babinec[a,b], Martin Polášek[c], Ján Urban[b], Pavel Mach[b], Jozef Mášik[b] and Jerzy Leszczyński[a]

[a]*Department of Chemistry, Jackson State University, 1400 Lynch Street, P.O. Box 17910, Jackson, MS 39217, USA*

[b]*Division of Chemical Physics, Faculty of Mathematics and Physics, Comenius University, Mlynská dolina F1, 842 15 Bratislava, Slovakia*

[c]*Institute of Physics, Faculty of Science, Silesian University, Bezručovo nám. 13, 746 01 Opava, Czech Republic*

Abstract

The coupling of electronic and vibrational motions is studied by two canonical transformations, namely, normal coordinate transformation and momentum transformation on molecular Hamiltonian. It is shown that by these transformations we can pass from crude approximation to adiabatic approximation and then to non-adiabatic (diabatic) Hamiltonian. This leads to renormalized fermions and renormalized diabatic phonons. Simple calculations on H_2, HD, and D_2 systems are performed and compared with previous approaches. Finally, the problem of reducing diabatic Hamiltonian to adiabatic and crude adiabatic is discussed in the broader context of electronic quasi-degeneracy.

1. Introduction

Many atomic systems (e.g. molecules, clusters and crystals) are systems with interacting electrons and nuclei and can be thus described by Schrödinger equation

$$H\Psi = E\Psi. \tag{1}$$

In general case the number of degrees of freedom of such systems with coulomb interaction is too large and even using high-performance computers to solve this equation becomes impossible. Therefore the only possibility to solve eq. (1) is to suggest some approximations [1–6]. The most important approximation and also most often used is the Born-Oppenheimer (BO) [7] and the adiabatic approximations. These approximations are based on the fact that masses of nuclei are 1870 × Z times (Z is the nuclear charge) heavier than the masses of electrons. This leads to the idea of potential energy surface. Beside the many attempts to go beyond the BO approximation and many different approaches, certain facts are not clear completely. In this paper we decided to study the coupling of electronic and vibrational motions by two canonical transformations, namely, normal coordinate and momentum transformations. Our approach is similar to quasiparticle transformations often done in solid state physics. In

*To whom correspondence should be addressed.

A. Hernández-Laguna et al. (eds.), Quantum Systems in Chemistry and Physics, Vol. 1: Basic Problems and Model Systems, 383–400.

order to make our approach more transparent we repeat here main features of adiabatic approximation. We follow the arguments of recent Kutzelnigg's paper [8].

Let us write the total molecular Hamiltonian as

$$H = T_N(R) + E_{NN}(R) + H_{EN}(r, R) + H_{EE}(r) \tag{2}$$

where $T_N(R)$ is the kinetic energy of nuclei, $E_{NN}(R)$ is the interaction between nuclei, R denotes nuclear coordinates, r denotes electronic coordinates, and $H_{EN}(r, R) + H_{EE}(r)$ is the electronic Hamiltonian

$$H_{EN}(r, R) + H_{EE}(r) = T_E(r) + U_{EN}(r, R) + H_{EE}(r) \tag{3}$$

where $T_E(r)$ is the kinetic energy of electrons, $U_{EN}(r, R)$ is the electron–nuclei interaction term and $H_{EE}(r)$ represents the electron-electron interaction. In Born and Huang [1,9] approach, the total wavefunction depending on the nuclear coordinates R and on the electronic coordinates r is expanded as

$$\Psi(r, R) = \sum_k \psi_k(r, R)\chi_k(R), \tag{4}$$

where ψ_k are a complete set of known functions of r that depend parametrically on the nuclear coordinates R and where the $\chi_k(R)$ are regarded as unknown. The ψ_k are conventionally chosen as a set of eigenfunctions of the clamped nuclei (CN) Hamiltonian.

Both the BO and adiabatic approximation can be based on choosing a single term in eq. (4)

$$\Psi(r, R) = \psi_k(r, R)\chi_k(R). \tag{5}$$

This is referred as BO ansatz. This ansatz is taken as a variational trial function. Terms beyond the leading order in m/M are neglected (m is the electronic and M is nuclear mass, respectively). The problem with expansion (4) is that functions $\psi(r, R)$ contain except bound states also continuum function since it includes the centre of mass (COM) motion. Variation principle does not apply to continuum states. To avoid this problem we can separate COM motion. The remaining Hamiltonian for the relative motion of nuclei and electrons has then bound state solution. But there is a problem, because this separation mixes electronic with nuclear coordinates and also there is a question how to define molecule-fixed coordinate system. This is in detail discussed by Sutcliffe [5] . In the recent paper by Kutzelnigg [8] this problem is also discussed and it is shown how to derive adiabatic corrections using, as he called it, the Born–Handy ansatz. There are few important steps to arrive at formula for a diabatic corrections. Firstly, one separates off COM motion. Secondly, (very important step) one does not specify the relative coordinates (which are to some extent arbitrary). In this way one arrives at relative Hamiltonian H_{rel} [8] with trial wavefunction Ψ_{rel}. If we make BO ansatz

$$\Psi_{rel} = \psi(\cdots \rho_{lk}, \cdots \rho_{\mu\nu} \cdots)\chi(\cdots \rho_{\mu\nu} \cdots) \tag{6}$$

where ρ_{lk}, $\rho_{\mu\nu}$ are non-specified relative coordinates and ψ is chosen as a solution of the CN Schrödinger equation. The adiabatic correction ΔE takes very simple form

$$\Delta E = -\frac{1}{2}\langle\Psi|\sum_{\mu} M_{\mu}\nabla_{\mu}^{2}|\Psi\rangle. \tag{7}$$

It was firstly introduced by Born in 1956 [1] and used by Handy [12]. It was used previously also by Sellers and Pulay [13]. (See also Davidov [9] or Kaplan [10] for its derivation). For practical calculation the identity

$$\int \psi \frac{\partial^2}{\partial Q_i^2} \psi \, \mathrm{d}r_i = -\int \left(\frac{\partial \psi}{\partial Q_i}\right)^2 \mathrm{d}r_i \tag{8}$$

can be used.

Note that in fact in every textbook of quantum chemistry [9,11] the validity of BO approximation is justified only when

$$\frac{\hbar\omega}{|E_n - E_m|} \ll 1 \tag{9}$$

when ω is the frequency of harmonic vibrations around the point R_0.

The aim of this paper is twofold:

i) We show how starting with molecular Hamiltonian (2) in crude adiabatic representation we arrive at adiabatic Hamiltonian by performing canonical transformations which mix together the electronic and vibrational motions (through normal coordinates). We derive simple formulae for adiabatic corrections.
ii) We generalize canonical transformations (through momenta) arriving at non–adiabatic Hamiltonian. We introduce the idea of quasiparticles (renormalized electrons and phonons) and present the formulae how to obtain the 'orbital energies', 'correlation corrections' and non–adiabatic frequencies for these quasiparticles. Finally, we perform some simple model calculations to demonstrate how the method works.

2. Theory

Let us start with electronic Hamiltonian (3) which we denote

$$H_{EN}(r, R) + H_{EE}(r) = h + v^0, \tag{10}$$

where h is the one-electron part representing the kinetic energy of the electrons and electron-nuclear attraction term, and v^0 is the two electron part of the Hamiltonian corresponding to electron-electron repulsion term. For the purpose of diagrammatic many-body perturbation theory it will be efficient to work in second quantization formalism. The electronic Hamiltonian (10) has the form

$$H_{EN} + H_{EE} = \sum_{PQ}\langle P|h|Q\rangle a_P^{+}a_Q + \frac{1}{2}\sum_{PQRS}\langle PQ|v^0|RS\rangle a_P^{+}a_Q^{+}a_S a_R \tag{11}$$

where $a_P^{+}(a_Q)$ is the creation (annihilation) operator for electrons in the spinorbital basis $|P\rangle, |Q\rangle, \ldots$. If we apply the Wick theorem to (11) we can write this equation as

$$H_{EN} + H_{EE} = \sum_{I} h_{II} + \frac{1}{2}\sum_{IJ}(v^0_{IJIJ} - v^0_{IJJI}) + \sum_{PQ} h_{PQ} N[a^+_P a_Q]$$

$$+ \sum_{PQI}(v^0_{PIQI} - v^0_{PIIQ}) N[a^+_P a_Q] + \frac{1}{2}\sum_{PQRS} v^0_{PQRS} N[a^+_P a^+_Q a_S a_R] \quad (12)$$

where $v^0_{ABAB}(v^0_{ABBA})$ denotes the coulomb (exchange) integral. One possibility is to work within crude representation in which the spinorbital basis $|P\rangle$, $|Q\rangle$, ... is determined at some fixed (equilibrium coordinate R_o) point. Note that Hamiltonian (12) has $3N - 6$ degrees of freedom (in fact 3N degrees of which 6 are zero). Hamiltonian (12) has only bound-state solutions. Let us divide individual terms of the Hamiltonian (12) into two parts. Namely calculated at point R_o and the terms which are shifted with respect to point R_o (we use prime to denote these terms). The electronic Hamiltonian (12) can be rewritten as

$$H_{EN} + H_{EE} = E^0_{SCF} + h'_{SCF} + \sum_{P} \varepsilon_P N[a^+_P a_P] + \sum_{PQ} h'_{PQ} N[a^+_P a_Q]$$

$$+ \frac{1}{2}\sum_{PQRS} v^0_{PQRS} N[a^+_P a^+_Q a_S a_R] \quad (13)$$

where E^0_{SCF} is the Hartree-Fock energy calculated at the point R_0, and h'_{SCF} is the shift in the Hartree-Fock energy with respect to the point other than R_0. The same is true for one-particle operator of eq. (13), where ε_P are the one-particle Hartree-Fock energies calculated at point R_0. The correlation operator is not changed because it does not depend on nuclear coordinates R. Let us perform the Taylor expansion for the energies E_{NN} and u_{SCF} around the point R_0. For the notation see [14].

$$E_{NN} = E^{(0)}_{NN} + E'_{NN} = \sum_{i=0}^{\infty} E^{(i)}_{NN} \quad (14)$$

and

$$u_{SCF} = u^{(0)}_{SCF} + u'_{SCF} = \sum_{i=0}^{\infty} u^{(i)}_{SCF} \quad (15)$$

Using (14) and (15) we can rewrite our Hamiltonian (13) in the form

$$H = E^{(0)}_{NN} + E^{(0)}_{SCF} + \sum_{r} \hbar\omega_r\left(b^+_r b_r + \frac{1}{2}\right) + u^{(2)}_{SCF} + \sum_{P} \varepsilon_P N[a^+_P a_P]$$

$$+ \frac{1}{2}\sum_{PQRS} v^0_{PQRS} N[a^+_P a^+_Q a_S a_R] + E'_{NN} - E^{(2)}_{NN} + u'_{SCF} - u^{(2)}_{SCF} + \sum_{PQ} u'_{PQ} N[a^+_P a_Q]$$

$$(16)$$

where

$$\sum_{r} \hbar\omega_r\left(b^+_r b_r + \frac{1}{2}\right) = T_N + E^{(2)}_{NN} + u^{(2)}_{SCF} \quad (17)$$

ω_r is the frequency of the harmonic oscillator and b^+ (b) are boson (phonon) creation (annihilation) operators. In order to use the perturbation theory we have to split the Hamiltonian (16) onto the unperturbed part H_0 and the perturbation H'

$$H = H_0 + H'. \tag{18}$$

Due to the crude approximation, we can partition the Hamiltonian (16) in the following way

$$H_0 = E_{NN}^{(0)} + E_{SCF}^{(0)} + \sum_P \varepsilon_P N[a_P^+ a_P] + \sum_r \hbar\omega_r \left(b_r^+ b_r + \frac{1}{2} \right) \tag{19}$$

and

$$H' = H'_E + H'_F + H'_I. \tag{20}$$

Where H' contains all terms in (16) except (19). In eq. (16) all quantities were defined through the cartesian coordinates. For further purposes it will be natural to work in normal coordinates $\{B_r\}$. The normal coordinate in second quantized formalism is given as

$$B_r = b_r + b_r^+. \tag{21}$$

If we transform Hamiltonian (16) into normal coordinates we arrive at the following expressions [14]

$$\begin{aligned} H = {} & E_{NN}^{(0)} + E_{SCF}^{(0)} + \sum_P \varepsilon_P N[a_P^+ a_P] + \sum_r \hbar\omega_r \left(b_r^+ b_r + \frac{1}{2} \right) \\ & + \frac{1}{2} \sum_{PQRS} v_{PQRS}^0 N[a_P^+ a_Q^+ a_S a_R] \quad \{\equiv H'_E\} \\ & + \sum_{n=1,(n\neq 2)}^{\infty} \sum_{k=0}^{[n/2]} (\mathbf{E}_{NN}^{(k,n-2k)} + \mathbf{u}_{SCF}^{(k,n-2k)}).\mathbf{B}^{(n-2k)} \quad \{\equiv H'_F\} \\ & + \sum_{n=1}^{\infty} \sum_{k=0}^{[n/2]} \sum_{PQ} \mathbf{u}_{PQ}^{(k,n-2k)}.\mathbf{B}^{(n-2k)} N[a_P^+ a_Q] \quad \{\equiv H'_I\}. \end{aligned} \tag{22}$$

The term H'_E is the electron correlation operator, the term H'_F corresponds to phonon-phonon interaction and H'_I corresponds to electron-phonon interaction. If we analyze the last term H'_I we see that when using crude approximation this corresponds to such phonons that force constant in eq. (17) is given as a second derivative of electron–nuclei interaction with respect to normal coordinates. Because we used crude adiabatic approximation in which minimum of the energy is at the point R_o, this is also reflected by basis set used. Therefore this approximation does not properly describes the physical vibrations i.e. if we move the nuclei, electrons are distributed according to the minimum of energy at point R_o and they do not feel correspondingly the R dependence. The perturbation term H'_I which corresponds to electron–phonon interaction is too large and thus perturbation theory based on splitting given by eq. (19, 20) will not converge [15]. Natural way to improve this situation will be to use basis set which is generally R dependent. We can do this in second–quantized formalism in a way that we pass from electron creation (annihilation) operators a_P^+ (a_Q) which act on R_o dependent basis set to a new fermion creation (annihilation) operators $\overline{a}_P^+$ $(\overline{a}_Q)$ which act on R dependent basis. Similar transformation was studied for solid state theory by Wagner [16], who also discuss the convergency properties of adiabatic approximation [17]. This we can achieve by canonical transformation passing from old electron operators a_P^+ (a_Q) to new operators $\overline{a}_P^+$ $(\overline{a}_P)$ through normal coordinates B_r. In this way we can pass from crude adiabatic Hamiltonian to what is called clamped nucleus Hamiltonian and corresponding clamped nucleus wavefunction $\Psi(r, R)$. The proof that this is a canonical transformation is in [14].

$$\overline{a}_P = a_P + \sum_Q \sum_{k=1}^{\infty} \frac{1}{k!} \sum_{r_1 \ldots r_k} C_{PQ}^{r_1 \ldots r_k} B_{r_1} \ldots B_{r_k} a_Q \tag{23}$$

$$\overline{a}_P^+ = a_P^+ + \sum_Q \sum_{k=1}^{\infty} \frac{1}{k!} \sum_{r_1 \ldots r_k} C_{PQ}^{r_1 \ldots r_k} B_{r_1} \ldots B_{r_k} a_Q^+, \tag{24}$$

where B_r are second quantized normal coordinates.

In short notation we can also write [14]

$$\begin{aligned} \overline{a}_P &= \sum_Q \sum_{k=0}^{\infty} \frac{1}{k!} \mathbf{C}_{PQ}^{(k)} \mathbf{B}^k a_Q \\ &= \sum_Q \sum_{k=0}^{\infty} C_{PQ}^{(k)} a_Q \\ &= \sum_Q C_{PQ} a_Q \end{aligned} \tag{25}$$

$$\begin{aligned} \overline{a}_P^+ &= \sum_Q \sum_{k=0}^{\infty} \frac{1}{k!} \mathbf{C}_{PQ}^{(k)*} \cdot \mathbf{B}^k a_Q^+ \\ &= \sum_Q \sum_{k=0}^{\infty} C_{PQ}^{(k)+} a_Q^+ \\ &= \sum_Q C_{PQ}^+ a_Q^+. \end{aligned} \tag{26}$$

We also perform analogous canonical transformation for phonons

$$\overline{b}_r = b_r + \sum_{PQ} \sum_{k=0}^{\infty} \frac{1}{k!} \sum_{s_1 \ldots s_k} d_{rPQ}^{s_1 \ldots s_k} B_{s_1} \ldots B_{s_k} a_P^{+} a_Q, \tag{27}$$

$$\overline{b}_r^{+} = b_r^{+} + \sum_{PQ} \sum_{k=0}^{\infty} \frac{1}{k!} \sum_{s_1 \ldots s_k} d_{rPQ}^{s_1 \ldots s_r^*} B_{s_1} \ldots B_{s_k} a_Q^{+} a_P. \tag{28}$$

The coefficients C_{PQ} (C_{PQ}^{+}) in eqs. (25, 26) are determined so that $\overline{a}_P$ $(\overline{a}_P^{+})$ satisfy fermion anticommutation relation. The coefficients d_{rPQ} (d_{rPQ}^{+}) in eqs. (27,28) are determined so that $\overline{b}_r$ $(\overline{b}_r^{+})$ satisfy boson commutation relation. Finally we ask fermions $\overline{a}_P$ $(\overline{a}_P^{+})$ to commute with bosons $\overline{b}_r$ $(\overline{b}_r^{+})$. This means that we can write similarly as in (5) the total wave function $\Psi(r, R)$ as a product of fermion wave function $\psi_k(r, R)$ and boson wave function χ_k as $\Phi(r, R)$

$$\Phi(r, R) = \psi_k(r, R)\chi_k(R). \tag{29}$$

It is easy to show that we have two invariants of transformations, namely number operator of fermions

$$\overline{N} = N. \tag{30}$$

and normal coordinate

$$\overline{\mathbf{B}} = \mathbf{B}. \tag{31}$$

The next step is that we find inverse transformations to (25–28) and substitute these inverse transformations into eq. (22) and then applying Wick theorem, we requantize the whole Hamiltonian (16) in a new fermions and bosons [14]. This leads to new V-E Hamiltonian (we omit sign on the second quantized operators)

$$H = H_A + H_B \tag{32}$$

where

$$\begin{aligned} H_A = {} & E_{NN}(B) - E_{NN}^{(2)}(B) - V_N^{(2)}(B) + \sum_{RSI} h_{RS}(B) C_{RI} C_{SI} \\ & + \frac{1}{2} \sum_{RSTUIJ} (v_{RTSU}^0 - v_{RSTU}^0) C_{RI} C_{SI} C_{TJ} C_{UJ} + \sum_{PQRS} h_{RS} C_{RP} C_{SQ} N[a_P^{+} a_Q] \\ & + \sum_{PQRSTUI} (v_{RTSU}^0 - v_{RSTU}^0) C_{RP} C_{SQ} C_{TI} C_{UI} N[a_P^{+} a_Q] \\ & + \sum_{PQRSTUVW} v_{TUVW}^0 C_{TP} C_{UQ} C_{VR} C_{WS} N[a_P^{+} a_Q^{+} a_S a_R] \end{aligned} \tag{33}$$

and

$$H_B = \sum_r \hbar\omega_r\left(b_r^+ b_r + \frac{1}{2}\right) + \sum_{PQr} \hbar\omega_r(b_r^+ d_{rPQ} + d_{rQP} b_r)N[a_P^+ a_Q]$$

$$+ \sum_{AIr} \hbar\omega_r (d_{rAI})^2 + \sum_{PQAIr} \hbar\omega_r(d_{rPA} d_{rQA} - d_{rPI} d_{rQI})N[a_P^+ a_Q]$$

$$+ \sum_{PQRSr} \hbar\omega_r d_{rPS} d_{rQR} N[a_P^+ a_Q^+ a_S a_R]. \tag{34}$$

If we introduce the following quantities

$$E_{SCF} = \sum_{RSI} h_{RS} C_{RI} C_{SI} + \frac{1}{2}\sum_{RSTUIJ}(v^0_{RTSU} - v^0_{RSTU}) C_{RI} C_{SI} C_{TJ} C_{UJ} \tag{35}$$

the new Hartree-Fock operator f with the matrix elements

$$f_{PQ} = \sum_{RS} h_{RS} C_{RP} C_{SQ} + \sum_{RTSUI}(v^0_{RTSU} - v^0_{RSTU}) C_{RP} C_{SQ} C_{TI} C_{UI} \tag{36}$$

and the new two-particle integral

$$v_{PQRS} = \sum_{TUVW} v^0_{TUVW} C_{TP} C_{UQ} C_{VR} C_{WS}. \tag{37}$$

We can rewrite our Hamiltonian H_A (33) into the form

$$H_A = E_{NN} - E_{NN}^{(2)} - V_N^{(2)} + E_{SCF} + \sum_{PQ} f_{PQ} N[a_P^+ a_Q] + \frac{1}{2}\sum_{PQRS} v_{PQRS} N[a_P^+ a_Q^+ a_S a_R] \tag{38}$$

Here in eq. (38) $\sum_{PQ} f_{PQ} N[a_P^+ a_Q]$ is new Hartree-Fock operator for a new fermions (25), (26), operator $\frac{1}{2}\sum_{PQRS} v_{PQRS} N[a_P^+ a_Q^+ a_S a_R]$ is a new fermion correlation operator and E_{SCF} is a new fermion Hartree-Fock energy. Our new basis set is obtained by diagonalizing the operator f from eq. (36). The new Fermi vacuum is renormalized Fermi vacuum and new fermions are renormalized electrons. The diagonalization of f operator (36) leads to coupled perturbed Hartree-Fock (CPHF) equations [18–20]. Similarly operators $\overline{b}_r$ ($\overline{b}_r^+$) corresponds to renormalized phonons. Using the quasiparticle canonical transformations (25–28) and the Wick theorem the V-E Hamiltonian takes the form

$$H = H_A + B_B \tag{39}$$

where

$$H_A = E_{NN}^0 + E_{SCF}^0 + \sum_P \epsilon_P N[a_P^+ a_P] + \sum_{n=1}^{\infty} \sum_{k=0}^{[n/2]} \mathbf{E}^{(k,n-2k)} . \mathbf{B}^{(n-2k)}$$

$$+ \sum_{n=1}^{\infty} \sum_{k=0}^{[n/2]} \sum_{PQ} \mathbf{f}_{PQ}^{(k,n-2k)} . \mathbf{B}^{(n-2k)} N[a_P^+ a_Q]$$

$$+ \frac{1}{2} \sum_{n=0}^{\infty} \sum_{k=0}^{[n/2]} \sum_{PQRS} \mathbf{v}_{PQRS}^{(k,n-2k)} . \mathbf{B}^{(n-2k)} N[a_P^+ a_Q^+ a_S a_R]. \tag{40}$$

and

$$H_B = \sum_r \hbar\omega_r \left(b_r^+ b_r + \frac{1}{2} \right) + \sum_{AIr} \hbar\omega_r (d_{rAI})^2 + \sum_{PQr} \hbar\omega_r (b_r^+ d_{rPQ} + d_{rQP} b_r) N[a_P^+ a_Q]$$

$$+ \sum_{PQAIr} \hbar\omega_r (d_{rPA} d_{rQA} - d_{rPI} d_{rQI}) N[a_P^+ a_Q]$$

$$+ \sum_{PQRSr} \hbar\omega_r d_{rPS} d_{rQR} N[a_P^+ a_Q^+ a_S a_R]. \tag{41}$$

As we have shown in [21,14] this quasiparticle transformation leads from crude adiabatic to adiabatic Hamiltonian. The Hamiltonian (39) is adiabatic Hamiltonian. Note that the force constant for harmonic oscillators is given as second derivative of E_{SCF} at point R_o . We shall call the corresponding phonons the adiabatic phonons.

3. Diabatic Canonical Transformation

In previous part we developed canonical transformation (through normal coordinates) by which we were able to pass from crude adiabatic to adiabatic Hamiltonian. We started with crude adiabatic molecular Hamiltonian on which we applied canonical transformation on second quantized operators

$$\overline{a}_P = \sum_Q C_{PQ}(B) a_Q \tag{42}$$

$$\overline{a}_P^+ = \sum_Q C_{PQ}(B)^+ a_Q^+ \tag{43}$$

$$\overline{b}_r = b_r + \sum_{PQ} d_{rPQ}(B) a_P^+ a_Q \tag{44}$$

$$\overline{b}_r^+ = b_r^+ + \sum_{PQ} d_{rQP}(B)^+ a_P^+ a_Q \tag{45}$$

where operators $\overline{a}_P$ ($\overline{a}_P^+$) corresponds to fermions and operators $\overline{b}_r$ ($\overline{b}_r^+$) to bosons and $B = b + b^+$ is the normal coordinate. The coefficients C_{PQ} (d_{rPQ}) can be found from the solution of CPHF equations. We also found that adiabatic corrections can be calculated as perturbation corrections, which mean that we expect that adiabatic corrections represents small perturbation. The situation can be more complex if we cannot treat non-adiabaticity as a perturbation. This is the case when non-adiabaticity can cause strong coupling between two or more electronic states. In order to treat such situations we can proceed in a way in which we generalize transformations (25)–(28). In these equations the expansion coefficients C and d were functions of normal coordinates $B = b + b^+$.

The generalization can be done in a way that these coefficients are some general functions $C_{PQ}(b, b^+)$ and $d_{rPQ}(b, b^+)$ of b and b^+ operators. We can expect that these coefficients will be not only the function of normal coordinate $B = b + b^+$ but also the function of momentum $\tilde{B} = b - b^+$. Therefore general transformations will have the form [21,22]

$$\bar{a}_P = \sum_Q C_{PQ}(B, \tilde{B}) a_Q \tag{46}$$

$$\bar{a}_P^+ = \sum_Q C_{PQ}(B, \tilde{B})^+ a_Q^+ \tag{47}$$

$$\bar{b}_r = b_r + \sum_{PQ} d_{rPQ}(B, \tilde{B}) a_P^+ a_Q \tag{48}$$

$$\bar{b}_r^+ = b_r^+ + \sum_{PQ} d_{rQP}(B, \tilde{B})^+ a_P^+ a_Q. \tag{49}$$

Such transformations would be rather complex, therefore we try simple approximation

$$C(B, \tilde{B}) = C(B).\tilde{C}(\tilde{B}). \tag{50}$$

Further we can proceed similarly as in the case of adiabatic approximation. We shall not present here the details, these are presented in [21,22]. We just mention the most important features of our transformation (46–50). Firstly, when passing from crude adiabatic to adiabatic approximation the force constant changed from second derivative of electron–nuclei interaction $u_{SCF}^{(2)}$ to second derivative of Hartree–Fock energy $E_{SCF}^{(2)}$. Therefore when performing transformation (46–50) we expect change of force constant and therefore change of the vibrational part of Hamiltonian

$$H_B = E_{kin}(\tilde{B}) + E_{pot}(B). \tag{51}$$

The potential energy is determined by the quadratic part of the nuclear energy $E_{NN}^{(2)}(B)$ as well as by some potential energy $V_N^{(2)}(B)$ which is a quadratic function of coordinate operators and has its origin in the interaction of the electrons with the vibrating nuclei. Therefore we have

$$E_{pot}(B) = E_{NN}^{(2)}(B) + V_N^{(2)}(B). \tag{52}$$

In the case of kinetic energy term this was identical with the kinetic energy of the nuclei in the case of adiabatic approximation. In the case of the breakdown of adiabatic approximation we have to remember the finite mass of electrons and therefore to introduce more general kinetic energy term. Therefore, we add to the kinetic energy of the nuclei $T_N(\tilde{B})$ some other yet unknown term which will be the quadratic function of momentum operator

$$E_{kin}(\tilde{B}) = T_N(\tilde{B}) + W_N^{(2)}(\tilde{B}). \tag{53}$$

The total vibrational–electronic Hamiltonian

$$H = H_A + H_B \tag{54}$$

will have the form

$$H_B = T_N^{(2)}(\tilde{B}) + W_N^{(2)}(\tilde{B}) + E_{NN}^{(2)}(B) + V_N^{(2)}(B) \tag{55}$$

and

$$H_A = E_{NN}(B) - E_{NN}^{(2)}(B) - V_N^{(2)}(B) - W_N^{(2)}(\tilde{B})$$
$$+ \sum_{P,Q} h_{PQ}(B) a_p^+ a_q + \frac{1}{2} \sum_{PQRS} V^0_{PQRS} a_P^+ a_Q^+ a_S a_R. \tag{56}$$

Secondly, coefficients C_{PQ} and $\tilde{C}_{PQ}$ are determined through equations [22]

$$u^r_{PQ} + (\varepsilon^0_P - \varepsilon^0_Q) C^r_{PQ} + \sum_{AI} [(V^0_{PIQA} - V^0_{PIAQ}) C^r_{AI} - (V^0_{PAQI} - V^0_{PAIQ}) C^r_{IA}] - \hbar\omega_r \tilde{C}^r_{PQ} = \varepsilon^r_P \delta_{PQ} \tag{57}$$

$$(\varepsilon^0_P - \varepsilon^0_Q) \tilde{C}^r_{PQ} + \sum_{AI} [(V^0_{PIQA} - V^0_{PIAQ}) \tilde{C}^r_{AI} - (V^0_{PAQI} - V^0_{PAIQ}) \tilde{C}^r_{IA}] - \hbar\omega_r C^r_{PQ} = \tilde{\varepsilon}^r_P \delta_{PQ} \tag{58}$$

where $\hbar\omega_r$ is the new non-adiabatic phonon given by

$$H_B = \sum_r \hbar\omega_r \left(b_r^+ b_r + \frac{1}{2} \right). \tag{59}$$

The expressions for extra terms $V_N^{(2)}(B)$ and $W_N^{(2)}(\tilde{B})$ in (55) are given as

$$(V_N^{(2)}(B)) \equiv V_N^{rs} = \sum_I u^{rs}_{II} + \sum_{IA} [(u^r_{AI} + \hbar\omega_r \tilde{C}^r_{AI}) C^s_{IA} + (u^r_{AI} + \hbar\omega_s \tilde{C}^r_{AI}) C^r_{IA}] \tag{60}$$

and

$$(W_N^{(2)}(\tilde{B})) \equiv W_N^{rs} = 2\hbar\omega_r \sum_{AI} C^r_{AI} \tilde{C}^s_{IA}. \tag{61}$$

This means that the resulting vibrational frequency ω depends explicitly on coefficients $\tilde{C}^r_{PQ}$ and C^r_{PQ}.

Finally fermion part of Hamiltonian will be given as

$$H_F = H_F^0 + H_{F'} + H_{F''} + H_{F'''}. \tag{62}$$

For the ground state energy we get

$$H_F^0 = E_{NN}^0 + E_{SCF}^0 + \sum_{AIr} \hbar\omega_r (|C_A^r|^2 - |\tilde{C}_{AI}^r|^2). \tag{63}$$

One-fermion part will be

$$\begin{aligned} H_{F'} = & \sum_P \varepsilon_P^0 N[a_P^+ a_P] + \sum_{PQr} \hbar\omega_r \left[\sum_A (C_{PA}^r C_{QA}^{r*} - \tilde{C}_{PA}^r \tilde{C}_{QA}^{r*}) \right. \\ & \left. - \sum_I (C_{PI}^r C_{QI}^{r*} - \tilde{C}_{PI}^r \tilde{C}_{QI}^{r*}) \right] N[a_P^+ a_Q] - 2 \sum_{PQr} E^{r*} \tilde{C}_{PQ}^r N[a_P^+ a_Q] \\ & + \sum_{PQr} [(h(P) - p(P))\varepsilon_P^{r*} + (h(Q) - p(Q))\varepsilon_Q^{r*}] \tilde{C}_{PQ}^r N[a_P^+ a_Q] \\ & - \sum_{PQAIr} [(v_{PIQA}^r - v_{PIAQ}^r) \tilde{C}_{IA}^{r*} + (v_{PAQI}^r - v_{PAIQ}^r) \tilde{C}_{AI}^{r*}] N[a_P^+ a_Q]. \end{aligned} \tag{64}$$

Two-fermion part will be

$$\begin{aligned} H_{F''} = & \frac{1}{2} \sum_{PQRS} v_{PQRS}^0 N[a_P^+ a_Q^+ a_S a_R] \\ & + \sum_{PQRSr} \hbar\omega_r (C_{PR}^r C_{SQ}^{r*} - \tilde{C}_{PR}^r \tilde{C}_{SQ}^{r*}) N[a_P^+ a_Q^+ a_S a_R] \\ & - 2 \sum_{PQRSr} \varepsilon_P^r \tilde{C}_{SQ}^{r*} N[a_P^+ a_Q^+ a_S a_R] \\ & + 2 \sum_{PQRSTr} \left\{ \sum_I [v_{PQRS}^0 C_{TI}^r - v_{PQTI}^0 C_{TS}^r + (v_{TQSI}^0 - v_{TQIS}^0) C_{PT}^r] \tilde{C}_{RI}^{r*} \right. \\ & + \sum_I [v_{TIRS}^0 C_{QT}^r - v_{TQRS}^0 C_{IT}^r + (v_{IQST}^0 - v_{IQST}^0) C_{TR}^r] \tilde{C}_{IP}^{r*} \\ & - \sum_A [v_{PQTS}^0 C_{TA}^r - v_{PQTA} C_{TS}^r + (v_{TQAS}^0 - v_{TQAS}^0) C_{PT}^r] \tilde{C}_{RA}^{r*}. \\ & \left. - \sum_A [v_{TARS}^0 C_{QT}^r - v_{TQRS} C_{AT}^r + (v_{AQTS}^0 - v_{AQST}^0) C_{TR}^r] \tilde{C}_{AP}^{r*} \right\} N[a_P^+ a_Q^+ a_S a_R]. \end{aligned} \tag{65}$$

Three-fermion part will be (as a result of transformation (50) the three fermion term appears)

$$H_{F'''} = -2 \sum_{PQRSTUVr} (v^0_{PQVT} C^r_{RS} - v^0_{VQST} C^r_{PV}) \tilde{C}^{r*}_{UR} N[a^+_P a^+_Q a^+_R a_U a_T a_S]. \quad (66)$$

The bosonic part of Hamiltonian H_B is not given in a diagonal form. To bring it to diagonal form as in eq. (59) we can proceede as follows.

$$H_B = T_N + E^{(2)}_{NN} + \sum_{r,s} \hat{E}^{rs}_{SCF} - \sum_I \left[\epsilon^0_1 \hat{S}^{rs}_{II} + \frac{1}{2}(\epsilon^r_I \hat{S}^s_{II} + \epsilon^s_I \hat{S}^r_{II}) \right]$$

$$+ \sum_{RI} [(\hat{f}^r_{RI} - \epsilon^0_I \hat{S}^r_{RI} + \hbar\omega_r \tilde{C}^r_{RI})\hat{C}^s_{RI} + (\hat{f}^s_{RI} - \epsilon^0_I \hat{S}^s_{RI} + \hbar\omega_s \tilde{C}^s_{RI})\hat{C}^r_{RI} B_r B_s$$

$$+ \sum_{r,s} 2\hbar\omega_r \sum_{AI} (\hat{C}^r_{AI} + \langle A|I^r\rangle)\tilde{C}^s_{AI} \tilde{B}_r \tilde{B}_s. \quad (67)$$

Our aim is to bring this Hamiltonian into diagonal form. We can extract adiabatic part ($\hbar\omega^a_r$) and we get

$$H_B = \sum_r \hbar\omega^a_r \left(b^+_r b_s + \frac{1}{2} \right) \delta_{rs} + F^1_{rs} + F^2_{rs} \quad (68)$$

where

$$F^1_{rs} = 2 \sum_{r,s} \left\{ \sum_{R,I} \left[(\hat{f}^s_{RI} - \epsilon^0_I)(\hat{C}^s_{RI} - \hat{\hat{C}}^r_{RI}) + \hbar\omega_r \tilde{C}^r_{RI} \tilde{C}^s_{RI} \right] \right\} B_r B_s \quad (68)$$

$$F^2_{rs} = 2 \sum_{r,s} \hbar\omega_r \left\{ \sum_{AI} (\hat{C}^r_{AI} + \langle A|I^r\rangle)\hat{C}^s_{AI} \right\} \tilde{B}_r \tilde{B}_s \quad (70)$$

where $\hat{\hat{C}}^r_{RI}$ is identical with $\hat{C}^r_{RI}$ coefficients from adiabatic transformation eqs. (23, 24). If we substitute in eq. (69) and (70) for $B_r = b^+_r + b_r$ and $\tilde{B}_r = b_r - b^+_r$ we get for (68) the expression

$$H_B = \sum_{rs} \left[A_{rs} b^+_r b_s + \frac{1}{2} B_{rs}(b^+_r b^+_s + b_r b_s) \right] \quad (71)$$

where

$$B_{rs} = 2(F^1_{rs} + F^2_{rs}) \quad (72)$$

$$A_{rs} = F^1_{rs} + F^1_{sr} - F^2_{rs} - F^2_{sr}. \quad (73)$$

Diagonalizing the above Hamiltonian we obtain diabatic frequencies

$$H_B = \sum_r \hbar\omega^d_r \left(b^+_r b_r + \frac{1}{2} \right). \quad (74)$$

Hamiltonian (71) has a form of quadratic Hamiltonian [23,24] and can be diagonalized by Bogoljubov transformation, which leads to the condition

$$det\begin{pmatrix} \hat{A}-\omega^d\hat{1} & \hat{B} \\ \hat{B} & \hat{A}+\omega^d\hat{1} \end{pmatrix} = 0. \tag{75}$$

Secular equation (75) gives us diabatic phonons $\hbar\omega_r^d$.

If we look at eq. (41) we see that we have corrections due to non–adiabaticity to one-particle part as well as to two-particle part. We see the hierarchical structure of our Hamiltonian. If the non–adiabatic coupling is small, i.e. $\tilde{C}$ goes to zero, we have the adiabatic Hamiltonian. If this coupling is strong we cannot use the adiabatic approximation but we have to work with full V-E diabatic Hamiltonian (54).

4. Calculations

In order to compare our approach with other approaches dealing with adiabatic corrections we perform simple model calculations for adiabatic corrections to ground state energy. We start with adiabatic Hamiltonian (32). We now perform the following approximation. We limit ourselves to finite orders of Taylor expansion of the operators H'_A and H'_B. We shall use similar approximation as in [25]. The diagrammatic representation of our approximate Hamiltonian will be

$$H = H_o + H' = E^o_{NN} + E^o_{SCF}\sum_P \epsilon_P N[a_P^+ a_P] + \sum_r \hbar\omega_r\left(b_r^+ b_r + \frac{1}{2}\right) \tag{76}$$

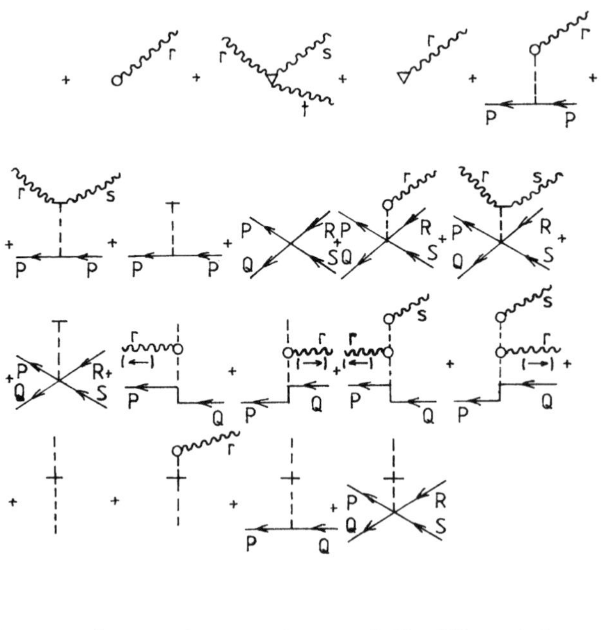

The adiabatic corrections to the ground state of H_2, HD, and D_2 we shall calculate using second–order Rayleigh–Schrödinger many-body perturbation theory (RS–

MBPT) and our Hamiltonian (76). If we assume that we know the solution of the unperturbed Schrödinger equation

$$H_0|\varphi_o\rangle = E_o|\varphi_o\rangle, \tag{77}$$

where H_0 is the unperturbed Hamiltonian $H_A^o + H_B^o$ where

$$H_A^o = E_{NN}^o + E_{SCF}^o + \sum_P \epsilon_P N[a_P^+ a_P] \tag{78}$$

and

$$H_B^o = \sum_r \hbar\omega_r \left(b_r^+ b_r + \frac{1}{2} \right). \tag{79}$$

The perturbed (exact) Schrödinger equation will have the form

$$H|\Psi\rangle = \mathcal{E}|\Psi\rangle, \tag{80}$$

where H will be our Hamiltonian (76). The perturbed energy E will be given through the RS–MBPT expansion as

$$\mathcal{E} = E_o + \langle\varphi_o|H'|\varphi_o\rangle + \langle\varphi_o|H'Q_oH'|\varphi_o\rangle + \ldots, \tag{81}$$

where H' is the perturbation and Q_o is the resolvent

$$Q_o = \sum_{i\neq 0} \frac{|\varphi_i\rangle\langle\varphi_i|}{E_0 - E_i}. \tag{82}$$

Since our sets of boson creation and annihilation operators and fermion creation and annihilation operators commute we can write our unperturbed wavefuntion $|\varphi_o\rangle$ as the product of the fermion state vector $|\psi_o\rangle$ and the boson state vector $|\chi_o\rangle$, i.e.

$$|\varphi_o\rangle = |\psi_o\rangle|\chi_o\rangle. \tag{83}$$

Further we want to study the nonadiabatic corrections to the ground state. Therefore $|\psi_o\rangle$ will be the unperturbed ground state wave function (we shall use Hartree–Fock ground state Slater determinant –Fermi vacuum) and $|\chi_o\rangle$ will be boson ground state–boson vacuum $|0\rangle$.

$$|\chi_0\rangle = |0\rangle. \tag{84}$$

The exact ground state energy will be given by perturbation expansion (up to the second order)

$$\mathcal{E} = \langle\psi_0|\langle 0|H_0|0\rangle|\psi_0\rangle + \langle\psi_0|\langle 0|H'|0\rangle|\psi_0\rangle + \langle\psi_0|\langle 0|H'Q_0H'|0\rangle|\psi_0\rangle + \ldots \tag{85}$$

Substituting for $H_o = H_A^o + H_B^o$ from (78, 79) into the first term in (85) we get

$$\langle\psi_0|\langle 0|H_o|0\rangle|\psi_0\rangle = E_{NN}^o + E_{SCF}^o + \sum_r \frac{1}{2}\hbar\omega_r. \tag{86}$$

We shall not present all terms for perturbation corrections from the right hand side of eq. (85). There are corrections which corresponds to electron correlation, anharmonicity corrections and adiabatic corrections [25]. We shall pay attention only to adiabatic corrections given through second-order term in eq. (85). Analyzing diagrammatic contributions through the Hamiltonian (76) we find that the adiabatic corrections are given through the second up to fourth term in eq. (41). From these terms we calculate only contributions from the second and the third term which are given through the first order of Taylor expansion and these terms are used in second order RS–MBPT. We obtain the following simple expressions

$$\begin{aligned}\langle\psi_0|\langle 0|H'|0\rangle|\psi_0\rangle &\sim \sum_{AIr}\hbar\omega_r(\hat{C}^r_{AI} + \langle A(0)|\hat{I}^r\rangle)^2 \\ &\quad + \sum_{AIr}\hbar^2\omega_r^2(\hat{C}^r_{AI} + \langle A(0)|\hat{I}^r\rangle)^2.(\varepsilon_I - \varepsilon_A - \hbar\omega_r)^{-1} \\ &\quad + \sum_{AIr}\hbar^2\omega_r^2(\hat{C}^r_{AI} + \langle A(0)|\hat{I}^r\rangle)^2.(\varepsilon_I - \varepsilon_A + \hbar\omega_r)^{-1} \\ &\approx \sum_{AIr}\hbar\omega_r(\hat{C}^r_{AI} + \langle A(0)|\hat{I}^r\rangle)^2 \\ &\quad + 2\sum_{AIr}\hbar^2\omega_r^2(\hat{C}^r_{AI} + \langle A(0)|\hat{I}^r\rangle)^2.(\varepsilon_I - \varepsilon_A)^{-1}. \end{aligned} \tag{87}$$

For the notation see [22,25]. We believe that these three terms on right hand side of (87) represents the dominant contributions to adiabatic corrections. The last approximate formula is valid due to the eq. (9). The formula (87) without approximation was used to calculate the adiabatic corrections to the ground state energy of the H_2, D_2, and HD, molecules. Results are presented in Table 1. We also used the same approach to calculate the adiabatic corrections to the energies of the first vibrational transitions for the same molecules [25]. Analyzing eq. (87), we can see that the first term on the right hand side of eq. (87) is always positive and correspond to largest contribution, while the second term on right hand side of eq. (87) is always negative and represents smaller contribution than the previous term. Therefore, we can expect that the eq. (87) should converge to the true value of adiabatic correction from above. This also explain the larger values for H_2, D_2, and HD adiabatic correction obtained through eq. (87) than true value obtained by Wolniewicz [26]. Another source of difference can be basis set

Table 1 *Adiabatic corrections (in cm^{-1}) for H_2, D_2 and HD molecules*

–	Wolfsberg [28]	Wolniewicz [26]	Kutzelnigg [29]	Handy [12]	Present method, eq. (87)
H_2	101.3	114.591	114.591	101	141.811
D_2	50.7	57.296*	57.296*	50.5*	70.97
HD	76.0	85.943*	85.943*	75.8*	106.36*

*These adiabatic corrections were obtained by reduced mass rescaling from H_2 corrections [28].

used and also the contribution from other terms in (41). In our calculation we have obtained for H_2 value of 141.81 cm^{-1} using Roos Augmented Triple Zeta ANO [27] basis set. Handy [12] using basis set of similar quality as used in our calculations obtained value 101 cm^{-1}. Similar value of adiabatic correction was obtained also in an older study by Wolfsberg [28]. Cencek and Kutzelnigg [29] in his paper using wave function expansion consisting of 1200 functions obtained Wolniewicz's [26] value 114.591 cm^{-1}. It is apparent that calculations of adiabatic corrections are strongly basis set dependent.

5. Conclusions

In this article we performed simple model calculations of adiabatic corrections for ground state energy of the H_2, HD, and D_2 molecules. The corrections were derived through canonical transformation applied to crude adiabatic molecular Hamiltonian. These transformations mix together electrons and phonons (normal coordinate canonical transformation) leading to adiabatic molecular Hamiltonian. Using second quantization formalism and many–body diagrammatic perturbation theory and splitting the adiabatic Hamiltonian into unperturbed part and perturbation we derived the formulae for adiabatic corrections. The results were compared with that obtained by different approaches by Wolniewicz [26] and recently by Kutzelnigg [29]. The quasiparticle canonical transformations were then generalized in a way that electrons and phonons are mixed not only through the normal coordinate but also through the momenta. This canonical transformation leads to non–adiabatic molecular Hamiltonian (motion of electrons does not follow the motion of nuclei, the electrons are phase shifted with respect to nuclei). One can clearly see that the electronic and vibrational motion cannot be separated. The mixed system behaves as one whole quasiparticle (the electrons and phonons coupled through the last (momentum) transformation behaves as a renormalized fermions and an another combination of electrons and phonons leads to renormalized bosons). This is in some extent analogous to the introduction of quasiparticles in the solid state theory, where the 'bare' electron interacting with quantized lattice vibrations is renormalized to 'absorb' some part of this interaction, and this quasiparticle is known as a polaron. We were able to derive equations for non–adiabatic $\tilde{C}$ coefficients, which permits us to calculate the so called mass polarization terms and thus non–adiabatic phonons. It is interesting that the quasiparticles preserves some interesting features known from pure electronic molecular Hamiltonian calculations e.g. we can speak about orbital energies of a new quasiparticles, correlation energies of a new quasiparticles, Hartree–Fock energy (holes and particles), etc. Further very important property which follow from the last canonical (momentum) transformation is that we clearly see that in the case of electronic quasidegeneracy when $\tilde{C}$ coefficients are non-negligible (non–adiabaticity is not a small correction) we should work with a full non–adiabatic Hamiltonian. In the case when non–adiabaticity is a small correction and $\tilde{C}$ coefficients are negligible, we can work with adiabatic Hamiltonian (we have only C coefficients through CPHF equations) and only if system is perfectly separable that even C coefficients are negligible we can work with purely electronic Hamiltonian.

Acknowledgements

This work was supported by the grants 1/4197/97 of the Slovak Grant Agency for Science and 202/98/1028 of the Grant Agency of the Czech Republic.

References

1. M. Born and K. Huang, *Dynamical Theory of Crystal Lattices*, London, Oxford University Press (1956).
2. H.C. Longuet–Higgins, *Adv. Spectrosc.* **2**, 429 (1961).
3. H. Köppel, W. Domcke and L.S. Cederbaum, *Adv. Chem. Phys.* **57**, 59 (1984).
4. S. Wilson, *Electron Correlation in Molecules*, Clarendon Press, Oxford (1984).
5. B.T. Sutcliffe, *Methods in Computational Chemistry* **4** (S. Wilson, Ed.), p. 33, Plenum Press, New York (1992).
6. D. Yarkony, *Rev. Mod. Phys* **68**, 985 (1996).
7. M. Born and R. Oppenheimer, *Ann. Physik (Leipzig)* **84**, 4357 (1927).
8. W. Kutzelnigg, *Mol. Phys.* **90**, 909 (1997).
9. A.S. Davidov, *Quantum Mechanics*, Addison Wesley, New York (1965).
10. I.G. Kaplan, *Theory of Molecular Interactions*, Elsevier, Amsterdam (1986).
11. I.B. Bersuker, *The Jahn-Teller Effect and Vibronic Interactions in Modern Chemistry*, Plenum Press, New York (1983).
12. N.C. Handy and A.M. Lee, *Chem. Phys. Lett.* **252**, 425 (1996).
13. H. Sellers and P. Pulay, *Chem. Phys. Lett.* **103 103**, 463 (1984).
14. I. Hubač and M. Svrček *Int. J. Quant. Chem.* **23**, 403 (1988).
15. I. Hubač, P. Čársky, unpublished results.
16. M. Wagner, *Phys. Stat. Sol. (b)* **107**, 617 (1981).
17. M. Wagner, *J. Chem. Phys.* **82**, 3207 (1985).
18. J.A. Pople, K. Raghavachari, H.B. Schlegel, J.S. Binkley *Int. J. Quant. Chem. Symp* **13**, 225 (1979).
19. J. Gerratt and J.M. Mills, *J. Chem. Phys.* **49**, 1719 (1968).
20. J. Gerratt and J.M. Mills, *J. Chem. Phys.* **49**, 1730 (1968).
21. M. Svrček, *PhD. Thesis*, Faculty of Mathematics and Physics, Bratislava (1986).
22. I. Hubač and M. Svrček, *Methods in Computational Chemistry* **4** (S. Wilson, Ed.), p.145, Plenum Press, New York (1992).
23. J.P. Blaizot and G. Ripka, *Quantum Theory of Finite Systems*, The MIT Press Cambridge, Massachusetts, London, England (1986).
24. I. Hubač, P. Babinec, J. Urban, P. Mach, J. Mášik, M. Polášek, J. Leszczynski, *Asian J. Spectroscopy*, **1** 181 (1997).
25. I. Hubač, M. Svrček, E.A. Salter, C. Sosa and R.J. Bartlett, *Lecture Notes in Chemistry.* Vol. 52, p. 95, Springer, Berlin (1988).
26. L. Wolniewicz, *J. Chem. Phys.* **99**, 1851 (1993).
27. P.O. Widmark, P.A. Malmqvist, B. Roos, *Theor. Chim. Acta.* **77**, 291 (1990).
28. L.I. Kleinman and M. Wolfsberg, *J. Chem. Phys.* **60**, 4740 (1974).
29. W. Cencek, W. Kutzelnigg, *Chem. Phys. Lett.* **266**, 383 (1997).

The Effect of Pseudopotential on the Torsional Energy Levels of Hydrogen Peroxide and Deuterium Peroxide

M.L. Senent and Y.G. Smeyers

Departamento de Física y Química Teóricas, I. Estructura de la Materia, C.S.I.C., Serrano 113b, Madrid 28006, Spain

Abstract

In this paper, the effect of the pseudopotential term, arising from the quantum mechanical correction to classical mechanism (V'), on the torsional levels of hydrogen peroxide and deuterium peroxide is evaluated. The V' operator, depends on the first and second derivatives with respect to the torsional coordinate of the determinant of the g inertia matrix and on the first derivatives of the B kinetic energy parameter of the vibrational Hamiltonian. V' has been determined for each nuclear conformation from the optimized coordinates obtained using MP2/AUG-cc-pVTZ ab initio calculations.

Three different algorithms are developed for the determination of the g and B derivatives: Two of them are analytical methods and the last one is a numerical method. For both species, H_2O_2 and D_2O_2, the consideration of V' goes down the energies to the experimental results. Finally, the effect of V' on the levels is compared with the corresponding effect of the zero vibrational energy correction of the neglected 3N-5 vibrations.

1. Introduction

As is well known, the vibrational Hamiltonian defined in internal coordinates may be written as the sum of three different terms: the kinetic energy operator, the Potential Energy Surface and the V' pseudopotential [1–3]. V' is a kinetic energy term that arises when the classic vibrational Hamiltonian in non-cartesian coordinates is transformed into the quantum-mechanical operator using the 'Podolsky trick' [4]. The determination of V' is a long process which requires the calculation of the molecular geometry and the derivatives of various structural parameters.

It is commonly accepted that the effect of the pseudopotential on the vibrational levels is very small [2]. In several cases, it could be of an order of magnitude close to some secondary effects that usually are neglected. In addition, the determination of the pseudopotencial term is laborious and many authors argue that the effort required is not worth it. For these reasons, it is usually neglected. Other effects, for example, the interactions with the neglected internal motions in calculations of molecules showing large amplitude motions, can be more significant. So, the variation of the vibrational energy with the basis set and the correlation energy corrections in ab initio calculations can be larger than the absolute values of the pseudopotential correction.

In a large series of studies of non-rigid molecules [6–11], we have omitted the pseudopotential correction on the basis of a previous estimation [12]. Since the aim of these papers was the analysis of vibrational structures, the frequencies and intensities, for $J = 0$, of molecules exhibiting one, two or three large amplitude motions were calculated. However, at the present time, we extend the previous methods to the study

A. Hernández-Laguna et al. (eds.), Quantum Systems in Chemistry and Physics, Vol. 1: Basic Problems and Model Systems, 401–413.

of the rotational structures of bands which include the roto-vibrational interactions [3]. The new purpose of the calculations requires a considerable precision and it is necessary to take into account all the effects that could influence the energies. These are the flexbility of the kinetic parameters [13], the zero vibrational [14] and the pseudopotential corrections.

The aim of this paper is to search for an optimum way to determine the pseudopotential. The V' have to be determined without great expense of personal effort since the order of magnitude of the correction is relatively small. For this purpose, we have developed three different algorithms. The first one requires the determination of the kinetic energy parameters B and the determinant of the inertia matrix in several selected conformations. The results are fitted to adapted Fourier series whose derivatives are calculated. The second way is a numerical algorithm and the third one is the rigorous analytical method. This last way could be laborious for molecules such as hydrogen peroxide and may be inadequate for large molecules.

In this paper, we apply the three algorithms to the determination of the V' effect of the torsional levels of hydrogen peroxide. The size of this molecule is optimum for the comparison of the three methods. In addition, a previous study shows that torsion of the central bond can be separated from the rest of the vibrational modes, thereby simplifying the problem. Finally, the torsional pseudopotential of D_2O_2 is also calculated for determining the isotopic substitution effects. The experimental data are from Camy-Peiret et al [15].

2. The Torsional Pseudopotential

The classical roto-vibrational kinetic energy can be defined in internal coordinates as a function of the angular momentum [2–3]:

$$2T = (P^t, p^t)\begin{pmatrix} I & X \\ X^t & Y \end{pmatrix}^{-1}\begin{pmatrix} P \\ p \end{pmatrix} = (P^t, p^t)G\begin{pmatrix} P \\ p \end{pmatrix} \tag{1}$$

where **P** is the vector of the overall angular momentum components corresponding to the molecular axis and **p** is the vector of the conjugate momenta of the internal coordinates. The **G** matrix is defined in equation (1).

By using internal coordinates, the quantum-mechanical operator may be obtained by the application of the 'Podolsky trick' [4]. For this purpose, the kinetic energy has to be multiplied by the determinant g of the $\mathbf{G}^{-1}$ matrix. The classical kinetic energy operator is:

$$2T = g^{-\frac{1}{4}}\sum_{\alpha=1}^{3}\sum_{\beta=1}^{3} P_\alpha g^{\frac{1}{2}} g_{\alpha\beta} P_\beta g^{-\frac{1}{4}} + g^{-\frac{1}{4}}\sum_{i=1}^{3N-6}\sum_{j=1}^{3N-6} p_i g^{\frac{1}{2}} g_{(i+3j+3)} p_j g^{-\frac{1}{4}}$$

$$+ g^{-\frac{1}{4}}\sum_{\alpha=1}^{3}\sum_{i=1}^{3N-6}(p_i g^{\frac{1}{2}} g_{(i+3\alpha)} P_\alpha + P_\alpha g^{\frac{1}{2}} g_{(\alpha i+3)} p_i) g^{-\frac{1}{4}} \tag{2}$$

where g_{ij} are the elements of **G**. The first term represents the rotational kinetic energy and the second one the vibrational energy. The last term represents the roto-vibrational interaction.

The Hamiltonian is defined as the sum of operators, $H = H_{vib} + H_{rot} + H_{ro\text{-}vib}$. The vibrational term is:

$$H_{vib} = \frac{\hbar^2}{2} \sum_{k=1}^{3N-6} \sum_{l=1}^{3N-6} \left(\frac{\partial}{\partial q_k}\right) g_{kl} \left(\frac{\partial}{\partial q_k}\right) + V' + V(q_1, q_2, \ldots) \tag{3}$$

where V is the potential energy surface and V' is the pseudo-potential given by,

$$V' = \frac{\hbar^2}{8} \sum_{k=1}^{3N-6} \sum_{l=1}^{3N-6} \left[\left(\frac{\partial}{\partial q_k}\right) g_{kl} \left(\frac{\partial \ln g}{\partial q_1}\right) - \left(\frac{\partial \ln g}{\partial q_k}\right) g_{kl} \left(\frac{\partial}{\partial q_1}\right)\right]$$

$$+ \frac{\hbar^2}{32} \sum_{k=1}^{3N-6} \sum_{l=1}^{3N-6} \left(\frac{\partial \ln g}{\partial q_k}\right) g_{kl} \left(\frac{\partial \ln g}{\partial q_1}\right) \tag{4}$$

where ln g is the logarithm of the determinant g.

In the case of hydrogen peroxide the torsional energy levels may be obtained accurately using a flexible one-dimensional model. In this case, the vibrational Hamiltonian depends on a single variable which is the torsional angle, γ. This coordinate depends on a single internal coordinate and can be defined as: $2\gamma = 180°$-dihedral angle (H1O1O2-O2O2H2). The rest of the internal coordinates are allowed to be relaxed during the torsional motion. They become functions of the coordinate γ [3]:

$$q_{j \neq \theta} = q_{j \neq \gamma}(\gamma) \tag{5}$$

and the **X** and **Y** sub-matrix elements have to be defined using the chain rule:

$$X_{i4} = \sum_{a=1}^{N} m_a \left[\vec{r}_a x \sum_{j=1}^{3N-6} \left(\frac{\partial \vec{r}_a}{\partial q_j}\right)\left(\frac{\partial q_j}{\partial \gamma}\right)\right]_i \tag{6}$$

$$Y_{44} = \sum_{a=1}^{N} m_a \left[\sum_{i=1}^{3N-6} \left(\frac{\partial \vec{r}_a}{\partial q_i}\right)\left(\frac{\partial q_i}{\partial \gamma}\right)\right]\left[\sum_{j=1}^{3N-6} \left(\frac{\partial \vec{r}_a}{\partial q_j}\right)\left(\frac{\partial q_j}{\partial \gamma}\right)\right] \tag{7}$$

These equations differ from the previous definitions of the **X** and **Y** matrix elements since the derivatives of the internal coordinates with respect to vibrational coordinate are considered.

The one-dimensional vibrational Hamiltonian is,

$$\hat{H}_{vib}(\gamma) = -\frac{\partial}{\partial \gamma} B(\gamma) \frac{\partial}{\partial \gamma} + V'(\gamma) + V(\gamma) \tag{8}$$

where $B = (\hbar^2/2) g_{44}$ and the one-dimensional pseudopotential is:

$$V' = \frac{1}{4}\left[B(\gamma)\left(\frac{\partial^2 \ln g}{\partial \gamma^2}\right) + \frac{B(\gamma)}{4}\left(\frac{\partial \ln g}{\partial \gamma}\right)^2 + \left(\frac{\partial B(\gamma)}{\partial \gamma}\right)\left(\frac{\partial \ln g}{\partial \gamma}\right)\right] \tag{9}$$

which depends on the first derivatives of the B kinetic energy parameters and on the first and second derivatives of ln g. In the present work this derivatives are determined numerically and analytically and the G^{-1} matrix elements are always determined analytically.

3. Ab Initio Calculations

In a previous paper [5], it has been demonstrated that the MP2/AUG-cc-pVTZ [16] level of ab initio calculations permit one to obtain very accurate torsional energies for hydrogen peroxide. For this reason, we employ here the same level of calculations. The surfaces were determined form the electronic energies of nuclear conformations selected for different values of the torsional angle. The energies were calculated with the Gaussian 94 package [17] and they were fitted to even x even Fourier series. The number of gaussian functions employed in the ebonite calculation was 206 which were contracted to form 138 atomic orbitals. Thus, the potential energy surface is:

$$V(\gamma) = 842.912 - 1080.571 \cos 2\gamma + 671.389 \cos 4\gamma - 48.817 \cos 6\gamma$$
$$+ 1.431 \cos 8\gamma + 0.749 \cos 10\gamma - 0.534 \cos 12\gamma \ (cm^{-1}) \tag{10}$$

The most stable structure of hydrogen peroxide has been determined to be a trans-gauche geometry. The $2\gamma_0$ angle of the minimum has been calculated to be 67.387° with MP2/AUG-cc-pVTZ which is in a good agreement with the experimental value of $68.1 \pm 0.4°$ [15]. The C-C and O-H bond distances were determined to be 1.447943 A and 0.964138 A, and the ⟨OOH angle was found to be 99.752°. The double trans-gauche minimum is separated by a barrier whose height is 386.5 cm^{-1}. The semi-empirical value of Johns et al. [14] determined using a one-dimensional model was 387.07 ± 0.2 cm^{-1} [15]. The cis barrier has been calculated to be 2643.8 cm^{-1} with the four sets whereas the corresponding experimental data is 2562.8 ± 60 cm^{-1} [15].

The geometry was fully optimized in all the nuclear conformations. The optimized coordinates [5] have been refined with the reduction of the quota imposed to the gradients in the optimization process. The dependence relation (eq. 7) of the internal coordinates on the vibrational coordinate determined with MP2/AUG-cc-pVTZ are the following:

$$\mathbf{d} = \mathbf{O1O2} = 1.4512 + 0.0013 \cos 2\gamma + 0.0065 \cos 4\gamma - 0.0003 \cos 6\gamma$$

$$\mathbf{R} = \mathbf{H1O1} = \mathbf{H2O2} = 0.9642 - 0.0007 \cos 2\gamma - 0.0005 \cos 4\gamma + 0.0003 \cos 6\gamma$$

$$\alpha = \mathbf{H1O1O2} = \mathbf{H2O2O1} = 100.928 - 2.924 \cos 2\gamma + 0.212 \cos 4\gamma - 0.126 \cos 6\gamma$$
$$+ 0.035 \cos 8\gamma + 0.0013 \cos 10\gamma + 0.0012 \cos 12\gamma - 0.0015 \cos 14\gamma$$

$$\beta = \mathbf{dih(H1O1O2\text{-}O1O2H2)} = 180° - 2\gamma$$

From these equations and the relation between cartesian and internal coordinates in Appendix I, the B kinetic parameter has been determined analytically using the KNP.F fortran program [13]. The results for each geometry were fitted to a Fourier series. The final surfaces are:

$$B_{HOOH}(\gamma) = 10.0946 - 0.3729\cos 2\gamma + 0.0755\cos 4\gamma - 0.0420\cos 6\gamma + 0.0255\cos 8\gamma - 0.0143\cos 10\gamma + 0.0098\cos 12\gamma (cm^{-1}) \quad (11)$$

for H_2O_2 and,

$$B_{DOOD}(\gamma) = 5.4509 - 0.3118\cos 2\gamma + 0.0508\cos 4\gamma - 0.0173\cos 6\gamma + 0.0109\cos 8\gamma - 0.0055\cos 10\gamma + 0.0035\cos 12\gamma (cm^{-1}) \quad (12)$$

for the case of the deuterium peroxide.

If the geometry is fully optimized in all the conformations, the 3N-5 neglected vibrations remain at $E = 0$ during the complete torsion instead of lying at the zero vibrational. For this reason, a zero vibrational correction V_{ZERO} can be added to the potential. For each selected conformation the simplest correction is:

$$V_{ZERO} = \frac{1}{2}\sum_{i=2}^{3N-6} \omega_i \quad (13)$$

where ω_i are the harmonic frequencies that can be determined in the ab initio calculations. The values can be fitted to a Fourier series. Figure 1 shows the zero vibrational correction of H_2O_2 as a function of the torsional angle. The correction for hydrogen peroxide is:

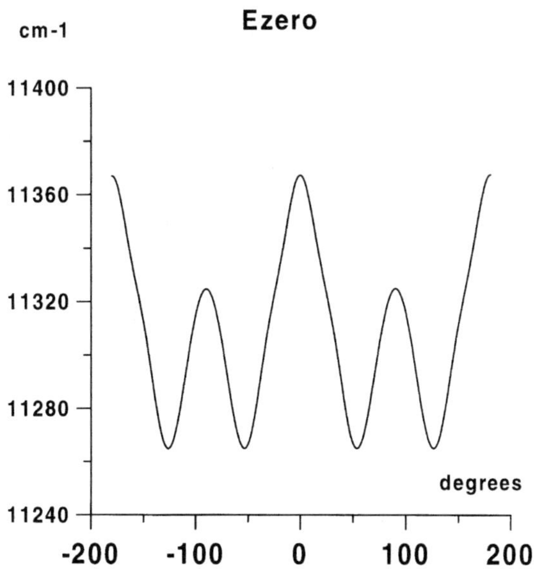

Fig. 1. The correction of the zero vibrational level of H_2O_2 determined with MP2/AUG-cc-pVTZ.

$$V_{ZERO}(\gamma) - V_{ZERO}(\gamma = 0) = -58.6380 + 25.1681\cos 2\gamma + 34.8010\cos 4\gamma - 6.8533 + 1.6868\cos 8\gamma + 2.8681\cos 10\gamma + 0.9672\cos 12\gamma (cm^{-1}) \quad (14)$$

4. Computation of the Pseudopotential

Tables 1 and 2 show the lowest torsional energy levels of hydrogen peroxide and deuterium peroxide which have been determined variationally using as basis functions the rigid rotor solutions. Experimental data are from Camy-Peiret et al [15]. The first set of leval data are from Camy-Peiret et al [15]. The first set of levels (SET I) has been calculated without including the pseudopotential ($V' = 0$). The levels corresponding to the other sets (SET II, SET III and SET IV) were obtained including pseudopotentials calculated with different numerical and analytical algorithms. Finally, the zero point vibration energy correction was introduced in the SET V [14].

Three different algorithms have been developed for the determination of the pseudopotential. Three corresponding subroutines (in Fortran 77) have been implemented in the KNP.F program [13]. These three are:

A) The first one, which was employed in [5], supposes the fitting of the values of B kinetic parameter and $\ln g$ for seven conformations with values of 2γ separated by 30°, to Fourier series. From the resulting $\ln g\ (\gamma)$ and $B(\gamma)$ functions, the first and second derivatives are easy determined. The corresponding levels are the SET II.

Table 1 *Torsional energy levels (cm^{-1}) of H_2O_2*

υ	τ	SET I	SET II	SET III–IV	SET V	exp.[ ]
0	1	0.0000	0.0000	0.0000	0.0000	0.0000
0	2	0.0000	0.0000	0.0000	0.0000	0.0000
0	3	12.0625	11.9564	12.4941	8.2094	11.4372
0	4	12.0625	11.9564	12.4941	8.2094	11.4372
1	1	256.6822	257.1041	254.2714	273.7169	254.5499
1	2	256.6822	257.1041	254.2715	273.7170	254.5499
1	3	377.0655	377.4650	375.6559	375.1448	370.8932
1	4	377.0655	377.4652	375.6561	375.1451	370.8932
2	1	579.1228	579.2210	577.3094	579.4482	569.7427
2	2	579.1228	579.2221	577.3106	579.4497	569.7442
2	3	790.5089	790.7029	788.5109	784.1644	776.1148
2	4	790.5152	790.7092	788.5173	784.1715	776.1215
3	1	1020.3494	1020.4942	1018.0758	1012.0283	1000.8820
3	2	1020.3814	1020.5261	1018.1083	1012.0624	1000.8820

Table 2 *Torsional energy levels (cm^{-1}) of H_2O_2*

v	τ	SET I	SET II	SET III–IV
0	1	0.0000	0.0000	0.0000
0	2	0.0000	0.0000	0.0000
0	3	1.8536	1.8432	1.9042
0	4	1.8536	1.8432	1.9042
1	1	213.3422	213.5867	212.5968
1	2	213.3422	213.5867	212.5968
1	3	255.4401	255.6701	254.7015
1	4	255.4401	255.6701	254.7015
2	1	394.6113	394.7061	393.6351
2	2	394.6113	394.7061	393.6351
2	3	519.7268	519.8817	518.8627
2	4	519.7268	519.8817	518.8627
3	1	668.3313	668.4317	667.4431
3	2	668.3313	668.4317	667.4431

B) In the second method, the derivatives of ln g and B are obtained numerically for each conformation. This method permits one to calculate V' in a single nuclear conformation. The corresponding energy levels are those of the SET III.

C) The derivatives of ln g and B are obtained analytically. This method is arduous since requires to determine first, second and third derivatives of all the structural parameters. It permits the use of a single nuclear conformation. The corresponding levels are called SET IV and the derivatives are shown in Table 3. Appendix I contains the formulas of the derivatives of the molecular cartesian coordinates.

In the three cases, the pseudopotential is calculated with the equation (9) and the values for each conformation have been fitted to a totally symmetric Fourier series.

Finally, the fifth set of levels on Table 1 (SET V) is obtained adding the zero vibrational correction (equation 13) which is neglected in the remaining calculations.

A algorithm: Fourier series

In the first case, ln g is determined from its values corresponding to the selected conformations. For hydrogen peroxide, it has been determined to be:

$$\ln g(\gamma) - 4\ln(1\,umaA) = 6.9575 + 0.3272\cos 2\gamma + 0.0085\cos 4\gamma + 0.0047\cos 6\gamma - 0.0034\cos 8\gamma - 0.0012\cos 12\gamma \quad (15)$$

where (1 uma A) is the unit of mass x distance and V':

Table 3 *Numerical and analitical derivatives of Ln g and B*

2γ	B′	(Ln g)′	(Ln g)″	V′
H_2O_2				
0	0.0	0.0	−0.21210425	−0.5183945
30	−0.09080722	−0.00126548	0.29288876	0.7175492
60	0.68824078	−0.06799784	2.19835990	5.4333647
90	1.30102890	−0.14799801	1.87555250	4.6707902
120	0.74943701	−0.05334943	−0.27748130	−0.7161390
150	1.34565330	−0.07009473	−0.36889385	−0.9813806
180	0.0	0.0	0.46389929	1.2333458
D_2O_2				
0	0.0	0.0	−0.40099892	−0.51944780
30	0.10533452	−0.00446667	−0.31480621	−0.40951502
60	0.38020683	−0.03300607	1.35229200	1.78381630
90	0.73571270	−0.11917699	1.14459144	1.93755750
120	0.68221352	−0.07825813	−0.15559284	−0.22768182
150	0.73232659	−0.05621637	−0.35505237	−0.51799593
180	0.0	0.0	0.33140294	0.48473619

$$V'(\gamma) = -0;00271 + 0.32621\cos 2\gamma + 0.32278\cos 4\gamma + 0.43666\cos 6\gamma - 0.56320\cos 8\gamma + 0.05989\cos 10\gamma - 0.45271\cos 12\gamma (cm^{-1}) \quad (16)$$

In the case of the deuterium peroxide is:

$$V'(\gamma) = -0.00116 + 0.18980\cos 2\gamma + 0.07988\cos 4\gamma + 0.15126\cos 6\gamma - 0.23025\cos 8\gamma - 0.01145\cos 10\gamma - 0.23148\cos 12\gamma (cm^{-1}) \quad (17)$$

B algorithm: Numerical derivatives

The derivatives of ln g and B of a single conformation which correspond to the γ_i torsional angle, have been calculated numerically using the double precision declaration as real *16. For this purpose, nine values of B and ln g around $\gamma_i(\gamma_i + \Delta\gamma, \gamma_i + 2\Delta\gamma \ldots$ and $\gamma_i - \Delta\gamma, \gamma_i - 2\Delta\gamma \ldots)$ have been calculated. The grid has been defined for $\Delta\gamma = 10^{-8}$ rad. The resulting nine values have been fitted to four termed Taylor series which were derived. The numerical derivatives are identical (up to 15 decimal numbers) to those obtained analytically with the C algorithm shown in Table 3.

C algorithm: analytical derivatives

The vibrational levels of hydrogen peroxide can be obtained in one-dimension. In this case, the B parameters is:

$$B = \frac{|I|}{g} \tag{18}$$

where $|I|$ is the determinant of the external rotation matrix. The two determinants $|I|$ and g are defined as:

$$|I| = Y_{44}I_x(I_yI_z - I_{yz}^2) - Y_{44}I_{xy}(I_{xy}I_z - I_{xz}I_{yz}) + Y_{44}I_{xz}(I_{xy}I_{yz} - I_{xz}I_y) \tag{19}$$

and g:

$$\begin{aligned} g = {} & [Y_{44}I_x - X_{14}^2][I_yI_z - I_{yz}^2] - X_{24}^2[I_xI_z - X_{xz}^2] \\ & - X_{34}^2[I_xI_y - I_{xy}^2] + [2X_{14}X_{24} - Y_{44}I_{xy}][I_{xy}I_z - I_{xz}I_{yz}] \\ & + [2X_{14}X_{34} - Y_{44}I_{xz}][I_{xz}I_y - I_{xy}I_{yz}] + 2X_{24}X_{34}[I_{yz}I_x - I_{xz}I_{xy}] \end{aligned} \tag{20}$$

The derivatives of these equations require one to obtain the first and second derivatives of the G^{-1} matrix elements. This, in turn, requires to obtain the first, second and third derivatives of the **d**, **R**, α and $\boldsymbol{\beta}$ (equations 5) internal coordinates with respect to γ, and the first, second and third derivatives of the cartesian coordinates with respect to the internal coordinates.

As an initial molecular system of reference a system centered on the O1 atom, has been selected. The x axis coincides with the O1-O2 bond and the three atoms O1, O2 and H1 lie in the *xy* plane. Appendix I shows the equations that connect cartesian and internal coordinates and their derivatives. From the initial cartesian coordinates, the X, $Y$and Z center-of-mass coordinates and its X', $Y' yZ'$ ... derivatives are calculated. The positions of the atoms have to be referred to the center of mass:

$$X = \frac{1}{M}\sum_{i=1}^{3N-6} m_i x_i;\ X_i = x_1 - X$$

$$X' = \frac{1}{M}\sum_{i=1}^{3N-6} m_i x_i';\ X_i' = x_1 - X' \ldots$$

$$X = (X, Y, Z) \ldots X' = (X', Y', Z') \ldots \tag{21}$$

The derivatives of B and ln g are easily determined:

$$B' = |I|'g - |I|g'$$

$$(\ln g)' = \frac{g'}{g}$$

$$(\ln g)'' = \frac{g''g - (g')^2}{g^2} \tag{22}$$

and the pseudopotential is obtained with equation (9). In the case of hydrogen peroxide, the analytical pseudopotential is:

$$V'(\gamma) = 1.58028 + 1.22340 \cos 2\gamma - 2.26795 \cos 4\gamma - 2.34179 \cos 6\gamma$$
$$+ 0.93386 \cos 8\gamma + 0.24252 \cos 10\gamma + 0.11129 \cos 12\gamma \ (cm^{-1}) \quad (23)$$

which is shown in Figure 2. V' for D_2O_2:

$$V'(\gamma) = 0.42480 + 0.19920 \cos 2\gamma - 1.06558 \cos 4\gamma - 0.83786 \cos 6\gamma$$
$$+ 0.53530 \cos 8\gamma + 0.13657 \cos 10\gamma + 0.08812 \cos 12\gamma \ (cm^{-1}) \quad (24)$$

5. Discussion and Conclusion

In a previous paper on hydrogen peroxide, the pseudopotential was determined with the A algorithm. The aim of this paper was the evaluation of the basis set effect on the levels. In the present paper, the old calculations have been repeated with the optimal basis set improving the optimization of the internal coordinates. It has to be remarked from the results in Tables, 1 and 2 that the B and C algorithms lead to similar results whereas the A algorithm gives the lowest results for the pseudopotential. V' is unappreciated when is determined with the A algorithm.

The best method is the C algorithm where the derivatives are calculated analytically. In this case, the possible error in the results only arises from the fit of the equations (5) and with the number of selected conformations for the ab initio calculations that could be insufficient. The method and the derivatives are extremely accurate. In addition, the same conclusion concern the B algorithm since the first 15 decimal number of derivatives are identical to the analytical ones. Differences between these two last

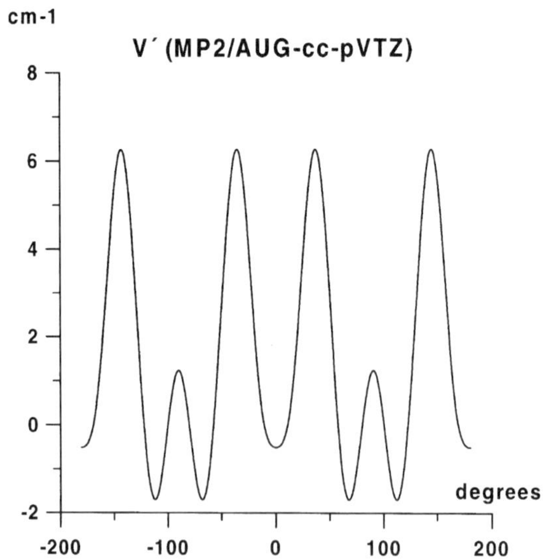

Fig. 2. The Torsional Pseudopotential of H_2O_2 determined with MP2/AUG-cc-pVTZ.

methods and the first one come from the calculations of the second derivatives of ln g which produces a series of errors arising form all the stages of the A method.

Since the B and C methods produce identical results, it may be concluded that the B numerical algorithm is the optimum. It has to be taken into consideration the personal effort required by the C method for a large molecule showing more that one independent vibration. In the C case, the equations shown in Appendix I have to be derived for each molecule. In addition, equation 20 increases with the number of large amplitude motions. On the other hand, expenditure of personal effort using the B numerical algorithm is independent of the size of the problem.

V' improves slightly the levels of hydrogen peroxide (Table 1). The energies go down to the experimental values of Camy-Peiret et al [15]. The $n = 1$, 2 and 3 levels decrease 2 cm^{-1}. The effect of the torsional staggering is very small. The splitting of $n = 0$ arising from the trans barrier is 12.0625 cm^{-1} with no V' and 12.4941 cm^{-1} if V' is added. The $n = 1$ level splitting is 120.4 cm^{-1} with no V' and changes 1 cm^{-1} when the pseudopotential is added. As was expected given the size of the cis barrier, V' has not significant effect on the cis staggering. In addition, the effect of V' on D_2O_2 is approximately half part of the effect on the hydrogen species.

Finally, the zero point vibration corrections (SET V) use to be much larger than the pseudopotential corrections. In the present case, these zero point corrections seems to give rise to unrrealistic values, probably because of the harmonic approximation used in the calculations. The torsion mode as well as its interactions with the remaining modes are indeed very anharmonic.

Appendix

Coordinates

$$x1 = x2 = x3 = y2 = y3 = z3 = 0$$
$$x2 = d$$
$$x3 = d - R\cos\alpha$$
$$y3 = R\sin\alpha$$
$$x4 = R\cos\alpha - d$$
$$y4 = R\sin\alpha\cos\beta$$
$$z4 = -R\sin\alpha\,\mathrm{sen}\,\beta$$

First derivatives

$$x1' = x2' = x3' = y2' = y3' = z3' = 0$$
$$x2' = d'$$
$$x3' = d' - R'\cos\alpha + R\sin\alpha(\alpha')$$
$$y3' = R'\sin\alpha + R\cos\alpha(\alpha')$$
$$x4' = R'\cos\alpha - R\sin\alpha(\alpha') - d'$$
$$y4' = R'\sin\alpha\cos\beta + R\cos\alpha(\alpha')\cos\beta - R\sin\alpha\sin\beta(\beta')$$
$$z4 = -R'\sin\alpha\sin\beta - R\cos\alpha(\alpha')\sin\beta - R\sin\alpha\cos\beta(\beta)'$$

Second derivatives

$$
\begin{aligned}
x1'' &= x2'' = x3'' = y2'' = y3'' = z3'' = 0\\
x2'' &= d''\\
x3'' &= d'' - R''\cos\alpha + 2R'\sin\alpha(\alpha') + R\cos\alpha(\alpha')^2 + R\sin\alpha(\alpha'')\\
y3'' &= R''\sin\alpha + 2R'\cos\alpha(\alpha') - R\sin\alpha(\alpha')^2 + R\cos\alpha(\alpha'')\\
x4'' &= R''\cos\alpha - 2R'\sin\alpha(\alpha') - R\cos\alpha(\alpha')^2 - R\sin\alpha(\alpha'') - d''\\
y4'' &= R''\sin\alpha\cos\beta + 2R'\cos\alpha(\alpha')\cos\beta - 2R'\sin\alpha\sin\beta(\beta')\\
&\quad - R\sin\alpha(\alpha')^2\cos\beta + R\cos\alpha(\alpha'')\cos\beta\\
&\quad - 2R\cos\alpha(\alpha')\sin\beta(\beta') - R\sin\alpha\cos\beta(\beta')^2 - R\sin\alpha\sin\beta(\beta'')\\
z4'' &= -R''\sin\alpha\sin\beta - 2R'\cos\alpha(\alpha')\sin\beta - 2R'\sin\alpha\cos\beta(\beta')\\
&\quad + R\sin\alpha(\alpha')^2\sin\beta - R\cos\alpha(\alpha'')\sin\beta - 2R\cos\alpha(\alpha')\cos\beta(\beta')\\
&\quad + R\sin\alpha\sin\beta(\beta')^2 - R\sin\alpha\cos\beta(\beta'')
\end{aligned}
$$

Third derivatives

$$
\begin{aligned}
x1''' &= x2''' = x3''' = y2''' = y3''' = z3''' = 0\\
x2''' &= d'''\\
x3''' &= d'' - R'''\cos\alpha + 3R''\sin\alpha(\alpha') + 3R'\cos\alpha(\alpha')^2 + 3R'\sin\alpha(\alpha'') - R\sin\alpha(\alpha')^3\\
&\quad + 3R\cos\alpha(\alpha')(\alpha'') + R\sin\alpha(\alpha''')\\
y3''' &= R'''\sin\alpha + 3R''\cos\alpha(\alpha') - 3R'\sin\alpha(\alpha')^2 + 3R'\cos\alpha(\alpha'') - R\cos\alpha(\alpha')^3\\
&\quad - 3R\sin\alpha(\alpha')(\alpha'') + R\cos\alpha(\alpha''')\\
x4''' &= R'''\cos\alpha - 3R''\sin\alpha(\alpha') - 3R'\cos\alpha(\alpha')^2 - 3R'\sin\alpha(\alpha'') + R\sin\alpha(\alpha')^3\\
&\quad - 3R\cos\alpha(\alpha')(\alpha'') - R\sin\alpha(\alpha''') - d'''.\\
y4''' &= R'''\sin\alpha\cos\beta + 3R''\cos\alpha(\alpha')\cos\beta - 3R''\sin\alpha\sin\beta(\beta')\\
&\quad - 3R'\sin\alpha(\alpha')^2\cos\beta + 3R'\cos\alpha(\alpha'')\cos\beta - 6R'\cos\alpha(\alpha')\sin\beta(\beta')\\
&\quad - 3R'\sin\alpha\cos\beta(\beta')^2 - 3R'\sin\alpha\sin\beta(\beta'') - R\cos\alpha(\alpha')^3\cos\beta\\
&\quad - 3R\sin\alpha(\alpha')(\alpha'')\cos\beta + 3R\sin\alpha(\alpha')^2\sin\beta(\beta') + R\cos\alpha(\alpha''')\cos\beta\\
&\quad - 3R\cos\alpha(\alpha'')\sin\beta(\beta') - 3R\cos\alpha(\alpha')\cos\beta(\beta')^2 - 3R\cos\alpha(\alpha')\sin\beta(\beta'')\\
z4''' &= -R'''\sin\alpha\sin\beta - 3R''\cos\alpha(\alpha')\sin\beta - 3R''\sin\alpha\cos\beta(\beta') + 3R'\sin\alpha(\alpha')^2\sin\beta\\
&\quad - 3R'\cos\alpha(\alpha'')\sin\beta - 6R'\cos\alpha(\alpha')\cos\beta(\beta') + 3R'\sin\alpha\sin\beta(\beta')^2\\
&\quad - 3R'\sin\alpha\cos\beta(\beta'') + R\cos\alpha(\alpha')^3\sin\beta + 3R\sin\alpha(\alpha')(\alpha'')\sin\beta\\
&\quad + 3R\sin\alpha(\alpha')^2\cos\beta(\beta') - R\cos\alpha(\alpha''')\sin\beta - 3R\cos\alpha(\alpha'')\cos\beta(\beta')\\
&\quad + 3R\cos\alpha(\alpha')\sin\beta(\beta')^2 - 3R\cos\alpha(\alpha')\cos\beta(\beta'') + R\sin\alpha\cos\beta(\beta')^3\\
&\quad + 3R\sin\alpha\sin\beta(\beta')(\beta'') - R\sin\alpha\cos\beta(\beta''')
\end{aligned}
$$

Acknowledgements

This work has been supported by the 'Consejería de Educación y Cultura' de la Comunidad de Castilla y León (BU07/97) and the 'Vicerectorado de Investigación y Relaciones Internacionales' of the University of Burgos. Y.G.S. and M.L.S. also acknowledge the financial assistance from the 'Comision Interministerial de Ciencias y Tecnologia' of Spain through grant no. PB 96-0882.

References

1. J.D. Lewis, T.B. Malloy Jr., T.H. Chao and J. Laane, J. Mol. Struct., **12**, 472 (1972).
2. M.A. Harthcock and J. Laane, J. Phys. Chem., **89**, 4231 (1985).
3. M.L. Senent J. Mol. Spectrosc., **191**, 265 (1998).
4. B. Podolsky, Phys. Rev, **32**, 812 (1928).
5. S. Fernández-Herrera and M.L. Senent J. Mol. Struct., **470**, 313 (1998).
6. Y.G. Smeyers, M.L. Senent, V. Botella and D.C. Moule, J. Chem. Phys., **98**, 2754 (1993).
7. M.L. Senent, D.C. Moule and Y.G. Semeyers, Can. J. Phys., **73**, 425 (1995).
8. M.L. Senent, D.C. Moule and Y.G. Smeyers, J. Chem. Phys., **102**, 5952 (1995).
9. M.L. Senent and Y.G. Smeyers, J. Chem. Phys., **105**, 2789 (1996).
10. M.L. Senent, Y.G. Smeyers and D.C. Moule, **102**, 6730 (1998) J. Phys Chem.
11. M.L. Senent, Y.G. Smeyers and D.C. Moule, **94**, 949 (1998) Mol. Phys.
12. Y.G. Smeyers and A. Hernández-Laguna, Int. J. Quant. Chem., **22**, 681 (1982).
13. M.L. Senent, Chem. Phys. Lett, **296**, 299 (1998).
14. R.S. Grev, B.J. Deleeuw, Y. Yamaguchi, S.-J. Kim and F. Schaefer III, in 'Structures and Conformations of Non-Rigid Molecules', NATO SCI Series, Kluwer Academic Publishers, pag. 325, 1992.
15. J.M. Flaud, C. Camy-Peyret, J.W.C. Johns and B. Carli, J. Chem. Phys., **91**, 1504 (1989).
16. D.E. Woon and T.H. Dunning, Jr., J. Chem. Phys., **98**, 1358 (1993); R.A.Kendall, T.H. Dunning, Jr and R.J. Harrison, J. Chem. Phys., **96**, 6796 (1992); T.H. Dunning, Jr, J. Chem. Phys., **90**, 1007 (1989).
17. Gaussian 94 (Revision A.1), M.J. Frisch, G.W. Trucks, H.B. Schlegel, P.M.W. Gill, B.G. Johnson, M.A. Robb, J.R. Cheeseman, T.A. Keith, G.A. Petersson, J.A. Mongomery, K. Raghavachari, M.A. Al-Laham, V.G. Zakrzewski, J.V. Ortiz, J.B. Foresman, J. Cioslowsky, B.B. Stefanov, A. Nanayakkara, M. Challacombe, C.Y. Peng, P.Y. Ayala, W. Chem, M.W. Wong, J.L. Andres, E.S. Repogle, R. Gomberts, R.L. Martin, D.J. Fox, J.S. Binkley, D.J. Defrees, J. Baker, J.J.P. Steward, M. Head-Gordon, C. Gonzalez and J.A. Pople, Gaussian, Inc., Pittsburgh PA, 1995.

Contents of Volume 2

Part VIII. Reactive Collisions and Chemical Reactions

Part IX. Computational Chemistry and Physics

Combined Index to Volumes 1 and 2

(Entries are in the form [volume number]:[page number].)

A. Hernández-Laguna et al. (eds.), Quantum Systems in Chemistry and Physics, Vol 1: Basic Problems and Model Systems, 419–421.

Progress in Theoretical Chemistry and Physics

1. S. Durand-Vidal, J.-P. Simonin and P. Turq: *Electrolytes at Interfaces.* 2000
ISBN 0-7923-5922-4
2. A. Hernandez-Laguna, J. Maruani, R. McWeeny and S. Wilson (eds.): *Quantum Systems in Chemistry and Physics.* Volume 1: Basic Problems and Model Systems, Granada, Spain, 1997. 2000 ISBN 0-7923-5969-0; Set 0-7923-5971-2
3. A. Hernandez-Laguna, J. Maruani, R. McWeeny and S. Wilson (eds.): *Quantum Systems in Chemistry and Physics.* Volume 2: Advanced Problems and Complex Systems, Granada, Spain, 1998. 2000 ISBN 0-7923-5970-4; Set 0-7923-5971-2
4. J.S. Avery: *Hyperspherical Harmonics and Generalized Sturmians.* 1999
ISBN 0-7923-6087-7

KLUWER ACADEMIC PUBLISHERS – DORDRECHT / LONDON / BOSTON